MW00845846

Finite Element Multidisciplinary Analysis Second Edition

Kajal K. Gupta
NASA Dryden Flight Research Center
Edwards, California

John L. Meek
University of Queensland
Brisbane, Australia

EDUCATION SERIES
Joseph A. Schetz
Series Editor-in-Chief
Virginia Polytechnic Institute and State University
Blacksburg, Virginia

Published by
American Institute of Aeronautics and Astronautics, Inc.
1801 Alexander Bell Drive, Reston, VA 20191

American Institute of Aeronautics and Astronautics, Inc., Reston, Virginia

1 2 3 4 5

Library of Congress Cataloging-in-Publication Data

Gupta, Kajal K.
Finite element multidisciplinary analysis / Kajal K. Gupta,
John L. Meek.—2nd ed.
p. cm.—(AIAA education series)
Includes bibliographical references and index.
ISBN 1-56347-580-4
1. Finite element method. I. Meek, J. L. II. Title. III. Series.

TA347.F5.G87 2003
620′.0042—dc22

In memory of our late parents

Foreword

Finite Element Multidisciplinary Analysis, second edition, by Kajal K. Gupta and John L. Meek, brings together several diverse disciplines used in the design of aircraft and space vehicles in which the powerful concept of finite elements is utilized. This second edition of the book fills a gap within the finite element literature by addressing the challenges and developments in structural mechanics, heat transfer, fluid mechanics, controls engineering, and propulsion technology and their interaction as encountered in many practical problems in aeronautical, mechanical, and aerospace engineering. The synergism of writing this text by the leading NASA expert in the development of finite element codes and by a university professor and author of one of the earliest textbooks on finite elements in structural mechanics has produced an outstanding textbook as well as a rich source of reference materials for practicing engineers. Kajal Gupta is well known for his development of NASA STARS (Structural Analysis Routines), a multidisciplinary, finite element based, graphics oriented, linear and nonlinear software that includes such disciplines as structural mechanics, heat transfer, linear aerodynamics, computational fluid dynamics (CFD), and controls engineering. A version of this program is available from the first author. John Meek is particularly known for his book, *Matrix Structural Analysis*, published in 1971, which was a comprehensive exposition of the state-of-the art of finite element methods of analysis for structural systems.

The present text covers several fields of engineering disciplines that can be separately analyzed using finite element techniques and leads to the concept of interdisciplinary analysis in which two or more disciplines are interdependent. This includes interactions of structures, materials, heat transfer, aerodynamics, propulsion, controls, and servomechanisms. A particularly welcome addition is the inclusion of numerical solutions to several examples for each chapter. This greatly helps the reader in understanding the practical applications of finite element techniques used in these examples. The second edition covers additional topics on CFD, aeroelasticity, aeroservoelasticity, optimization, and sparse matrix storage and decomposition using current frontal techniques.

This text is recommended for senior level undergraduate or graduate level courses in aeronautical engineering and related disciplines. It also can be used for practicing engineers by providing them with an overview of the multidisciplinary analysis based on finite element concepts, including solutions to practical problems.

The AIAA Education Series of textbooks and monographs embraces a broad spectrum of theory and application of different disciplines in aeronautics and astronautics, including aerospace design practice. The series also includes texts on defense science, engineering, and management. The books serve as teaching texts for students as well as reference materials for practicing engineers, scientists, and managers. The complete list of textbooks published in the series (over 80 titles) can be found on the end pages of this volume.

J. S. Przemieniecki
Editor-in-Chief, Retired
AIAA Education Series

Table of Contents

Preface to the Second Edition

Preparation of this second edition was motivated by the desire to continuously update the book in keeping with the rapid development of technology in the area of multidisciplinary modeling and simulation of aerospace vehicles. Primary emphasis has been retained in the development of novel numerical solution procedures and associated finite element based software that enables rapid, efficient solution of complex practical problems encompassing a number of distinct engineering disciplines such as fluids, structures and controls, among others, and their interaction.

Thus, Chapter 4 now includes details of a torsion finite element. Chapter 7 on bandwidth minimization has been revised and additional references included. Significant additions have been implemented in Chapter 8, which deals with the efficient solution of sparse matrices that require minimal storage and significantly less solution time. These techniques include minimum fill methods as well as multifrontal solution procedures. Detailed numerical examples are also provides so that the student has better understanding of the intricacies of these numerical procedures.

A new section on optimization has been added in Chapter 12 with many new references. Structural optimization and numerical examples have been added to illustrate the efficacy of the associated finite element code, which also caters to aerodynamic shape optimization. Student exercises are also included in the chapter.

Significant revisions have been made in Chapter 15 in the area of CFD-based aeroelastic and aeroservoelastic analysis. A new section has been added on finite element numerical formulation of viscous flow involving solution of the Navier–Stokes equation. Detailed numerical example problems are also provided that testify to the accuracy of the current solution schemes. Also provided are details of revised and upgraded numerical schemes for aeroelastic and aeroservoelastic analysis of complex practical problems. Extensive numerical examples are also included showing results of these analysis techniques.

A new version of the upgraded STARS program is available for teaching purposes. As a textbook, this volume is suitable for the senior undergraduate or graduate level. For practicing engineers it should prove useful for solving practical, large-scale problems as well as to serve as a valuable reference in the area of multidisciplinary analysis and design.

The authors gratefully acknowledges the assistance provided by C. L. Lawson, S. F. Lung, A. Arena, R. Kolar, A. Ibrahim, and P. Nithiarasu, as well as C. Bach, E. Hahn, and T. Doyle for their help in the preparation of this second edition.

K. K. Gupta
J. L. Meek
March 2003

Preface to the First Edition

The motivation for writing this book arose from the development of a multidisciplinary solution methodology and associated finite element based numerical analysis software for aerospace vehicle modeling and simulation. During the development of the theory and software it was realized that although there are many excellent tools available to solve problems in the individual disciplines, there is a scarcity in the area of multidisciplinary analysis. The current developments affect areas of structural mechanics, heat transfer, fluid mechanics, controls engineering, and propulsion technology, and their interaction as encountered in many practical problems in aeronautical, aerospace, and mechanical engineering, among other disciplines. These topics are reflected in the 15 chapter titles of the book. Numerical examples are provided in these chapters to illustrate the applicability of the techniques. Exercises are given that may be solved either manually or by using suitable computer software. For this purpose, a version of the multidisciplinary analysis program STARS is available for teaching purposes. As a text the book should be useful at the senior undergraduate or graduate level. For the practicing engineer, it will serve as a guide for solving full-scale practical problems, as well as being a general reference in the various subject areas.

A brief survey of the book is as follows. In Chapter 1, the subject areas are delineated and a brief history of the beginning of the finite element method is given. Chapter 2 provides an overview of the various analysis methods that can be used in a finite element formulation. Chapter 3 includes the fundamentals of the finite element technology along with material characteristics pertaining to structural mechanics. In Chapter 4, a range of useful finite elements in one, two, and three dimensions are developed in detail. These elements provide the capabilities for solving a variety of problems encountered in practice. Chapter 5 deals with the dynamics of gyroscopic systems in which different structural components may be subjected to differing spin rates. A dynamic element methods that is characterized by frequency-dependent relevant matrices is described in Chapter 6. Chapter 7 deals with matrix assembly, bandwidth minimization techniques, imposition of deflection boundary conditions, and also storage schemes for sparse matrices. Various methods for the solution of the system equations pertaining to static problems are discussed in Chapter 8. Structural eigenproblem solution techniques are discussed in Chapter 9 for conventional as well as spinning structures, while Chapter 10 describes methods for computation of dynamic response of elastic structures. Chapter 11 continues with the solution of nonlinear static and dynamic problems. Chapter 12 provides a summary of stress computations in elements once the nodal displacements have been obtained. The subject matter of heat conduction in solids is addressed in Chapter 13. Multidisciplinary aerostructural–controls analysis of linear systems is described in Chapter 14. In Chapter 15 computational fluid dynamics–based nonlinear multidisciplinary simulation is addressed in detail.

In the Appendix, 52 exercises are given to assist students in assimilating the subject material given in the text. Because of the vast literature in the subject area it is not possible to include all of them in the lists of references; however, a number of textbooks have been quoted that include many of these references.

We wish to thank those people who helped us with so many aspects of the book. The work on the multidisciplinary analysis and development of related software, which is the core of this book, was due to the inspiration and support provided by colleagues at NASA Dryden Flight Research Center, including L. Voelker, R. Knight, K. Petersen, K. Szalai, and M. Brenner. Thanks are due to S. F. Lung for his tireless effort in the preparation of the manuscript and C. L. Lawson, A. Arena, and R. Kolar for many useful suggestions. Thanks are also given to E. Hahn, R. Truax, C. Bach, T. Doyle, T. Cowan, T. Walsh, Z. Duron, and A. Ahmadi, members of the STARS engineering group at NASA Dryden Flight Research Center, for preparation of example problems and continued support in this effort over many years. Sincere thanks are given to our families for their support, patience, and understanding during the years of preparing the manuscript.

K. K. Gupta
J. L. Meek
April 2000

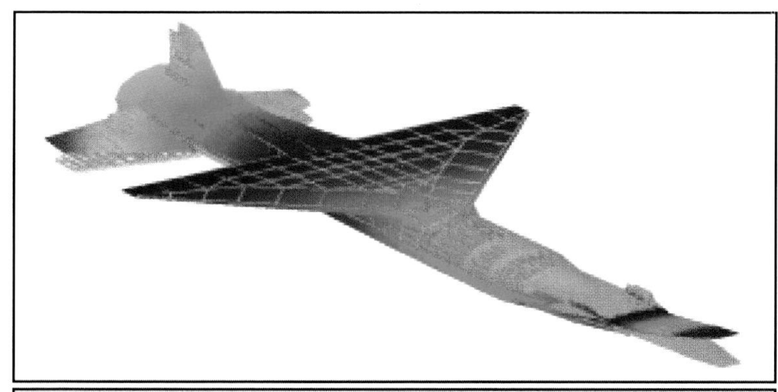

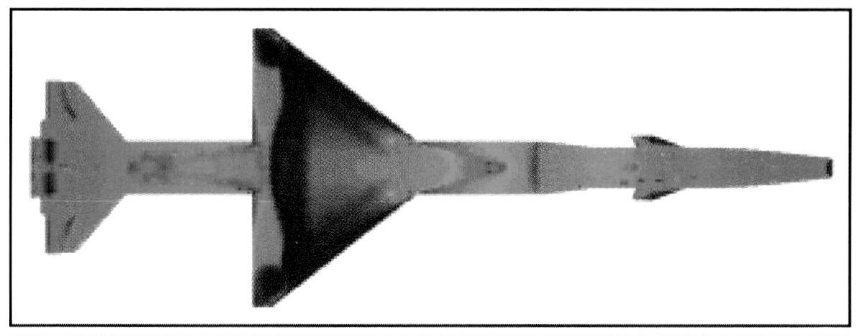

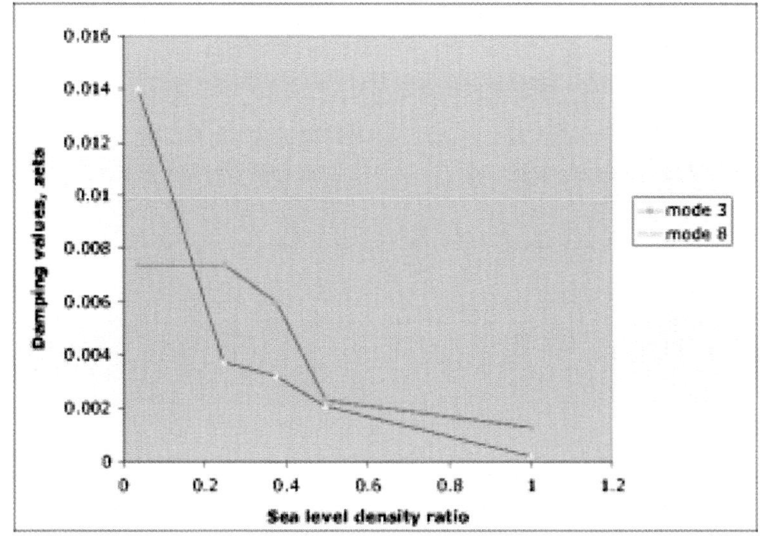
0.016
0.014
0.012
0.01
0.008
0.006
0.004
0.002
0
Damping values, zeta
0
0.2
0.4
0.6
0.8
1
1.2
Sea level density ratio
mode 3
mode 8

Integrated Systems Analysis

NASA
AD89-513a

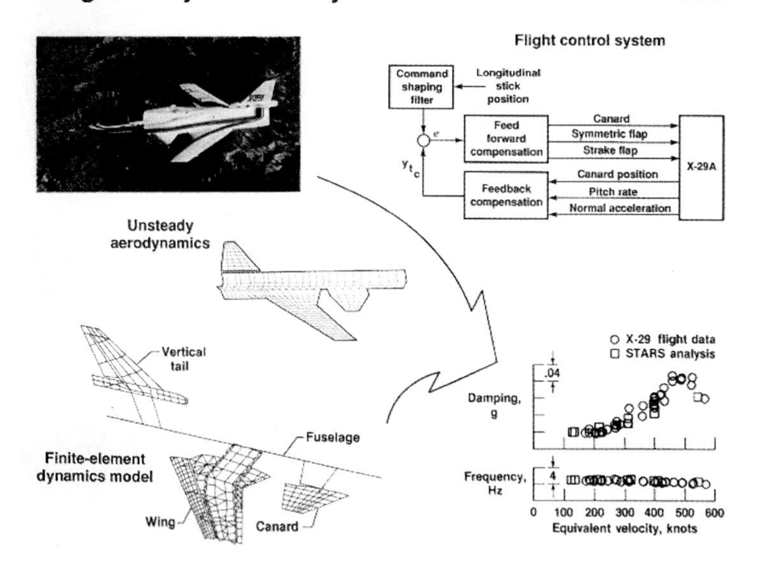

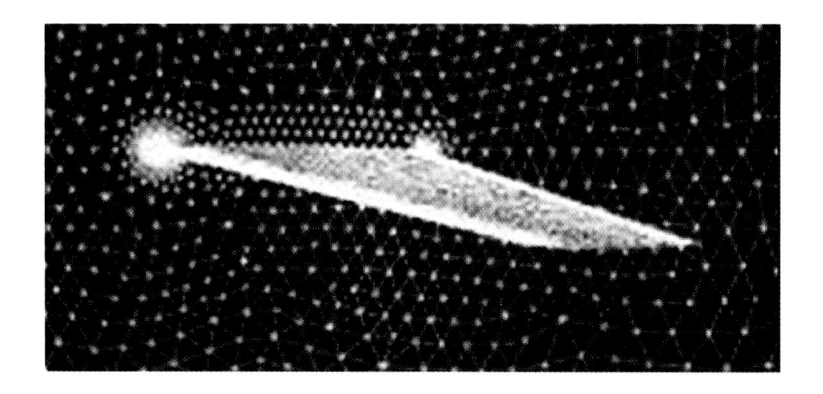

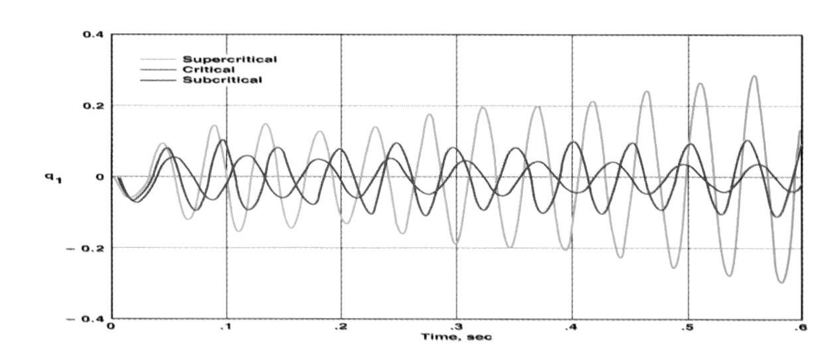

Nomenclature

Some of the major symbols referred to in this book are given below; others are defined in the text as needed. In general, a matrix or row vector is represented by a bold, upper case letter whereas a bold, lower case letter is indicative of a column vector.

$\boldsymbol{A}$	=	plant dynamics matrix in body-fixed coordinates
$A_e(k)$	=	aerodynamic influence coefficient matrix
A_j	=	coefficient matrices of aerodynamic approximation, $j = 0, 1 \ldots$
$\hat{\boldsymbol{A}}$	=	plant dynamics matrix in inertial frame of reference
$\tilde{\boldsymbol{A}}$	=	solution to aerodynamic approximations $\boldsymbol{A}_j$
$\boldsymbol{B}$	=	control influence matrix in body-fixed coordinates
$\hat{\boldsymbol{B}}$	=	control influence matrix in inertial frame of reference
$\boldsymbol{C}$	=	output state matrix in body-fixed coordinates
$\boldsymbol{C}_c$	=	Coriolis acceleration matrix
$\boldsymbol{C}_d$	=	elastic damping matrix
$\hat{\boldsymbol{C}}$	=	generalized damping matrix
$\bar{c}$	=	mean aerodynamic chord
$\boldsymbol{D}$	=	output control matrix in body-fixed coordinates
$\boldsymbol{D}_E$	=	elastic stress–strain relation matrix in structural analysis; also $\boldsymbol{D}$
$\boldsymbol{f}$	=	fluid flux vector
E	=	Young's modulus
$\boldsymbol{G}$	=	feedback controller matrix
g	=	aeroelastic or structural damping
$\boldsymbol{H}$	=	open-loop transfer function matrix of plant and all analog elements
$\hat{\boldsymbol{H}}$	=	closed-loop transfer function matrix
$\boldsymbol{H}^*$	=	hybrid loop gain matrix
$\boldsymbol{I}$	=	identity matrix
Im	=	imaginary part
$\boldsymbol{i}^*$	=	$\sqrt{-1}$, imaginary number
$\boldsymbol{K}$	=	elastic stiffness matrix
$\boldsymbol{K}_G$	=	geometric stiffness matrix
$\boldsymbol{K}'$	=	centripetal acceleration matrix
$\hat{\boldsymbol{K}}$	=	generalized stiffness matrix
$\boldsymbol{K}_2$	=	dynamic element correction stiffness matrix
k	=	reduced frequency
k_i	=	discrete set of reduced frequencies
$\boldsymbol{M}$	=	elastic mass or inertia matrix
$\hat{\boldsymbol{M}}$	=	generalized mass matrix

$\boldsymbol{M}_2$	=	dynamic element correction mass matrix
$\boldsymbol{p}$	=	forcing function for elastic dynamics
$\hat{\boldsymbol{p}}$	=	generalized forcing function
$\boldsymbol{Q}$	=	generalized aerodynamic force matrix
$\hat{\boldsymbol{Q}}$	=	approximation of $\boldsymbol{Q}$
$\hat{\boldsymbol{Q}}_R, \hat{\boldsymbol{Q}}_I$	=	real and imaginary parts of $\hat{\boldsymbol{Q}}$
$\bar{\boldsymbol{Q}}$	=	$\boldsymbol{Q}$ without the rigid air loads
$\boldsymbol{q}$	=	displacement vector
$\boldsymbol{q}_s$	=	modal displacements at sensor location
$\bar{q}$	=	dynamic pressure
$\dot{q}, \ddot{q}$	=	velocity and accelerations
R_e	=	real part
s	=	Laplace variable
$\boldsymbol{T}$	=	thermal load in structural analysis
$\tilde{\boldsymbol{T}}_i$	=	inertial to body-fixed coordinate transformation matrices for combined rigid, elastic, and aerodynamic lag states
$\boldsymbol{T}_s$	=	sensor interpolation matrix
T	=	sample time for digital controller
$\boldsymbol{u}$	=	system input vector
$W1B, F1B, \ldots$	=	vibration modes
$\boldsymbol{v}$	=	fluid solution vector
$\boldsymbol{x}$	=	system state vector in body-fixed reference frame
$\hat{\boldsymbol{x}}$	=	system state vector in inertial frame
$\boldsymbol{x}_j$	=	jth lag state vector
$\boldsymbol{y}$	=	system output vector
$x, y, z, (X, Y, Z)$	=	coordinate system
α	=	angle of attack
α_1	=	trim angle of attack
β	=	angle of sideslip
β_i	=	aerodynamic lag terms
δ	=	control surface deflection
$\boldsymbol{\eta}$	=	generalized coordinate vector
ζ	=	damping
λ	=	natural frequencies
$\boldsymbol{\lambda}$	=	structural direction cosine matrix
ν	=	Poisson's ratio
ρ	=	density
τ	=	system time delay
$\boldsymbol{\phi}$	=	structural modes of vibration
$\boldsymbol{\Omega}, \tilde{\boldsymbol{\Omega}}$	=	angular velocity matrices
ω	=	frequency

Subscripts

e, E	=	elastic
I	=	inertial frame of reference

r = rigid
x, y, z = coordinate system, also X, Y, Z
p, P = plastic

Superscripts

T = matrix transpose
-1 = matrix inverse
e = elementwise

1
Introduction

1.1 Introduction

A variety of engineering problems encountered in practice exhibit behavior that is predicated by the interaction of complex physical phenomena, i.e., their response is determined not by a single phenomena but by multidisciplinary interaction. This real behavior is in contrast to much of the current design practice in which analysis is frozen in each separate discipline. The situation where this static approach is inadequate is in the design of modern high performance aerospacecraft that are characterized by complex aerothermal-structural-controls-propulsion interaction. Hypersonic vehicles, in particular, are expected to exhibit unprecedented levels of multidisciplinary interaction among the disciplines of structural dynamics, propulsion, aerodynamics, heat transfer, and controls engineering that may impose considerable constraint on the dynamic stability and controls performance margins stipulated for flight safety. The accurate prediction of flight characteristics is thus of utmost importance to ensure safe vehicle design and acceptable levels of performance.

Multidisciplinary analysis requires the capability to model the simultaneous influences of the fluids/aerodynamics, structural dynamics, propulsive forces, control system dynamics, and actuation in a coupled and integrated computational environment. All potential couplings between separate disciplines demand time-accurate solutions of the thermofluid–electromechanical interactions in a computational simulation of adequate fidelity for stability analysis.

The analysis methods are beyond the capabilities of simple closed-form solutions and it is necessary to develop numerical techniques to obtain approximate solutions. The choice of an appropriate methodology for modeling of the various associated solution domains and the accurate simulation of the relevant forces at their respective boundaries is of utmost importance to ensure effective interaction of relevant interdisciplinary phenomena. This must be done in a routine fashion if the methods are to be applied to practical design problems. The finite element method (FEM) has proven to be an effective and viable technique to model both the fluid and the solid continua and is the basis of the multidisciplinary analysis developed in this book. The analysis methodology, using the unifying finite element approach, enables the effective integration of each discipline in a simple and consistent manner, thus providing the necessary compact and coherent analysis framework. The primary elements of this multidisciplinary approach include structural dynamics and stress analysis, heat transfer, computational fluid dynamics, controls, and propulsion engineering. To encapsulate these within the FEM, it is necessary also to provide a three-dimensional unstructured grid (or element) generation capability and also a solids–fluids interface pressure interpolation module.

Examples of multidisciplinary problems include flutter and divergence of aircraft, traffic and wind loading on bridges and buildings, motion of turbine blades in fluids, and aerostructural-controls interaction in high performance aircraft. It is the purpose of this book to delineate the theory necessary for the development and use of software packages capable of undertaking some of these tasks. These software packages consist of a suite of integrated general purpose finite element codes for the solution of either linear or nonlinear multidisciplinary problems.

1.2 Areas of Analysis

There are a number of major interacting physical phenomena that provide the focus of our analysis: e.g., structural or solid mechanics (both static and dynamic), heat transfer (steady state and transient), and fluid mechanics and controls engineering. These topics[1,2,3] are introduced in the following sections of this chapter.

1.2.1 Structural Mechanics

The discipline of structural mechanics is diverse in nature. Both linear and nonlinear solutions are sought as appropriate for the determination of displacements and stress distributions as well as related stability characteristics under prescribed loads and boundary conditions. Typically the loads are mechanical and thermal in nature. Primary areas of analysis are as follows: static; vibration of conventional and rotating structures (with or without damping); stability; dynamic response; visco-elasticity, plasticity, and creep; fatigue and crack propagation; thermal stresses and displacements; and optimization. Figure 1.1 depicts a typical vibration mode shape of a generic hypersonic vehicle.

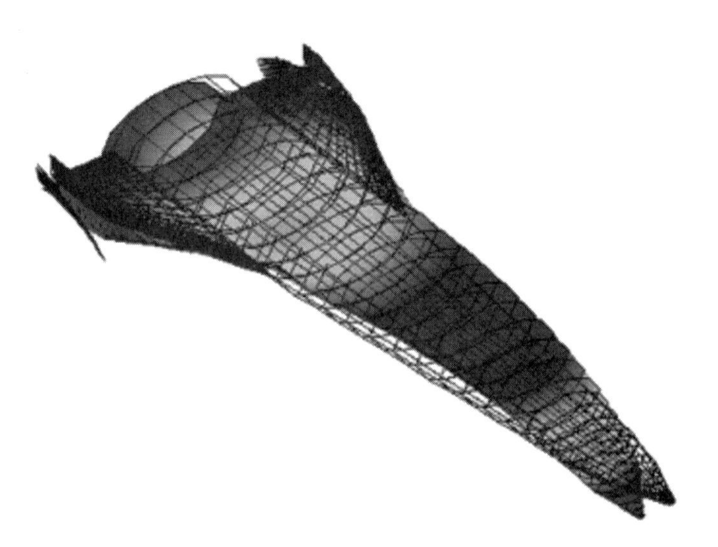

Fig. 1.1 Typical mode shape of a generic hypersonic vehicle.

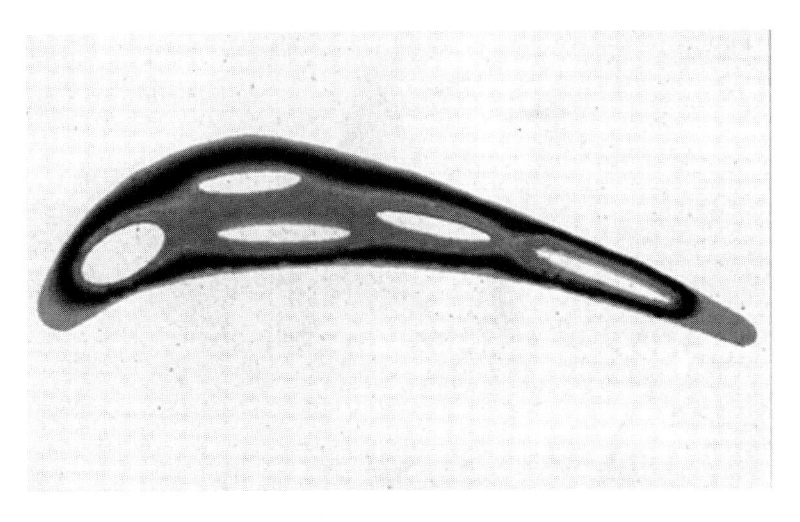

Fig. 1.2 Temperature contours in turbine blade cross section.

1.2.2 Heat Transfer

Heat transfer studies provide one of the interconnecting links in multidisciplinary analysis. Heat transfer analysis (linear or nonlinear) is concerned with heat conduction in structures and may involve either steady-state or transient solutions. Nonlinearities are caused by radiation boundary conditions as well as temperature-dependent material properties. Other heat transfer analyses include convection and radiation phenomena. Important analysis areas are steady state, transient, linear and nonlinear, conduction with various boundary conditions, convection, and radiation. A typical heat conduction analysis of a turbine blade with convection boundary conditions is shown in Fig. 1.2.

1.2.3 Fluid Mechanics

Fluid flow analysis by its very nature involves a wide variety of subject areas. Some of these topics are as follows: incompressible flow; compressible flow; inviscid flow; viscous flow; Newtonian fluid; non-Newtonian fluid; laminar, transition, and turbulent flow; steady-state flow; unsteady transient flow; internal and external flows; flowfield analysis; and chemically reactive flows. Figure 1.3 depicts the Mach distribution on the external surface of the SR-71, and Fig. 1.4 shows similar results for the internal flow in a vertical takeoff aircraft engine.

1.2.4 Controls

Suitable controllers are needed to ensure controlled motion as well as stability of engineering systems such as aerospace vehicles. Relevant topics are control law design, analog controller, digital controller, hybrid systems, state-space formulation, damping and frequencies, and phase and gain margins.

Fig. 1.3 Mach distribution on the surface of the SR-71.

1.2.5 Multidisciplinary Analysis

Multidisciplinary analysis involves interactions among a number of disciplines. Figure 1.5 shows a relevant block diagram encountered in aerospace engineering; suitable combination of the individual subject areas pertain to structural dynamics, aeroelastic, and aeroservoelastic analysis topics.

1.2.5.1 Aeroelasticity. This subject area concerns the interaction of fluids and structures, and some of the key analyses are unsteady aerodynamics-panel methods, unsteady aerodynamics–computational fluid dynamics (CFD) methods, flutter and divergence, and damping and frequencies. Figure 1.6 presents the results of a flutter analysis of a wing using finite element structural and CFD solutions (see also the color example at the front of this book).

1.2.5.2 Aeroservoelasticity. In the area of aeroservoelasticity involving aerostructural-controls interaction, state-space formulation, unsteady aerodynamics (panel and CFD methods), damping and frequencies, digital and analog

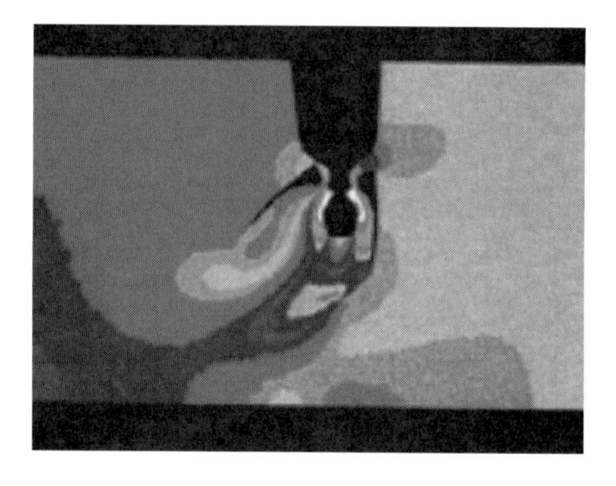

Fig. 1.4 Flow analysis for an aircraft engine.

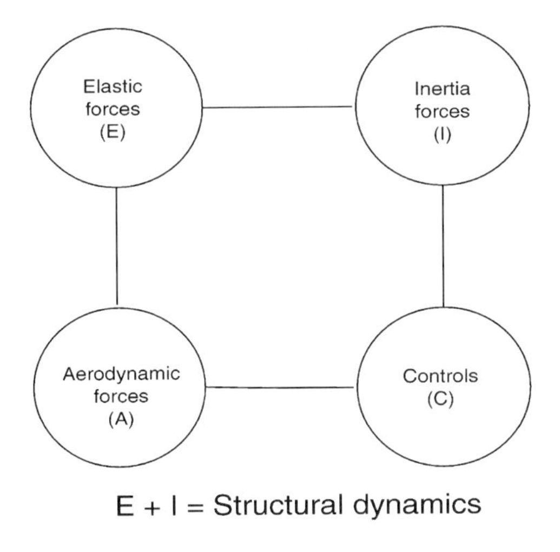

Fig. 1.5 Multidisciplinary analysis block diagram.

controllers, and frequency response gain and phase margins are of primary importance. Results of a typical aeroservoelastic analysis are summarized in Fig. 1.7, which depicts half of a symmetric aircraft structural and panel linear aerodynamic model, as well as the flight control system of the aircraft (see also the color example at the front of this book). The resulting damping and frequency values, representative of the stability characteristics of the aircraft, are also shown in the figure.

1.3 Methods of Analysis

Engineering problems arising in practice usually may be characterized by ordinary or partial differential equations with their associated boundary conditions.[1] Solving these problems may be achieved by either difference or integral formulations. The first approach replaces the derivatives with finite difference approximations. The later formulation is more versatile because it is also capable of accurate discretization of irregular solution domains and is based on the use of trial functions with undetermined parameters. In this approach either the weighted residual method or the stationary functional method may be used for the solution.

In the weighted residual method a trial solution, expressed in terms of undetermined parameters, is chosen that satisfies all boundary conditions, and when inserted in the differential equation yields a residual. The magnitude of the residual is required to be small in any of the four available procedures (see Fig. 1.8), and this in turn depends on the choice of the trial solution.

The stationary functional method is based on the variational approach and involves the determination of the extremum value of the chosen functional. The trial function that gives the minimum value of the functional is the approximate solution

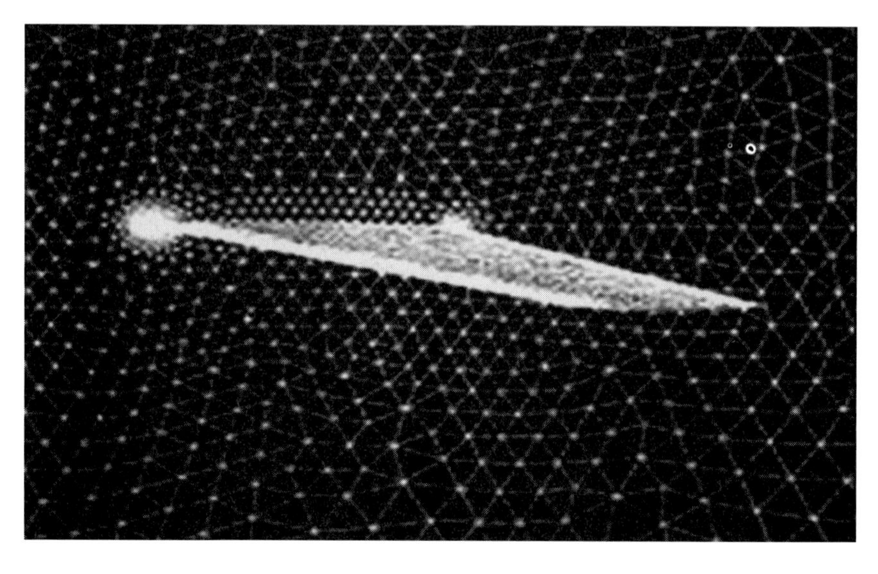

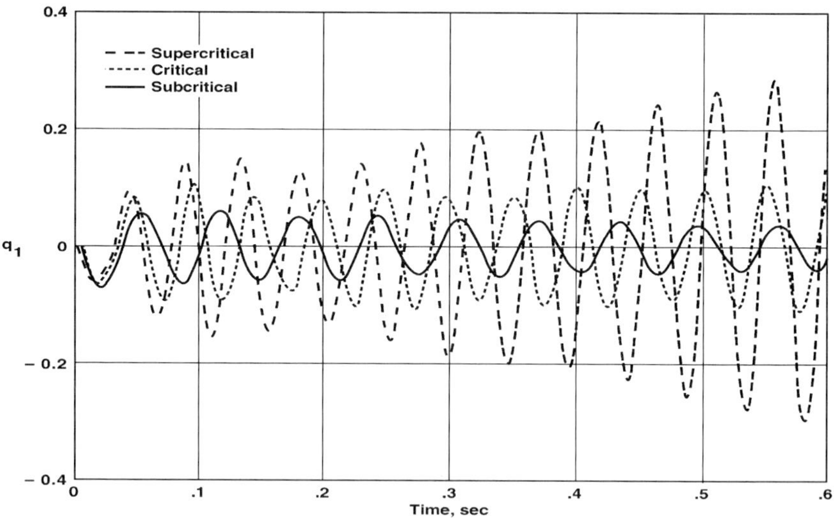

Fig. 1.6 Flutter analysis of an aircraft wing.

of the governing differential equation. In solid mechanics the functional used is the total potential energy or some variant of this quantity.

In the FEM, a continuum is discretized into smaller domains or elements. Trial solutions are sought within each element for which, in the variational method, the total potential energy is minimized. The idea is that the piecewise discretization will have a similar result for the entire continuum when summed over all elements. Similarly in the weighted residual method the procedure sets out to minimize the residual error for the entire continuum.

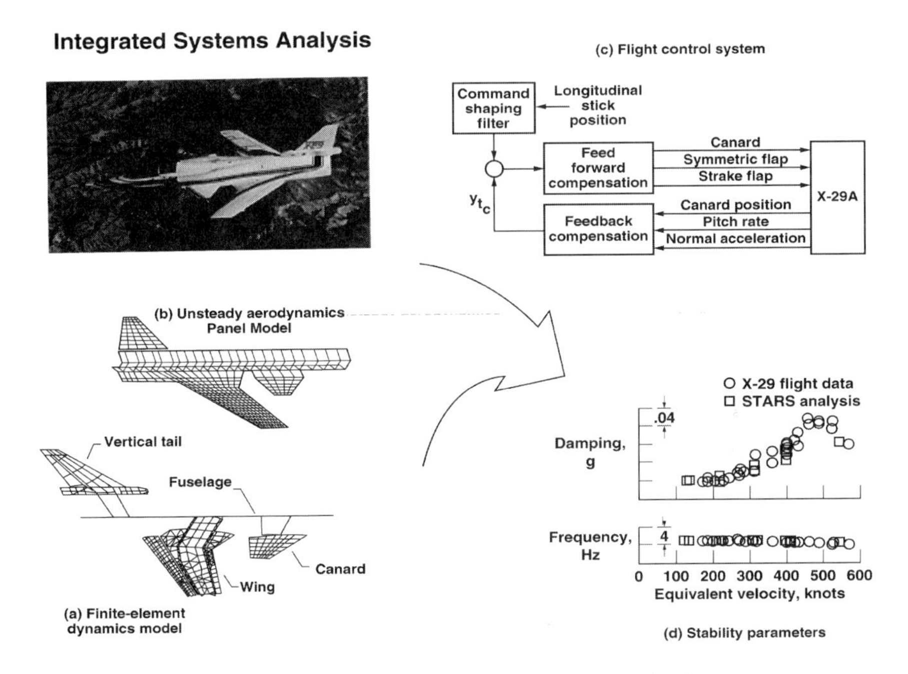

Fig. 1.7 Aircraft linear aeroservoelastic analysis.

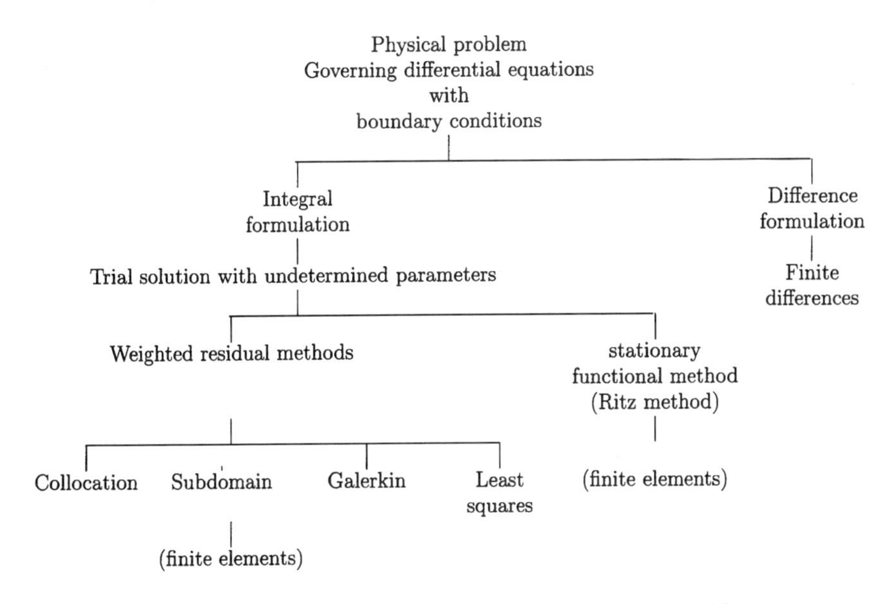

Fig. 1.8 Approximate solution methods of engineering problems.

The trial functions for the field variables within an element are chosen as a suitable interpolation of these values at the nodes of the element, and application of any one of the two integral formulations provides the desired element property matrices. Appropriate assembly of these element matrices yields the global system of equations that are modified to incorporate the boundary conditions. The solution gives the unknown nodal function values. From these nodal values other useful solution data can be obtained. For example, in the area of structural mechanics, nodal displacements are the calculated unknowns and the derived variables are the element stresses. The element relations are the stiffness matrices and the typical boundary conditions may be prescribed displacements and forces. Element stresses are calculated from the nodal displacements once they are determined from the solution of the system equations. In heat transfer analysis, nodal temperatures are the unknown quantities, and in fluid mechanics, density, velocity, and total energy are the common unknowns.

Physical engineering problems, based on their expected pattern of behavior may be classified as follows:

1) Equilibrium problems refer to time-independent, steady-state phenomena, e.g., static stresses in structures, steady fluid flow, and steady-state temperature distribution. These problems are characterized by ordinary or partial differential equations of the elliptic type with closed boundary conditions.

2) Eigenvalue problems are concerned with determination of critical parameters in addition to the associated steady-state configurations and are characterized by similar differential equations and boundary conditions as the equilibrium problem. Relevant examples include free vibration and stability of structures and resonance in acoustics.

3) Propagation problems refer to time-dependent, transient, and unsteady-state phenomena and are often termed initial value problems. They are characterized by partial differential equations of the hyperbolic or parabolic type with open boundary conditions. Thus, dynamic response of structures, unsteady pressure waves in fluids, and propagation of heat in a continuum are example problems in this category.

Numerical solution of practical engineering problems consist of two distinct yet related steps.

1) Discretize the continuum with the FEM, defining nodal geometry and element topology, and perform the following steps:

a) Define a piecewise continuous smooth solution trial function for each element and apply either the stationary functional or the weighted residual methods to obtain the element equations in terms of unknown nodal parameters.
b) Assemble system equations, the order of which will be the total number of nodal unknowns.
c) Apply the desired displacement boundary conditions to the set of system equations as well as applied mechanical and thermal loads.

2) Solve the appropriate system equations by a suitable numerical technique.

Subsequently, calculations are made of other required information derived from the basic solution of the physical problem being studied. In structural analysis, the element stresses are computed after obtaining element nodal displacements in the previous step.

For complex practical problems, a suitable discretization yields rather large sets of algebraic equations that occur in sparse, banded form. Economical solution of such equations is in itself a viable area of research as is the topic of the continuum discretization.

The variational method is applicable only for linear self-adjoint problems, whereas no such limitations exist for the weighted residual method. Also the existence of the variational principle for a given differential equation can be verified by the use of Frechet differentials. However, the variational method has the advantage that the functional contains derivatives of an order lower than that of the differential operator, and solutions may be sought for a larger class of functions. Accuracy of the two procedures is dependent on the choice of the trial functions. In general, the trial function may be constructed from polynomials. These functions must satisfy certain criteria for convergence of the solution, e.g., the constant strain state should be able to be modeled in elasticity problems. This being the case, higher-order functions usually give more accurate solutions for the same density of nodal points than those obtained from the use of simple linear functions. However, because of increased matrix bandwidth, there may not be an advantage in solution time for the same degree of accuracy.

1.4 Computer Software

In an effort to ensure accurate interaction of various disciplines in the context of multidisciplinary modeling and simulation of engineering systems, e.g., aerospace vehicles, the finite element discretization procedure is commonly employed for both solids and fluids continua. To achieve this objective, general purpose finite element codes are developed in each critical discipline area that are then suitably combined to yield an integrated multidisciplinary analysis tool. In addition, it is necessary to develop interfacing routines to facilitate transfer of analysis results from one discipline to another. Thus, in the area of fluids–structure interaction for an aerospace vehicle, results of the CFD analysis are interpolated from aerodynamic grid data on the vehicle surface to the structural nodes on the same surface; alternatively modal data are interpolated from structural to the fluids nodes. For aeroservoelastic problems involving aerostructural controls, interactions interface routines are also written that simulate patterns of interaction among the three disciplines. The optimization module in a code relates to the design and synthesis of an engineering system. In structural mechanics it involves, for example, the weight as the optimization parameter, whereas in aerodynamics, shape optimization, for maximum lift and minimum drag, is the usual criterion. Numerical analysis routines for manipulation of matrices derived from finite element discretization form an important, integral part of any multidisciplinary analysis program. These routines are designed to effect accurate and efficient solution of the relevant large sparse matrices.

A versatile preprocessor is also an essential module of the program. This routine must include automatic grid generation capability, particularly in connection with solid and fluid mechanics disciplines. Similarly a postprocessor module that displays various analysis results (including animation) in color is a crucial feature of the software. Finally such a system should be designed so that it is entirely user-friendly. In this regard the graphical user interface (GUI) is an integral part of the software. The various modules of an existing typical multidisciplinary analysis program are as shown in Fig. 1.9.

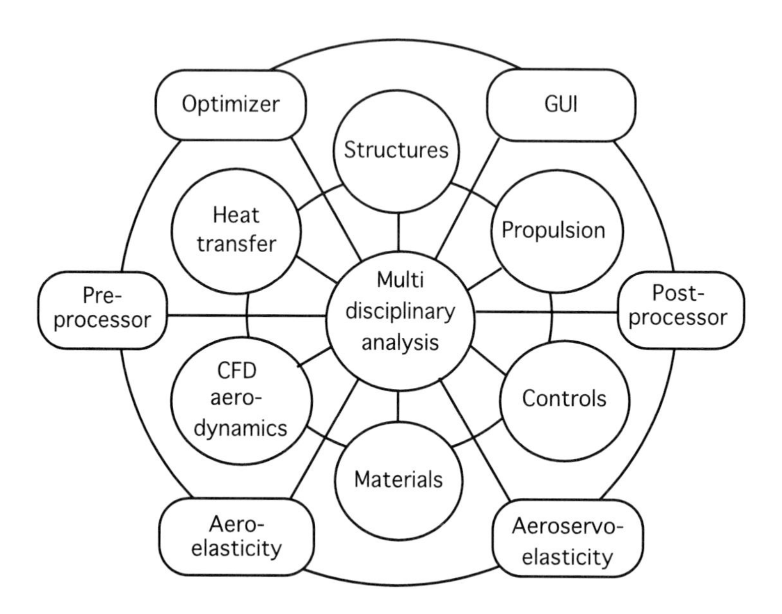

Fig. 1.9 Modules of a multidisciplinary simulation program.

1.5 Brief History of the Finite Element Method

This section traces the development of the ideas that led to the finite element method (FEM) of analysis in which the field equations of mathematical physics are approximated over simple regions (triangular, quadrilateral domains). They are then assembled together, so that equilibrium or continuity is satisfied at the interconnecting nodal points of the domains.

There are five groups of papers that may be considered in the development of FEM: Courant,[4] Argyris,[5] Turner et al.,[6] Clough,[7–9] and Zienkiewicz and Cheung.[10] Here the contribution of each of these five groups of papers is examined and we attempt to put them in their place in the history of the FEM. In this survey of the papers, they are examined by the following criteria:

1) Is the discretization technique clearly explained so that the reader will be capable of implementing it to model any practical problem with complex geometry and loading?

2) Has the paper explained how the solution convergence may be achieved?

3) Can the presentation in the paper be adapted routinely to automatic computation?

1.5.1 Courant

Courant[4] developed the idea of the minimization of a functional using linear approximation over subregions with the values being specified at discrete points that in essence become the node points of a mesh of elements. In his paper, Courant showed clearly the mesh subdivision used with one, two, three, and five approximating points to solve the St Venant's torsion of a square hollow box of

(2 × 2) with wall thickness of 1/4. Courant noted that St Venant's torsion problem can be solved using a stress function ϕ that has a zero value on the external boundary and constant value on any closed internal boundary. The shear stresses in the shaft are given as the first derivatives of the stress function ϕ. The condition of stationary potential energy is expressed. If ϕ is described in terms of a number of discrete parameters a_i, the stationary condition leads to a set of linear equations.

In the Appendix of his paper, Courant applied the preceding theory to the solution for the stress function ϕ for the (2 × 2) hollow square shaft, and introduced the idea of linear approximation to ϕ over a number of triangular areas. The meshes chosen show little understanding of the physics of the problem and certainly do not represent a study of convergence of the solution. Courant did not give any of the mathematical details of his piecewise linear approximation to the ϕ surface, but he did indicate a procedure that apparently could be used in the minimization of the total potential energy of the torsion problem.

1.5.2 Argyris

The series of papers under the title *Energy Theorems and Structural Analysis* by Argyris[5] is perhaps one of the most significant landmarks in structural mechanics of all time. This publication completely developed the matrix theory of structures for the discrete elements and then went on to show that this is only a particular case of the general continuum in which stresses and strains have been specified. This breakthrough led to the concept of flexibility and stiffness (as given by Argyris).

In these equations developed for both stiffness and flexibility, Argyris must be described as the father of modern structural mechanics. Argyris recognized completely the concept of duality between equilibrium and compatibility. The paper determines the stiffnesses of a rectangular panel for unit displacements in the local coordinate directions of the panel. The stiffness is an (8 × 8) matrix. Argyris noted, "Assume that the displacements vary linearly between nodal points. This assumption offends against the equilibrium conditions but its effect upon stiffness is not pronounced as long as we keep the unit panels reasonably small."

From this quotation we see that Argyris anticipated convergence with mesh refinement, although his examples do not explicitly prove the point. Argyris wrote, "Naturally the grid does not have to be restricted to this definition and we can always choose a finer one if the stiffeners are widely spaced so that the assumption of linear variation between adjacent nodal points can represent adequately the displacement pattern."

Thus Argyris developed the rectangular panel stiffness matrix in the state of plane stress from the point of view of element interpolation functions in terms of nodal displacements; i.e., he used orthogonal interpolation functions.

1.5.3 Turner

The third pioneering paper is that by Turner, Clough, Martin, and Topp.[6] After some discussion on rectangular plate elements it then turns to triangular elements: "The triangle is not only simpler to handle than the rectangle but later it will be used as the basic 'building bloc' for calculating stiffness matrices for plates of arbitrary shape."

The Turner paper uses the triangular element in the study of the deflections of a plate showing that for an irregular mesh, composed of triangles, the errors tend to disappear as the mesh is refined. Thus the Turner paper addresses the question of convergence.

1.5.4 Clough

In his most interesting paper,[7] Clough outlined the research program undertaken at the Boeing Company in 1952–1953 for the calculation of flexibility coefficients for low aspect ratio wing structures for dynamic analysis. Clough gave full credit to Turner for the invention of the triangular plane stress element in a state of constant stress. In his own work at Berkeley from 1957 onward, Clough's own contribution is significant because of his extension of Turner's work to the calculation of stresses and the verification that for known geometries and loading these stresses converged to the corresponding analytic solution. Clough outlined how he first invented the name *finite element method* in Refs. 8 and 9 because he wished to show the distinction between the continuum analysis and the matrix methods of structural analysis.

1.5.5 Zienkiewicz and Cheung

The development of the nonstructural applications by means of minimization of the total potential energy of a system is taken up again by Zienkiewicz and Cheung.[10] They analyze heat transfer and St Venant's torsion of prismatic shafts. The same functional as given in Ref. 4 is studied and the approximation to the function in terms of the nodal values of the triangular domain, into which the region is subdivided, is set up. Following the paper by Clough,[8] the approximation is now referred to as the finite element method. Because no details of the integration procedure are given by Courant,[4] we must nevertheless assume that he followed the same methods for the triangular domains given in Ref. 10, as any other method would have been too laborious. These details were provided by Zienkiewicz and Cheung.[10] Because each triangle contributes to only three nodes, the summation evidently leads to sparse equations, which as was noted by Zienkiewicz, can be highly banded. Thus the notion of the topology of the system was introduced. The concept of continuity of the function along common element boundaries is also discussed.

1.6 Concluding Remarks

In summary, a careful examination of the paper by Courant[4] shows that he apparently used a finite element type of procedure in a potential energy minimization of a functional for the torsion stress function using grid point values as the unknown parameters. The word "apparently" is used with caution because no details of the calculations were given. There is no indication that the calculation of the mesh integrals could be made in a repetitive fashion using the mesh topology, and the convergence study was inconclusive.

The Argyris paper[5] is in a different category. Here, for the first time, the numerical techniques necessary for the application of the Gauss divergence theorem (principles of virtual displacements and forces) are set out. These matrix methods can become the basis for nearly all stress analysis applications of the FEM. Argyris successfully demonstrated how the method may be applied to a rectangular planar elasticity element for which interelement displacement compatibility is assured by the choice of linear functions along element edges. This is a fundamental paper in the development of FEM and, because it was the first published, should be considered as the originating point of FEM. The publication *Energy Theorems and Structural Analysis*[5] is a classic work in the development of structural mechanics. However Argyris did not undertake convergence studies on his rectangular element, although the notion is anticipated in his paper.

The paper by Turner et al.[6] was written apparently without knowledge of the Argyris work, and develops independently the theory for the stiffness of a planar triangular element in a state of constant stress. In this paper the study of the convergence characteristics of planar elements is given, comparing the new approach with simple theory and the relaxation method for the deflections of a cantilever beam. This pioneering paper, which was published a little later than the Argyris work, was apparently developed at about the same time. Clough in Ref. 7 acknowledged the Turner contribution when he wrote, "Also it should be recognized that the principal credit for conceiving the procedure should go to M. J. Turner, who not only led the developmental effort for the two critical years of 1952–1953, but who also provided the inspiration." Clough's contribution was to continue convergence studies on stress components and to popularize the ideas by giving them the name *finite element method*. Clough also gave lectures on the method in the spring of 1958.

The function minimization techniques, referred to so obliquely by Courant,[4] were finally clarified by Zienkiewicz and Cheung[10] in 1965 and opened the way to the analysis of field problems by the FEM. It is also quite apparent that the FEM technique was essentially conceived and developed in the industry.

References

[1]Crandall, S. H., *Engineering Analysis*, McGraw–Hill, New York, 1955.

[2]Streeter, V. L., *Fluid Dynamics*, McGraw–Hill, New York, 1984.

[3]Nelson, R. C., *Flight Stability and Automatic Control*, McGraw–Hill, New York, 1989.

[4]Courant, R., "Methods for the Solution of Problems of Equilibrium and Vibrations," *Transactions of the American Mathematical Society*, June 1942, pp. 1–23.

[5]Argyris, J. H., "Energy Theorems and Structural Analysis Part 1," *Aircraft Engineering*, Vols. 26, 27, Oct. 1954, May 1955.

[6]Turner, M. J., Clough, R. W., Martin, H. C., and Topp, L. T., "Stiffness and Deflection Analysis of Complex Structures," *Journal of Aeronautical Sciences*, Vol. 25, No. 9, 1956, pp. 805–823.

[7]Clough, R. W., "Original Formulation of the Finite Element Method," *Proceeding of the ASCE Structures Congress Session on Computer Utilization in Structural Engineering*, American Society of Civil Engineers, New York, 1989, pp. 1–10.

[8]Clough, R. W., "The Finite Element Method in Plane Stress Analysis," *Proceedings of the 2nd ASCE Conference on Electronic Comp.* American Society of Civil Engineers, New York, 1960.

[9]Clough, R. W., and Wilson, E. W., "Stress Analysis of a Gravity Dam by the Finite Element Method," *Proceedings of the Symposium on the Use of Computers in Civil Engineering*, Lab. Nacional de Engenharia Civil, Lisbon, Portugal, 1962.

[10]Zienkiewicz, O. C., and Cheung, Y. K., "Finite Elements in the Solution of Field Problems," *The Engineer*, Sept. 1965, pp. 507–510.

2
Finite Element Discretization of Physical Systems

2.1 Introduction

A field problem in engineering science is associated with a domain of specified geometry within which the field variables are required to be investigated. The physical behavior represented by these field variables is characterized by differential equations and must satisfy the given boundary conditions. Except in simple cases, closed-form solutions are not available and recourse is made to approximate solutions obtained from numerical analysis techniques. Approximate solutions of the differential equations are obtained by several techniques including Fourier analysis, finite difference techniques, Rayleigh–Ritz functions over the whole region, or by functions prescribed over discretized finite regions of the original domain, to name but a few of the methods available. All of these have their particular appeal for different applications.

The method developed in the present book is, of course, the last and perhaps latest of these techniques and has come to be known as the finite element method. In this technique an approximate numerical solution is obtained by satisfying the conditions of the differential equations (equilibrium, continuity, etc.) at discrete points. The accuracy being improved by successive refinement of the size of the contributing domains and the consequent increase in the points at which the field conditions are satisfied. In setting up the nodal equations for approximation of the differential equations, a variety of techniques can be used.

For example, in the weighted residual method, an approximate solution of the differential equation is sought within each element by choosing a set of trial functions exclusive to that element. Depending on the parameters chosen, there can be various families of possible approximations within an element. For one such family, any one of four suitable criteria (given later in this chapter) may be prescribed in the weighted residual method. The element characteristic relationships are obtained by applying the weighted residual method directly to the differential equation within the element and applying the element boundary conditions.

In the stationary functional method, a functional is first identified that relates to the phenomena being studied in the differential equation. A trial solution with undetermined parameters that are chosen to minimize the value of the functional is then taken to be the approximate solution of the differential equation. Accuracy depends on the correct choice of functional as well as the trial functions used. For this variational approach a necessary requirement is the existence of a functional. The weighted residual method that does not have this restriction is more versatile in its application.

In the application of the finite element method (FEM) to structural mechanics, either the variational approach or the direct application of the Gauss divergence theorem via the dual principles of virtual displacements and virtual forces leads to elegant methods for setting up the element properties and are the preferred methods. For field problems the method of weighted residuals proves to be both versatile and convenient over a wide range of applications. Once the individual element matrices relating nodal variables are obtained, the topology of the system (element node numbering) is used to construct the set of algebraic equations for the entire region. When the appropriate boundary conditions are applied, the solution of the equations yields the piecewise approximation to the original differential equations. In this chapter details of the two procedures are given. A simple example demonstrates their application. References 1–20 provide detailed information on the various relevant techniques.

2.2 Finite Element Solutions

Equilibrium problems are characterized by differential equations with closed boundary conditions, specified on the entire boundary of the domain. Mathematically, the problem is to determine the function ψ that satisfies the differential equation,

$$L_{2m}(\psi) - f = 0 \tag{2.1}$$

within the given domain and subject to the given boundary conditions,

$$B(\psi) - g = 0 \tag{2.2}$$

in which the differential operator $L_{2m}(\psi)$ contains ψ and its derivatives up to the order $2m$, and f is a known function of ψ. Similarly, the eigenvalue and propagation problems may be formulated in the same manner. In the finite element solution technique, trial functions with undetermined parameters are chosen to approximate the function that is given by $\hat{\psi}$ in Eq. (2.3):

$$\psi \approx \hat{\psi} = \boldsymbol{N}\hat{\boldsymbol{\psi}} \tag{2.3}$$

In Eq (2.3), $\boldsymbol{N}$ is the row vector of known shape functions within the element and $\hat{\boldsymbol{\psi}}$ are the corresponding undetermined parameters. Substitution in the differential equation (2.1) results in a residual quantity e that indicates the error in the approximation to the solution. This is expressed by

$$L_{2m}(\hat{\psi}) - f = e \tag{2.4}$$

It is the intention of the FEM to choose a solution that attempts to minimize e within each element in some viable way. The various methods are discussed in the next two sections.

2.2.1 Residual Minimization Techniques

The residual minimization techniques presume that a "trial" function, e.g., a polynomial series that has been used to approximate the dependent variable of a problem in mathematical physics, does not in general satisfy the relevant differential equation and its boundary conditions exactly at every point of the domain. Substitution into the differential equation will result in a residual error. The objective of the numerical analysis is to find a best-fit solution. To obtain the best-fit solution, the value of the residual is minimized in some way and because the method is approximate, as mentioned in Sec. 1.3, there are at least four ways to achieve such an approximation. These are classified under the headings *collocation*, *subdomain integration*, *Galerkin*, and *least squares*.

Each method must produce slightly different results for a given subdivision of the domain, but it is a necessary requirement of any suitable method that for successive refinement of the domain, convergence to a common solution is obtained. This condition is checked by analyzing problems for which analytic solutions are available.

Let e be the error obtained when a trial function is substituted into the differential equation. Then the four methods may be summarized as

$$\text{collocation} \qquad \sum_{i=1}^{n} e_i = 0 \tag{2.5}$$

$$\text{subdomain integration} \qquad \int_V e \, \mathrm{d}V = 0 \tag{2.6}$$

$$\text{Galerkin} \qquad \int_V \boldsymbol{N}^T e \, \mathrm{d}V = 0 \tag{2.7}$$

$$\text{least squares} \qquad \int_V e^2 \, \mathrm{d}V = 0 \tag{2.8}$$

2.2.1.1 Collocation method. The sum of the residuals at r points chosen within each element is set to zero. In this way a set of simultaneous algebraic equations for the nodal unknowns is generated. Obviously the choice of points in the elements is critical to the accuracy of the method.

2.2.1.2 Subdomain integration. In this technique the element is subdivided into r cells, each of simple defined geometry. Now the integral of the residual over each subdomain is set to zero resulting in r equations.

2.2.1.3 Galerkin method. In the Galerkin method, the integral over the element (subdomain) of the weighted residual of the error is set equal to zero. There will be as many equations per element as there are weighted residuals, i.e., one equation for each nodal variable. In applications in structural mechanics e will be a force residual of the equations of equilibrium. Then the integral in Eq. (2.7) represents the work done by the generalized displacement N_i, i.e., although equilibrium may not be satisfied point by point throughout the element, it is satisfied in the generalized (shape function) modes. Hopefully, the more modes taken for the shape functions, the better will be the point-by-point convergence to equilibrium.

If the differential equation describing the field variable ψ is written symbolically as $L_{2m}(\hat{\psi}) - f$, then the Galerkin condition equation (2.7) for the approximate solution $\hat{\psi}$ for the element is written as

$$\int_V \boldsymbol{N}^T [L_{2m}(\boldsymbol{N}\hat{\boldsymbol{\psi}}) - f] \, \mathrm{d}V = 0 \tag{2.9}$$

2.2.1.4 Least-squares method. As given in Eq. (2.8), in this method the integral of the square of the residual obtained by substituting the trial function in the differential equation is set equal to zero, again leading to as many element equations as there are trial functions.

2.2.2 Stationary Functional Method

In this procedure, based on the variational approach, a functional $\boldsymbol{\Pi}$ is chosen such that a stationary value of the functional yields an approximate solution of the governing differential equation. A trial solution of the system is substituted into the functional, which is minimized to yield the desired solution. This trial solution of Eq. (2.1) may be expressed as

$$\psi = \boldsymbol{c}^T \boldsymbol{\phi} \tag{2.10}$$

where $\boldsymbol{c}$ are the undetermined parameters and $\boldsymbol{\phi}$ are linearly independent known functions within the element. In the Ritz variational method these coefficients are determined by setting the derivatives of the total potential energy $\boldsymbol{\Pi}$ with respect to $\boldsymbol{c}$ to zero resulting in the r equations:

$$\frac{\mathrm{d}\boldsymbol{\Pi}}{\mathrm{d}\boldsymbol{c}} = 0 \tag{2.11}$$

Solving for $\boldsymbol{c}$ and substitution in Eq. (2.10) gives the approximate solution of the system.

2.3 Application of the Galerkin Method

In this section an example is given of the application of the Galerkin method [Eq. (2.9)]. This is a simple linear problem in nondimensional form, applicable to such physical problems as the deformation of a bar under axial force and the steady-state heat transfer in the same bar with given temperature boundary conditions. In this problem several steps in the analysis become evident: 1) the element property matrix derivation, 2) the element boundary conditions, 3) the assembly of the element property matrices, 4) implementation of boundary conditions, and 5) solution of the linear algebraic equations.

Each of these items has an important role in the overall numerical solution of the finite element approximation. In this example, the Gauss divergence theorem will be applied to Eq. (2.9) to derive the element property matrices. The Gauss divergence theorem relates the rate of change of a field variable within a body to the flux across its external boundary. That is, for a continuous function $\boldsymbol{\psi}$ components (ψ_i), differentiable and single valued in the region V and bounded by the surface S, it can be shown that

$$\int_V \psi_{i,i}\, \mathrm{d}V = \int_S \psi_i n_i\, \mathrm{d}S \tag{2.12}$$

In Eq. (2.12), i takes the values 1, 2, or 3 depending on whether the problem domain is in one-, two-, or three-dimensional space. The notation $\psi_{i,i}$ implies differentiation with respect to the i independent variable, and a repeated index, unless noted otherwise, means a summation over the range of the variable dimension. The term n_i is the component of the surface normal vector.

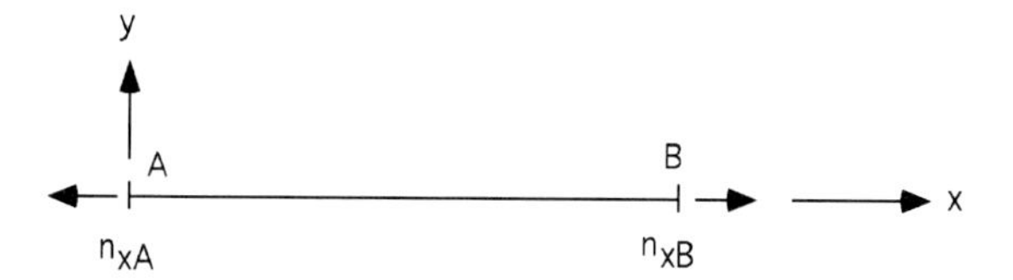

Fig. 2.1 Line element *A-B*.

Example 2.1: Line Element

For the line element A-B shown in Fig. 2.1, the Eq. (2.12) reduces to Eq. (2.14), since, from Fig. 2.1,

$$n_{xB} = 1 = -n_{xA} \tag{2.13}$$

i.e.,

$$\int_A^B \psi_{,x} = n_{xA}\psi_A + n_{xB}\psi_B \tag{2.14}$$

Consider now the second-order differential equation,

$$D\psi_{,ii} + Q = 0 \tag{2.15}$$

In the most general case both D and Q can be functions of x but for this example they are assumed to be constant. A solution to this differential equation is required for ψ over a length RS of which AB is one segment. At the extremities of RS there will be some boundary conditions on ψ, and because of Q there will be internal variation over the length of RS, which is divided into equal segments, each of length l by the points 1–6. Define the ψ values at these points; AB is one region of RS.

In any element AB, the approximate solution $\hat{\psi}$ is given by

$$\hat{\psi} = [N_1 \quad N_2]\begin{bmatrix} \hat{\psi}_A \\ \hat{\psi}_B \end{bmatrix} = \boldsymbol{N}\hat{\boldsymbol{\psi}} \tag{2.16}$$

The linear functions N_1, N_2 are shown in Fig. 2.2. From Fig. 2.2, the interpolation functions are expressed as

$$N_1 = 1 - \frac{x}{l} = \zeta_1; \qquad N_2 = \frac{x}{l} = \zeta_2 \tag{2.17}$$

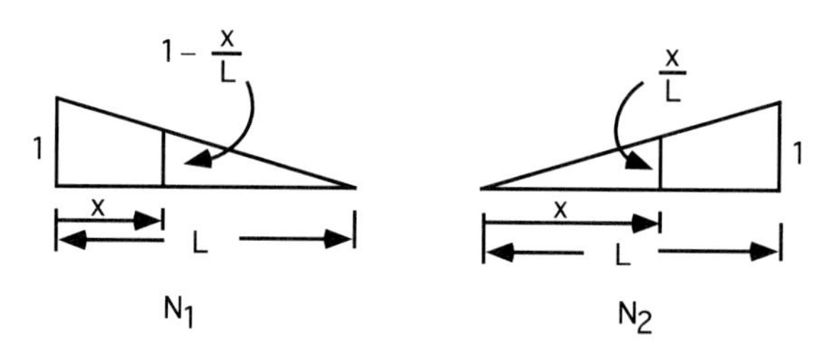

Fig. 2.2 Linear line element shape functions.

Consider the function of x,

$$DN_i \hat{\psi}_{,x} \qquad i = 1 \text{ or } 2 \tag{2.18}$$

and integrate over AB. Then Gauss divergence theorem equation (2.12) gives

$$\int_A^B (DN_i \hat{\psi}_{,x})_{,x} \, \mathrm{d}x = n_A (DN_i \hat{\psi}_{,x})_A + n_B (DN_i \hat{\psi}_{,x})_B \tag{2.19}$$

The first term on the right-hand side of Eq. (2.19) is equal to zero if $N_i = N_2$, and the second term is equal to zero if $N_i = N_1$. The left-hand side of Eq. (2.19) is expanded, so that

$$\int_A^B (DN_i \hat{\psi}_{,x})_{,x} \, \mathrm{d}x = \int_A^B DN_{i,x} \, \mathrm{d}x + \int_A^B DN_i \hat{\psi}_{,xx} \, \mathrm{d}x \tag{2.20}$$

Now the Galerkin method is used, in which the weighted residual is set equal to zero, i.e.,

$$\int_A^B N_i (D\hat{\psi}_{,xx} + Q) \, \mathrm{d}x = 0 \tag{2.21}$$

Then recalling the boundary conditions on A-B, the two equations obtained from Eqs. (2.15) and (2.16) are written as

$$\int_A^B DN_{1,x} \hat{\psi}_{,x} \, \mathrm{d}x = \int_A^B N_1 Q \, \mathrm{d}x - D\hat{\psi}_{,x}|_A \tag{2.22}$$

$$\int_A^B DN_{2,x} \hat{\psi}_{,x} \, \mathrm{d}x = \int_A^B N_2 Q \, \mathrm{d}x - D\hat{\psi}_{,x}|_B \tag{2.23}$$

Now, from Eq. (2.16),

$$\hat{\psi}_{,x} = [N_{1,x} \quad N_{2,x}] \begin{Bmatrix} \hat{\psi}_A \\ \hat{\psi}_B \end{Bmatrix} \tag{2.24}$$

Hence Eqs. (2.22) and (2.23) can be combined with Eq. (2.24)

$$\int_A^B D \, \boldsymbol{B}\boldsymbol{B}^T \, \mathrm{d}x \begin{Bmatrix} \hat{\psi}_A \\ \hat{\psi}_B \end{Bmatrix} = \int_A^B \begin{Bmatrix} N_1 Q \\ N_2 Q \end{Bmatrix} \mathrm{d}x + \begin{Bmatrix} -D\hat{\psi}_{,x}|_A \\ D\hat{\psi}_{,x}|_B \end{Bmatrix} \tag{2.25}$$

in which $\boldsymbol{B}$ is given by the coefficient matrix in Eq. (2.24). Equation (2.25) is written symbolically as

$$\boldsymbol{K}^{(e)} \begin{Bmatrix} \hat{\psi}_A \\ \hat{\psi}_B \end{Bmatrix} = \boldsymbol{f}^{(e)} + \begin{Bmatrix} -D\hat{\psi}_{,xA} \\ D\hat{\psi}_{,xB} \end{Bmatrix} \tag{2.26}$$

Usually the superscript e, indicating the element level, will be omitted for clarity because it will be evident when element-level matrices are being considered. Written in terms of all its components, Eq. (2.26) is shown in Eq. (2.27). This is a useful form when the element-level matrices are being added into the global equations of the whole system:

$$\begin{bmatrix} K_{AA} & K_{AB} \\ K_{BA} & K_{BB} \end{bmatrix} \begin{Bmatrix} \hat{\psi}_A \\ \hat{\psi}_B \end{Bmatrix} = \begin{Bmatrix} f_A \\ f_B \end{Bmatrix} + \begin{Bmatrix} -D\hat{\psi}'_A \\ D\hat{\psi}'_B \end{Bmatrix} \tag{2.27}$$

In Eq. (2.27) the prime indicates differentiation with respect to x and thus gives an end slope of $\hat{\psi}$. In the example, from Fig. 2.2, the interpolation functions

$$\begin{Bmatrix} N_1 \\ N_2 \end{Bmatrix} = \begin{Bmatrix} \zeta_1 \\ \zeta_2 \end{Bmatrix} \tag{2.28}$$

and their derivatives are given as

$$\begin{Bmatrix} N_{1,x} \\ N_{2,x} \end{Bmatrix} = \frac{1}{l} \begin{Bmatrix} -1 \\ 1 \end{Bmatrix} \tag{2.29}$$

and its substitution in Eq. (2.25) gives

$$\boldsymbol{K}_{AB} = \frac{D}{l} \begin{bmatrix} 1 & -1 \\ -1 & 1 \end{bmatrix} \tag{2.30}$$

If Q is constant over AB,

$$\int_A^B N_1 Q \, \mathrm{d}x = \int_A^B N_2 Q \, \mathrm{d}x = Ql/2 \tag{2.31}$$

Consider now an interior point 2 in Fig. 2.3. For segment 1,

$$\frac{D}{l}(-1)\hat{\psi}_1 + \frac{D}{l}(1)\hat{\psi}_2 - \frac{Ql}{2} = D\hat{\psi}'_{(1)-2} \tag{2.32}$$

$$\frac{D}{l}(-1)\hat{\psi}_2 + \frac{D}{l}(1)\hat{\psi}_3 - \frac{Ql}{2} = D\hat{\psi}'_{(2)-1} \tag{2.33}$$

For continuity of the function at node 2,

$$D\hat{\psi}'_{(1)-2} = D\hat{\psi}'_{(2)-1} \tag{2.34}$$

Adding Eqs. (2.32) and (2.33) gives the approximate continuity equation at node 2,

$$\frac{D}{l}[-1 \quad 2 \quad -1] \begin{Bmatrix} \hat{\psi}_1 \\ \hat{\psi}_2 \\ \hat{\psi}_3 \end{Bmatrix} = Ql \tag{2.35}$$

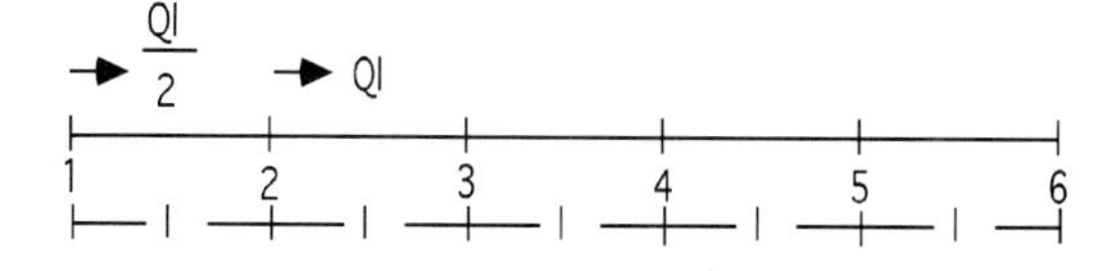

Fig. 2.3 Domain subdivision.

For all of the node points 1–6 we see that the combined equations are written as

$$\begin{bmatrix} 1 & -1 & \cdot & \cdot & \cdot & \cdot \\ -1 & 2 & -1 & \cdot & \cdot & \cdot \\ \cdot & -1 & 2 & -1 & \cdot & \cdot \\ \cdot & \cdot & -1 & 2 & -1 & \cdot \\ \cdot & \cdot & \cdot & -1 & 2 & -1 \\ \cdot & \cdot & \cdot & \cdot & -1 & 1 \end{bmatrix} \begin{Bmatrix} \hat{\psi}_1 \\ \hat{\psi}_2 \\ \hat{\psi}_3 \\ \hat{\psi}_4 \\ \hat{\psi}_5 \\ \hat{\psi}_6 \end{Bmatrix} = \frac{Ql^2}{D} \begin{Bmatrix} \frac{1}{2} \\ 1 \\ 1 \\ 1 \\ 1 \\ \frac{1}{2} \end{Bmatrix} + \begin{Bmatrix} -l\hat{\psi}'_{(1)-1} \\ 0 \\ 0 \\ 0 \\ 0 \\ l\hat{\psi}'_{(5)-2} \end{Bmatrix} \tag{2.36}$$

These equations cannot be solved because the coefficient matrix on the left-hand side is singular. The boundary conditions must now be applied. Choose $\hat{\psi}_1 = 0$ and let

$$\hat{\psi}'_{(5)-2} = Ql/D \tag{2.37}$$

The resulting equations to be solved are

$$\begin{bmatrix} 2 & -1 & \cdot & \cdot & \cdot \\ -1 & 2 & -1 & \cdot & \cdot \\ \cdot & -1 & 2 & -1 & \cdot \\ \cdot & \cdot & -1 & 2 & -1 \\ \cdot & \cdot & \cdot & -1 & 1 \end{bmatrix} \begin{Bmatrix} \hat{\psi}_2 \\ \hat{\psi}_3 \\ \hat{\psi}_4 \\ \hat{\psi}_5 \\ \hat{\psi}_6 \end{Bmatrix} = \frac{Ql^2}{D} \begin{Bmatrix} 1 \\ 1 \\ 1 \\ 1 \\ \frac{1}{2} \end{Bmatrix} \tag{2.38}$$

Solving these equations makes the $\hat{\psi}$ values:

$$\hat{\psi}_2 = 5.5$$
$$\hat{\psi}_3 = 10.0$$
$$\hat{\psi}_4 = 13.5$$
$$\hat{\psi}_5 = 16.0$$
$$\hat{\psi}_6 = 17.5$$

A second solution is obtained for $Q = 0$ and $\hat{\psi}'_{(5)-2} = 20.0$. These results are

$$\hat{\psi}_2 = 20.0$$
$$\hat{\psi}_3 = 40.0$$
$$\hat{\psi}_4 = 60.0$$
$$\hat{\psi}_5 = 80.0$$
$$\hat{\psi}_6 = 100.0$$

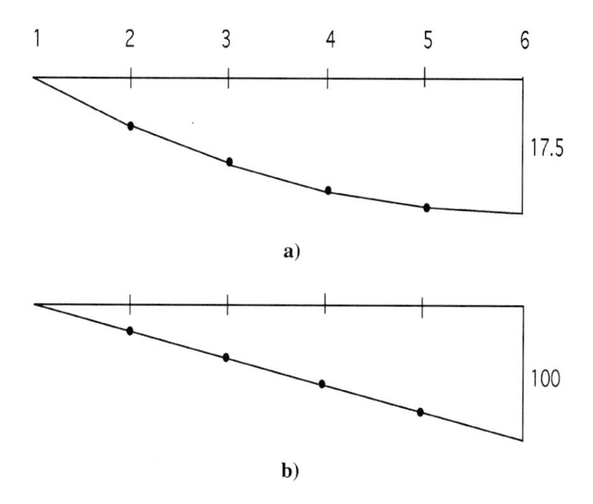

Fig. 2.4 Approximate solutions for $\hat{\psi}$.

The values of $l\hat{\psi}'_{(1)-1}$, obtained by substituting in Eq. (2.35), are 5.5 and 20.0, respectively. The values of $\hat{\psi}$ are plotted in Fig. 2.4.

It is seen in Fig. 2.4a that the effect on $\hat{\psi}$ is parabolic in nature because the right-hand side of Eq. (2.38) represents a constant second derivative. In Fig. 2.4b, the concentrated value of 20 on the unconstrained point 6 produces a linear variation of $\hat{\psi}$, i.e., its second derivative is equal to zero.

2.4 Concluding Remarks

In this chapter the basic idea of the discretization of a domain into a number of finite and distinct regions was introduced. In each of these regions the field variables (displacements, temperature, stress functions, and so forth), were approximated by interpolation functions using a given number of controlling nodal values. That is, it was recognized that closed form analytic solutions were not available and numerical techniques must be devised for an approximate solution. How the relationships are set up depends on the method by which the error function is minimized and also on a number of concepts that were introduced in this chapter (e.g., collocation, Galerkin, least squares). In structural problems this can be achieved by the minimization of the total potential energy function or alternatively, very succinctly, through the use of the principles of virtual displacements and virtual forces.

References

[1]Argyris, J. H., *Energy Theorems and Structural Analysis*, Butterworths, London, 1960 (reprinted from *Aircraft Engineering*, Vols. 26, 27, 1954–55).

[2]Bathe, K. J., and Wilson, E. L., *Numerical Methods in Finite Element Analysis*, Prentice–Hall, Englewood Cliffs, NJ, 1976.

[3]Crandall, S. H., *Engineering Analysis*, McGraw–Hill, New York, 1955.

[4]Chung, T. J., *Finite Element Analysis in Fluid Mechanics*, McGraw–Hill, New York, 1978.

[5]Connor, J. J., and Brebbia, C. A., *Finite Element Technique for Fluid Flow*, Butterworths, London, 1976.

[6]Cook, R. D., Malkus, D. S., and Plesha, M. E., *Concepts and Applications of Finite Element Analysis*, 3rd ed., Wiley, New York, 1988.

[7]Desai, C. S., and Abel, J. F., *Introduction to the Finite Element Method*, Van Nostrand Reinhold, New York, 1972.

[8]Dhatt, G., and Touzot, G., *The Finite Element Displayed*, Wiley, New Delhi, India, 1984.

[9]Finlayson, B. A., *The Method of Weighted Residuals and Variational Principles*, Academic, New York, 1972.

[10]Gallagher, R. H., *Finite Element Analysis Fundamentals*, Prentice–Hall, Englewood Cliffs, NJ, 1975.

[11]Huebner, K. H., *The Finite Element Method for Engineers*, Wiley, New York, 1975.

[12]Irons, B. M., and Ahmad, S., *Techniques of Finite Elements*, Ellis Harwood, Chichester, England, U.K. 1978.

[13]Meek, J. L., *Computer Methods in Structural Analysis*, E&F Sponn, London, 1990.

[14]Oden, J. T., *Finite Elements of Non-Linear Continua*, McGraw–Hill, New York, 1972.

[15]Przemieniecki, J. S., *Theory of Matrix Structural Analysis*, McGraw–Hill, New York, 1968.

[16]Segerlind, L. J., *Applied Finite Element Analysis*, Wiley, New York, 1976.

[17]Weaver, W., Jr., and Johnston, P. R., *Finite Elements for Structural Analysis*, Prentice–Hall, Englewood Cliffs, NJ, 1984.

[18]Washizu, *Variational Mehods in Elasticity and Plasticity*, 2nd ed., Pergamon, New York, 1975.

[19]Yang, T. Y., *Finite Element Structural Analysis*, Prentice–Hall, Englewood Cliffs, NJ, 1986.

[20]Zienkiewicz, O. C., *The Finite Element Method in Engineering Science*, 4th ed., McGraw–Hill, New York, 1989.

3
Structural Mechanics—Basic Theory

3.1 Introduction

Analysis of a structural system involves computation of deformations and stresses due to externally applied forces such as mechanical and thermal loads, the magnitudes of which are indicators of a safe design. This analysis in the present context involves first a finite element discretization of the continuum, e.g., for a static problem, resulting in a set of simultaneous algebraic equations that can be solved to yield the required unknown variables. The relevant analysis procedure is composed of the following basic steps:

1) Idealize the continuum as a set of smaller regions known as finite elements.

2) Select nodes at interelement boundaries and element interiors for the purpose of setting up of interpolating functions (see Fig. 3.1).

3) Use interpolation functions to express displacement values at element interior points in terms of nodal variables.

4) Develop element force–displacement matrices by applying either the variational principles or the weighted residual method.

5) Assemble equilibrium matrices for the entire region in global coordinates for all of the element matrices, and solve the resulting set of algebraic equations for the unknown nodal values.

6) Calculate element stresses and strains from the calculated nodal displacements.

In the following chapters the analysis procedures will be presented in detail. First, however, some of the fundamental precepts of the theory of elasticity are set out followed by details of structural modeling and simulation procedures.

3.2 Modeling of Material Behavior

Materials in structural mechanics, either manufactured or occurring naturally, tend to exhibit complex patterns of behavior that may be linear or nonlinear, elastic or elastoplastic, a function of temperature, loading rate, or time, and have a dependence on the past strain history. Constitutive models of such materials may be constructed from laboratory experiments and field tests and the resulting relationships simplified for use in the subsequent structural analyses.

3.2.1 Basic Equations of the Theory of Elasticity

For the numerical methods of structural analysis it is necessary to develop the matrix equations of the theory of elasticity.[1,2] In the linear, small strain theory presented herein, displacements and rotations are considered to be small and the material is assumed to exhibit linear elastic behavior between the applied stresses

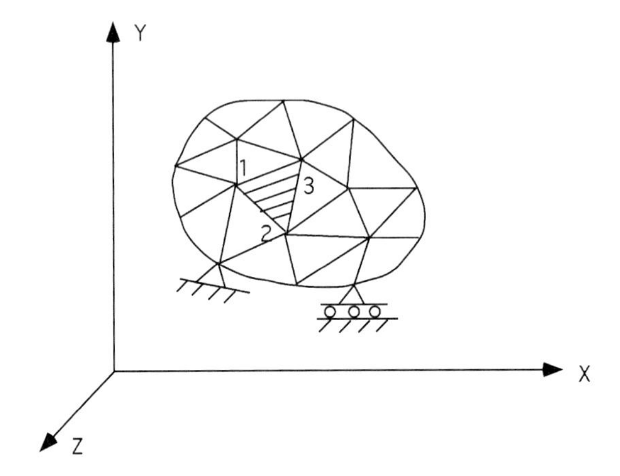

Fig. 3.1 Finite element idealization.

and the resulting strains. That is, for a linear anisotropic material, the elastic strains expressed in a rectangular Cartesian coordinate system are related to the corresponding stresses by the three-dimensional formulation of Hooke's law, and the total strain may be expressed as

$$\begin{aligned} \boldsymbol{\epsilon} &= \boldsymbol{\epsilon}_E + \boldsymbol{\epsilon}_T + \boldsymbol{\epsilon}_I \\ &= \boldsymbol{C}\boldsymbol{\sigma} + \boldsymbol{\epsilon}_T + \boldsymbol{\epsilon}_I \end{aligned} \tag{3.1}$$

in which

$$\boldsymbol{\epsilon}_E = \{\epsilon_{xx} \quad \epsilon_{yy} \quad \epsilon_{zz} \quad \epsilon_{xy} \quad \epsilon_{yz} \quad \epsilon_{zx}\} \tag{3.2}$$

$$\boldsymbol{\sigma} = \{\sigma_{xx} \quad \sigma_{yy} \quad \sigma_{zz} \quad \sigma_{xy} \quad \sigma_{yz} \quad \sigma_{zx}\} \tag{3.3}$$

are the elastic strain and stress vectors, $\boldsymbol{C}$ is the constitutive relationship, and $\boldsymbol{\epsilon}_T$ and $\boldsymbol{\epsilon}_I$ are the thermal and initial strain vectors, respectively. Equation (3.1) may be solved for the inverse relationship of stresses expressed in terms of the strains

$$\boldsymbol{\sigma} = \boldsymbol{D}\boldsymbol{\epsilon} - \boldsymbol{D}\boldsymbol{\epsilon}_T - \boldsymbol{D}\boldsymbol{\epsilon}_I \tag{3.4}$$

where $\boldsymbol{D}$ is the inverse of $\boldsymbol{C}$.

The stress–strain matrix $\boldsymbol{D}$ is symmetric and has the following form with 21 independent constants for an anisotropic material:

$$\boldsymbol{D} = \begin{bmatrix} d_{11} & d_{12} & \dots & d_{16} \\ & d_{22} & \dots & d_{26} \\ \vdots & \vdots & \ddots & \vdots \\ & \text{sym} & \dots & d_{66} \end{bmatrix} \tag{3.5}$$

For an orthotropic material having three orthogonal planes of symmetry, this relationship reduces to the following form:

$$\boldsymbol{D} = \begin{bmatrix} \boldsymbol{D}_{11} & \boldsymbol{0} \\ \boldsymbol{0} & \boldsymbol{D}_{22} \end{bmatrix} \tag{3.6}$$

having nine independent constants in which the Cartesian xyz axes are the planes of the material symmetry. In Eq. (3.6) there are six independent constants in $\boldsymbol{D}_{11}$ and three in the diagonal matrix $\boldsymbol{D}_{22}$. An isotropic material is characterized by an independent choice of reference frame and the stress–strain matrix is defined in terms of two constants for which a variety of options is possible. One choice is Young's modulus E and Poisson's ratio ν.

The material constants can be measured using simple uniaxial tension and pure torsion (shear) tests that are then related to the elements of the $\boldsymbol{C}$ matrix of Eq. (3.1) as follows[3]:

$$\begin{gathered} c_{11} = 1/E_{11}, \quad c_{12} = -\nu_{12}/E_{11}, \quad c_{13} = -\nu_{13}/E_{11}, \quad c_{14} = \eta_{14}/E_{11} \\ c_{22} = 1/E_{22}, \quad c_{23} = -\nu_{23}/E_{22}, \quad c_{24} = \eta_{24}/E_{22} \\ c_{33} = 1/E_{33}, \quad c_{34} = \eta_{34}/E_{33}, \quad c_{44} = 1/G_{12} \\ c_{55} = 1/G_{23}, \quad c_{66} = 1/G_{13} \end{gathered} \tag{3.7}$$

where

E_{ij} = Young's moduli in the 1, 2, and 3 (x, y, and z) directions, respectively
ν_{ij} = Poisson's ratio; ratio of the strain in the j direction over the strain in the i direction caused by a stress in the latter direction and multiplied by -1
G_{ij} = shear moduli in i, j plane
η_{ij} = shear coupling ratios

The $\boldsymbol{D}$ matrix can be then obtained by inverting $\boldsymbol{C}$. For an orthotropic material these relationships simplify to the following expressions:

$$\begin{gathered} d_{11} = (1 - \nu_{23}\nu_{32})\nu E_{11}, \quad d_{12} = (\nu_{12} + \nu_{13}\nu_{32})\nu E_{22}, \quad d_{13} = (\nu_{13} + \nu_{23}\nu_{12})\nu E_{33} \\ d_{22} = (1 - \nu_{31}\nu_{13})\nu E_{22}, \quad d_{23} = (\nu_{23} + \nu_{21}\nu_{13})\nu E_{33} \\ d_{33} = (1 - \nu_{12}\nu_{21})\nu E_{33}, \quad d_{44} = G_{12}, \quad d_{55} = G_{23}, \quad d_{66} = G_{31} \end{gathered} \tag{3.8}$$

in which

$$\nu = 1/(1 - \nu_{12}\nu_{21} - \nu_{23}\nu_{32} - \nu_{31}\nu_{13} - 2\nu_{12}\nu_{23}\nu_{31}) \tag{3.9}$$

For an isotropic material the preceding expressions reduce to the following values:

$$\begin{aligned} d_{11} = d_{22} = d_{33} &= \frac{(1-\nu)E}{(1+\nu)(1-2\nu)} \\ d_{12} = d_{13} = d_{23} &= \frac{\nu E}{(1+\nu)(1-2\nu)} \\ d_{44} = d_{55} = d_{66} = G &= \frac{E}{2(1+\nu)} \end{aligned} \tag{3.10}$$

For shell and plate problems the stress–strain relationship for the participating stresses and strains for the general anisotropic case is written as

$$\begin{bmatrix} \sigma_{xx} \\ \sigma_{yy} \\ \sigma_{xy} \\ \sigma_{yz} \\ \sigma_{zx} \end{bmatrix} = \begin{bmatrix} d_{11} & d_{12} & d_{14} & 0 & 0 \\ & d_{22} & d_{24} & 0 & 0 \\ & 0 & d_{44} & 0 & 0 \\ & & & d_{55} & d_{56} \\ & & \text{sym} & & d_{66} \end{bmatrix} \begin{bmatrix} \epsilon_{xx} \\ \epsilon_{yy} \\ \epsilon_{xy} \\ \epsilon_{yz} \\ \epsilon_{zx} \end{bmatrix} \tag{3.11}$$

and for the orthotropic case the stiffness coefficients are expressed as

$$\begin{gathered} d_{11} = \frac{E_{11}}{1 - \nu_{12}\nu_{21}}, \qquad d_{12} = \frac{\nu_{12}E_{11}}{1 - \nu_{12}\nu_{21}} \\ d_{22} = \frac{E_{22}}{1 - \nu_{12}\nu_{21}}, \qquad d_{44} = G_{12} \\ d_{55} = G_{23}, \qquad d_{66} = G_{31} \\ d_{14} = d_{24} = d_{56} = 0 \end{gathered} \tag{3.12}$$

and for the isotropic material ($\sigma_{zz} = \sigma_{yz} = \sigma_{zx} = 0$) from Eqs. (3.11) and (3.12),

$$\begin{gathered} d_{11} = d_{22} = \frac{E}{(1 - \nu^2)}, \; d_{12} = \nu d_{11} \\ d_{44} = G = \frac{E}{2(1 + \nu)}, \; d_{55} = d_{66} = G \end{gathered} \tag{3.13}$$

For plane stress problems the constitutive matrix is reduced to the first (3×3) submatrix of Eq. (3.11). For the plane strain case in which $\epsilon_{zz} = \epsilon_{zx} = \epsilon_{zy} = 0$, the constitutive matrix takes the following forms.

Orthotropic case:

$$\begin{bmatrix} \sigma_{xx} \\ \sigma_{yy} \\ \sigma_{xy} \end{bmatrix} = \begin{bmatrix} \dfrac{E'_{xx}}{\left(1 - n'\nu'^2_{yx}\right)} & \dfrac{\nu'_{yx}E'_{xx}}{\left(1 - n'\nu'^2_{yx}\right)} & 0 \\ \text{sym} & \dfrac{E'_{yy}}{\left(1 - n'\nu'^2_{yx}\right)} & 0 \\ & & G_{xy} \end{bmatrix} \begin{bmatrix} \epsilon_{xx} \\ \epsilon_{yy} \\ \epsilon_{xy} \end{bmatrix} \tag{3.14}$$

in which

$$\begin{gathered} E'_{xx} = \frac{E_{xx}}{1 - n\nu^2_{zx}}, \qquad E'_{yy} = \frac{E_{yy}}{1 - n\nu^2_{zx}} \\ n = E_{xx}/E_{yy}, \qquad n' = E'_{xx}/E'_{yy} \\ \nu'_{yx} = \frac{\nu_{yx} + \nu_{zx}\nu_{zy}(E_{yy}/E_{zz})}{1 - \nu^2_{zy}(E_y/E_z)} \end{gathered} \tag{3.15}$$

Isotropic case:

$$d_{11} = d_{22} = \frac{E(1-\nu)}{(1+\nu)(1-2\nu)}, \quad d_{12} = \frac{\nu}{(1-\nu)} d_{11}, \quad d_{33} = \frac{E}{2(1+\nu)} = G \tag{3.16}$$

and

$$\sigma_{zz} = \nu(\sigma_{xx} + \sigma_{yy}) \tag{3.17}$$

The thermal stress term in Eq. (3.4) may be written as

$$\boldsymbol{D}\boldsymbol{\epsilon_T} = \boldsymbol{D}\boldsymbol{S}_T \boldsymbol{\alpha} T \tag{3.18}$$

in which

$$\boldsymbol{\alpha} = [\alpha_{xx} \quad \alpha_{yy} \quad \alpha_{zz} \quad 0 \quad 0 \quad 0]^T \tag{3.19}$$

and the vector $\boldsymbol{S}_T$ is a diagonal vector for the anisotropic case that reduces to the unit vector

$$\boldsymbol{S}_T = [1 \quad 1 \quad 1 \quad 0 \quad 0 \quad 0]^T \tag{3.20}$$

for the isotropic material. The complete stress–strain relationship of Eq. (3.4) can be written as

$$\boldsymbol{\sigma} = \boldsymbol{D}\boldsymbol{\epsilon} - \boldsymbol{D}\boldsymbol{S}_T \boldsymbol{\alpha} T - \boldsymbol{D}\boldsymbol{\epsilon_I} \tag{3.21}$$

in which

$$\boldsymbol{\epsilon_I} = [\epsilon_{Ixx} \quad \epsilon_{Iyy} \quad \epsilon_{Izz} \quad \epsilon_{Ixy} \quad \epsilon_{Iyz} \quad \epsilon_{Izx}]^T \tag{3.22}$$

The external forces applied on any surface of a body must be in equilibrium with the internal stresses. If $\hat{\boldsymbol{n}}$ is the surface normal, $\tilde{\boldsymbol{p}}$ the surface pressure vector, and $[\boldsymbol{\sigma}]$ the stress tensor, then equilibrium of the surface element requires that

$$[\boldsymbol{\sigma}]^T \hat{\boldsymbol{n}} = \tilde{\boldsymbol{p}} \tag{3.23}$$

The stress tensor in Eq. (3.23) is written as

$$[\boldsymbol{\sigma}] = \begin{bmatrix} \sigma_{xx} & \sigma_{xy} & \sigma_{xz} \\ \sigma_{yx} & \sigma_{yy} & \sigma_{yz} \\ \sigma_{zx} & \sigma_{zy} & \sigma_{zz} \end{bmatrix} \tag{3.24}$$

whereas the surface normal is

$$\hat{\boldsymbol{n}} = [l \quad m \quad n]^T \tag{3.25}$$

in which l, m, and n are direction cosines of the outward normal η, drawn at the point on the surface, and are given as

$$l = \cos(\eta, x), \qquad m = \cos(\eta, y), \qquad n = \cos(\eta, z) \tag{3.26}$$

The surface pressure can be written as

$$\tilde{\boldsymbol{p}} = [\tilde{p}_x \quad \tilde{p}_y \quad \tilde{p}_z]^T \tag{3.27}$$

The strain tensor corresponding to the stress tensor is given for small displacements and strains in terms of the first partial derivatives of the displacement components u, v, w; i.e., using the same notation as in Eq. (3.24),

$$[\epsilon] = \begin{bmatrix} \epsilon_{xx} & \epsilon_{xy} & \epsilon_{xz} \\ \epsilon_{yx} & \epsilon_{yy} & \epsilon_{yz} \\ \epsilon_{zx} & \epsilon_{zy} & \epsilon_{zz} \end{bmatrix} \tag{3.28}$$

In Eq. (3.28), the terms of the strain matrix are defined as

$$\epsilon_{xx} = \frac{\partial u}{\partial x}; \qquad \epsilon_{yy} = \frac{\partial v}{\partial y}; \qquad \epsilon_{zz} = \frac{\partial w}{\partial z}$$

$$\epsilon_{xy} = \frac{1}{2}\left(\frac{\partial u}{\partial y} + \frac{\partial v}{\partial x}\right); \qquad \epsilon_{yz} = \frac{1}{2}\left(\frac{\partial v}{\partial z} + \frac{\partial w}{\partial y}\right); \qquad \epsilon_{zx} = \frac{1}{2}\left(\frac{\partial w}{\partial x} + \frac{\partial u}{\partial z}\right) \tag{3.29}$$

and these relationships are used extensively in a finite element formulation.

For layered anisotropic materials it is desirable to formulate a stress transformation matrix to achieve stresses in any desirable coordinate system from that calculated in a reference coordinate system (Fig. 3.2). Such a relationship can be expressed in vector form as

$$\boldsymbol{\sigma}' = \boldsymbol{T}_\sigma \boldsymbol{\sigma} \tag{3.30}$$

where

$$\boldsymbol{T}_\sigma = \begin{bmatrix} l_1^2 & m_1^2 & n_1^2 & 2l_1m_1 & 2m_1n_1 & 2n_1l_1 \\ l_2^2 & m_2^2 & n_2^2 & 2l_2m_2 & 2m_2n_2 & 2n_2l_2 \\ l_3^2 & m_3^2 & n_3^2 & 2l_3m_3 & 2m_3n_3 & 2n_3l_3 \\ l_1l_2 & m_1m_2 & n_1n_2 & l_1m_2 + l_2m_1 & m_1n_2 + m_2n_1 & n_1l_2 + n_2l_1 \\ l_2l_3 & m_2m_3 & n_2n_3 & l_2m_3 + l_3m_2 & m_2n_3 + m_3n_2 & n_2l_3 + n_3l_2 \\ l_3l_1 & m_3m_1 & n_3n_1 & l_3m_1 + l_1m_3 & m_3n_1 + m_1n_3 & n_3l_1 + n_1l_3 \end{bmatrix} \tag{3.31}$$

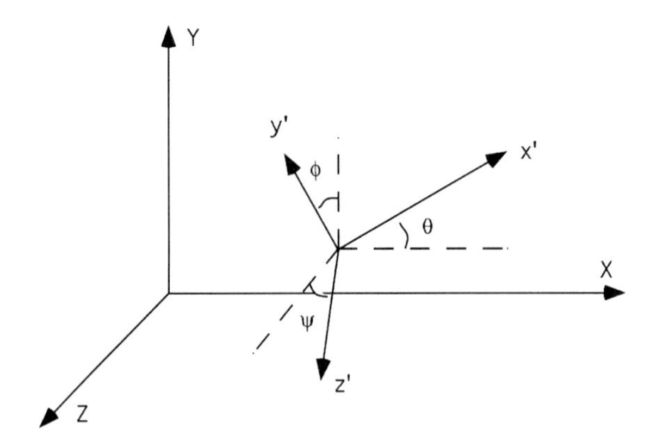

Fig. 3.2 Global and material coordinate axes.

between the coordinate system (X, Y, Z) and (x', y', z') and in which the direction cosine matrix is defined as

$$\tilde{\boldsymbol{\lambda}} = \begin{bmatrix} l_1 & m_1 & n_1 \\ l_2 & m_2 & n_2 \\ l_3 & m_3 & n_3 \end{bmatrix} = \begin{bmatrix} \cos(X, x') & \cos(Y, x') & \cos(Z, x') \\ \cos(X, y') & \cos(Y, y') & \cos(Z, y') \\ \cos(X, z') & \cos(Y, z') & \cos(Z, z') \end{bmatrix} \tag{3.32}$$

in which typically $\cos(X, x')$ is the angle measured from X to x' axis. Alternatively in matrix–tensor notation

$$[\sigma'] = [\boldsymbol{\lambda}]^T [\sigma][\boldsymbol{\lambda}] \tag{3.33}$$

in which

$$\boldsymbol{\lambda} = \begin{bmatrix} \tilde{\boldsymbol{\lambda}} & 0 \\ 0 & \tilde{\boldsymbol{\lambda}} \end{bmatrix}$$

Equations (3.30) and (3.33) are exactly equivalent.

3.2.2 *Elastoplasticity*

Figure 3.3 depicts typical elastic and elastoplastic stress–strain curves for a ductile material in which the elastic part is linear (Fig. 3.3a) or nonlinear (Fig. 3.3b). When the material is subjected to repeated loading and unloading such a relationship takes the form shown in Fig. 3.4a and its linearized elastic version, often adapted for routine analysis, is shown in Fig. 3.4b. Any nonlinear elastic behavior in rubberlike materials up to the yield point may be approximately analyzed by piecewise linear representation of the stress–strain curve.

Beyond the elastic range the material behavior[4–6] may be characterized as being either time independent such as elastoplasticity or time dependent like creep and viscoelastoplasticity. For an elastoplastic material, material deformation is irreversible beyond the yield point. The deformation may be divided into elastic and plastic parts and then the total strain may be expressed as a combination of elastic and plastic strains:

$$\epsilon = \epsilon_E + \epsilon_P \tag{3.34}$$

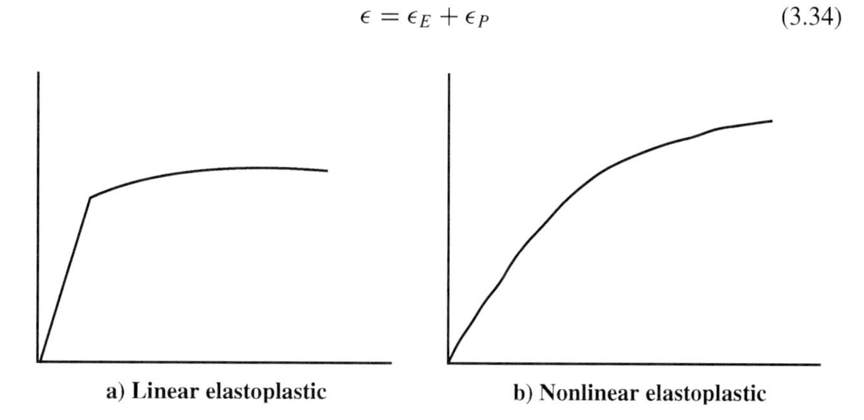

Fig. 3.3 Material stress–strain curve.

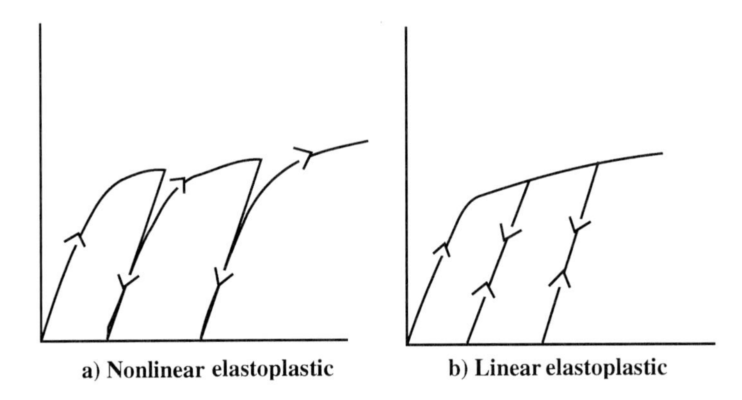

Fig. 3.4 Elastoplastic stress–strain curve for repeated loading.

To determine the stress state at which yielding occurs, it is possible to define a yield criterion as a function of the stresses,

$$f(\boldsymbol{\sigma}) = f(J_1, J_2, J_3) = 0 \tag{3.35}$$

where J_1, J_2, and J_3 are stress invariants, and Eq. (3.35) defines the yield surface with the principal stresses $\sigma_1, \sigma_2, \sigma_3$ as the coordinate axes. The principal stresses may be calculated from the determinantal equation,

$$|\boldsymbol{\sigma} - \boldsymbol{\sigma}_i \delta_{ij}| = 0 \tag{3.36}$$

where $\boldsymbol{\sigma}_i$ is the diagonal matrix of principal stresses. The stress invariants are the coefficients of terms obtained from the cubic equation in Eq. (3.36) defined as

$$\begin{aligned}
J_1 &= \sigma_1 + \sigma_2 + \sigma_3 \\
&= \sigma_x + \sigma_y + \sigma_z \\
J_2 &= -(\sigma_1\sigma_2 + \sigma_2\sigma_3 + \sigma_3\sigma_1) \\
&= \sigma_x\sigma_y + \sigma_y\sigma_z + \sigma_z\sigma_x + \tau_{xy}^2 + \tau_{yz}^2 + \tau_{zx}^2 \\
J_3 &= \sigma_1\sigma_2\sigma_3 \\
&= \sigma_x\sigma_y\sigma_z + 2\tau_{xy}\tau_{yz}\tau_{zx} - \sigma_x\tau_{yz}^2 - \sigma_y\tau_{zx}^2 - \sigma_z\tau_{xy}^2
\end{aligned} \tag{3.37}$$

A simplification of the yield criterion involves the principal components $(\sigma_1', \sigma_2', \sigma_3')$ of the deviatoric or reduced stress tensor

$$\sigma_{ij}' = \sigma_{ij} - \tfrac{1}{3}\sigma_{ij}\delta_{ij} \tag{3.38}$$

in which $\sigma_1' + \sigma_2' + \sigma_3' = 0$. The yield criterion reduces to

$$f(J_2', J_3') = 0 \tag{3.39}$$

where

$$\begin{aligned}
J_2' &= -(\sigma_1'\sigma_2' + \sigma_2'\sigma_3' + \sigma_3'\sigma_1') \\
&= \tfrac{1}{2}\left(\sigma_1'^2 + \sigma_2'^2 + \sigma_3'^2\right)
\end{aligned} \tag{3.40}$$

$$J_3' = \sigma_1'\sigma_2'\sigma_3' \tag{3.41}$$

Thus the Von Mises yield criterion states that yielding occurs when the effective stress $\sqrt{3J_2'}$ equals the yield stress $\tilde{\sigma}_{yp}$ measured in a uniaxial tension test, and the yield surface is defined as

$$f = \sqrt{3J_2'} - \tilde{\sigma}_{yp} \tag{3.42}$$

In terms of the principal stresses, the Von Mises yield criterion is given by

$$f = (\sigma_1 - \sigma_2)^2 + (\sigma_2 - \sigma_3)^2 + (\sigma_3 - \sigma_1)^2 - 2(\tilde{\sigma}_{yp})^2 = 0 \tag{3.43}$$

This formula is generally applicable to a number of metals.

Using the incremental theory of plasticity, the total strain is the sum of the elastic and plastic parts,

$$\mathrm{d}\epsilon = \mathrm{d}\epsilon_E + \mathrm{d}\epsilon_P \tag{3.44}$$

and the increment in stress may be written as

$$\mathrm{d}\boldsymbol{\sigma} = \boldsymbol{D}_{EP}\,\mathrm{d}\epsilon \tag{3.45}$$

Computation of the elastoplastic matrix $\boldsymbol{D}_{EP}$ is dependent on evaluation of plastic strain increment $\mathrm{d}\epsilon_P$, which is achieved by combining the Von Mises yield criterion with the Prandtl–Reuss equation for the plastic strain components. The final form of the matrix for isotropic material is obtained as

$$\mathrm{d}\boldsymbol{\sigma} = \left[[\boldsymbol{D}] - \frac{[\boldsymbol{D}]\left[\dfrac{\partial \sigma_e}{\partial \sigma}\right]\left[\dfrac{\partial \sigma_e}{\partial \sigma}\right]^{\boldsymbol{T}}[\boldsymbol{D}]}{\boldsymbol{H}' + \left[\dfrac{\partial \sigma_e}{\partial \sigma}\right]^{\boldsymbol{T}}[\boldsymbol{D}]\left[\dfrac{\partial \sigma_e}{\partial \sigma}\right]} \right] \mathrm{d}\epsilon \tag{3.46}$$

in which

$$\left[\frac{\partial \sigma_e}{\partial \sigma}\right] = \frac{3}{2\sigma_e} \begin{bmatrix} \sigma'_{xx} \\ \sigma'_{yy} \\ \sigma'_{zz} \\ 2\tau_{xy} \\ 2\tau_{yz} \\ 2\tau_{zx} \end{bmatrix} \tag{3.47}$$

and the numerator in Eq. (3.46) is defined by S, where

$$S = H' + \left[\frac{\partial \sigma_e}{\partial \sigma}\right]^{T}[\boldsymbol{D}]\left[\frac{\partial \sigma_e}{\partial \sigma}\right] \tag{3.48}$$

and

$$H' = \frac{\mathrm{d}\sigma_e}{\mathrm{d}\varepsilon_{ep}} \tag{3.49}$$

being the slope of the curve relating effective stress σ_e to the equivalent plastic strain ε_{ep}, obtained from a uniaxial test.

3.3 Finite Element Formulation Based on the Stationary Functional Method

In the finite element method the solution domain is divided into a number of discrete elements. The displacements within an element are generally the unknown field variables that are expressed in terms of unknown nodal values. This may be achieved by first expressing each displacement component in terms of trial coordinate functions, usually expressed as polynomials, the number of unknown coefficients depending on the number of nodal variables defined in the element as shown:

$$\boldsymbol{u} = \boldsymbol{A}\boldsymbol{c} \tag{3.50}$$

in which $\boldsymbol{u} = [u_x\ u_y\ u_z]^T$, $\boldsymbol{c}$ is a vector whose scalars are element spatial coordinates and are unity, whereas $\boldsymbol{A}$ is a matrix of unknown coefficients. These coefficients are determined from element boundary conditions yielding the relationship that expresses displacements within the element to their unknown nodal values:

$$\boldsymbol{u} = \boldsymbol{N}\boldsymbol{u}^e \tag{3.51}$$

where the superscript e refers to elementwise values and $\boldsymbol{N}$ are the shape functions, being functions of the position coordinates. For three-dimensional elements the shape function matrix $\boldsymbol{N}$ has three rows and its number of columns is equal to the total degrees of freedom of the element. This matrix also may be obtained directly by employing suitable interpolation functions.

The strain–displacement relationships of Eq. (3.29) are next generated using Eq. (3.51):

$$\epsilon = \boldsymbol{B}\boldsymbol{u}^e \tag{3.52}$$

by differentiating the appropriate displacement components. The matrix $\boldsymbol{B}$ has six rows and its number of columns is equal to the number of degrees of freedom of the element.

The principle of stationary total potential energy V can be simply stated as that of all displacement states satisfying compatibility and boundary conditions; those that also satisfy equilibrium make the total potential energy assume a stationary value.[7–10] For a stable structure, the value of V is always a minimum, which also is expressed as

$$V = U - W \tag{3.53}$$

in which U is the internal strain energy and W is the potential of the external forces. Assuming that the body is subjected to time-varying external forces, the displacements, strains, and stresses within a finite element all will be functions of time. Then the strain energy of an element is given as

$$U^e = \frac{1}{2}\int_V \sigma^T \epsilon_E \, \mathrm{d}V \tag{3.54}$$

which also can be expressed as

$$U^e = \frac{1}{2}\left[\int_V \epsilon^T \boldsymbol{D}\epsilon \, \mathrm{d}V - \int_V \epsilon_T^T \boldsymbol{D}\epsilon \, \mathrm{d}V - \int_V \epsilon_I^T \boldsymbol{D}\epsilon \, \mathrm{d}V\right] \tag{3.55}$$

using Eq. (3.4) and further expanding using Eq. (3.1) as shown in the following:

$$U^e = \frac{1}{2}\left[\int_V \epsilon^T D(\epsilon - \epsilon_T - \epsilon_I)\,\mathrm{d}V - \int_V \epsilon_T^T D(\epsilon - \epsilon_T - \epsilon_I)\,\mathrm{d}V \right.$$
$$\left. - \int_V \epsilon_I^T D(\epsilon_T - \epsilon_I)\,\mathrm{d}V\right] \tag{3.56}$$

which reduces to

$$U^e = \frac{1}{2}\int_V \epsilon^T D\epsilon\,\mathrm{d}V - \int_V \epsilon^T D\epsilon_T\,\mathrm{d}V - \int_V \epsilon^T D\epsilon_I\,\mathrm{d}V \tag{3.57}$$

neglecting terms independent of elastic displacements and noting that $\epsilon^T D\epsilon_T = \epsilon_T^T D\epsilon$, and so forth. Equation (3.57) may finally be written in terms of element nodal displacements by using Eq. (3.52) that relates to total strain

$$\begin{aligned} U^e &= \frac{1}{2}\int_V u^{eT} B^T D B u^e\,\mathrm{d}V - \int_V u^{eT} B^T D\epsilon_T\,\mathrm{d}V - \int_V u^{eT} B^T D\epsilon_I\,\mathrm{d}V \\ &= \frac{1}{2} u^{eT} K^e u^e - u^{eT}\int_V B^T D\epsilon_T\,\mathrm{d}V - u^{eT}\int_V B^T D\epsilon_I\,\mathrm{d}V \end{aligned} \tag{3.58}$$

where K^e is the element stiffness matrix, defined as

$$K^e = \int_V B^T D B\,\mathrm{d}V \tag{3.59}$$

The work done by concentrated loads p, body force p_B, and surface forces p_S in an element may be expressed as

$$\begin{aligned} W^e &= u^{eT} p(t) + \int_V u^T p_B(t)\,\mathrm{d}V + \int_S u^T p_S(t)\,\mathrm{d}S - \int_V u^T \rho\ddot{u}\,\mathrm{d}V \\ &= u^{eT} p(t) + \int_V u^{eT} N^T p_B(t)\,\mathrm{d}V + \int_S u^{eT} N^T p_S(t)\,\mathrm{d}S - \int_V u^{eT} N^T \rho N\ddot{u}^e\,\mathrm{d}V \\ &= u^{eT}\left(p(t) + \int_V N^T p_B(t)\,\mathrm{d}V + \int_S N^T p_S(t)\,\mathrm{d}S - M^e\ddot{u}^e\right) \end{aligned} \tag{3.60}$$

because $\ddot{u}^e = N u^{eT}$ and M^e is the element mass matrix,[11] defined as

$$M^e = \int_V N^T \rho N\,\mathrm{d}V \tag{3.61}$$

The expression for the total potential energy of an element may now be written as

$$V^e = U^e - W^e \tag{3.62}$$

For the entire structure, defining q as the nodal unknowns,

$$\begin{aligned} V &= \Sigma V^e = \Sigma U^e - \Sigma W^e \\ &= \tfrac{1}{2} q^T K q - q^T p_T(t) - q^T p_I(t) - q^T(p(t) + p_B(t) + p_S(t) - M\ddot{q}) \end{aligned} \tag{3.63}$$

Then the requirement of minimum total potential energy

$$\frac{\mathrm{d}V}{\mathrm{d}\boldsymbol{q}} = 0 \tag{3.64}$$

yields the equation of motion

$$\begin{aligned} \boldsymbol{Kq} + \boldsymbol{M\ddot{q}} &= \boldsymbol{p}_T(t) + \boldsymbol{p}_I(t) + \boldsymbol{p}(t) + \boldsymbol{p}_B(t) + \boldsymbol{p}_S(t) \\ &= \boldsymbol{f}(t) \end{aligned} \tag{3.65}$$

where the matrices and vectors refer to the entire structure and where

$\boldsymbol{K}$ = stiffness matrix, $\boldsymbol{K}_E$ (elastic stiffness matrix)
$\boldsymbol{M}$ = mass or inertia matrix
$\boldsymbol{p}_T = \int_V \boldsymbol{B}^T \boldsymbol{D}\epsilon_T \, \mathrm{d}V$ = thermal load
$\boldsymbol{p}_I = \int_V \boldsymbol{B}^T \boldsymbol{D}\epsilon_I \, \mathrm{d}V$ = inertia load
$\boldsymbol{p}$ = concentrated load
$\boldsymbol{p}_B = \int_V \boldsymbol{N}^T \boldsymbol{p}_B(t) \, \mathrm{d}V$ = body forces
$\boldsymbol{p}_S = \int_S \boldsymbol{N}^T \boldsymbol{p}_S(t) \, \mathrm{d}S$ = surface forces

Equation (3.65) may also be derived by using the principle of virtual work for dynamic systems, also known as Hamilton's principle, first at element level and then analyzing the same for the entire structure.

Most structures are characterized by the presence of structural as well as viscous damping, in which case Eq. (3.65) takes the following form[7,12]:

$$\boldsymbol{K}(1 + ig)\boldsymbol{q} + \boldsymbol{C}_D\dot{\boldsymbol{q}} + \boldsymbol{M}\ddot{\boldsymbol{q}} = \boldsymbol{f}(t) \tag{3.66}$$

in which $\boldsymbol{C}_D$ is viscous damping and equals $\alpha\boldsymbol{K} + \beta\boldsymbol{M}$, if the damping is proportional. For spinning structures with viscous damping the dynamic equations of motion can be written as[13]

$$\boldsymbol{Kq} + \boldsymbol{C\dot{q}} + \boldsymbol{M\ddot{q}} = \boldsymbol{f}(t) \tag{3.67}$$

or

$$(\boldsymbol{K}_E + \boldsymbol{K}_G + \boldsymbol{K}')\boldsymbol{q} + (\boldsymbol{C}_C + \boldsymbol{C}_D)\dot{\boldsymbol{q}} + \boldsymbol{M}\ddot{\boldsymbol{q}} = \boldsymbol{f}(t) \tag{3.68}$$

where

$\boldsymbol{K}_G$ = geometric stiffness matrix, incorporating the effect of in-plane stretching on out-of-plane motion
$\boldsymbol{K}'$ = centrifugal stiffness matrix
$\boldsymbol{C}_C$ = Coriolis matrix

All matrices in the preceding formulation except $\boldsymbol{C}_C$ are symmetric and usually highly banded. The matrix $\boldsymbol{C}_C$ is skew symmetric being similarly banded.

The associated matrix equation of free vibration may be written in the general form as

$$\boldsymbol{Kq} + \boldsymbol{C\dot{q}} + \boldsymbol{M\ddot{q}} = 0 \tag{3.69}$$

where the definitions for $\boldsymbol{K}$, $\boldsymbol{C}$, and $\boldsymbol{M}$ depend on the problem type. Similarly, the undamped equation of free vibration takes the form

$$\boldsymbol{K}\boldsymbol{q} + \boldsymbol{M}\ddot{\boldsymbol{q}} = 0 \tag{3.70}$$

and similar equations are encountered for structural instability or buckling problems. For static problems, the matrix equation reduces simply to

$$\boldsymbol{K}\boldsymbol{q} = \boldsymbol{f} \tag{3.71}$$

Clearly, the numerical analysis of a structural system consists of two distinct yet related solution procedures. First, a finite element model of the system yields a set of algebraic equations that are then solved by employing a suitable numerical procedure. Because of the very nature of finite element discretization, the resulting equations tend to be rather large in size, as well as highly banded for many practical problems. An economical solution of such problems poses as much a challenge to an analyst as the process of discretization itself.

3.4 Concluding Remarks

In this chapter the basic tools of the finite element method used in structural mechanics were developed. First, the material constructive relationships for three dimensional and planar elasticity were given, together with the definition of thermal and initial strain vectors. The latter may be used in material nonlinear analysis to accumulate plastic strains. Plasticity was discussed and the Von Mises yield criterion developed. From this, using the Prandtl–Reuss assumptions for the plastic strain components, the elastoplastic constitutive matrix was derived with strain hardening depending on an experimentally obtained stress–strain curve. In Sec. 3.3 the finite element equations were developed for a body with mass, damping, and stiffness subjected to various time-dependent load effects. The stiffness and mass matrices ($\boldsymbol{K}$ and $\boldsymbol{M}$) will be the subject of Chapter 4.

References

[1]Sokolnikoff, I. S., *Mathematical Theory of Elasticity*, 2nd ed., McGraw–Hill, New York, 1956.

[2]Timoshenko, S., and Goodier, N., *Theory of Elasticity*, McGraw–Hill, New York, 1951.

[3]Jones, R. M, *Mechanics of Composite Materials*, McGraw–Hill, New York, 1975.

[4]Fung, Y. C., *Foundations of Solid Mechanics*, Prentice–Hall, Englewood Cliffs, NJ, 1965.

[5]Hill, R., *Mathematical Theory of Plasticity*, Clarendon Press, Oxford, England, U.K., 1950.

[6]Prager, W., *An Introduction to Plasticity*, Addison-Wesley, Reading, MA, 1959.

[7]Argyris, J. H., *Energy Theorems and Structural Analysis*, Butterworths, London, 1960 (reprinted from *Aircraft Engineering*, Vols. 26, 27, 1954–55).

[8]Crandall, S. H., *Engineering Analysis*, McGraw–Hill, New York, 1955.

[9]Przemieniecki, J. S., *Theory of Matrix Structural Analysis*, McGraw–Hill, New York, 1968.

[10]Washizu, K., *Variational Methods in Elasticity and Plasticity*, Pergamon Press, 2nd ed., 1975.

[11]Archer, J. S. "Consistent Matrix Formulations for Structural Analysis Using Finite-Element Techniques," *AIAA Journal*, Vol. 3, No. 10, 1965, pp. 1910–1918.
[12]Hurty, W. C., and Rubinstein, M. F., *Dynamics of Structures*, Prentice–Hall, Englewood Cliffs, NJ, 1964.
[13]Dokainish, M. A., and Rawtani, S., "Vibration Analysis of Rotating Cantilever Plates," *International Journal for Numerical Methods in Engineering*, Vol. 3, 1971, pp. 233–248.

4 Structural Mechanics—Finite Elements

4.1 Introduction

Any structural domain may be classified into one of three categories and suitably discretized by finite elements as follows. One-dimensional space: line elements, two-dimensional space: triangular and quadrilateral elements, and three-dimensional space: tetrahedral, prismatic, and hexahedral elements, or any combination thereof. The displacement values within an element are interpolated in terms of nodal values that may be displacements, or their derivatives. Depending on the pattern of the displacement distribution, problems in structural mechanics may be broadly categorized as beam, truss, and cable; plane stress and strain; axisymmetric solids; plate bending; and shells and three-dimensional solids.

Finite element force-displacement characteristics may be derived for each class of problem by applying variational or weighted-residual techniques or simply using the Gauss divergence theorem. The generation of stiffness, inertia, and other associated element matrices is considered in this chapter. The fundamental step in developing these matrices involves choosing appropriate shape functions, either in the local coordinate system or using a natural coordinate system that enables convenient use of standard numerical integration procedures. With the availability of symbolic manipulation software packages it is also quite straightforward to generate these matrices using the local coordinate system. A displacement expression must satisfy rigid-body and constant strain requirements as well as interelement compatibility. The latter requirement is satisfied if the displacement field is continuous up to the derivative one order lower than the highest derivative occurring in the strain displacement equations. Satisfaction of rigid-body and constant strain states are achieved when the displacement expansion is chosen as a complete polynomial of order equal to the highest derivative occurring in the strain–displacement relations.

4.2 One-Dimensional Line Elements

A one-dimensional beam element with six displacement degrees of freedom at each node is shown in Fig. 4.1. Structural characteristics of the straight element with uniform cross section A may be obtained from assumed axial, torsional, and flexural deformation components, combining these individual relationships as described next.[1,2] The local coordinate system is assumed such that the x axis corresponds to the axial direction of the element. The six displacement degrees of freedom at any point on the element are shown in Fig. 4.1.

Thus for the axial element the displacement interpolation in the local x direction assumes a linear relationship:

$$u_x = c_1 + c_2 x \tag{4.1}$$

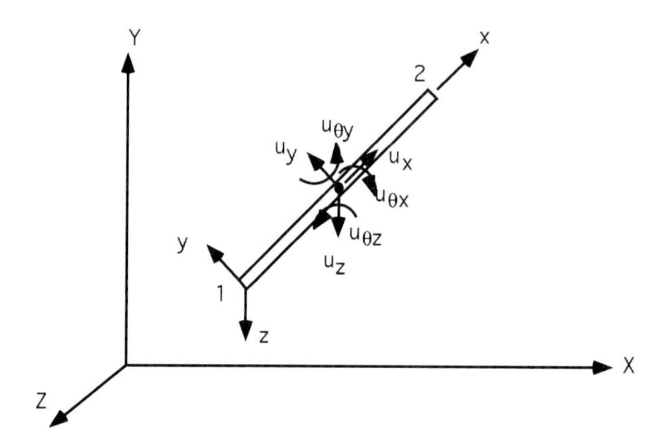

Fig. 4.1 One-dimensional line element.

The coefficients c_1 and c_2 are evaluated in terms of nodal displacements by setting the boundary conditions $u_x = u_{x_1}$ at $x = 0$ and $u_x = u_{x_2}$ at $x = l$, so that

$$\begin{aligned} u_x &= \begin{bmatrix} 1 - \dfrac{x}{l} & \dfrac{x}{l} \end{bmatrix} \begin{Bmatrix} u_{x_1} \\ u_{x_2} \end{Bmatrix} \\ &= \boldsymbol{N} \boldsymbol{u}_x^e \end{aligned} \tag{4.2}$$

where $\boldsymbol{N}$ is the shape function row vector pertaining to the x direction. The strain is given by differentiation of u_x with respect to x,

$$\begin{aligned} \epsilon_{xx} &= \frac{\mathrm{d}u_x}{\mathrm{d}x} = \frac{1}{l}[-1 \quad 1] \begin{Bmatrix} u_{x_1} \\ u_{x_2} \end{Bmatrix} \\ &= \boldsymbol{B} \boldsymbol{u}_x^e \end{aligned} \tag{4.3}$$

and the stiffness and inertia matrices are obtained as

$$\begin{aligned} \boldsymbol{K}_x^e &= \int_V \boldsymbol{B}^T \boldsymbol{D} \boldsymbol{B} \, \mathrm{d}V \\ &= \frac{EA}{l} \begin{bmatrix} 1 & -1 \\ -1 & 1 \end{bmatrix} \end{aligned} \tag{4.4}$$

$$\begin{aligned} \boldsymbol{M}_x^e &= \int_V \rho \boldsymbol{N}^T \boldsymbol{N} \, \mathrm{d}V \\ &= \frac{\rho A l}{6} \begin{bmatrix} 2 & 1 \\ 1 & 2 \end{bmatrix} \end{aligned} \tag{4.5}$$

using Eqs. (3.59) and (3.61) derived in Chapter 3. In these equations, E is the Young's modulus of elasticity, ρ the mass density, A the area of the beam cross section, and l the length of the beam element.

For a torsional distortion $u_{\theta x}$, the interpolation can also be expressed by Eq. (4.1) and has the same shape function $\boldsymbol{N}$ as in Eq. (4.2). The shearing strain within a cross section is assumed to vary linearly along the radius r and is expressed as

$$\begin{aligned}\gamma_{xy} &= r\frac{\mathrm{d}u_{\theta x}}{\mathrm{d}x} \\ &= \frac{r}{l}[-1 \quad 1]\begin{Bmatrix} u_{\theta x_1} \\ u_{\theta x_2} \end{Bmatrix} \\ &= \boldsymbol{B}\boldsymbol{u}^e_{\theta x}\end{aligned} \tag{4.6}$$

The stress–strain relationship is given as

$$\sigma_{xy} = G\gamma_{xy} \tag{4.7}$$

and thus $\boldsymbol{D} = G$, the shear modulus. From these expressions the stiffness and inertia matrices are derived as

$$\begin{aligned}\boldsymbol{K}^e_{\theta x} &= \int_V \boldsymbol{B}^T\boldsymbol{D}\boldsymbol{B}\,\mathrm{d}V \\ &= \int_V \frac{G}{l^2}r^2\begin{Bmatrix} -1 \\ 1 \end{Bmatrix}[-1 \quad 1]\,r\,\mathrm{d}\theta\,\mathrm{d}r \\ &= \frac{GJ}{l}\begin{bmatrix} 1 & -1 \\ -1 & 1 \end{bmatrix}\end{aligned} \tag{4.8}$$

and

$$\begin{aligned}\boldsymbol{M}^e_{\theta x} &= \int_V \rho J\boldsymbol{N}^T\boldsymbol{N}\,\mathrm{d}V \\ &= \frac{\rho J l}{6}\begin{bmatrix} 2 & 1 \\ 1 & 2 \end{bmatrix}\end{aligned} \tag{4.9}$$

in which J is the torsional constant of the cross section, being the polar moment of inertia for a circular section. In a flexural element the displacement u_y is the unknown and can be expressed in the x–y principal plane of the cross section as the cubic polynomial

$$u_y = c_1 + c_2x + c_3x^2 + c_4x^3 \tag{4.10}$$

The unknown coefficients are evaluated in terms of the associated nodal displacements. Thus four boundary conditions need to be imposed that also can be obtained from the solution of the homogeneous part of the governing differential equation for beam,

$$\frac{\mathrm{d}^4u_y}{\mathrm{d}x^4} = 0 \tag{4.11}$$

$$\begin{aligned} u_y(0) &= u_{y_1} \qquad & u_y(l) &= u_{y_2} \\ \frac{\mathrm{d}u_y}{\mathrm{d}x}(0) &= u_{\theta z_1} \qquad & \frac{\mathrm{d}u_y}{\mathrm{d}x}(l) &= u_{\theta z_2}\end{aligned} \tag{4.12}$$

Substituting in Eq. (4.10) and solving the resulting simultaneous equations yields the coefficients as follows:

$$
\begin{aligned}
c_1 &= u_{y_1} \\
c_2 &= u_{\theta z_1} \\
c_3 &= -\frac{1}{l}\left(2u_{\theta z_1} + u_{\theta z_2}\right) + \frac{3}{l^2}\left(u_{y_2} - u_{y_1}\right) \\
c_4 &= \frac{1}{l^2}\left(u_{\theta z_1} + u_{\theta z_2}\right) + \frac{2}{l^3}\left(u_{y_1} - u_{y_2}\right)
\end{aligned}
\tag{4.13}
$$

Equation (4.10) may then be rewritten in standard form

$$
\begin{aligned}
u_y &= [N_1 \quad N_2 \quad N_3 \quad N_4]\boldsymbol{u}^e \\
&= \boldsymbol{N}\boldsymbol{u}^e
\end{aligned}
\tag{4.14}
$$

in which the shape function $\boldsymbol{N}$ for the y direction has the following components:

$$
\begin{aligned}
N_1 &= \frac{1}{l^3}(l^3 - 3x^2l + 2x^3) \qquad N_2 = \frac{1}{l^2}(xl^2 - 2x^2l + x^3) \\
N_3 &= \frac{1}{l^3}(3x^2l - 2x^3) \qquad N_4 = \frac{1}{l^2}(-x^2l + x^3)
\end{aligned}
\tag{4.15}
$$

and

$$
\boldsymbol{u}^e = \left[u_{y_1} \quad u_{\theta z_1} \quad u_{y_2} \quad u_{\theta z_2}\right]^T \tag{4.16}
$$

The flexural strain is related to the curvature by the expression

$$
\epsilon_x = -y\frac{\mathrm{d}^2 u_y}{\mathrm{d}x^2} \tag{4.17}
$$

Using Eq. (4.14), the strain displacement matrix may be obtained as

$$
\boldsymbol{B} = [B_1 \quad B_2 \quad B_3 \quad B_4] \tag{4.18}
$$

with

$$
\begin{aligned}
B_1 &= \frac{1}{l^3}(-6l + 12x) \qquad B_2 = \frac{1}{l^2}(-4l + 6x) \\
B_3 &= \frac{1}{l^3}(6l - 12x) \qquad B_4 = \frac{1}{l^2}(-2l + 6x)
\end{aligned}
\tag{4.19}
$$

The element stiffness matrix that includes the effect of shear deformation may then be calculated from

$$\boldsymbol{K}^e_{u_y} = \int_V \boldsymbol{B}^T \boldsymbol{D}\boldsymbol{B}\,\mathrm{d}V$$

$$= \frac{EI_z}{l^3}\begin{bmatrix} \frac{12}{\psi_y} & \frac{6l}{\psi_y} & \frac{-12}{\psi_y} & \frac{6l}{\psi_y} \\ & 4l^2 & -6l & 2l^2 \\ & \text{sym} & \frac{12}{\psi_y} & \frac{-6l}{\psi_y} \\ & & & 4l^2 \end{bmatrix} \tag{4.20}$$

in which $I_z = \int y^2\,\mathrm{d}A$ is the moment of inertia about z axis; and

$$\psi_y = 1 + \frac{12EI_z}{GA_{sy}l^2} \tag{4.21}$$

A_{sy} being the beam area effective in shear. The inertia matrix is obtained as

$$\boldsymbol{M}^e_{u_y} = \int_V \rho \boldsymbol{N}^T \boldsymbol{N}\,\mathrm{d}V + \int_0^l \rho I_z \boldsymbol{N}^T_{,x}\boldsymbol{N}_{,x}\,\mathrm{d}x$$

$$= \rho Al \begin{bmatrix} \frac{13}{35} + \frac{6I_z}{5Al^2} & \frac{11l}{210} + \frac{I_z}{10Al} & \frac{9}{70} - \frac{6I_z}{5Al^2} & -\frac{13l}{420} + \frac{I_z}{10Al} \\ & \frac{l^2}{105} + \frac{2I_z}{15A} & \frac{13l}{420} - \frac{I_z}{10Al} & -\frac{l^2}{140} - \frac{I_z}{30A} \\ & \text{sym} & \frac{13}{35} + \frac{6I_z}{5Al^2} & -\frac{11l}{210} - \frac{I_z}{10Al} \\ & & & \frac{l^2}{105} + \frac{2I_z}{15A} \end{bmatrix} \tag{4.22}$$

if the rotary inertia effect is included and shear deformation effects excluded in the formulation. The formulation when shear deformations are included is rather cumbersome. Similarly, expressions equivalent to Eqs. (4.20) and (4.22) are obtained for the flexural element involving the u_z displacement in terms of nodal displacements $\boldsymbol{u}^e$ and I_y the moment of inertia about y-principal axis

$$\boldsymbol{u}^e = \begin{bmatrix} u_{z_1} & u_{\theta y_1} & u_{z_2} & u_{\theta y_2} \end{bmatrix}^T \tag{4.23}$$

Thus the stiffness and inertia matrices of the two flexural elements and axial and torsional elements are combined to yield matrices for the beam element with 12 degrees-of-freedom such that

$$\boldsymbol{u}^e = \begin{bmatrix} u_{x_1} & u_{y_1} & u_{z_1} & u_{\theta x_1} & u_{\theta y_1} & u_{\theta z_1} & u_{x_2} & u_{y_2} & u_{z_2} & u_{\theta x_2} & u_{\theta y_2} & u_{\theta z_2} \end{bmatrix}^T \tag{4.24}$$

as shown in Fig. 4.1. This is easily achieved by adding each element of the individual matrices to the corresponding degree-of-freedom location in the final (12×12) matrix.

For completeness the stiffness matrix for the 12 degree-of-freedom beam element is given in Eqs. (4.25–4.28):

$$\boldsymbol{K}^e = \begin{bmatrix} \boldsymbol{K}_{11} & \boldsymbol{K}_{12} \\ \text{sym} & \boldsymbol{K}_{22} \end{bmatrix} \tag{4.25}$$

in which

$$\boldsymbol{K}_{11} = \begin{bmatrix} \frac{EA}{l} & 0 & 0 & 0 & 0 & 0 \\ & \frac{12EI_z}{l^3\psi_y} & 0 & 0 & 0 & \frac{6EI_z}{l^2\psi_y} \\ & & \frac{12EI_y}{l^3\psi_z} & 0 & \frac{-6EI_y}{l^2\psi_z} & 0 \\ & & & \frac{GJ}{l} & 0 & 0 \\ & \text{sym} & & & \frac{\psi_z' EI_y}{l} & 0 \\ & & & & & \frac{\psi_y' EI_z}{l} \end{bmatrix} \tag{4.26}$$

$$\boldsymbol{K}_{12} = \begin{bmatrix} -\frac{EA}{l} & 0 & 0 & 0 & 0 & 0 \\ 0 & -\frac{12EI_z}{l^3\psi_y} & 0 & 0 & 0 & \frac{6EI_z}{l^2\psi_y} \\ 0 & 0 & -\frac{12EI_y}{l^3\psi_z} & 0 & -\frac{6EI_y}{l^2\psi_z} & 0 \\ 0 & 0 & 0 & -\frac{GJ}{l} & 0 & 0 \\ 0 & 0 & \frac{6EI_y}{l^2\psi_z} & 0 & \frac{(3-\psi_z)EI_y}{l\psi_z} & 0 \\ 0 & -\frac{6EI_z}{l^2\psi_y} & 0 & 0 & 0 & \frac{(3-\psi_y)EI_z}{l\psi_y} \end{bmatrix} \tag{4.27}$$

$$
\boldsymbol{K}_{22} = \begin{bmatrix}
\frac{EA}{l} & 0 & 0 & 0 & 0 & 0 \\
 & \frac{12EI_z}{l^3\psi_y} & 0 & 0 & 0 & -\frac{6EI_z}{l^2\psi_y} \\
 & & \frac{12EI_y}{l^3\psi_z} & 0 & \frac{6EI_y}{l^2\psi_z} & 0 \\
 & & & \frac{GJ}{l} & 0 & 0 \\
 & \text{sym} & & & \frac{\psi_z' EI_y}{l} & 0 \\
 & & & & & \frac{\psi_y' EI_z}{l}
\end{bmatrix} \tag{4.28}
$$

In these equations, the following substitutions are to be made:

$$
\begin{aligned}
\psi_y &= 1 + 12EI_z/GA_{sy}l^2, \qquad \psi_y' = (3+\psi_y)/\psi_y \\
\psi_z &= 1 + 12EI_y/GA_{sz}l^2, \qquad \psi_z' = (3+\psi_z)/\psi_z
\end{aligned} \tag{4.29}
$$

where A_{s_y} and A_{s_z} are the beam effective shear cross-sectional areas in the y and z axes, respectively. Similarly the inertia matrices are written, accounting for both translational and rotational inertia:

$$
\boldsymbol{M}^e = \rho A l \begin{bmatrix} \boldsymbol{M}_{11} & \boldsymbol{M}_{12} \\ \text{sym} & \boldsymbol{M}_{22} \end{bmatrix} \tag{4.30}
$$

in which

$$
\boldsymbol{M}_{11} = \begin{bmatrix}
\frac{1}{3} & 0 & 0 & 0 & 0 & 0 \\
 & \frac{13}{35} + \frac{6I_z}{5Al^2} & 0 & 0 & 0 & \frac{11l}{210} + \frac{I_z}{10Al} \\
 & & \frac{13}{35} + \frac{6I_y}{5Al^2} & 0 & -\frac{11l}{210} - \frac{I_y}{10Al} & 0 \\
 & & & \frac{J_x}{3A} & 0 & 0 \\
 & \text{sym} & & & \frac{l^2}{105} + \frac{2I_y}{15A} & 0 \\
 & & & & & \frac{l^2}{105} + \frac{2I_z}{15A}
\end{bmatrix} \tag{4.31}
$$

$$
\boldsymbol{M}_{12} = \begin{bmatrix}
\frac{1}{6} & 0 & 0 & 0 & 0 & 0 \\
0 & \frac{9}{70} - \frac{6I_z}{5Al^2} & 0 & 0 & 0 & -\frac{13l}{420} + \frac{I_z}{10Al} \\
0 & 0 & \frac{9}{70} - \frac{6I_y}{5Al^2} & 0 & \frac{13l}{420} - \frac{I_y}{10Al} & 0 \\
0 & 0 & 0 & \frac{J_x}{6A} & 0 & 0 \\
0 & 0 & -\frac{13l}{420} + \frac{I_y}{10Al} & 0 & -\frac{l^2}{140} - \frac{I_y}{30A} & 0 \\
0 & \frac{13l}{420} - \frac{I_z}{10Al} & 0 & 0 & 0 & -\frac{l^2}{140} - \frac{I_z}{30A}
\end{bmatrix} \tag{4.32}
$$

$$
\boldsymbol{M}_{22} = \begin{bmatrix}
\frac{1}{3} & 0 & 0 & 0 & 0 & 0 \\
 & \frac{13}{35} + \frac{6I_z}{5Al^2} & 0 & 0 & 0 & -\frac{11l}{210} - \frac{I_z}{10Al} \\
 & & \frac{13}{35} + \frac{6I_y}{5Al^2} & 0 & \frac{11l}{210} + \frac{I_y}{10Al} & 0 \\
 & & & \frac{J_x}{3A} & 0 & 0 \\
 & \text{sym} & & & \frac{l^2}{105} + \frac{2I_y}{15A} & 0 \\
 & & & & & \frac{l^2}{105} + \frac{2I_z}{15A}
\end{bmatrix} \tag{4.33}
$$

In these expressions for mass matrix, the effect of shear deformation[2] has been neglected. The preceding element may be modified to include rigid links and springs. The effect of temperature change, lack of fit, etc., is to produce restraint forces on the nodes in the zero displacement state. These forces, reversed in sign, become the nodal forces to be applied to the nodes of the structure. The thermal nodal forces for the line element for a uniform temperature change T may be calculated from Eq. (3.65) as

$$
\begin{aligned}
\boldsymbol{p}_T &= \int_V \boldsymbol{B}^T \boldsymbol{D} \epsilon_T \, \mathrm{d}V \\
&= \frac{A}{l} \int_0^l [-1 \quad 1]^T E\alpha T \, \mathrm{d}x \\
&= EA\alpha T [-1 \quad 1]^T
\end{aligned} \tag{4.34}
$$

Similarly, effects of initial strains caused by lack of fit or shrinkage are calculated replacing the thermal strains by these values, using Eq. (3.65):

$$\begin{aligned} \boldsymbol{p}_I &= \int_V \boldsymbol{B}^T \boldsymbol{D} \epsilon_I \, \mathrm{d}V \\ &= EA\epsilon_I [-1 \quad 1]^T \end{aligned} \tag{4.35}$$

4.3 Two-Dimensional Plane Elements

4.3.1 *Constant Strain Triangle*

A triangular plane element[3,4] with three nodes is shown in Fig. 4.2. The two displacement functions (u_x, u_y) at any point within the triangle, expressed in its local coordinate system, have linear variations:

$$\begin{aligned} u_x &= [c_1 \quad c_2 \quad c_3] \begin{Bmatrix} 1 \\ x \\ y \end{Bmatrix} \\ &= \boldsymbol{a}_1^T \boldsymbol{\phi} \\ u_y &= [c_4 \quad c_5 \quad c_6] \begin{Bmatrix} 1 \\ x \\ y \end{Bmatrix} \\ &= \boldsymbol{a}_2^T \boldsymbol{\phi} \end{aligned} \tag{4.36}$$

or

$$\begin{aligned} \boldsymbol{u} &= \begin{bmatrix} c_1 & c_2 & c_3 \\ c_4 & c_5 & c_6 \end{bmatrix} \begin{Bmatrix} 1 \\ x \\ y \end{Bmatrix} \\ &= \boldsymbol{C\phi} \end{aligned} \tag{4.37}$$

The unknown coefficients c_i are evaluated in terms of nodal values of the unknown displacements by applying appropriate boundary conditions at the nodes, i.e., at

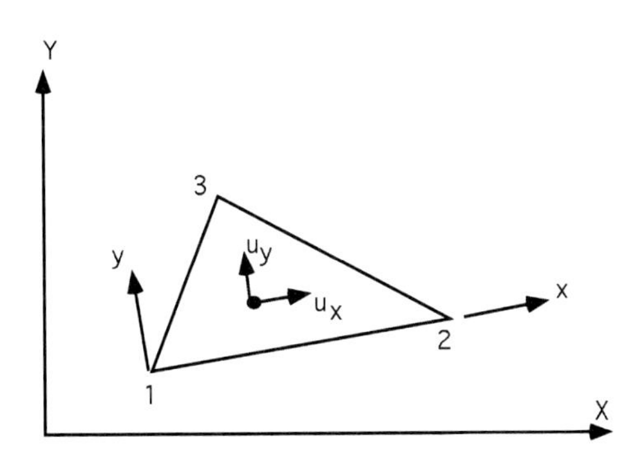

Fig. 4.2 Triangular plane element.

$x = 0$, $y = 0$, $u_x = u_{x_1}$, and $u_y = u_{y_1}$, etc., yielding

$$\boldsymbol{u}_x^e = \boldsymbol{A}\boldsymbol{a}_1, \qquad \boldsymbol{u}_y^e = \boldsymbol{A}\boldsymbol{a}_2 \tag{4.38}$$

Further,

$$\boldsymbol{u} = \begin{Bmatrix} u_x \\ u_y \end{Bmatrix} = \begin{bmatrix} N_1 & 0 & N_2 & 0 & N_3 & 0 \\ 0 & N_1 & 0 & N_2 & 0 & N_3 \end{bmatrix} \begin{Bmatrix} u_{x_1} \\ u_{y_1} \\ u_{x_2} \\ u_{y_2} \\ u_{x_3} \\ u_{y_3} \end{Bmatrix} \tag{4.39}$$

or

$$\boldsymbol{u} = \boldsymbol{N}\boldsymbol{u}^e \tag{4.40}$$

where the interpolation functions N_1, N_2, N_3 in Eq. (4.39) are obtained from the expressions

$$\begin{aligned} \boldsymbol{N}_x &= \boldsymbol{N}_y \\ &= [N_1 \quad N_2 \quad N_3] \\ &= \boldsymbol{\phi}^T \boldsymbol{A}^{-1} \end{aligned} \tag{4.41}$$

and

$$N_1 = \frac{1}{2A}\{(x_2 - x)y_3 - (x_2 - x_3)y\} \qquad N_2 = \frac{1}{2A}(xy_3 - x_3 y) \qquad N_3 = x_2 y \tag{4.42}$$

where A is the area of the triangle. To obtain the strain–displacement matrix, the planar strain vector is expressed as

$$\boldsymbol{\epsilon} = [\epsilon_x \quad \epsilon_y \quad \gamma_{xy}]^T \tag{4.43}$$

in which

$$\epsilon_x = \frac{\partial u_x}{\partial x} \qquad \epsilon_y = \frac{\partial u_y}{\partial y} \qquad \gamma_{xy} = \frac{\partial u_x}{\partial y} + \frac{\partial u_y}{\partial x} \tag{4.44}$$

and the required matrix is obtained from Eqs. (4.39), (4.42), and (4.44) as

$$\begin{aligned} \boldsymbol{\epsilon} &= \frac{1}{2A} \begin{bmatrix} -y_3 & 0 & y_3 & 0 & 0 & 0 \\ 0 & -(x_2 - x_3) & 0 & -x_3 & 0 & x_2 \\ -(x_2 - x_3) & -y_3 & -x_3 & y_3 & x_2 & 0 \end{bmatrix} \begin{Bmatrix} u_{x_1} \\ u_{y_1} \\ u_{x_2} \\ u_{y_2} \\ u_{x_3} \\ u_{y_3} \end{Bmatrix} \\ &= \boldsymbol{B}\boldsymbol{u}^e \end{aligned} \tag{4.45}$$

For the plane stress problem the stress–strain relationship for a general material can be written as

$$\boldsymbol{\sigma} = \boldsymbol{D}(\boldsymbol{\epsilon} - \boldsymbol{\epsilon}_T - \boldsymbol{\epsilon}_I) \tag{4.46}$$

in which

$$\boldsymbol{\sigma} = [\sigma_x \;\; \sigma_y \;\; \sigma_{xy}]^T, \qquad \boldsymbol{\epsilon}_T = T[\alpha_x \;\; \alpha_y \;\; 0]^T \tag{4.47}$$

and where $\boldsymbol{\epsilon}_I$ is the initial strain vector. For homogeneous, isotropic material, $\alpha_x = \alpha_y = \alpha$, so that

$$\boldsymbol{\epsilon}_T = T\alpha[1 \;\; 1 \;\; 0]^T \tag{4.48}$$

Then the stiffness and inertia matrices and thermal and initial load vectors are derived from standard relationships such as

$$\boldsymbol{K}^e = \int_V \boldsymbol{B}^T \boldsymbol{D}\boldsymbol{B} \,\mathrm{d}V = \int_A \boldsymbol{B}^T \boldsymbol{D}\boldsymbol{B} t \,\mathrm{d}A \tag{4.49}$$

$$\boldsymbol{M}^e = \int_V \rho \boldsymbol{N}^T \boldsymbol{N} \,\mathrm{d}V = \int_A \rho \boldsymbol{N}^T \boldsymbol{N} t \,\mathrm{d}A \tag{4.50}$$

$$\boldsymbol{p}_T = \int_V \boldsymbol{B}^T \boldsymbol{D} \boldsymbol{\epsilon}_T \,\mathrm{d}V = \int_A \boldsymbol{B}^T \boldsymbol{D} \boldsymbol{\epsilon}_T t \,\mathrm{d}A \tag{4.51}$$

$$\boldsymbol{p}_I = \int_V \boldsymbol{B}^T \boldsymbol{D} \boldsymbol{\epsilon}_I \,\mathrm{d}V = \int_A \boldsymbol{B}^T \boldsymbol{D} \boldsymbol{\epsilon}_I t \,\mathrm{d}A \tag{4.52}$$

and also

$$\begin{aligned} \boldsymbol{p}_T &= \int_V \boldsymbol{B}^T \boldsymbol{D} \boldsymbol{\epsilon}_T \,\mathrm{d}V \\ &= T\alpha t \int_A \boldsymbol{B}^T \boldsymbol{D} \,\mathrm{d}A \end{aligned} \tag{4.53}$$

For the constant strain triangle (CST), the thermal nodal forces are calculated to be

$$\boldsymbol{p}_T = T\alpha t A \boldsymbol{B}^T \boldsymbol{D} \boldsymbol{\epsilon}_T \tag{4.54}$$

In the local coordinate system,

$$\boldsymbol{p}_T = \frac{T\alpha t}{2} \begin{bmatrix} -y_3 & 0 & -(x_2 - x_3) \\ 0 & -(x_2 - x_3) & -y_3 \\ y_3 & 0 & -x_3 \\ 0 & -x_3 & y_3 \\ 0 & 0 & x_2 \\ 0 & x_2 & 0 \end{bmatrix} \boldsymbol{D} \boldsymbol{\epsilon}_T \tag{4.55}$$

in which

$$\boldsymbol{D} \boldsymbol{\epsilon}_T = \frac{TE}{1 - \nu^2} \begin{bmatrix} \alpha_x + \nu \alpha_y \\ \nu \alpha_x + \alpha_y \\ 0 \end{bmatrix} \tag{4.56}$$

and for homogeneous, isotropic material $\alpha_x = \alpha_y = \alpha$,

$$p_T = \frac{T\alpha t E}{2(1-\nu)} \begin{bmatrix} -y_3 \\ -(x_2 - x_3) \\ y_3 \\ -x_3 \\ 0 \\ x_2 \end{bmatrix} \tag{4.57}$$

For initial strains ϵ_I for the case of an isotropic material, Eq. (4.52) can be used to calculate the load vector in the local coordinate system:

$$p_I = \frac{\epsilon_I t E}{2(1-\nu)} \begin{bmatrix} -y_3 \\ -(x_2 - x_3) \\ y_3 \\ -x_3 \\ 0 \\ x_2 \end{bmatrix} \tag{4.58}$$

The assumption of linear variation of displacements within the element results in constant strains and stresses. A linear distribution of displacements along any edge of the triangle assures compatibility of displacements between two adjoining elements. Furthermore, choice of a complete first order polynomial in conformity with the highest derivative of the strain-displacement relation assures rigid-body and constant strain requirements. For elements using higher order functions integration may be carried out with a symbolic algebraic manipulation package yielding FORTRAN expressions for each matrix element. For the numerically integrated isoparametric elements Gaussian integration may also be used.

4.3.2 Linear and Quadratic Strain Triangles

The preceding constant strain plane triangle needs a large number of elements to achieve an accurate solution for a number of practical problems where linear strain distribution is predominant. Thus considerable effort has been expended toward deriving refined, higher-order elements.[5–7] Figure 4.3 shows a six-noded triangular element that assumes quadratic variation of displacements within the element

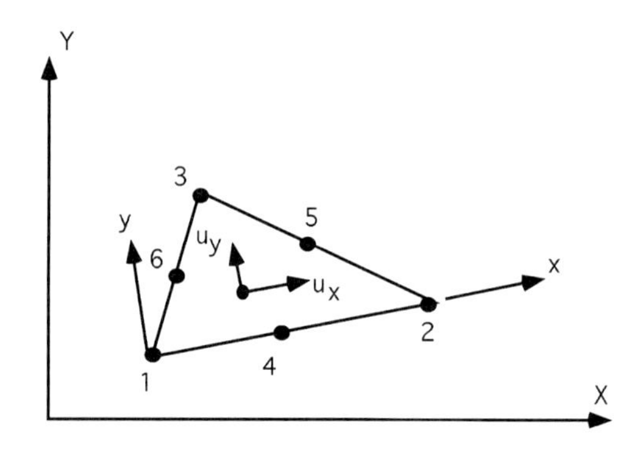

Fig. 4.3 Linear strain triangle.

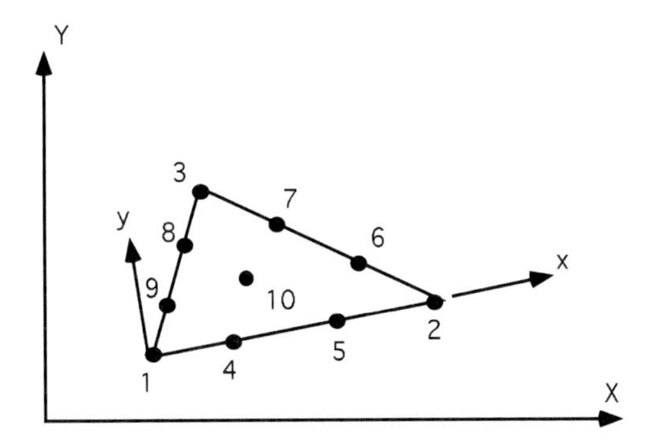

Fig. 4.4 Quadratic strain triangle.

resulting in linear distribution of strain. An element displacement component can be expressed in terms of a complete second-order polynomial, so that

$$\begin{aligned} u_x &= c_1 + c_2x + c_3y + c_4x^2 + c_5xy + c_6y^2 \\ u_y &= c_7 + c_8x + c_9y + c_{10}x^2 + c_{11}xy + c_{12}y^2 \end{aligned} \tag{4.59}$$

or

$$\boldsymbol{u} = \begin{bmatrix} c_1 & c_2 & c_3 & c_4 & c_5 & c_6 \\ c_7 & c_8 & c_9 & c_{10} & c_{11} & c_{12} \end{bmatrix} \begin{Bmatrix} 1 \\ x \\ y \\ x^2 \\ xy \\ y^2 \end{Bmatrix}$$

$$= \boldsymbol{C\phi} \tag{4.60}$$

The coefficients c_1–c_{12} may again be evaluated, as done earlier. The standard procedures may then be applied as before to yield the relevant element matrices defined in Eqs. (4.49–4.52). Again the algebraic symbolic manipulation can be used to derive corresponding FORTRAN instructions to be incorporated into a structural analysis software. In a similar fashion it is also convenient to derive a quadratic strain triangle (in Fig. 4.4), that requires a complete cubic polynomial:

$$\begin{aligned} u_x &= c_1 + c_2x + c_3y + c_4x^2 + c_5xy + c_6y^2 + c_7x^3 + c_8x^2y + c_9xy^2 + c_{10}y^3 \\ u_y &= c_{11} + c_{12}x + c_{13}y + c_{14}x^2 + c_{15}x^2 + c_{16}y^2 + c_{17}x^3 \\ &\quad + c_{18}x^2y + c_{19}xy^2 + c_{20}y^3 \end{aligned}$$

This polynomial requires an extra node, number 10, located at the triangle centroid for determination of the coefficients c_1–c_{20}, and the required element matrices are derived from equations (4.49–4.52).

With increasing mesh subdivision a finite element solution is expected to converge to the corresponding analytic solution. The analysis effort is a function of $(n \times m^2)$, where n is the order and m is the half bandwidth of the stiffness matrix for the entire structure. Adoption of an element type is a compromise between

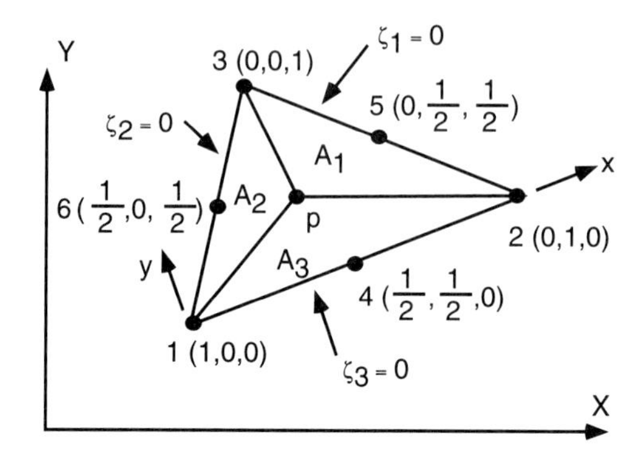

Fig. 4.5 Natural coordinates for a triangular element.

available computing resources, modeling effort, and desired solution accuracy. Although higher-order elements may require fewer elements and thus less effort for data preparation, they may involve increased demand on computing time because of increased bandwidth.

4.3.2.1 Element matrix using natural coordinate system. An alternative, commonly used procedure for derivation of element matrices using the natural coordinate system[8] is described next. Thus Fig. 4.5 shows the natural coordinate system for the element. The relationship between the element local and the natural coordinates may be written as

$$\begin{Bmatrix} 1 \\ x \\ y \end{Bmatrix} = \begin{bmatrix} 1 & 1 & 1 \\ x_1 & x_2 & x_3 \\ y_1 & y_2 & y_3 \end{bmatrix} \begin{Bmatrix} \zeta_1 \\ \zeta_2 \\ \zeta_3 \end{Bmatrix} \tag{4.61}$$

This equation can be inverted to give the following inverse relationship:

$$\begin{Bmatrix} \zeta_1 \\ \zeta_2 \\ \zeta_3 \end{Bmatrix} = \frac{1}{2A} \begin{bmatrix} x_2 y_3 - x_3 y_2 & y_2 - y_3 & x_3 - x_2 \\ x_3 y_1 - x_1 y_3 & y_3 - y_1 & x_1 - x_3 \\ x_1 y_2 - x_2 y_1 & y_1 - y_2 & x_2 - x_1 \end{bmatrix} \begin{Bmatrix} 1 \\ x \\ y \end{Bmatrix} \tag{4.62}$$

In these equations $\zeta_i = A_i/A$ is referred to as an area coordinate. Equation (4.62) reduces to the following when expressed in the local element coordinate system:

$$\begin{Bmatrix} \zeta_1 \\ \zeta_2 \\ \zeta_3 \end{Bmatrix} = \frac{1}{2A} \begin{bmatrix} x_2 y_3 - x_3 y_2 & -y_3 & x_3 - x_2 \\ 0 & y_3 & -x_3 \\ 0 & 0 & x_2 \end{bmatrix} \begin{Bmatrix} 1 \\ x \\ y \end{Bmatrix} \tag{4.63}$$

Now the displacement functions can be written as polynomials of the natural coordinates:

$$u_x(\zeta_1, \zeta_2, \zeta_3) = \alpha_1 \zeta_1^2 + \alpha_2 \zeta_2^2 + \alpha_3 \zeta_3^2 + \alpha_4 \zeta_1 \zeta_2 + \alpha_5 \zeta_2 \zeta_3 + \alpha_6 \zeta_3 \zeta_1 \tag{4.64}$$

$$u_y(\zeta_1, \zeta_2, \zeta_3) = \alpha_7 \zeta_1^2 + \alpha_8 \zeta_2^2 + \alpha_9 \zeta_3^2 + \alpha_{10} \zeta_1 \zeta_2 + \alpha_{11} \zeta_3 \zeta_3 + \alpha_{12} \zeta_3 \zeta_1 \tag{4.65}$$

thus, e.g., Eqs. (4.64) and (4.65) can be written as

$$u_x = \boldsymbol{a}_1^T \boldsymbol{\phi}, \qquad u_y = \boldsymbol{a}_2^T \boldsymbol{\phi} \tag{4.66}$$

where

$$\boldsymbol{a}_1 = [\alpha_1 \quad \alpha_2 \quad \alpha_3 \quad \alpha_4 \quad \alpha_5 \quad \alpha_6]^T \qquad \boldsymbol{\phi} = \left[\zeta_1^2 \quad \zeta_2^2 \quad \zeta_3^2 \quad \zeta_1\zeta_2 \quad \zeta_2\zeta_3 \quad \zeta_3\zeta_1\right]^T$$

Evaluating the polynomial at the nodal points, the relationship is obtained,

$$\begin{Bmatrix} u_{x_1} \\ u_{x_2} \\ u_{x_3} \\ u_{x_4} \\ u_{x_5} \\ u_{x_6} \end{Bmatrix} = \begin{bmatrix} 1 & 0 & 0 & 0 & 0 & 0 \\ 0 & 1 & 0 & 0 & 0 & 0 \\ 0 & 0 & 1 & 0 & 0 & 0 \\ \frac{1}{4} & \frac{1}{4} & 0 & \frac{1}{4} & 0 & 0 \\ 0 & \frac{1}{4} & \frac{1}{4} & 0 & \frac{1}{4} & 0 \\ \frac{1}{4} & 0 & \frac{1}{4} & 0 & 0 & \frac{1}{4} \end{bmatrix} \begin{Bmatrix} \alpha_1 \\ \alpha_2 \\ \alpha_3 \\ \alpha_4 \\ \alpha_5 \\ \alpha_6 \end{Bmatrix} \tag{4.67}$$

or

$$\boldsymbol{u}_x^e = \boldsymbol{A}\boldsymbol{a}_1 \tag{4.68}$$

Inverting this relationship the unknown coefficients are obtained as

$$\boldsymbol{a}_1 = \boldsymbol{A}^{-1}\boldsymbol{u}_x^e \tag{4.69}$$

where

$$\boldsymbol{A}^{-1} = \begin{bmatrix} 1 & 0 & 0 & 0 & 0 & 0 \\ 0 & 1 & 0 & 0 & 0 & 0 \\ 0 & 0 & 1 & 0 & 0 & 0 \\ -1 & -1 & 0 & 4 & 0 & 0 \\ 0 & -1 & -1 & 0 & 4 & 0 \\ -1 & 0 & -1 & 0 & 0 & 4 \end{bmatrix}$$

Then the shape function is derived using Eq. (4.66) as

$$\begin{aligned} u_x &= \boldsymbol{a}_1^T \boldsymbol{\phi} \\ &= \left(\boldsymbol{u}_x^e\right)^T (\boldsymbol{A}^{-1})^T \boldsymbol{\phi} \\ &= \left(\boldsymbol{u}_x^e\right)^T \boldsymbol{N}_x^T \\ &= \boldsymbol{N}_x \boldsymbol{u}_x^e \end{aligned} \tag{4.70}$$

The shape function in the individual directions is given by

$$\boldsymbol{N}_x = \boldsymbol{N}_y = \boldsymbol{\phi}^T(\boldsymbol{A}^{-1}) \tag{4.71}$$

Performing appropriate algebraic operations gives the quadratic form of the shape function:

$$\begin{aligned} \boldsymbol{N}_x &= [\zeta_1(2\zeta_1 - 1)\ \zeta_2(2\zeta_2 - 1)\ \zeta_3(2\zeta_3 - 1)\ 4\zeta_1\zeta_2\ 4\zeta_2\zeta_3\ 4\zeta_3\zeta_1] \\ &= [N_1 \quad N_2 \quad N_3 \quad N_4 \quad N_5 \quad N_6] \end{aligned} \tag{4.72}$$

The same shape function is used to interpolate the u_y displacement and thus together these are written as

$$\begin{Bmatrix} u_x \\ u_y \end{Bmatrix} = \begin{bmatrix} N_1 & 0 & N_2 & 0 & N_3 & 0 & N_4 & 0 & N_5 & 0 & N_6 & 0 \\ 0 & N_1 & 0 & N_2 & 0 & N_3 & 0 & N_4 & 0 & N_5 & 0 & N_6 \end{bmatrix} \boldsymbol{u}^e \tag{4.73}$$

or

$$\boldsymbol{u} = \boldsymbol{N}\boldsymbol{u}^e \tag{4.74}$$

The strain–displacement relations may be obtained by applying the chain rule of differentiation, e.g., in the x direction

$$\frac{\partial u_x}{\partial x} = \left(\frac{\partial}{\partial \zeta_1}\frac{\partial \zeta_1}{\partial x} + \frac{\partial}{\partial \zeta_2}\frac{\partial \zeta_2}{\partial x} + \frac{\partial}{\partial \zeta_3}\frac{\partial \zeta_3}{\partial x} \right) u_x \tag{4.75}$$

and noting that $u_x = \boldsymbol{N}_x \boldsymbol{u}_x^e$, the expression for ϵ_x is obtained:

$$\begin{aligned} \epsilon_x &= \frac{\partial u_x}{\partial x} \\ &= \frac{1}{2A}[b_1(4\zeta_1 - 1)\ b_2(2\zeta_2 - 1)\ b_3(4\zeta_3 - 1)\ 4(\zeta_1 b_2 + \zeta_2 b_1) \\ &\quad 4(\zeta_2 b_3 + \zeta_3 b_2)\ 4(\zeta_1 b_3 + \zeta_3 b_1)]\boldsymbol{u}_x^e \\ &= \boldsymbol{B}_x \boldsymbol{u}_x^e \end{aligned} \tag{4.76}$$

and similarly for ϵ_y,

$$\begin{aligned} \epsilon_y &= \frac{\partial u_y}{\partial y} = \boldsymbol{B}_y u_y^e \\ \gamma_{xy} &= \frac{\partial u_x}{\partial y} + \frac{\partial u_y}{\partial x} = \boldsymbol{B}_y \boldsymbol{u}_x^e + \boldsymbol{B}_x \boldsymbol{u}_y^e \end{aligned} \tag{4.77}$$

Together these expressions are written as

$$\begin{Bmatrix} \epsilon_x \\ \epsilon_y \\ \gamma_{xy} \end{Bmatrix} = \begin{bmatrix} \boldsymbol{B}_x & 0 \\ 0 & \boldsymbol{B}_y \\ \boldsymbol{B}_y & \boldsymbol{B}_x \end{bmatrix} \begin{Bmatrix} \boldsymbol{u}_x^e \\ \boldsymbol{u}_y^e \end{Bmatrix} \tag{4.78}$$

or

$$\epsilon = \boldsymbol{B}\boldsymbol{u}^e \tag{4.79}$$

It should be noted that $\boldsymbol{B}_y$ is obtained from $\boldsymbol{B}_x$ simply by substituting a_1, a_2, a_3 for b_1, b_2, and b_3, respectively, and where in the local triangle system

$$\begin{aligned} a_1 &= x_3 - x_2 \qquad & b_1 &= -y_3 \\ a_2 &= -x_3 \qquad & b_2 &= y_3 \\ a_3 &= x_2 \qquad & b_3 &= 0 \end{aligned} \tag{4.80}$$

and noting that these values constitute the third and second columns, respectively, of the coefficient transformation matrix in Eq. (4.63).

Because the strains are linear functions of the apex nodal values, it is advantageous to obtain these nodal strains from apex nodal displacements first. This is achieved by substituting ξ nodal coordinate values in Eq. (4.78), so that

$$\begin{Bmatrix} \varepsilon_{x1} \\ \varepsilon_{x2} \\ \varepsilon_{x3} \\ \varepsilon_{y1} \\ \varepsilon_{y2} \\ \varepsilon_{y3} \\ \gamma_{xy1} \\ \gamma_{xy2} \\ \gamma_{xy3} \end{Bmatrix} = \begin{bmatrix} \boldsymbol{\Phi}_1 & 0 \\ 0 & \boldsymbol{\Phi}_2 \\ \boldsymbol{\Phi}_2 & \boldsymbol{\Phi}_1 \end{bmatrix} \begin{Bmatrix} \boldsymbol{u}_x^e \\ \boldsymbol{u}_y^e \end{Bmatrix} = \boldsymbol{\Phi} \begin{Bmatrix} \boldsymbol{u}_x^e \\ \boldsymbol{u}_y^e \end{Bmatrix} \tag{4.81}$$

or

$$\boldsymbol{\epsilon}_c = \boldsymbol{\Phi} \begin{Bmatrix} \boldsymbol{u}_x^e \\ \boldsymbol{u}_y^e \end{Bmatrix} \tag{4.82}$$

in which

$$\boldsymbol{\Phi}_1 = \frac{y_3}{2A} \begin{bmatrix} -3 & -1 & 0 & 4 & 0 & 0 \\ 1 & 3 & 0 & -4 & 0 & 0 \\ 1 & -1 & 0 & 0 & 4 & -4 \end{bmatrix}$$

$$\boldsymbol{\Phi}_2 = -\frac{x_3}{2A} \begin{bmatrix} -3 & -1 & 0 & 4 & 0 & 0 \\ 1 & 3 & 0 & -4 & 0 & 0 \\ 1 & -1 & 0 & 0 & 4 & -4 \end{bmatrix} + \frac{x_2}{2A} \begin{bmatrix} -3 & 0 & -1 & 0 & 0 & 4 \\ 1 & 0 & -1 & -4 & 4 & 0 \\ 1 & 0 & 3 & 0 & 0 & -4 \end{bmatrix} \tag{4.83}$$

The strains at any point in the triangle are expressed in terms of the apex node values,

$$\begin{Bmatrix} \varepsilon_x \\ \varepsilon_y \\ \gamma_{xy} \end{Bmatrix} = \begin{bmatrix} \boldsymbol{N}_1 & 0 & 0 \\ 0 & \boldsymbol{N}_1 & 0 \\ 0 & 0 & \boldsymbol{N}_1 \end{bmatrix} \begin{Bmatrix} \boldsymbol{\epsilon}_{x_c} \\ \boldsymbol{\epsilon}_{y_c} \\ \boldsymbol{\gamma}_{xy_c} \end{Bmatrix} = \boldsymbol{N}\boldsymbol{\epsilon}_c \tag{4.84}$$

$$\boldsymbol{N}_1 = [\zeta_1 \quad \zeta_2 \quad \zeta_3]$$

hence

$$\begin{Bmatrix} \varepsilon_x \\ \varepsilon_y \\ \gamma_{xy} \end{Bmatrix} = \boldsymbol{N}\boldsymbol{\Phi} \begin{Bmatrix} \boldsymbol{u}_x^e \\ \boldsymbol{u}_y^e \end{Bmatrix} = \boldsymbol{B} \begin{Bmatrix} \boldsymbol{u}_x^e \\ \boldsymbol{u}_y^e \end{Bmatrix} \tag{4.85}$$

Now the element stiffness matrix is given by the integral

$$\boldsymbol{K}^e = \int_V \boldsymbol{B}^T \boldsymbol{D}\boldsymbol{B}\,\mathrm{d}V$$
$$= t\int_A \boldsymbol{B}^T \boldsymbol{D}\boldsymbol{B}\,\mathrm{d}A \tag{4.86}$$

or

$$\boldsymbol{K}^e = \boldsymbol{\Phi}^T t\left[\int_A \boldsymbol{N}^T \boldsymbol{D}\boldsymbol{N}\,\mathrm{d}A\right]\boldsymbol{\Phi} \tag{4.87}$$

The element stiffness matrix $\boldsymbol{K}^e$ can now be simply derived by noting that

$$\int_A \boldsymbol{N}^T \boldsymbol{D}\boldsymbol{N}\,\mathrm{d}A = \begin{bmatrix} d_{11}\boldsymbol{\Lambda} & d_{12}\boldsymbol{\Lambda} & d_{13}\boldsymbol{\Lambda} \\ & d_{22}\boldsymbol{\Lambda} & d_{23}\boldsymbol{\Lambda} \\ \text{sym} & & d_{33}\boldsymbol{\Lambda} \end{bmatrix} \tag{4.88}$$

where d_{ij} are the elements of the stress–strain matrix $\boldsymbol{D}$ and

$$\boldsymbol{\Lambda} = \int_A \boldsymbol{N}_1^T \boldsymbol{N}_1\,\mathrm{d}A = \frac{A}{12}\begin{bmatrix} 2 & 1 & 1 \\ 1 & 2 & 1 \\ 1 & 1 & 2 \end{bmatrix} \tag{4.89}$$

because

$$\boldsymbol{N}_1^T \boldsymbol{N}_1 = \begin{bmatrix} \zeta_1^2 & \zeta_1\zeta_2 & \zeta_1\zeta_3 \\ \zeta_1\zeta_2 & \zeta_2^2 & \zeta_2\zeta_3 \\ \zeta_1\zeta_3 & \zeta_2\zeta_3 & \zeta_3^2 \end{bmatrix} \tag{4.90}$$

can easily be integrated using the known formulation

$$\int_A \zeta_1^p\,\zeta_2^q\,\zeta_3^r\,\mathrm{d}A = 2A\frac{p!\,q!\,r!}{(p+q+r+2)!} \tag{4.91}$$

Then using Eq. (4.88) the stiffness matrix may be derived from Eq. (4.87).

Likewise the inertia matrix may be derived as

$$\boldsymbol{M}^e = \rho\int_V \boldsymbol{N}^T \boldsymbol{N}\,\mathrm{d}V$$
$$= \rho t\int_A \boldsymbol{N}^T \boldsymbol{N}\,\mathrm{d}A \tag{4.92}$$

for constant thickness t. In the x direction

$$\boldsymbol{M}_x^e = \rho t\int \boldsymbol{N}_x^T \boldsymbol{N}_x\,\mathrm{d}A = \frac{At\rho}{180}\begin{bmatrix} 6 & -1 & -1 & 0 & -4 & 0 \\ & 6 & -1 & 0 & 0 & -4 \\ & & 6 & -4 & 0 & 0 \\ & \text{sym} & & 32 & 16 & 16 \\ & & & & & 32 \end{bmatrix} \tag{4.93}$$

and

$$\boldsymbol{M}_y^e = \boldsymbol{M}_x^e \tag{4.94}$$

Finally, by suitable interchange of rows and columns the stiffness and inertia matrices are rearranged to conform to nodal variables expressed nodewise. It may also be noted that the stiffness matrix is derived in the local (x, y) coordinate system as defined in Fig. 4.5, whereas the inertia matrix is invariant under coordinate transformation.

Nodal release forces produced by thermal strains caused by a temperature change of T in the linear strain triangle (LST) are calculated using Eq. (4.51):

$$\boldsymbol{p}_T = \int_V \boldsymbol{B}^T \boldsymbol{D} \epsilon_T \, \mathrm{d}V \tag{4.95}$$

The $\boldsymbol{B}$ matrix for the LST is given in Eq. (4.85) as

$$\boldsymbol{B} = \boldsymbol{N\Phi} \tag{4.96}$$

For isotropic material making this substitution, and noting that

$$\epsilon_T = T\alpha[1 \quad 1 \quad 0]^T; \qquad \epsilon_T = \frac{ET\alpha}{1-\nu}[1 \quad 1 \quad 0]^T \tag{4.97}$$

the expression for $\boldsymbol{p}_T$ becomes,

$$\boldsymbol{p}_T = \frac{ET\alpha t}{1-\nu}\boldsymbol{\Phi}^T \int_{area} \boldsymbol{N}^T \, \mathrm{d}A \begin{bmatrix} 1 \\ 1 \\ 0 \end{bmatrix} \tag{4.98}$$

and in this case noting that $\boldsymbol{N}^T$ is composed of $\boldsymbol{N}_1$ matrices which each integrate to, $\frac{A}{3}[1 \quad 1 \quad 0]$ so that,

$$\boldsymbol{p}_T = \frac{ET\alpha t}{1-\nu}\boldsymbol{\Phi}^T \frac{A}{3} \begin{bmatrix} \boldsymbol{e}_3 \\ \boldsymbol{e}_3 \\ 0 \end{bmatrix}; \quad \boldsymbol{e}_3 = [1 \quad 1 \quad 0] \tag{4.99}$$

This expression is easily evaluated using the definition of $\boldsymbol{\Phi}$, see Eq. (4.82). For initial strains $T\alpha$ is replaced by ϵ_I.

4.3.2.2 Curved isoparametric elements. For some practical applications it is advantageous to use elements with curved edges. The isoparametric elements fit into this category. In such an element the geometry as well as displacements are defined by the same shape functions. Proceeding with the six-noded triangular plane element, the following relationships are defined:

$$[x \quad y] = \boldsymbol{NX} \tag{4.100}$$

where $\boldsymbol{N} = \boldsymbol{N}_x$ or $\boldsymbol{N}_y$ and

$$\boldsymbol{X} = \begin{bmatrix} x_1 & x_2 & x_3 & x_4 & x_5 & x_6 \\ y_1 & y_2 & y_3 & y_4 & y_5 & y_6 \end{bmatrix}^T \tag{4.101}$$

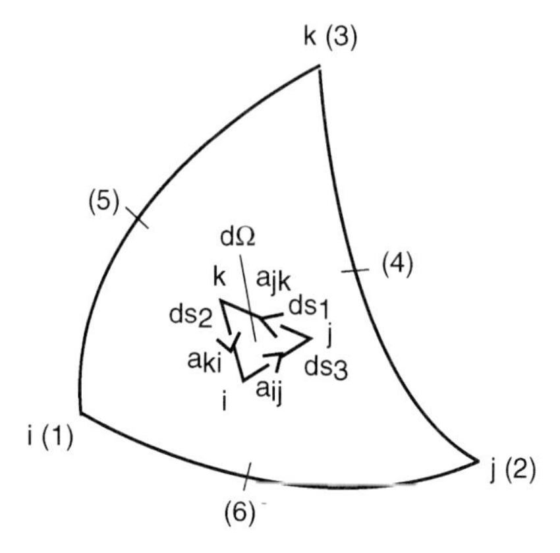

Fig. 4.6 Curved triangle: local coordinate vectors.

contains the nodal coordinates. Figure 4.6 shows an infinitesimal area $\mathrm{d}\Omega$ in the six-node isoparametric triangle in which

$$2\,\mathrm{d}\Omega = |\boldsymbol{a}_{ki} \times \boldsymbol{a}_{ij}|\,\mathrm{d}\zeta_i\,\mathrm{d}\zeta_j \tag{4.102}$$

where the unit vector, e.g., $\boldsymbol{a}_{ki}$, is given as

$$\boldsymbol{a}_{ki} = \begin{bmatrix} \dfrac{\partial x}{\partial \zeta_i} - \dfrac{\partial x}{\partial \zeta_k} \\[2ex] \dfrac{\partial y}{\partial \zeta_i} - \dfrac{\partial y}{\partial \zeta_k} \end{bmatrix} \tag{4.103}$$

Now the Jacobian for the triangle $\boldsymbol{J}_A$ is found as

$$\boldsymbol{J}_A = \left[\boldsymbol{e}_3 \quad \frac{\partial \boldsymbol{N}}{\partial \zeta_i} \boldsymbol{X} \right], \qquad i = 1, 2, 3 \tag{4.104}$$

then

$$2\,\mathrm{d}\Omega = \det \boldsymbol{J}_A\,\mathrm{d}\zeta_i\,\mathrm{d}\zeta_j \tag{4.105}$$

Equation (4.105) can be obtained by expanding Eq. (4.102) and evaluating the determinant of $\boldsymbol{J}_A$ given in Eq. (4.104). The derivatives of the shape function $\boldsymbol{N}$ with respect to $\zeta_1, \zeta_2, \zeta_3$ are written as

$$\frac{\partial \boldsymbol{N}}{\partial \boldsymbol{\xi}} = \boldsymbol{N}_{,\xi} = \begin{bmatrix} 4\zeta_1 - 1 & 0 & 0 & 4\zeta_2 & 0 & 4\zeta_3 \\ 0 & 4\zeta_2 - 1 & 0 & 4\zeta_1 & 4\zeta_3 & 0 \\ 0 & 0 & 4\zeta_3 - 1 & 0 & 4\zeta_2 & 4\zeta_1 \end{bmatrix} = \boldsymbol{\Phi}_\epsilon \tag{4.106}$$

Using the chain rule for differentiation, the x and y derivatives of a function f can be expressed in terms of ζ_1, ζ_2, and ζ_3 derivatives:

$$\begin{Bmatrix} \dfrac{\partial f}{\partial x} \\ \dfrac{\partial f}{\partial y} \end{Bmatrix} = \begin{bmatrix} \dfrac{\partial \zeta_1}{\partial x} & \dfrac{\partial \zeta_2}{\partial x} & \dfrac{\partial \zeta_3}{\partial x} \\ \dfrac{\partial \zeta_1}{\partial y} & \dfrac{\partial \zeta_2}{\partial y} & \dfrac{\partial \zeta_3}{\partial y} \end{bmatrix} \begin{Bmatrix} \dfrac{\partial f}{\partial \zeta_1} \\ \dfrac{\partial f}{\partial \zeta_2} \\ \dfrac{\partial f}{\partial \zeta_3} \end{Bmatrix} \tag{4.107}$$

or

$$\begin{aligned} \boldsymbol{f}_{,c} &= [\boldsymbol{L}_R]\{\boldsymbol{f}_{,\xi}\} \\ &= [\boldsymbol{L}_R][\boldsymbol{\Phi}_\epsilon]\{\boldsymbol{f}\} \end{aligned} \tag{4.108}$$

in which the suffix c stands for coordinate x or y and $\boldsymbol{f}$ contains the six nodal values of the function. It can be proven that $[\boldsymbol{L}_R]$ is composed of rows two and three of $\boldsymbol{J}_A^{-1}$. The stiffness matrix may then be calculated by numerical integration, using the seven-point Gauss integration scheme in Fig. 4.7. The scheme for this numerical evaluation of the stiffness matrix is described next. Form $\boldsymbol{X}$ matrix and then perform the following steps.

Step 1: For each point $(\zeta_1, \zeta_2, \zeta_3)_i$, obtain weight function w_i.
Step 2: Calculate $\boldsymbol{\Phi}_\epsilon$.
Step 3: Calculate $\boldsymbol{J}_A$, det $\boldsymbol{J}_A$, and $\boldsymbol{J}_A^{-1}$.
Step 4: Hence form $[\boldsymbol{L}_R][\boldsymbol{\Phi}_\epsilon]$.
Step 5: From the product in step 4,

$$\begin{Bmatrix} \epsilon_x \\ \epsilon_y \\ \gamma_{xy} \end{Bmatrix} = \begin{bmatrix} \boldsymbol{B}_x & 0 \\ 0 & \boldsymbol{B}_y \\ \boldsymbol{B}_y & \boldsymbol{B}_x \end{bmatrix} \begin{Bmatrix} \boldsymbol{u}_x \\ \boldsymbol{u}_y \end{Bmatrix} \tag{4.109}$$

$$= [\boldsymbol{B}]\begin{Bmatrix} \boldsymbol{u}_x \\ \boldsymbol{u}_y \end{Bmatrix} \tag{4.110}$$

$\boldsymbol{B}_x$, $\boldsymbol{B}_y$ being the first and second rows of the product.

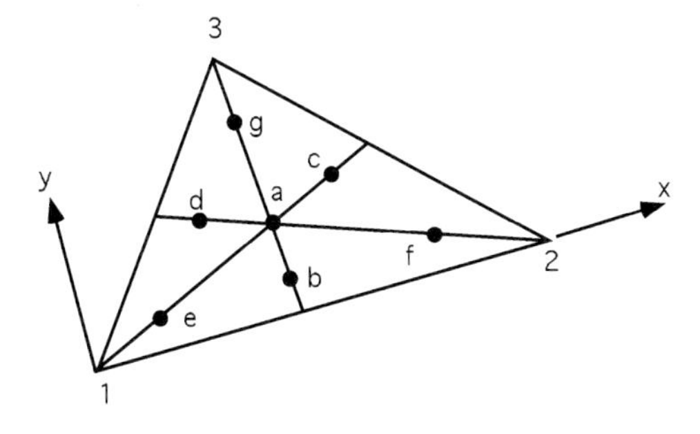

Fig. 4.7 Gauss integration points.

Table 4.1 Seven-point integration scheme

Point	$\zeta_1\ \zeta_2\ \zeta_3$	Weight	ζ values
1	$\alpha\ \alpha\ \alpha$	0.225	
2	$\beta_1\ \alpha_1\ \beta_1$	0.13239415	$\alpha = 0.33333333$
3	$\beta_1\ \beta_1\ \alpha_1$	0.13239415	$\alpha_1 = 0.05971587$
4	$\alpha_1\ \beta_1\ \beta_1$	0.13239415	$\beta_1 = 0.47014206$
5	$\beta_2\ \beta_2\ \alpha_2$	0.12593918	$\alpha_2 = 0.79742699$
6	$\alpha_2\ \beta_2\ \beta_2$	0.12593918	$\beta_2 = 0.10128651$
7	$\beta_2\ \alpha_2\ \beta_2$	0.12593918	

Step 6: Calculate

$$\boldsymbol{K}_i^e = \left(w\boldsymbol{B}^T\boldsymbol{D}\boldsymbol{B}\ \det \boldsymbol{J}_A\right)_i \tag{4.111}$$

which is accumulated to yield the total stiffness matrix, w being the integration weight.

Step 7: Repeat steps 1–6 for each point.

The expression for $\boldsymbol{K}^e$ in Eq. (4.111) contains quintic terms, and for this order at least seven Gauss points are required for integration. Table 4.1 provides details of Gauss integration points for the element.

The inertia matrix, likewise, may be obtained by numerical integration. In this process the matrix is evaluated for each Gauss point as

$$\boldsymbol{M}^e = \left(w\boldsymbol{N}^T\boldsymbol{N}\frac{\det \boldsymbol{J}_A}{2}\right)_i \rho t \tag{4.112}$$

and is summed over all integration points. The preceding integration procedure can be extended to any higher-order triangular element involving appropriate shape functions and nodal coordinates as in Sec. 4.3.2. Any quadrilateral element may be conveniently generated by subdividing it into four triangles, the apex of each triangle being the center of area of the element. Element matrices are combined from individual triangles and the apex point eliminated by static condensation. In the case of higher-order elements, all internal nodes are condensed out. Figure 4.8 shows such an arrangement for constant, linear, and quadratic strain elements with straight as well as curved edges. The thermal nodal release forces for the isotropic material are obtained by numerical integration as follows:

$$\boldsymbol{p}_T = \int_V \boldsymbol{B}^T\boldsymbol{D}\boldsymbol{\epsilon}_T\ \mathrm{d}V \tag{4.113}$$

$$= \frac{ET\alpha t}{1-\nu}\sum_{i=1}^{n}\frac{1}{2}(w\boldsymbol{B}^T\ \det \boldsymbol{J}_A)_i\begin{bmatrix}1\\1\\0\end{bmatrix} \tag{4.114}$$

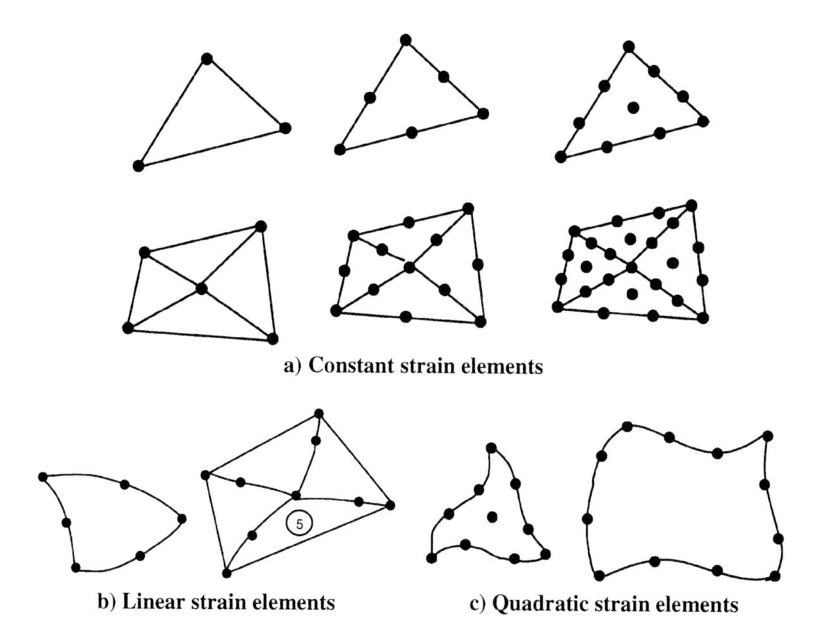

Fig. 4.8 Triangular and rectangular plane elements.

4.3.3 Axisymmetric Solids

The conditions of axisymmetry are satisfied when the body, its supports, and loading can be described by functions that are independent of the angle about the axis of rotational symmetry. The condition is shown in Fig. 4.9. In such a situation the cross section may be subdivided into finite elements, each of which now forms a volume of revolution about the axis of symmetry. In this case the deformations of the body produce four strain components,[9,10] ϵ_x, ϵ_y, γ_{xy}, for the two-dimensional

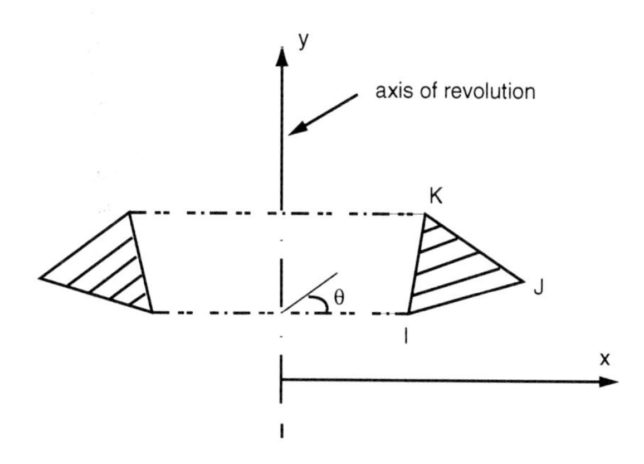

Fig. 4.9 Axisymmetric solid finite element.

case and hoop strains, ϵ_θ produced by the radial displacement u_x such that

$$\epsilon_\theta = u_x / x \tag{4.115}$$

If interpolation functions are assumed for both u_x and x, then this hoop strain is expressed as

$$\epsilon_\theta = \boldsymbol{N}\boldsymbol{u}_x^e / \boldsymbol{N}\boldsymbol{x}^e \tag{4.116}$$

where $\boldsymbol{x}^e$ is the vector of the nodal coordinates. Furthermore if the nodal displacements are grouped coordinatewise, then the strain transformation matrix is given as

$$\begin{Bmatrix} \epsilon_x \\ \epsilon_y \\ \epsilon_\theta \\ \gamma_{xy} \end{Bmatrix} = \begin{bmatrix} \boldsymbol{N}_{,x} & 0 \\ 0 & \boldsymbol{N}_{,y} \\ \dfrac{\boldsymbol{N}}{\boldsymbol{N}\boldsymbol{x}^e} & 0 \\ \boldsymbol{N}_{,y} & \boldsymbol{N}_{,x} \end{bmatrix} \begin{Bmatrix} \boldsymbol{u}_x^e \\ \boldsymbol{u}_y^e \end{Bmatrix} = \boldsymbol{B}\boldsymbol{u}^e \tag{4.117}$$

For the isotropic, elastic material, the constitutive matrix now includes the effect of the three direct strains, so that $\boldsymbol{D}$ is given by the expression

$$[\boldsymbol{D}] = \frac{E}{(1+\nu)(1+2\nu)} \begin{bmatrix} 1-\nu & \nu & \nu & 0 \\ \nu & 1-\nu & \nu & 0 \\ \nu & \nu & 1-\nu & 0 \\ 0 & 0 & 0 & \dfrac{1-2\nu}{2} \end{bmatrix} \tag{4.118}$$

For the CST, the $\boldsymbol{B}$ matrix may be expressed as

$$\boldsymbol{B} = \frac{1}{2A} \begin{bmatrix} b_1 & b_2 & b_3 & 0 & 0 & 0 \\ 0 & 0 & 0 & a_1 & a_2 & a_3 \\ \dfrac{2A\zeta_1}{x} & \dfrac{2A\zeta_2}{x} & \dfrac{2A\zeta_3}{x} & 0 & 0 & 0 \\ a_1 & a_2 & a_3 & b_1 & b_2 & b_3 \end{bmatrix} \tag{4.119}$$

The stiffness matrix is then calculated from the surface of revolution,

$$\begin{aligned} \boldsymbol{K}^e &= 2\pi \int_0^1 \int_{\zeta_2=0}^{1-\zeta_1} \boldsymbol{B}^T \boldsymbol{D}\boldsymbol{B} x \frac{|\boldsymbol{J}|}{2} \, \mathrm{d}\zeta_2 \, \mathrm{d}\zeta_1 \\ &= 2\pi A \int_0^1 \int_{\zeta_2=0}^{1-\zeta_1} \boldsymbol{B}^T \boldsymbol{D}\boldsymbol{B} (\boldsymbol{N}\boldsymbol{x}^e) \, \mathrm{d}\zeta_2 \, \mathrm{d}\zeta_1 \\ &= 2\pi A \sum_{i=1}^{I} w_i \boldsymbol{B}_i^T \boldsymbol{D}\boldsymbol{B}_i \left(\boldsymbol{N}_i \boldsymbol{x}^e\right) \end{aligned} \tag{4.120}$$

where I is the number of integration points. Similarly the mass matrix is obtained:

$$\boldsymbol{M}^e = 2\pi A\rho \sum_{i=1}^{I} w_i \boldsymbol{N}_i^T \boldsymbol{N}_i (\boldsymbol{N}_i \boldsymbol{x}^e) \tag{4.121}$$

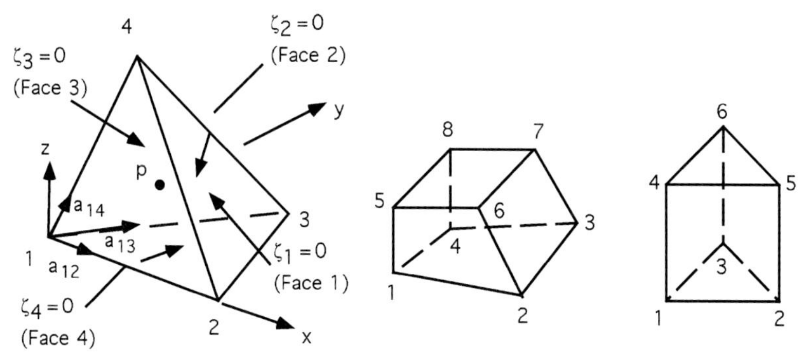

a) Tetrahedron element b) Hexahedron element c) Pentahedron element

Fig. 4.10 Four-noded tetrahedron and other solid elements.

4.4 Three-Dimensional Solid Elements

4.4.1 Constant Strain Tetrahedron

Figure 4.10 shows the natural coordinates system for a four-noded tetrahedron element with the face 1-2-3 on the local x–y plane.[11,12] The relationship between element local and natural coordinate systems may be expressed as

$$\begin{Bmatrix} 1 \\ x \\ y \\ z \end{Bmatrix} = \begin{bmatrix} 1 & 1 & 1 & 1 \\ x_1 & x_2 & x_3 & x_4 \\ y_1 & y_2 & y_3 & y_4 \\ z_1 & z_2 & z_3 & z_4 \end{bmatrix} \begin{Bmatrix} \zeta_1 \\ \zeta_2 \\ \zeta_3 \\ \zeta_4 \end{Bmatrix} \tag{4.122}$$

which, referring to Fig. 4.10, reduces to

$$\begin{Bmatrix} 1 \\ x \\ y \\ z \end{Bmatrix} = \begin{bmatrix} 1 & 1 & 1 & 1 \\ 0 & x_2 & x_3 & x_4 \\ 0 & 0 & y_3 & y_4 \\ 0 & 0 & 0 & z_4 \end{bmatrix} \begin{Bmatrix} \zeta_1 \\ \zeta_2 \\ \zeta_3 \\ \zeta_4 \end{Bmatrix} \tag{4.123}$$

In the volume coordinate system $\zeta_i = V_i / V$, in which V_i is the tetrahedron bound by the center of volume point P and surface opposite to the node i of the element. The volume V of the element is defined as

$$\begin{aligned} 6V &= \boldsymbol{a}_{14} \cdot (\boldsymbol{a}_{12} \times \boldsymbol{a}_{13}) \\ &= \det \boldsymbol{J} \end{aligned} \tag{4.124}$$

where

$$\boldsymbol{J} = \begin{bmatrix} x_4 & y_4 & z_4 \\ x_2 & y_2 & z_2 \\ x_3 & y_3 & z_3 \end{bmatrix} \tag{4.125}$$

The inverse transformation of Eq. (4.103) is obtained as

$$\begin{Bmatrix} \zeta_1 \\ \zeta_2 \\ \zeta_3 \\ \zeta_4 \end{Bmatrix} = \frac{1}{6V} \begin{bmatrix} a_1 & b_1 & c_1 & d_1 \\ a_2 & b_2 & c_2 & d_2 \\ a_3 & b_3 & c_3 & d_3 \\ a_4 & b_4 & c_4 & d_4 \end{bmatrix} \begin{Bmatrix} 1 \\ x \\ y \\ z \end{Bmatrix}$$
$$= \frac{1}{6V}[\boldsymbol{a} \quad \boldsymbol{b} \quad \boldsymbol{c} \quad \boldsymbol{d}] \tag{4.126}$$

or

$$\begin{Bmatrix} \zeta_1 \\ \zeta_2 \\ \zeta_3 \\ \zeta_4 \end{Bmatrix} = \frac{1}{6V} \begin{bmatrix} x_2 y_3 z_4 & -y_3 z_4 & z_4(x_3 - x_2) & y_3(x_4 - x_2) + y_4(x_2 - x_3) \\ 0 & y_3 z_4 & -x_3 z_4 & x_3 y_4 - x_4 y_3 \\ 0 & 0 & x_2 z_4 & -x_2 y_4 \\ 0 & 0 & 0 & x_2 y_3 \end{bmatrix} \begin{Bmatrix} 1 \\ x \\ y \\ z \end{Bmatrix} \tag{4.127}$$

A linear displacement distribution is assumed within the element. Thus typically

$$u_x = \alpha_1 \zeta_1 + \alpha_2 \zeta_2 + \alpha_3 \zeta_3 + \alpha_4 \zeta_4 \tag{4.128}$$

which with substitution of nodal boundary conditions yields

$$u_x = [\zeta_1 \quad \zeta_2 \quad \zeta_3 \quad \zeta_4]\boldsymbol{u}_x^e$$
$$= \boldsymbol{N}_x \boldsymbol{u}_x^e$$
$$= \boldsymbol{N} \boldsymbol{u}_x^e \tag{4.129}$$

and similarly for u_y and u_z displacements, $\boldsymbol{N}_y = \boldsymbol{N}_z = \boldsymbol{N}_x = \boldsymbol{N}$. Using the chain rule of differentiation, e.g., in the x direction for a function f

$$\frac{\partial f}{\partial x} = \frac{\partial f}{\partial \zeta_1}\frac{\partial \zeta_1}{\partial x} + \frac{\partial f}{\partial \zeta_2}\frac{\partial \zeta_2}{\partial x} + \frac{\partial f}{\partial \zeta_3}\frac{\partial \zeta_3}{\partial x} + \frac{\partial f}{\partial \zeta_4}\frac{\partial \zeta_4}{\partial x} \tag{4.130}$$

Now for $f = u_x$, it follows that

$$\frac{\partial u_x}{\partial x} = \left[\frac{\partial \boldsymbol{N}}{\partial \zeta_1}\frac{\partial \zeta_1}{\partial x} + \frac{\partial \boldsymbol{N}}{\partial \zeta_2}\frac{\partial \zeta_2}{\partial x} + \frac{\partial \boldsymbol{N}}{\partial \zeta_3}\frac{\partial \zeta_3}{\partial x} + \frac{\partial \boldsymbol{N}}{\partial \zeta_4}\frac{\partial \zeta_4}{\partial x} \right] \boldsymbol{u}_x^e$$
$$= \frac{1}{6V}\left[\frac{\partial \boldsymbol{N}}{\partial \zeta_1} b_1 + \frac{\partial \boldsymbol{N}}{\partial \zeta_2} b_2 + \frac{\partial \boldsymbol{N}}{\partial \zeta_3} b_3 + \frac{\partial \boldsymbol{N}}{\partial \zeta_4} b_4 \right] \boldsymbol{u}_x^e$$
$$= \frac{1}{6V}\boldsymbol{b}^T \boldsymbol{u}_x^e \tag{4.131}$$

because $\partial N/\partial \zeta$ is a unit matrix of order four. Similar expressions can be obtained for other displacement derivatives. The strain–displacement relationships for the three-dimensional case are

$$
\begin{gathered}
\epsilon_x = \frac{\partial u_x}{\partial x} \qquad \epsilon_y = \frac{\partial u_y}{\partial y} \qquad \epsilon_z = \frac{\partial u_z}{\partial z} \\
\gamma_{xy} = \frac{\partial u_y}{\partial x} + \frac{\partial u_x}{\partial y} \qquad \gamma_{yz} = \frac{\partial u_y}{\partial z} + \frac{\partial u_z}{\partial y} \qquad \gamma_{zx} = \frac{\partial u_z}{\partial x} + \frac{\partial u_x}{\partial z}
\end{gathered}
\tag{4.132}
$$

which can be written in matrix form:

$$
\begin{Bmatrix} \epsilon_x \\ \epsilon_y \\ \epsilon_z \\ \gamma_{xy} \\ \gamma_{yz} \\ \gamma_{zx} \end{Bmatrix} = \frac{1}{6V} \begin{bmatrix} \boldsymbol{b}^T & 0 & 0 \\ 0 & \boldsymbol{c}^T & 0 \\ 0 & 0 & \boldsymbol{d}^T \\ \boldsymbol{c}^T & \boldsymbol{b}^T & 0 \\ 0 & \boldsymbol{d}^T & \boldsymbol{c}^T \\ \boldsymbol{d}^T & 0 & \boldsymbol{b}^T \end{bmatrix} \begin{Bmatrix} \boldsymbol{u}_x^e \\ \boldsymbol{u}_y^e \\ \boldsymbol{u}_z^e \end{Bmatrix}
\tag{4.133}
$$

or

$$
\epsilon = \boldsymbol{B}\boldsymbol{u}^e \tag{4.134}
$$

Then the stiffness matrix is given by

$$
\begin{aligned}
\boldsymbol{K}^e &= \int_V \boldsymbol{B}^T \boldsymbol{D} \boldsymbol{B} \, \mathrm{d}V \\
&= V \boldsymbol{B}^T \boldsymbol{D} \boldsymbol{B}
\end{aligned}
\tag{4.135}
$$

because $\boldsymbol{B}$ is constant. The inertia matrix is given by

$$
\begin{aligned}
\boldsymbol{M}_x^e &= \int_V \rho \boldsymbol{N}^T \boldsymbol{N} \, \mathrm{d}V \\
&= \frac{\rho V}{20} \begin{bmatrix} 2 & 1 & 1 & 1 \\ 1 & 2 & 1 & 1 \\ 1 & 1 & 2 & 1 \\ 1 & 1 & 1 & 2 \end{bmatrix} \\
&= \boldsymbol{M}_y^e = \boldsymbol{M}_z^e
\end{aligned}
\tag{4.136}
$$

which for all degrees of freedom can be assembled into a (12×12) matrix for the element.

For the tetrahedron the thermal release forces are calculated as

$$
\boldsymbol{p}_T = \int_V \boldsymbol{B}^T \boldsymbol{D} \epsilon_T \, \mathrm{d}V
$$

which for the isotropic case reduces to

$$\boldsymbol{p}_T = \frac{ET\alpha}{1-2\nu}\int_V \boldsymbol{B}^T \,\mathrm{d}V \begin{bmatrix} 1 \\ 1 \\ 1 \\ 0 \\ 0 \\ 0 \end{bmatrix} \tag{4.137}$$

Using Eq. (4.133) this becomes

$$\boldsymbol{p}_T = \frac{ET\alpha}{6(1-2\nu)} \begin{bmatrix} \boldsymbol{b} \\ \boldsymbol{c} \\ \boldsymbol{d} \end{bmatrix} \tag{4.138}$$

The sequence of forces in $\boldsymbol{p}_T$ is $(\boldsymbol{p}_x \; \boldsymbol{p}_y \; \boldsymbol{p}_z)^T$. The stiffness and inertia matrices are finally rearranged in nodal sequence to yield the final element matrices. A number of complex, useful elements may be easily derived from simple elements. Thus three-dimensional hexahedron and pentahedron elements in Fig. 4.10 may be suitably subdivided into a number of constant strain tetrahedra and their individual stiffness and mass matrices are combined by the usual procedure to yield desired element matrices.

Thus a simple eight-node hexahedron element may be subdivided to 10 tetrahedra defined by appropriate nodal combinations:

1	2	3	6	2	7	5	6
6	3	8	7	4	5	7	8
1	3	4	8	1	2	4	5
1	3	8	6	2	3	4	7
1	6	8	5	2	4	5	7

The averages of the stiffness and mass matrices of these elements are accepted as the hexahedron element matrices. Similarly, for a six-node pentahedron element the following tetrahedra are assumed to constitute the element:

1	2	3	5	2	3	1	6	3	1	2	4
1	3	4	5	2	1	5	6	3	2	6	4
4	6	5	3	5	4	6	1	6	5	4	2
1	2	3	4	2	3	1	5	3	1	2	6
2	3	4	5	5	3	1	6	6	1	2	4
4	6	5	3	5	4	6	1	6	5	4	2

Both of these elements are known to produce accurate solutions for practical problems.

4.4.2 Linear Strain Tetrahedron

A quadratic displacement distribution is assumed within the element in Fig. 4.11 and, typically, for u_x, is given as

$$\begin{aligned} u_x(\zeta_1, \zeta_2, \zeta_3, \zeta_4) = {} & \alpha_1\zeta_1^2 + \alpha_2\zeta_2^2 + \alpha_3\zeta_3^2 + \alpha_4\zeta_4^2 + \alpha_5\zeta_1\zeta_2 + \alpha_6\zeta_2\zeta_3 \\ & + \alpha_7\zeta_3\zeta_1 + \alpha_8\zeta_1\zeta_4 + \alpha_9\zeta_2\zeta_4 + \alpha_{10}\zeta_3\zeta_4 \end{aligned} \tag{4.139}$$

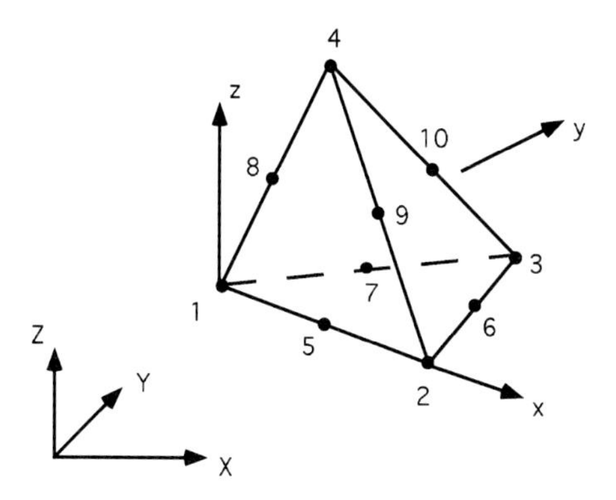

Fig. 4.11 Ten-noded linear strain tetrahedron element.

and following the procedure in Sec. 4.3.2, the shape function may simply be written as

$$
\begin{aligned}
\boldsymbol{N}_x &= [\zeta_1(2\zeta_1 - 1)\, \zeta_2(2\zeta_2 - 1)\, \zeta_3(2\zeta_3 - 1)\, \zeta_4(2\zeta_4 - 1) \\
&\quad 4\zeta_1\zeta_2\; 4\zeta_2\zeta_3\; 4\zeta_3\zeta_1\; 4\zeta_1\zeta_4\; 4\zeta_2\zeta_4\; 4\zeta_3\zeta_4] \\
&= \boldsymbol{N}_y = \boldsymbol{N}_z = \boldsymbol{N}_2
\end{aligned} \tag{4.140}
$$

Following the standard procedure of Sec. 4.3.2 the strain–displacement relationship can be obtained by chain differentiation. For example, for ϵ_x,

$$
\begin{aligned}
\epsilon_x &= \frac{\partial u_x}{\partial x} \\
&= \frac{1}{6V}[b_1(4\zeta_1 - 1)\, b_2(4\zeta_2 - 1)\, b_3(4\zeta_3 - 1)\, b_4(4\zeta_4 - 1) \\
&\quad 4(b_1\zeta_2 + b_2\zeta_1)\, 4(b_2\zeta_3 + b_3\zeta_2)\, 4(b_3\zeta_1 + 4b_1\zeta_3) \\
&\quad 4(b_1\zeta_4 + 4b_4\zeta_1)\, 4(b_2\zeta_4 + b_4\zeta_2)\, 4(b_3\zeta_4 + b_4\zeta_3)]\boldsymbol{u}_x^e
\end{aligned} \tag{4.141}
$$

and similarly for derivatives with respect to y and z in which $\boldsymbol{b}$ is replaced by $\boldsymbol{c}$ and $\boldsymbol{d}$, respectively. This leads to the complete strain–displacement relationship

$$
\begin{Bmatrix} \epsilon_x \\ \epsilon_y \\ \epsilon_z \\ \gamma_{xy} \\ \gamma_{yz} \\ \gamma_{zx} \end{Bmatrix} = \begin{bmatrix} \boldsymbol{B}_x & 0 & 0 \\ 0 & \boldsymbol{B}_y & 0 \\ 0 & 0 & \boldsymbol{B}_z \\ \boldsymbol{B}_y & \boldsymbol{B}_x & 0 \\ 0 & \boldsymbol{B}_z & \boldsymbol{B}_y \\ \boldsymbol{B}_z & 0 & \boldsymbol{B}_x \end{bmatrix} \begin{Bmatrix} \boldsymbol{u}_x^e \\ \boldsymbol{u}_y^e \\ \boldsymbol{u}_z^e \end{Bmatrix} \tag{4.142}
$$

This equation is now used to obtain the apex strains in terms of the element nodal displacements:

$$\begin{Bmatrix} \epsilon_{x_c} \\ \epsilon_{y_c} \\ \epsilon_{z_c} \\ \gamma_{xy_c} \\ \gamma_{yz_c} \\ \gamma_{zx_c} \end{Bmatrix} = \begin{bmatrix} \boldsymbol{\Phi}_1 & 0 & 0 \\ 0 & \boldsymbol{\Phi}_2 & 0 \\ 0 & 0 & \boldsymbol{\Phi}_3 \\ \boldsymbol{\Phi}_2 & \boldsymbol{\Phi}_1 & 0 \\ 0 & \boldsymbol{\Phi}_3 & \boldsymbol{\Phi}_2 \\ \boldsymbol{\Phi}_3 & 0 & \boldsymbol{\Phi}_1 \end{bmatrix} \begin{Bmatrix} \boldsymbol{u}_x^e \\ \boldsymbol{u}_y^e \\ \boldsymbol{u}_z^e \end{Bmatrix} \tag{4.143}$$

or

$$\boldsymbol{\epsilon}_c = \boldsymbol{\Phi} \begin{Bmatrix} \boldsymbol{u}_x^e \\ \boldsymbol{u}_y^e \\ \boldsymbol{u}_z^e \end{Bmatrix} \tag{4.144}$$

where $\boldsymbol{\epsilon}_{xc} = [\epsilon_{x1}\ \epsilon_{x2}\ \epsilon_{x3}\ \epsilon_{x4}]^T$, etc. In Eq. (4.143) the matrices $\boldsymbol{\Phi}_1$, $\boldsymbol{\Phi}_2$, $\boldsymbol{\Phi}_3$ are defined as follows:

$$\boldsymbol{\Phi}_1 = \frac{1}{6V} \begin{bmatrix} 3b_1 & -b_2 & -b_3 & -b_4 & 4b_2 & 0 & 4b_3 & 4b_4 & 0 & 0 \\ -b_1 & 3b_2 & -b_3 & -b_4 & 4b_1 & 4b_3 & 0 & 0 & 4b_4 & 0 \\ -b_1 & -b_2 & 3b_3 & -b_4 & 0 & 4b_2 & 4b_1 & 0 & 0 & 4b_4 \\ -b_1 & -b_2 & -b_3 & 3b_4 & 0 & 0 & 0 & 4b_1 & 4b_2 & 4b_3 \end{bmatrix} \tag{4.145}$$

and $\boldsymbol{\Phi}_2$ and $\boldsymbol{\Phi}_3$ may simply be obtained from the preceding equation by substituting for $\boldsymbol{b}$ with $\boldsymbol{c}$ and $\boldsymbol{d}$, respectively. The linear strains at any point within the tetrahedron may now be interpolated in terms of the element apex nodal values using the linear shape function:

$$\begin{Bmatrix} \epsilon_x \\ \epsilon_y \\ \epsilon_z \\ \gamma_{xy} \\ \gamma_{yz} \\ \gamma_{zx} \end{Bmatrix} = \begin{bmatrix} \boldsymbol{N}_1 & & & & & \\ & \boldsymbol{N}_1 & & & & \\ & & \boldsymbol{N}_1 & & & \\ & & & \boldsymbol{N}_1 & & \\ & & & & \boldsymbol{N}_1 & \\ & & & & & \boldsymbol{N}_1 \end{bmatrix} \begin{Bmatrix} \boldsymbol{\epsilon}_{x_c} \\ \boldsymbol{\epsilon}_{y_c} \\ \boldsymbol{\epsilon}_{z_c} \\ \boldsymbol{\gamma}_{xy_c} \\ \boldsymbol{\gamma}_{yz_c} \\ \boldsymbol{\gamma}_{zx_c} \end{Bmatrix} \tag{4.146}$$

or

$$\boldsymbol{\epsilon} = \boldsymbol{N}\boldsymbol{\epsilon}_c = \boldsymbol{N}\boldsymbol{\Phi} \begin{Bmatrix} \boldsymbol{u}_x^e \\ \boldsymbol{u}_y^e \\ \boldsymbol{u}_z^e \end{Bmatrix} = \boldsymbol{B} \begin{Bmatrix} \boldsymbol{u}_x^e \\ \boldsymbol{u}_y^e \\ \boldsymbol{u}_z^e \end{Bmatrix} \tag{4.147}$$

where

$$\boldsymbol{N}_1 = [\zeta_1 \quad \zeta_2 \quad \zeta_3 \quad \zeta_4]$$

The element stiffness matrix is next obtained as

$$\boldsymbol{K}^e = \int_V \boldsymbol{B}^T \boldsymbol{D} \boldsymbol{B}\, \mathrm{d}V$$

$$= \boldsymbol{\Phi}^T \left[\int_V \boldsymbol{N}^T \boldsymbol{D} \boldsymbol{N}\, \mathrm{d}V \right] \boldsymbol{\Phi} \tag{4.148}$$

This integral can be evaluated by noting that it contains submatrix values of the type

$$d_{ij} \int_V \boldsymbol{N}_1^T \boldsymbol{N}_1\, \mathrm{d}V = d_{ij} \frac{V}{20} \begin{bmatrix} 2 & 1 & 1 & 1 \\ 1 & 2 & 1 & 1 \\ 1 & 1 & 2 & 1 \\ 1 & 1 & 1 & 2 \end{bmatrix} \tag{4.149}$$

$$= d_{ij} \boldsymbol{\Lambda} \tag{4.150}$$

which are suitably assembled into a (24 × 24) apex nodal stiffness matrix. This is finally transformed to the usual (30 × 30) element stiffness matrix, using the formulation of Eq. (4.148). Similarly the inertia matrix is given by

$$\boldsymbol{M}_x^e = \int_V \rho \boldsymbol{N}_2^T \boldsymbol{N}_2\, \mathrm{d}V$$

$$= \frac{\rho V}{420} \begin{bmatrix} 6 & 1 & 1 & 1 & -4 & -6 & -4 & -4 & -6 & -6 \\ & 6 & 1 & 1 & -6 & -6 & -4 & -4 & -6 & -4 \\ & & 6 & 1 & -4 & -6 & -6 & -4 & -4 & -6 \\ & & & 6 & -4 & -4 & -4 & -6 & -6 & -6 \\ & & & & 32 & 16 & 16 & 16 & 16 & 8 \\ & & \text{sym} & & & 32 & 16 & 8 & 16 & 16 \\ & & & & & & 32 & 16 & 8 & 16 \\ & & & & & & & 32 & 16 & 16 \\ & & & & & & & & 32 & 16 \\ & & & & & & & & & 32 \end{bmatrix}$$

$$= \boldsymbol{M}_y^e = \boldsymbol{M}_z^e \tag{4.151}$$

which is assembled into the (30 × 30) element inertia matrix. Both the stiffness and inertia matrices are finally rearranged in nodal sequence to yield the final form of the corresponding matrices. For the linear strain tetrahedron, using Eqs. (4.123) and (4.125),

$$\boldsymbol{p}_T = \boldsymbol{\Phi}^T \int_V \boldsymbol{N}^T\, \mathrm{d}V \frac{ET\alpha}{1 - 2\nu} \begin{bmatrix} 1 \\ 1 \\ 1 \\ 0 \\ 0 \\ 0 \end{bmatrix} \tag{4.152}$$

$$= \frac{ET\alpha V}{4(1-2\nu)} \boldsymbol{\Phi}^T \begin{bmatrix} \boldsymbol{e}_4 \\ \boldsymbol{e}_4 \\ \boldsymbol{e}_4 \\ \boldsymbol{0}_4 \\ \boldsymbol{0}_4 \\ \boldsymbol{0}_4 \end{bmatrix} \tag{4.153}$$

in which $\boldsymbol{e}_4 = (1\ 1\ 1\ 1)^T$ and $\boldsymbol{0}_4 = (0\ 0\ 0\ 0)^T$. For initial strains using Eq. (4.52),

$$\boldsymbol{p}_I = \frac{E\epsilon_I V}{4(1-2\nu)} \boldsymbol{\Phi}^T \begin{bmatrix} \boldsymbol{e}_4 \\ \boldsymbol{e}_4 \\ \boldsymbol{e}_4 \\ \boldsymbol{0}_4 \\ \boldsymbol{0}_4 \\ \boldsymbol{0}_4 \end{bmatrix} \tag{4.154}$$

4.4.3 Isoparametric Tetrahedron

An isoparametric linear strain element is considered next. For the 10-noded tetrahedron, the geometric relationship may be written as

$$[x \quad y \quad z] = \boldsymbol{N}_2 \boldsymbol{X} \tag{4.155}$$

in which

$$\boldsymbol{N}_2 = \boldsymbol{N}_x = \boldsymbol{N}_y = \boldsymbol{N}_z$$

and

$$\boldsymbol{X} = \begin{bmatrix} x_1 & x_2 & x_3 & x_4 & x_5 & x_6 & x_7 & x_8 & x_9 & x_{10} \\ y_1 & y_2 & y_3 & y_4 & y_5 & y_6 & y_7 & y_8 & y_9 & y_{10} \\ z_1 & z_2 & z_3 & z_4 & z_5 & z_6 & z_7 & z_8 & z_9 & z_{10} \end{bmatrix}^T \tag{4.156}$$

containing the nodal coordinates. With reference to Fig. 4.12 the volume of an infinitesimal element may be written as

$$6\,\mathrm{d}V = \det \boldsymbol{J}_A \,\mathrm{d}\zeta_j \,\mathrm{d}\zeta_k \,\mathrm{d}\zeta_l \tag{4.157}$$

in which

$$\boldsymbol{J}_A = \begin{bmatrix} \boldsymbol{e}_4 & \dfrac{\partial \boldsymbol{N}_2}{\partial \zeta} X \end{bmatrix} \tag{4.158}$$

The derivatives of $\boldsymbol{N}_2$ with respect to the natural coordinates $\zeta_i, \zeta_j, \zeta_k, \zeta_l$ can be expressed as

$$\frac{\partial \boldsymbol{N}_2}{\partial \zeta} = \begin{bmatrix} 4\zeta_1 - 1 & 0 & 0 & 0 & 4\zeta_2 & 0 & 4\zeta_3 & 4\zeta_4 & 0 & 0 \\ 0 & 4\zeta_2 - 1 & 0 & 0 & 4\zeta_1 & 4\zeta_3 & 0 & 0 & 4\zeta_4 & 0 \\ 0 & 0 & 4\zeta_3 - 1 & 0 & 0 & 4\zeta_2 & 4\zeta_1 & 0 & 0 & 4\zeta_4 \\ 0 & 0 & 0 & 4\zeta_4 - 1 & 0 & 0 & 0 & 4\zeta_1 & 4\zeta_2 & 4\zeta_3 \end{bmatrix}$$

$$= \boldsymbol{\Phi}_\epsilon \tag{4.159}$$

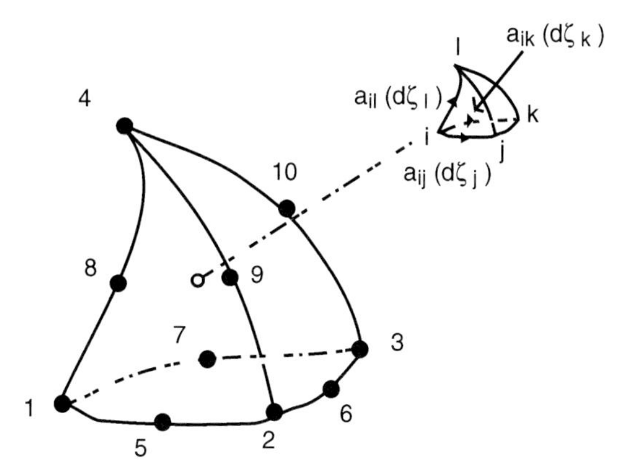

Fig. 4.12 Isoparametric tetrahedron element.

The rule for chain differentiation, in this case relating Cartesian and natural coordinates, can be written as

$$\begin{Bmatrix} \dfrac{\partial f}{\partial x} \\ \dfrac{\partial f}{\partial y} \\ \dfrac{\partial f}{\partial z} \end{Bmatrix} = \begin{bmatrix} \dfrac{\partial \zeta_1}{\partial x} & \dfrac{\partial \zeta_2}{\partial x} & \dfrac{\partial \zeta_3}{\partial x} & \dfrac{\partial \zeta_4}{\partial x} \\ \dfrac{\partial \zeta_1}{\partial y} & \dfrac{\partial \zeta_2}{\partial y} & \dfrac{\partial \zeta_3}{\partial y} & \dfrac{\partial \zeta_4}{\partial y} \\ \dfrac{\partial \zeta_1}{\partial z} & \dfrac{\partial \zeta_2}{\partial z} & \dfrac{\partial \zeta_3}{\partial z} & \dfrac{\partial \zeta_4}{\partial z} \end{bmatrix} \begin{Bmatrix} \dfrac{\partial f}{\partial \zeta_1} \\ \dfrac{\partial f}{\partial \zeta_2} \\ \dfrac{\partial f}{\partial \zeta_3} \\ \dfrac{\partial f}{\partial \zeta_4} \end{Bmatrix} \tag{4.160}$$

or

$$\begin{aligned} f_{,c} &= \boldsymbol{L}_R\{f_{,\zeta}\} \\ &= \boldsymbol{L}_R \boldsymbol{\Phi}_\epsilon \boldsymbol{f} \end{aligned} \tag{4.161}$$

The vector $\boldsymbol{f}$ contains the 10 nodal values of the function and $\boldsymbol{L_R}$ is composed of rows two, three, and four of matrix $\boldsymbol{J}_A^{-1}$. Also

$$\begin{Bmatrix} \boldsymbol{B}_x \\ \boldsymbol{B}_y \\ \boldsymbol{B}_z \end{Bmatrix} = \boldsymbol{L}_R \boldsymbol{\Phi}_\epsilon \tag{4.162}$$

and the strain–displacement relationship is given as

$$\epsilon = \boldsymbol{B}\boldsymbol{u}^e = \begin{bmatrix} \boldsymbol{B}_x & 0 & 0 \\ 0 & \boldsymbol{B}_y & 0 \\ 0 & 0 & \boldsymbol{B}_z \\ \boldsymbol{B}_y & \boldsymbol{B}_x & 0 \\ 0 & \boldsymbol{B}_z & \boldsymbol{B}_y \\ \boldsymbol{B}_z & 0 & \boldsymbol{B}_x \end{bmatrix} \boldsymbol{u}^e$$

The vector $\boldsymbol{u}^e$ contains 30 nodal displacements in the order of the global degrees of freedom. The stiffness matrix is given by

$$\boldsymbol{K}^e = \int_V \boldsymbol{B}^T \boldsymbol{D}\boldsymbol{B}\,\mathrm{d}V \tag{4.163}$$

Again this is obtained by numerical integration as the weighted contribution of Gauss points in a process similar to that described in Sec. 4.3.2 for the isoparametric six-node plane triangular element. The process involving evaluation of

$$\boldsymbol{K}^e = \sum (w\boldsymbol{B}^T\boldsymbol{D}\boldsymbol{B}\det \boldsymbol{J}_A)_i \tag{4.164}$$

which also contains quintic terms, needs nine points for Gaussian integration. The consistent mass matrix is obtained from Eq. (4.50):

$$\boldsymbol{M}^e = \rho t \sum (w\,\boldsymbol{N}^T\boldsymbol{N}\det \boldsymbol{J}_A)_i \tag{4.165}$$

4.5 Isoparametric Quadrilateral and Hexahedron Elements

4.5.1 Quadrilateral Element

Having developed triangular elements in detail and showing how they may be combined to form quadrilateral elements of arbitrary shape, their direct formulation of isoparametric elements[13,14] will be discussed in this section. A family is shown in Fig. 4.13 in which the x, y coordinates are expressed in terms of curvilinear coordinates ξ, η. The quadratic and cubic elements may be interpolated using products of linear space Lagrange functions in which the presence of internal nodes is required. Alternatively it is possible to use the serendipity functions that have

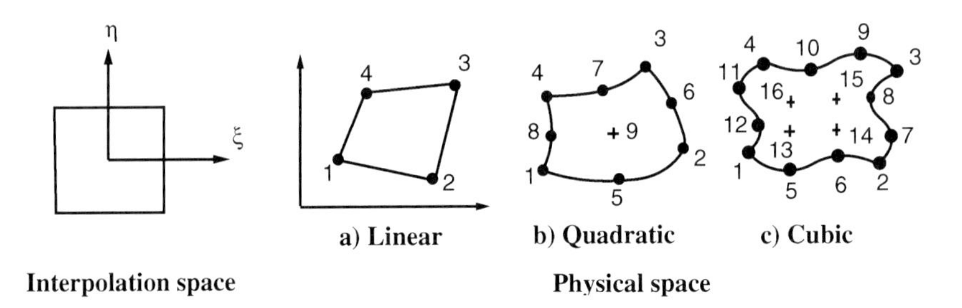

Fig. 4.13 Planar quadrilateral isoparametric elements.

nodes embedded in the sides only. Development of the element stiffness and inertia matrices follows directly from the procedure adopted for the plane triangle and tetrahedron elements. The interpolation functions are given in (ξ, η) space that spans $(+1, -1)$ in each coordinate direction, and the physical space is interpolated using the usual $\boldsymbol{N}$ shape functions,

$$[x \quad y] = \boldsymbol{N}\boldsymbol{X} \tag{4.166}$$

where $\boldsymbol{N} = \boldsymbol{N}_x = \boldsymbol{N}_y$, and $\boldsymbol{X}$ contains columns of the x and y nodal coordinate values. Then the derivatives of any function f with respect to ξ, η are written in terms of x, y derivatives as

$$\begin{Bmatrix} \dfrac{\partial f}{\partial \xi} \\ \dfrac{\partial f}{\partial \eta} \end{Bmatrix} = \begin{bmatrix} \dfrac{\partial x}{\partial \xi} & \dfrac{\partial y}{\partial \xi} \\ \dfrac{\partial x}{\partial \eta} & \dfrac{\partial y}{\partial \eta} \end{bmatrix} \begin{Bmatrix} \dfrac{\partial f}{\partial x} \\ \dfrac{\partial f}{\partial y} \end{Bmatrix} \tag{4.167}$$

$$= \boldsymbol{J} \begin{Bmatrix} \dfrac{\partial f}{\partial x} \\ \dfrac{\partial f}{\partial y} \end{Bmatrix} \tag{4.168}$$

The preceding coefficient matrix is the Jacobian of the coordinate transformation and from Eq. (4.166) can be derived as

$$\boldsymbol{J} = \begin{bmatrix} \dfrac{\partial x}{\partial \xi} & \dfrac{\partial y}{\partial \xi} \\ \dfrac{\partial x}{\partial \eta} & \dfrac{\partial y}{\partial \eta} \end{bmatrix} = \begin{bmatrix} \dfrac{\partial \boldsymbol{N}^T}{\partial \xi} \\ \dfrac{\partial \boldsymbol{N}^T}{\partial \eta} \end{bmatrix} [\boldsymbol{X}] \tag{4.169}$$

Inverting Eq. (4.168), the x and y derivatives are obtained in terms of ξ, η derivatives, i.e.,

$$\begin{Bmatrix} \dfrac{\partial f}{\partial x} \\ \dfrac{\partial f}{\partial y} \end{Bmatrix} = \boldsymbol{J}^{-1} \begin{Bmatrix} \dfrac{\partial f}{\partial \xi} \\ \dfrac{\partial f}{\partial \eta} \end{Bmatrix} \tag{4.170}$$

Substituting u_x for f results in

$$\begin{Bmatrix} \dfrac{\partial u_x}{\partial x} \\ \dfrac{\partial u_x}{\partial y} \end{Bmatrix} = \boldsymbol{J}^{-1} \begin{Bmatrix} \dfrac{\partial f}{\partial \xi} \\ \dfrac{\partial f}{\partial \eta} \end{Bmatrix} \tag{4.171}$$

Noting that the same interpolation function $\boldsymbol{N}$ is used for u_x and u_y, it is possible to write

$$\begin{bmatrix} \dfrac{\partial u_x}{\partial x} & \dfrac{\partial u_y}{\partial x} \\ \dfrac{\partial u_x}{\partial y} & \dfrac{\partial u_y}{\partial y} \end{bmatrix} = \boldsymbol{J}^{-1} \begin{bmatrix} \dfrac{\partial u_x}{\partial \xi} & \dfrac{\partial u_y}{\partial \xi} \\ \dfrac{\partial u_x}{\partial \eta} & \dfrac{\partial u_y}{\partial \eta} \end{bmatrix} \tag{4.172}$$

$$= \boldsymbol{J}^{-1} \begin{bmatrix} \dfrac{\partial \boldsymbol{N}}{\partial \xi} \\ \dfrac{\partial \boldsymbol{N}}{\partial \eta} \end{bmatrix} [\boldsymbol{u}_x \quad \boldsymbol{u}_y] \tag{4.173}$$

$$= \begin{bmatrix} \boldsymbol{B}_x \\ \boldsymbol{B}_y \end{bmatrix} [\boldsymbol{u}_x \quad \boldsymbol{u}_y] \tag{4.174}$$

where $\boldsymbol{u}_x$ and $\boldsymbol{u}_y$ are the nodal values of the x and y displacements, respectively.

The strain–displacement relationship is then obtained from the preceding equations and is written for all strain components:

$$\begin{Bmatrix} \epsilon_x \\ \epsilon_y \\ \gamma_{xy} \end{Bmatrix} = \begin{bmatrix} \boldsymbol{B}_x & 0 \\ 0 & \boldsymbol{B}_y \\ \boldsymbol{B}_y & \boldsymbol{B}_x \end{bmatrix} \begin{Bmatrix} \boldsymbol{u}_x \\ \boldsymbol{u}_y \end{Bmatrix} \tag{4.175}$$

The element stiffness matrix is obtained by numerical integration as

$$\boldsymbol{K}^e = \sum_i \sum_j w_i w_j (\boldsymbol{B}^T \boldsymbol{D} \boldsymbol{B})_{ij} \det \boldsymbol{J}_{ij} \tag{4.176}$$

The preceding procedure can be applied for any of the chosen shape functions, and these are summarized next for the serendipity family (see Fig. 4.14).

Linear:

$$N_i = \tfrac{1}{4}(1 + \xi_i \xi)(1 + \eta_i \eta) \qquad i = 1, 2, 3, 4 \tag{4.177}$$

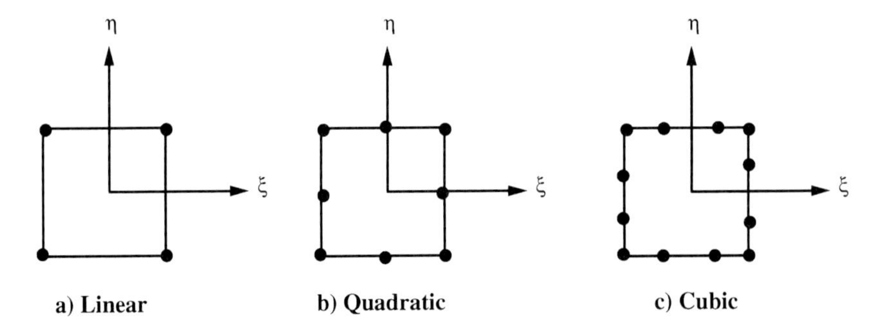

Fig. 4.14 Isoparametric plane element nodal points in (ξ, η) space.

Quadratic:

$$
\begin{aligned}
N_i &= \tfrac{1}{4}(1+\xi_i\xi)(1+\eta_i\eta)(\xi_i\xi+\eta_i\eta-1) && \text{at corner nodes } 1, 2, 3, 4 \\
&= \tfrac{1}{2}(1-\xi^2)(1+\eta_i\eta) && \text{at side nodes } 5, 7 \\
&= \tfrac{1}{2}(1+\xi_i\xi)(1-\eta^2) && \text{at side nodes } 6, 8 \qquad (4.178)
\end{aligned}
$$

Cubic:

$$
\begin{aligned}
N_i &= \frac{1}{32}(1+\xi_i\xi)(1+\eta_i\eta)[9(\xi^2+\eta^2)-10] && \text{at corner nodes } 1, 2, 3, 4 \\
&= \frac{9}{32}(1-\xi^2)(1+9\xi\xi_i)(1+\eta\eta_i) && \text{all side nodes} \qquad (4.179)
\end{aligned}
$$

The inertia matrix is given as

$$
\boldsymbol{M}^e = \rho A t \int \boldsymbol{N}^T \boldsymbol{N} \det \boldsymbol{J} \, \mathrm{d}A \tag{4.180}
$$

which is evaluated as

$$
\boldsymbol{M}^e = \rho t \sum_i \sum_j w_i w_j (\boldsymbol{N}^T \boldsymbol{N})_{ij} \det \boldsymbol{J}_{ij} \tag{4.181}
$$

at the Gauss integration points. Alternatively, all numerical integrations can also be achieved by employing symbolic manipulation.

4.5.2 Hexahedron Elements

A family of three-dimensional solid elements[13,14] is shown in Fig. 4.15. Other elements with cubic and higher displacement fields follow naturally. The interpolation function in the ξ, η, ζ coordinate system are related to the physical x, y, z space as follows:

$$
[x \quad y \quad z] = \boldsymbol{N}\boldsymbol{X} \tag{4.182}
$$

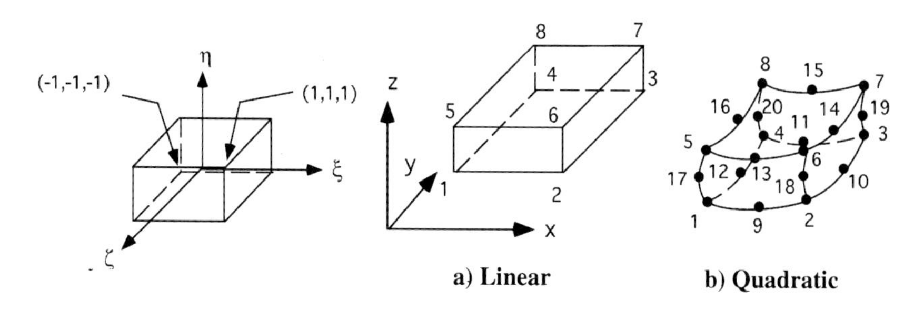

Fig. 4.15 Hexahedron isoparametric elements.

in which $\boldsymbol{N} = \boldsymbol{N}_x = \boldsymbol{N}_y = \boldsymbol{N}_z$, with matrix $\boldsymbol{X}$ containing element nodal coordinates. As before, the relationship between derivatives of the two coordinate systems is written as

$$\begin{Bmatrix} \dfrac{\partial f}{\partial \xi} \\ \dfrac{\partial f}{\partial \eta} \\ \dfrac{\partial f}{\partial \zeta} \end{Bmatrix} = \begin{bmatrix} \dfrac{\partial x}{\partial \xi} & \dfrac{\partial y}{\partial \xi} & \dfrac{\partial z}{\partial \xi} \\ \dfrac{\partial x}{\partial \eta} & \dfrac{\partial y}{\partial \eta} & \dfrac{\partial z}{\partial \eta} \\ \dfrac{\partial x}{\partial \zeta} & \dfrac{\partial y}{\partial \zeta} & \dfrac{\partial z}{\partial \zeta} \end{bmatrix} \begin{Bmatrix} \dfrac{\partial f}{\partial x} \\ \dfrac{\partial f}{\partial y} \\ \dfrac{\partial f}{\partial z} \end{Bmatrix} = \boldsymbol{J} \begin{Bmatrix} \dfrac{\partial f}{\partial x} \\ \dfrac{\partial f}{\partial y} \\ \dfrac{\partial f}{\partial z} \end{Bmatrix} \tag{4.183}$$

and the coefficient matrix $\boldsymbol{J}$ may also be written as

$$\boldsymbol{J} = \begin{bmatrix} \dfrac{\partial \boldsymbol{N}^T}{\partial \xi} \\ \dfrac{\partial \boldsymbol{N}^T}{\partial \eta} \\ \dfrac{\partial \boldsymbol{N}^T}{\partial \zeta} \end{bmatrix} [\boldsymbol{X}] \tag{4.184}$$

Inverting Eq. (4.183) the transformation from (ξ, η, ζ) to (x, y, z) derivatives is obtained,

$$\begin{Bmatrix} \dfrac{\partial f}{\partial x} \\ \dfrac{\partial f}{\partial y} \\ \dfrac{\partial f}{\partial z} \end{Bmatrix} = \boldsymbol{J}^{-1} \begin{Bmatrix} \dfrac{\partial f}{\partial \xi} \\ \dfrac{\partial f}{\partial \eta} \\ \dfrac{\partial f}{\partial \zeta} \end{Bmatrix} \tag{4.185}$$

which may be written in terms of three displacements as

$$\begin{bmatrix} \dfrac{\partial u_x}{\partial x} & \dfrac{\partial u_y}{\partial x} & \dfrac{\partial u_z}{\partial x} \\ \dfrac{\partial u_x}{\partial y} & \dfrac{\partial u_y}{\partial y} & \dfrac{\partial u_z}{\partial y} \\ \dfrac{\partial u_x}{\partial z} & \dfrac{\partial u_y}{\partial z} & \dfrac{\partial u_z}{\partial z} \end{bmatrix} = \boldsymbol{J}^{-1} \begin{bmatrix} \dfrac{\partial u_x}{\partial \xi} & \dfrac{\partial u_y}{\partial \xi} & \dfrac{\partial u_z}{\partial \xi} \\ \dfrac{\partial u_x}{\partial \eta} & \dfrac{\partial u_y}{\partial \eta} & \dfrac{\partial u_z}{\partial \eta} \\ \dfrac{\partial u_x}{\partial \zeta} & \dfrac{\partial u_y}{\partial \zeta} & \dfrac{\partial u_z}{\partial \zeta} \end{bmatrix} \tag{4.186}$$

$$\begin{bmatrix} \frac{\partial u_x}{\partial x} & \frac{\partial u_y}{\partial x} & \frac{\partial u_z}{\partial x} \\ \frac{\partial u_x}{\partial y} & \frac{\partial u_y}{\partial y} & \frac{\partial u_z}{\partial y} \\ \frac{\partial u_x}{\partial z} & \frac{\partial u_y}{\partial z} & \frac{\partial u_z}{\partial z} \end{bmatrix} = \boldsymbol{J}^{-1} \begin{bmatrix} \frac{\partial \boldsymbol{N}^T}{\partial \xi} \\ \frac{\partial \boldsymbol{N}^T}{\partial \eta} \\ \frac{\partial \boldsymbol{N}^T}{\partial \zeta} \end{bmatrix} [\boldsymbol{u}_x \quad \boldsymbol{u}_y \quad \boldsymbol{u}_z] \tag{4.187}$$

$$= \begin{bmatrix} \boldsymbol{B}_x \\ \boldsymbol{B}_y \\ \boldsymbol{B}_z \end{bmatrix} [\boldsymbol{u}_x \quad \boldsymbol{u}_y \quad \boldsymbol{u}_z] \tag{4.188}$$

where $\boldsymbol{u}_x, \boldsymbol{u}_y, \boldsymbol{u}_z$ are the nodal values of x, y, z displacements, respectively.

Next, the strain–displacement relationship is written as

$$\begin{Bmatrix} \epsilon_x \\ \epsilon_y \\ \epsilon_z \\ \epsilon_{xy} \\ \epsilon_{yz} \\ \epsilon_{zx} \end{Bmatrix} = \begin{bmatrix} \boldsymbol{B}_x & 0 & 0 \\ 0 & \boldsymbol{B}_y & 0 \\ 0 & 0 & \boldsymbol{B}_z \\ \boldsymbol{B}_y & \boldsymbol{B}_x & 0 \\ 0 & \boldsymbol{B}_z & \boldsymbol{B}_y \\ \boldsymbol{B}_z & 0 & \boldsymbol{B}_x \end{bmatrix} \begin{Bmatrix} \boldsymbol{u}_x^e \\ \boldsymbol{u}_y^e \\ \boldsymbol{u}_z^e \end{Bmatrix} \tag{4.189}$$

As before, the stiffness matrix may be obtained by using $(3 \times 3 \times 3)$, 27 Gauss integration points as

$$\boldsymbol{K}^e = \sum_i \sum_j \sum_k w_i w_j w_k [\boldsymbol{B}^T \boldsymbol{D} \boldsymbol{B}]_{ijk} \det \boldsymbol{J}_{ijk} \tag{4.190}$$

$$\boldsymbol{K}^e = \int_V \boldsymbol{B}^T \boldsymbol{D} \boldsymbol{B} \, \mathrm{d}V$$

which is the standard expression for the element stiffness matrix, and can be evaluated by using any standard symbolic manipulation program.

Using appropriate shape functions the preceding formulation will represent any pattern of displacement distribution in the element. Some currently used shape functions are given next.

Linear:

$$N_i = \tfrac{1}{8}(1 + \xi_i \xi)(1 + \eta_i \eta)(1 + \zeta_i \zeta) \qquad i = 1, \ldots, 8 \tag{4.191}$$

Quadratic:

$$\begin{aligned} N_i &= \tfrac{1}{8}(1 + \xi_i \xi)(1 + \eta_i \eta)(1 + \zeta_i \zeta)(\xi_i \xi + \eta_i \eta + \zeta_i \zeta - 2) \text{ apex nodes; } i = 1, \ldots, 8 \\ &= \tfrac{1}{4}(1 - \xi^2)(1 + \eta_i \eta)(1 + \zeta_i \zeta) \qquad i = 10, 12, 14, 16 \\ &= \tfrac{1}{4}(1 - \eta^2)(1 + \xi_i \xi)(1 + \zeta_i \zeta) \qquad i = 9, 11, 13, 15 \\ &= \tfrac{1}{4}(1 - \zeta^2)(1 + \xi_i \xi)(1 + \eta_i \eta) \qquad i = 17, \ldots, 20 \end{aligned} \tag{4.192}$$

and other higher-order shape functions can be created in the same way. The inertia matrix is calculated as

$$M^e = \rho \int_V N^T N \det J \, \mathrm{d}V \tag{4.193}$$

which may be evaluated by numerical integration, summing weighted values at the Gauss points

$$M^e = \rho \sum_i \sum_j \sum_k w_i w_j w_k (N^T N)_{ijk} \det J_{ijk} \tag{4.194}$$

4.6 Torsion of Prismatic Shafts

4.6.1 Introduction: St. Venant's Theory

From the strength of materials, it is known that for the torsion of a prismatic shaft of a circular cross section, the assumption that cross sections at right angles to the axis of the shaft remain plane in the deformed state leads to a correct solution for the shear stress distribution according to the theory of elasticity. The shear stress at any point of the cross section is perpendicular to the radius of the circle. It is easily demonstrated that this assumption is invalid for shafts of noncircular cross section, for example the rectangle shown in Fig. 4.16. From Fig. 4.16 the shear stress τ_r has components τ_s and τ_n, and because τ_n is nonzero for equilibrium at the surface, longitudinal traction forces are necessary for equilibrium. St. Venant's torsion can be solved as an initial strain problem by first making the assumption that the elementary theory remains valid for the displacements (u, v) and that the surface stresses that would be induced are removed by allowing the cross section to warp in the z direction. All sections are assumed to warp alike, so that longitudinal σ_z stresses are not induced. Thus if θ is the twist per unit length of the shaft at the cross section distance z from the fixed end, then the three components of the displacements (see Fig. 4.16) are given by

$$u = -y\theta z \qquad v = x\theta z \qquad w = \theta\psi(x, y) \tag{4.195}$$

The function $\psi(x, y)$ is to be determined by the analysis. Given the displacement assumptions in Eq. (4.195), the in-plane shear strains are

$$\begin{Bmatrix} \gamma_{xz} \\ \gamma_{yz} \end{Bmatrix} = \begin{Bmatrix} w_{,x} + u_{,z} \\ w_{,y} + v_{,z} \end{Bmatrix} = \begin{Bmatrix} w_{,x} \\ w_{,y} \end{Bmatrix} + \begin{Bmatrix} -y \\ x \end{Bmatrix} \theta \tag{4.196}$$

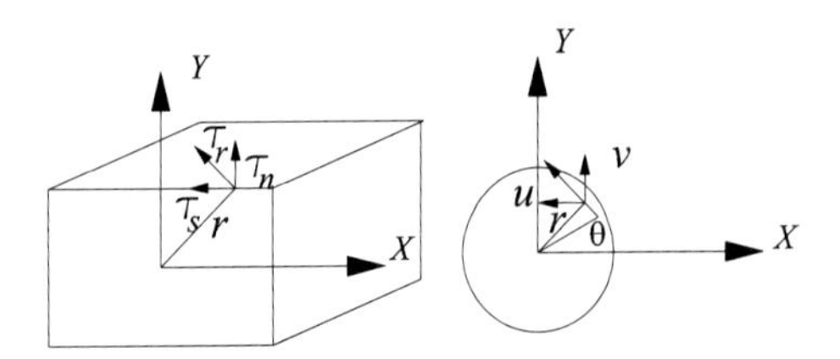

Fig. 4.16 Shaft of rectangular cross section.

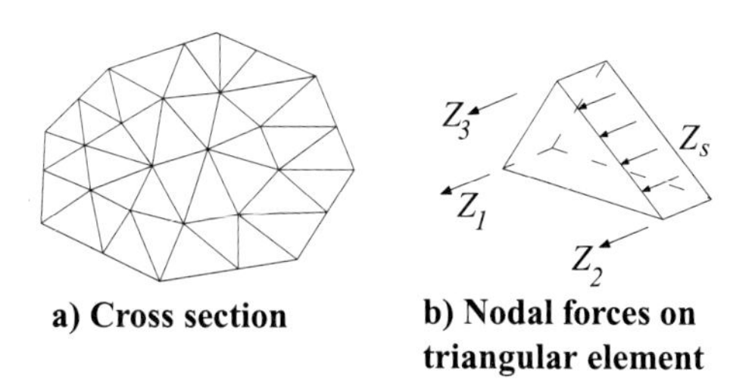

Fig. 4.17 Cross section idealization.

4.6.2 Torsion–Finite Element Approximation

The shaft cross section is subdivided into finite elements,[52] with triangles as shown in Fig.4.17a. Within each finite element express the longitudinal warping function approximation $\tilde{w}$ in terms of its nodal value $\boldsymbol{w}$, using the interpolation polynomial $\boldsymbol{N}$:

$$\tilde{w} = \boldsymbol{N}\boldsymbol{w} \tag{4.197}$$

Then the x and y derivatives are given by

$$\begin{Bmatrix} \tilde{w}_{,x} \\ \tilde{w}_{,y} \end{Bmatrix} = \begin{Bmatrix} \boldsymbol{N}_{,x} \\ \boldsymbol{N}_{,y} \end{Bmatrix} \boldsymbol{w} \tag{4.198}$$

Thence, from Eq.(4.196), the shear strains are expressed as

$$\begin{Bmatrix} \gamma_{xz} \\ \gamma_{yz} \end{Bmatrix} = \begin{Bmatrix} \boldsymbol{N}_{,x} \\ \boldsymbol{N}_{,y} \end{Bmatrix} \{\boldsymbol{w}\} + \begin{bmatrix} 0 & -\theta \\ \theta & 0 \end{bmatrix} \begin{Bmatrix} \tilde{x} \\ \tilde{y} \end{Bmatrix} \tag{4.199}$$

Evidently $\tilde{x}$ and $\tilde{y}$ also can be interpolated in terms of their nodal values, and for the straight sided triangle,

$$[\tilde{x} \quad \tilde{y}] = \boldsymbol{N}_1[\boldsymbol{X} \quad \boldsymbol{Y}] \tag{4.200}$$

Substituting these values for $(\tilde{x}, \tilde{y})$ in Eq. (4.199) gives

$$\begin{Bmatrix} \gamma_{xz} \\ \gamma_{yz} \end{Bmatrix} = \begin{Bmatrix} \boldsymbol{N}_{,x} \\ \boldsymbol{N}_{,y} \end{Bmatrix} \boldsymbol{w} + \begin{bmatrix} 0 & -\theta \boldsymbol{N}_1 \\ \theta \boldsymbol{N}_1 & 0 \end{bmatrix} \begin{Bmatrix} \boldsymbol{X} \\ \boldsymbol{Y} \end{Bmatrix} \tag{4.201}$$

Equation (4.201) is then written as

$$\tilde{\boldsymbol{\epsilon}} = [\boldsymbol{T}]\boldsymbol{w} + \boldsymbol{\epsilon}_0 \tag{4.202}$$

It is noted that the second term on the right-hand side of Eq. (4.202) is independent of the warped shape and that $\boldsymbol{\epsilon}_0$ is the initial strain term produced by the distortions that violate the stress boundary conditions. The constitutive equation relating shear stresses and strains is

$$\begin{Bmatrix} \tau_{xz} \\ \tau_{yz} \end{Bmatrix} = \begin{bmatrix} G & 0 \\ 0 & G \end{bmatrix} \begin{Bmatrix} \gamma_{xz} \\ \gamma_{yz} \end{Bmatrix} \tag{4.203}$$

or symbolically,

$$\tilde{\boldsymbol{\tau}} = \boldsymbol{k}\tilde{\boldsymbol{\epsilon}} \tag{4.204}$$

Thence, combining Eqs. (4.202) and (4.204),

$$\tilde{\tau} = \boldsymbol{kTw} + \boldsymbol{k}\tilde{\epsilon}_0 \tag{4.205}$$

The nodal forces that are to be considered are the shears complementary to those on the x, y plane of the cross section and for a single element, as in Fig. 4.17b. The contribution to these nodal forces for an infinitesimal area, applying the contragredient law, is

$$\Delta\{\boldsymbol{R}\} = [\boldsymbol{T}]^T\{\tilde{\tau}\} \tag{4.206}$$

For the whole element, integrating over the area of an element,

$$\{\boldsymbol{R}\} = \int_A [\boldsymbol{T}]^T[\boldsymbol{k}][\boldsymbol{T}]\,\mathrm{d}A\{\boldsymbol{w}\} + \int_A \{\tilde{\epsilon_0}\}\,\mathrm{d}A \tag{4.207}$$

For the whole configuration of elements, contributions to node forces from all elements (triangles in the present case), are added. At both internal and external nodes the sum of the forces must be equal to zero for equilibrium. These equations of equilibrium take the form

$$\boldsymbol{Kw} = -\boldsymbol{R}_0 \tag{4.208}$$

in which

$$\boldsymbol{K} = \sum \int_A \boldsymbol{T}^T\boldsymbol{kT}\,\mathrm{d}A \tag{4.209}$$

and

$$\boldsymbol{R}_0 = \int_A \boldsymbol{T}^T\boldsymbol{k}\tilde{\epsilon}_0\,\mathrm{d}A \tag{4.210}$$

For the general cross section without axes of symmetry, one node value of w is set equal to zero. On an axis of symmetry, $w = 0$, and only a portion of the cross section need be analyzed. The preceding theory is based on the approximate solution to the Poisson equation,

$$\nabla^2 w = \text{const} \tag{4.211}$$

An alternative approach to the solution of St Venant's torsion problem is to use a stress function ϕ that leads to the solution of the Laplace equation:

$$\nabla^2\phi = 0 \tag{4.212}$$

In the finite element method analysis, the solution to Eq. (4.211) is to be preferred because in Eq. (4.212) voids must be meshed whereas in the former they are not.

4.6.3 Calculation of Torsion Stiffness and Shear Stress Distribution

From Eq. (4.209) written for all node points, the solution is obtained for $\boldsymbol{w}$, the warping function in terms of the twist per unit length θ. It is seen from Eq. (4.195) that $\psi(x, y) = w_1$ for $\theta = 1$. Once the approximation has been obtained for w, the element shear stresses are calculated from Eq. (4.205); i.e.,

$$\begin{Bmatrix} \tilde{\tau}_{xz} \\ \tilde{\tau}_{yz} \end{Bmatrix} = G\theta \begin{Bmatrix} \boldsymbol{N}_{,x} \\ \boldsymbol{N}_{,y} \end{Bmatrix} \boldsymbol{w} + \begin{bmatrix} 0 & -\theta \boldsymbol{N}_1 \\ \theta \boldsymbol{N}_1 & 0 \end{bmatrix} \begin{Bmatrix} \boldsymbol{X} \\ \boldsymbol{Y} \end{Bmatrix} \tag{4.213}$$

The twisting moment M_T may be obtained as function of θ using the expression

$$M_T = \int A(\tilde{x}\tilde{\tau}_{yz} - \tilde{y}\tilde{\tau}_{xz})\,\mathrm{d}A \tag{4.214}$$

For one finite element the contribution to M_T is then given as

$$\Delta M_T = \int_A [\tilde{y}\ \tilde{x}] \begin{Bmatrix} -\tilde{\tau}_{xz} \\ \tilde{\tau}_{yz} \end{Bmatrix} \mathrm{d}A \tag{4.215}$$

Then, substituting for $\tilde{x}$, $\tilde{y}$, and $\tilde{\tau}_{xz}$, $\tilde{\tau}_{yz}$,

$$\Delta M_T = G\theta[\boldsymbol{X}^T\ \boldsymbol{Y}^T] \int_A \begin{bmatrix} \boldsymbol{0} & \boldsymbol{N}_1^T \\ \boldsymbol{N}_1^T & \boldsymbol{0} \end{bmatrix} \left[\begin{Bmatrix} -\boldsymbol{N}_{,x} \\ \boldsymbol{N}_{,y} \end{Bmatrix} \boldsymbol{w}_1 + \begin{bmatrix} \boldsymbol{0} & \boldsymbol{N}_1 \\ \boldsymbol{N}_1 & \boldsymbol{0} \end{bmatrix} \begin{Bmatrix} \boldsymbol{X} \\ \boldsymbol{Y} \end{Bmatrix} \right] \mathrm{d}A \tag{4.216}$$

Then, for all elements,

$$M_T = \sum_{1 \text{ to } n}^{i} \Delta M_T = [\boldsymbol{K}_S] G\theta \tag{4.217}$$

and twist per unit length

$$\theta = \frac{M_T}{G\boldsymbol{K}_S} \tag{4.218}$$

Having calculated θ, the shear stress distribution may be obtained in terms of M_T.

Example 4.1: Cross Section Modeled with Three-Node Triangles

In this case,

$$\boldsymbol{N} = \boldsymbol{N}_1 = [\zeta_1\ \zeta_2\ \zeta_3] \tag{4.219}$$

The x and y derivatives are given:

$$\begin{Bmatrix} \boldsymbol{N}_{1,x} \\ \boldsymbol{N}_{1,y} \end{Bmatrix} = \frac{1}{2A} \begin{bmatrix} b_1\ b_2\ b_3 \\ a_1\ a_2\ a_3 \end{bmatrix} = \frac{1}{2A} \begin{bmatrix} \boldsymbol{b}^T \\ \boldsymbol{a}^T \end{bmatrix} \tag{4.220}$$

and then, from Eq. (4.201),

$$[\mathbf{T}] = \frac{1}{2A} \begin{bmatrix} \boldsymbol{b}^T \\ \boldsymbol{a}^T \end{bmatrix} \tag{4.221}$$

Then

$$[\boldsymbol{T}]^T[k][\boldsymbol{T}] = \frac{G}{4A^2}[\boldsymbol{b}\ \boldsymbol{a}] \begin{bmatrix} \boldsymbol{b}^T \\ \boldsymbol{a}^T \end{bmatrix} \tag{4.222}$$

Integrating over the area of the triangle and simplifying,

$$\int_A [\boldsymbol{T}]^T [k][\boldsymbol{T}]\, \mathrm{d}A = \frac{G}{4A}[\boldsymbol{b}^T \boldsymbol{b} + \boldsymbol{a}^T \boldsymbol{a}] \tag{4.223}$$

The term $\{R_0\}$ is given by

$$\int_A [\boldsymbol{T}]^T [\boldsymbol{k}]\{\tilde{\epsilon}_0\}\, \mathrm{d}A = \frac{G\theta}{2A}\int_A [\boldsymbol{b}\ \boldsymbol{a}]\begin{bmatrix} \mathbf{0} & -\boldsymbol{N}_1 \\ \boldsymbol{N}_1 & \mathbf{0} \end{bmatrix}\begin{Bmatrix} \boldsymbol{X} \\ \boldsymbol{Y} \end{Bmatrix} \mathrm{d}A = \frac{G\theta}{2}[\boldsymbol{a}\bar{x} - \boldsymbol{b}\bar{y}] \tag{4.224}$$

In this equation $(\bar{x},\ \bar{y})$ are the coordinates of the centroid of the triangle, equal to

$$\frac{1}{3}\left(\sum_{i \text{ to } 3}^{i} x_i, \quad \frac{1}{3}\sum_{1 \text{ to } 3}^{i} y_1\right)$$

Hence, for one triangle,

$$\{\boldsymbol{R}\} = \begin{Bmatrix} R_{z1} \\ R_{z2} \\ R_{z3} \end{Bmatrix} = \frac{G}{4A}[\boldsymbol{b}\boldsymbol{b}^T + \boldsymbol{a}\boldsymbol{a}^T]\boldsymbol{w}_1\theta + \frac{G}{2}[\boldsymbol{b}\ \boldsymbol{a}]\begin{Bmatrix} -\bar{y} \\ \bar{x} \end{Bmatrix}\theta \tag{4.225}$$

Note that w_1 is the nodal z displacement for $\theta = 1$. The torsion constant $\boldsymbol{K}_S$ is now calculated from Eq. (4.216). In this case

$$\Delta \boldsymbol{K}_S = [\boldsymbol{X}^T\ \boldsymbol{Y}^T]\frac{1}{2A}\int_A \begin{bmatrix} \mathbf{0} & \boldsymbol{N}_1^T \\ \boldsymbol{N}_1^T & \mathbf{0} \end{bmatrix}\left[\begin{Bmatrix} -\boldsymbol{b} \\ \boldsymbol{a} \end{Bmatrix}\boldsymbol{w}_1 + \begin{bmatrix} \mathbf{0} & \boldsymbol{N}_1 \\ \boldsymbol{N}_1 & \mathbf{0} \end{bmatrix}\begin{Bmatrix} \boldsymbol{X} \\ \boldsymbol{Y} \end{Bmatrix}\right] \mathrm{d}A \tag{4.226}$$

It can be shown that the first term on the right-hand side is equal to

$$1/2[-\bar{y}\boldsymbol{b} + \bar{x}\boldsymbol{a}]\{\boldsymbol{w}_1\} \tag{4.227}$$

and the second term is equal to

$$\frac{1}{12}[\boldsymbol{X}^T\ \boldsymbol{Y}^T]\begin{bmatrix} \boldsymbol{\Lambda} & \mathbf{0} \\ \mathbf{0} & \boldsymbol{\Lambda} \end{bmatrix} \tag{4.228}$$

in which $(\bar{x},\ \bar{y})$ are the coordinates of the triangle centroid, and $\boldsymbol{\Lambda}$ is the matrix:

$$\boldsymbol{\Lambda} = \begin{bmatrix} 2 & 1 & 1 \\ 1 & 2 & 1 \\ 1 & 1 & 2 \end{bmatrix} \tag{4.229}$$

Then, for the one element,

$$\boldsymbol{K}_S = \frac{1}{12}[\boldsymbol{X}^T\ \boldsymbol{Y}^T] + \frac{1}{12}[\boldsymbol{X}^T\ \boldsymbol{Y}^T]\begin{bmatrix} \boldsymbol{\Lambda} & \mathbf{0} \\ \mathbf{0} & \boldsymbol{\Lambda} \end{bmatrix} \tag{4.230}$$

and, once $\boldsymbol{w}_1$ has been calculated, the shear stresses in an element are given by

$$\begin{Bmatrix} \tilde{\tau}_{xz} \\ \tilde{\tau}_{yz} \end{Bmatrix} = \frac{\boldsymbol{M}_T}{2A}\left[\begin{Bmatrix} \boldsymbol{b}^T \\ \boldsymbol{a}^T \end{Bmatrix}\boldsymbol{w}_1 + \begin{bmatrix} \mathbf{0} & -\boldsymbol{N}_1 \\ \boldsymbol{N}_1 & \mathbf{0} \end{bmatrix}\begin{Bmatrix} \boldsymbol{X} \\ \boldsymbol{Y} \end{Bmatrix}\right] \tag{4.231}$$

4.6.4 Center of Twist–Shear Center

The displacements at a point in the shaft cross section are calculated as

$$[u\ v] = \theta z[-y\ x\ ] \tag{4.232}$$

In this equation the assumption has been made that the displacement of the origin of coordinates is zero. In fact, the displacement is zero at the shear center or center of twist, coordinates (x_T, y_T). Then, using the center of twist,

$$[u\ v] = \theta z[-(y - y_T)\ (x - x_T)] \tag{4.233}$$

Now examine the cross section displacements,

$$[u\ v] = \theta[zy_T\ \ -zx_T] \tag{4.234}$$

These are rigid-body rotations of a cross section, and hence,

$$\gamma_{xz} = \gamma_{yz} = 0 \tag{4.235}$$

For γ_{xz}

$$\gamma_{xz} = \frac{\partial u}{\partial z} + \frac{\partial w}{\partial x} = 0 \tag{4.236}$$

That is

$$\frac{\partial w}{\partial x} = -\theta y_T \tag{4.237}$$

and integrating

$$w = -\theta y_T x \tag{4.238}$$

Similarly for γ_{yz}

$$w = \theta x_T y \tag{4.239}$$

so that the warping function is written as

$$w = \theta\psi(x, y) + \theta x_T y - \theta y_T x \tag{4.240}$$

To locate (x_T, y_T) it is noted that the w function should be such that if σ_z stress exists representing pure bending about either the x or y axis, the generalized force corresponding to the w displacement produced by pure twist should be zero. For example for an M_x generalized force,

$$\sigma_z = cy \tag{4.241}$$

Calculate the axial force and equate to zero,

$$R_z = \int_A \sigma_z w \,\mathrm{d}A = \int_A cy\theta(\psi(x, y) + x_T y - y_T x)\,\mathrm{d}A = 0 \tag{4.242}$$

If the origin is the centroid and the principal axes are used,

$$\int_A y^2 \,\mathrm{d}A = I_x \quad \text{and} \quad \int_A xy \,\mathrm{d}A = 0 \tag{4.243}$$

and Eq. (4.242) locates x_T:

$$x_T = -\frac{\int y\psi(x, y)\,\mathrm{d}A}{I_x} \tag{4.244}$$

Similarly,

$$y_T = \frac{\int x\psi(x, y)\,\mathrm{d}A}{I_y} \tag{4.245}$$

Write

$$\psi(x, y) = \boldsymbol{N}w; \qquad y = \boldsymbol{N}_1\boldsymbol{Y}; \qquad x = \boldsymbol{N}_1\boldsymbol{X}$$

The coordinates of the shear center are then located at

$$x_T = -\frac{\sum \boldsymbol{Y}^T \int \boldsymbol{N}_1\boldsymbol{N}_1\,\mathrm{d}A\,\boldsymbol{w}}{\sum \boldsymbol{Y}^T \int \boldsymbol{N}_1\boldsymbol{N}_1\,\mathrm{d}A\,\boldsymbol{Y}}$$

$$y_T = \frac{\sum \boldsymbol{X}^T \int \boldsymbol{N}_1\boldsymbol{N}_1\,\mathrm{d}A\,\boldsymbol{w}}{\sum \boldsymbol{X}^T \int \boldsymbol{N}_1\boldsymbol{N}_1\,\mathrm{d}A\,\boldsymbol{X}} \tag{4.246}$$

4.7 Plate Bending Elements

In a typical plate element the nodal unknowns consist of the transverse deflection u_z and the nodal rotations θ_x, θ_y. The transverse deflection u_z is expressed in a polynomial form and should satisfy compatibility conditions requiring continuity of u_z, $u_{\theta x}$, and $u_{\theta y}$ within the element. Also, the normal slope $u_{z,n}$ should be continuous across interelement boundaries if C^1 compatibility between elements is to be satisfied. Furthermore, the polynomial expansion must satisfy the rigid-body displacement state and reproduce a constant strain state. All of these conditions can be satisfied for a triangular element by selection of a complete quintic polynomial[15,16] in Fig. 4.18, in which nodal unknowns are as follows:

Nodes 1,2,3: $u_z, u_{z,x}, u_{z,y}, u_{z,xx}, u_{z,yy}, u_{z,xy}$

Nodes: 4,5,6: $u_{z,n}$

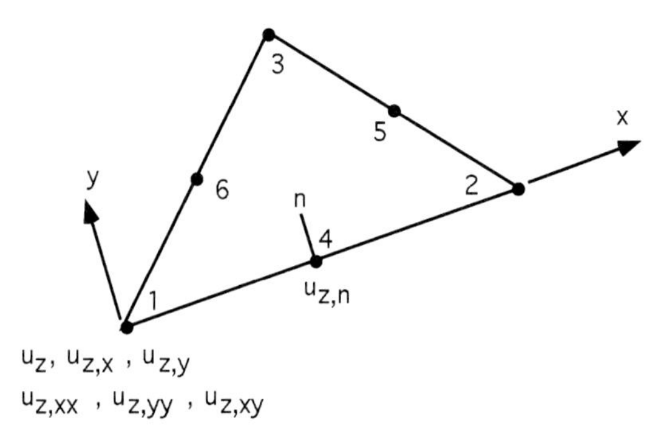

Fig. 4.18 Triangular plate bending elements using quintic interpolation displacement function.

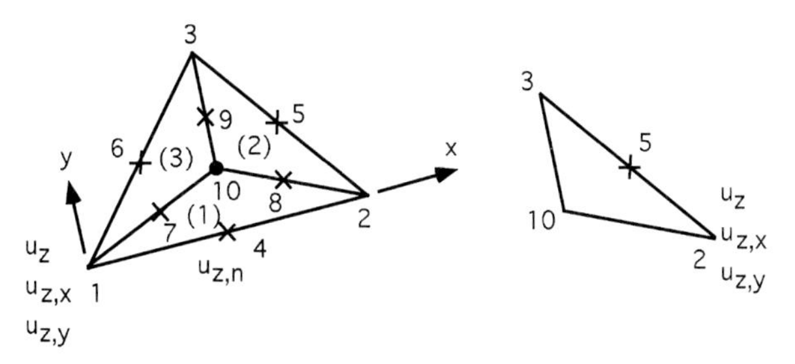

a) Triangular element **b) Triangular subelement**

Fig. 4.19 Triangular plate bending element.

It may be noted that, although u_z has quintic variation within the element, its first derivative is quartic and its second derivative is cubic, all being continuous in nature. The slope normal to the interface, along a typical edge 1-2, $u_{z,n}$, also has quartic variation and is uniquely defined by its values $(u^1_{z,n}, u^2_{z,n}, u^4_{z,n})$ and derivatives $(u^1_{z,nn}, u^2_{z,nn})$ embedded in nodal values along that edge. Thus the element satisfies all compatibility conditions up to level C^1. Also, because the polynomial is complete, the constant strain state can be achieved and further it satisfies the rigid-body displacement state. The midside nodes may be eliminated by expressing the $u_{z,n}$ in cubic rather than quartic form. This results in a three-noded, 18-degrees-of-freedom (DOF) element with little loss of accuracy.[15]

In practice, from the point of view of computational efficiency, it is expedient to have the nodal degrees of freedom not exceeding three. In this regard a fully compatible triangular element has been devised using cubic polynomial interpolation in each of the three subtriangles formed with the centroid of the original element as the common internal node (Fig. 4.19). For each subtriangle a complete cubic polynomial is used involving 10 nodal degrees of freedom that ensure normal slope continuity on the outer edge. By imposing internal compatibility conditions on these polynomials a 12-DOF compatible plate bending element is achieved. The element can be further simplified by imposing linear normal slope variation along the external edges resulting in a three-noded, nine-DOF plate bending element. This element,[17] although involving considerable derivation, results in a fully compatible, complete element that gives satisfactory results.

A number of nonconforming, simple, yet effective elements have been derived previously.[18–20] Figure 4.20 shows the deflected form of the element described in Ref. 20. Its derivation is included herein to illustrate the plate bending type of element. The deformation at any point within the element can be obtained in terms of six relative nodal slopes that in turn can be expressed in terms of nine nodal values, each node having three degrees of freedom, namely, u_z, θ_x, and θ_y. In this element the effect of shear deformations is not included. The rigid-body motion is given by

$$u_z^R = \boldsymbol{N}_1 \boldsymbol{u}_z^e = [\zeta_1 \quad \zeta_2 \quad \zeta_3] \begin{bmatrix} u_{zi} \\ u_{zj} \\ u_{zk} \end{bmatrix} \tag{4.247}$$

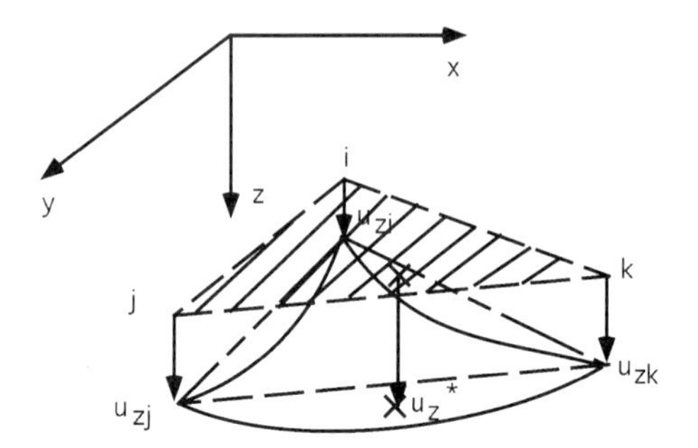

Fig. 4.20 Triangular plate bending element.

and the plate middle surface deflection can be written as

$$u_z = u_z^R + u_z^* \tag{4.248}$$

If $(\theta_x, \theta_y)_i$ are the typical nodal rotations, then the relative nodal rotations are

$$\theta_x^* = u_{z,y}^* = \theta_x - u_{z,y}^R; \qquad \theta_y^* = -u_{z,x}^* = \theta_y + u_{z,x}^R \tag{4.249}$$

The coordinate derivatives of u_z^R are obtained by differentiating Eq. (4.169) with respect to x, y, and are

$$\begin{Bmatrix} -u_{z,y}^R \\ -u_{z,x}^R \end{Bmatrix} = \frac{1}{2A}\begin{bmatrix} a_1 & -a_2 & -a_3 \\ b_1 & b_2 & b_3 \end{bmatrix}\begin{Bmatrix} u_{zi} \\ u_{zj} \\ u_{zk} \end{Bmatrix} \tag{4.250}$$

Defining the relative nodal rotation vector as

$$\boldsymbol{r}^{*T} = \begin{bmatrix} \theta_{xi}^* & \theta_{yi}^* & \theta_{xj}^* & \theta_{yj}^* & \theta_{xk}^* & \theta_{yk}^* \end{bmatrix} \tag{4.251}$$

then from Eqs. (4.249) and (4.256), it is shown that

$$\{\boldsymbol{r}^*\} = \frac{1}{2A}\begin{bmatrix} -a_1 & 2A & 0 & -a_2 & 0 & -a_3 & 0 & 0 \\ b_1 & 0 & 2A & b_2 & 0 & b_3 & 0 & 0 \\ -a_1 & 0 & 0 & -a_2 & 2A & -a_3 & 0 & 0 \\ b_1 & 0 & 0 & b_2 & 0 & b_3 & 0 & 0 \\ -a_1 & 0 & 0 & -a_2 & 0 & -a_3 & 2A & 0 \\ b_1 & 0 & 0 & b_2 & 0 & b_3 & 0 & 2A \end{bmatrix}\begin{Bmatrix} \boldsymbol{r}_i^e \\ \boldsymbol{r}_j^e \\ \boldsymbol{r}_k^e \end{Bmatrix} \tag{4.252}$$

where

$$\boldsymbol{r}_i^e = [u_{z1} \quad \theta_{x1} \quad \theta_{y1}]^T \tag{4.253}$$

and so on. Equation (4.252) is written in the form

$$\boldsymbol{r}^* = \boldsymbol{T}\boldsymbol{r}^e \tag{4.254}$$

Then the nodal forces are obtained as

$$\boldsymbol{R}^e = \boldsymbol{T}^T\boldsymbol{R}^* \tag{4.255}$$

in which

$$\left\{\boldsymbol{R}_l^*\right\}^T = [\boldsymbol{M}_{xl} \quad \boldsymbol{M}_{yl}] \qquad l = i, j, k \tag{4.256}$$

and

$$\left\{\boldsymbol{R}_l^T\right\} = [\boldsymbol{R}_{zl} \quad \boldsymbol{M}_{xl} \quad \boldsymbol{M}_{yl}] \qquad l = i, j, k \tag{4.257}$$

If the stiffness relationship between $\boldsymbol{R}^*$ and $\boldsymbol{r}^*$ is written as

$$\{\boldsymbol{R}^*\} = \{\boldsymbol{K}^*\}\{\boldsymbol{r}^*\} \tag{4.258}$$

then the global stiffness matrix relationship is found to be

$$\begin{aligned} \boldsymbol{R}^e &= \boldsymbol{T}^T\boldsymbol{K}^*\boldsymbol{T}\boldsymbol{r}^e \\ &= \boldsymbol{K}^e\boldsymbol{r}^e \end{aligned} \tag{4.259}$$

To obtain $\boldsymbol{K}^*$ the polynomial $\boldsymbol{\phi}_3$ is used such that

$$\boldsymbol{N}_3 = \begin{bmatrix} \zeta_1^2(-\zeta_2 b_3 + \zeta_3 b_2) \\ \zeta_1^2(-\zeta_2 a_3 + \zeta_3 a_2) \\ \zeta_2^2(-\zeta_3 b_1 + \zeta_1 b_3) \\ \zeta_2^2(-\zeta_3 a_1 + \zeta_1 a_3) \\ \zeta_3^2(-\zeta_1 b_2 + \zeta_2 b_1) \\ \zeta_3^2(-\zeta_1 a_2 + \zeta_2 a_1) \end{bmatrix} + \tfrac{1}{2}\zeta_1\zeta_2\zeta_3 \begin{bmatrix} b_3 - b_2 \\ a_3 - a_2 \\ b_1 - b_3 \\ a_1 - a_3 \\ b_2 - b_1 \\ a_2 - a_1 \end{bmatrix} \tag{4.260}$$

Thus, given $\boldsymbol{N}_3$ the generation of $\boldsymbol{K}_c$ follows the same procedure as for any finite element analysis. Differentiating Eq. (4.260) twice with respect to x or y and once with respect to x and y and substituting in ζ nodal values to obtain the curvatures at any point in the triangle yields

$$\chi_c = \boldsymbol{\Phi}\boldsymbol{r}^* \tag{4.261}$$

Then

$$\boldsymbol{K}_c = \int_A \boldsymbol{N}^T\boldsymbol{k}\boldsymbol{N}\,\mathrm{d}A \tag{4.262}$$

where

$$\boldsymbol{N} = \begin{bmatrix} \boldsymbol{N}_1 & 0 & 0 \\ 0 & \boldsymbol{N}_1 & 0 \\ 0 & 0 & \boldsymbol{N}_1 \end{bmatrix} \tag{4.263}$$

and

$$\boldsymbol{K}^* = \boldsymbol{\Phi}^T\boldsymbol{K}_c\boldsymbol{\Phi} \tag{4.264}$$

$$\boldsymbol{K}_c = \frac{Et^3 A}{144(1-\nu^2)} \begin{bmatrix} \boldsymbol{\Lambda} & \nu\boldsymbol{\Lambda} & 0 \\ \nu\boldsymbol{\Lambda} & \boldsymbol{\Lambda} & 0 \\ 0 & 0 & \dfrac{1-\nu}{2}\boldsymbol{\Lambda} \end{bmatrix} \tag{4.265}$$

in which

$$\boldsymbol{\Lambda} = \begin{bmatrix} 2 & 1 & 1 \\ 1 & 2 & 1 \\ 1 & 1 & 2 \end{bmatrix} \tag{4.266}$$

The material constitutive relationship is $\boldsymbol{k}$, and $\boldsymbol{\Phi}$ is the transformation matrix relating nodal curvature

$$\left(\frac{\partial^2 u_z}{\partial x^2}, \frac{\partial^2 u_z}{\partial y^2}, 2\frac{\partial^2 u_z}{\partial x \partial y}\right)_l$$

to nodal displacement $(u_z, \theta_x, \theta_z)_l$. The final matrix $\boldsymbol{K}^e$ is obtained from Eqs. (4.254) and (4.255). The inertia matrix may be calculated as

$$\boldsymbol{M}^e = \boldsymbol{T}^T \rho t \int_A \boldsymbol{N}^T \boldsymbol{N} \, \mathrm{d}A \boldsymbol{T} \tag{4.267}$$

This element is complete, although nonconforming, producing good convergence characteristics for displacement solution results, including structural dynamics.

A similar element was also developed in Ref. 18, in which three linear moment fields parallel with the three sides were used. This is a forcefield element for which the (9×9) stiffness matrix is obtained by first the inversion of a (6×6) natural flexibility matrix. Any of the three-noded, nine-DOF triangular elements,[17,18,20] described may be conveniently used to yield a quadrilateral plate bending element by subdividing it into four triangles with the common apex point being at the center of the area of the quadrilateral. Then stiffness and inertia matrices of each triangular element are combined as before, eliminating the internal node by static condensation.

4.8 Shell Elements

The fundamental difference between plate and shell elements lies in the fact that, in the latter, coupling between membrane and bending actions occurs because of shell curvature. The development of an effective curved-shell element is then complicated by the necessity to incorporate the effects of curvature into the element behavior requiring analytical description of the shell geometry as well as high-order representation of displacements.

Problems relating to the geometry representation of the shell surface are reduced somewhat by formulating the curved element in terms of a shallow shell theory. The necessary mathematical manipulations are then performed in a base reference plane, and it is sufficient to assume constant geometric curvature over the element. This, however, introduces geometric discontinuities into the approximation of the shell surface because adjacent elements are portions of different parabolic surfaces that will not match exactly. To ensure that shallow shell finite element approximations converge to the deep shell solution, it is essential that the shallowness assumption be enforced relative to a local base plane, rather than to the global horizontal plane. Regardless of whether or not the shallowness assumption is used, a proper description of the rigid-body modes in elements based on curvilinear shell theory is possible only with the inclusion of transcendental functions in the assumed displacement expressions. This violates interelement compatibility. Alternatively, one may use higher-order polynomials for the displacement fields, which leads to the introduction of additional nodal degrees of freedom, namely, second-order derivatives. The introduction of second-order derivatives complicates not only shell intersections but also skew boundaries in which the transformation of the stress-free boundary condition into these coordinates is not straightforward.

In classical shell theory, assumptions are imposed on the three-dimensional equations of elasticity, and these together with integration through the shell thickness reduce the problem from three to two dimensions. Because of the various assumptions and approximations that have been devised, there are a variety of shell theories.[21–25] Thus, a shell theory combined with additional assumptions of geometry and the discretization of the field variables forms the basis of the curved elements in finite element approximations to shells. The isoparametric formulation, which has been successfully applied to both two- and three-dimensional analysis, obviously has the potential to provide an alternative approach to analyzing shells. It was shown that the degenerated isoparametric concept was successful in the application to analysis of plates. In historical sequence the degenerated isoparametric concept was actually first introduced in Ref. 26 for the analysis of shells. However, for practical analysis of shells where computational cost-effectiveness is paramount, there is considerable appeal in the flat-facet formulation because of its simplicity and low cost of element stiffness matrix generation. The advantages in using an assemblage of flat-facet elements to model a shell are 1) simplicity and ease of formulation, 2) correct representation of rigid-body motions, and 3) convergence to the deep shell solution.

4.8.1 Flat-Facet Elements

Because of the relative ease of their formulation, many such elements have been developed by suitably combining membrane and bending elements (some of which have been described in earlier sections).

The concept of using flat-facet elements for analyzing the behavior of shell structures was suggested in Ref. 27 as early as 1960. Satisfactory bending stiffness matrices were first developed for the rectangular element. They were utilized in deriving the earliest successful flat shell elements for the analysis of arch dams and cylindrical roofs.[28] Using an assumed stress field for the membrane action and a nonconforming 12-term polynomial for the transverse displacement, a flat quadrilateral shell element was derived in Ref. 29. Another quadrilateral faceted shell element was formulated in Ref. 30 using the subdomain approach. Each of the four triangular elements consists of a superposition of a partly constrained linear strain triangle (LST) plane element with the Hsieh, Clough, and Tocher (HCT) bending element. An improved version of this quadrilateral was later proposed in Ref. 31. Both the rectangular and quadrilateral faceted shell elements are, however, restricted in use due to the need for all of the four nodes to lie in the same plane.

To model an arbitrary doubly curved shell with flat elements requires the use of triangular elements, although warped quadrilaterals are possible. Initial attempts at formulating faceted triangular shell elements in Refs. 32–34 were not particularly satisfactory due to the lack of suitable triangular bending elements. This situation was rectified with the development of other plate bending elements.[17,20] These plate bending elements could then be combined with the plane stress elements to yield suitable triangular faceted shell elements. This was accomplished in Ref. 35 where the constant strain triangle (CST) was used in conjunction with the bending element of Ref. 17, and in Ref. 36 with the plate bending element of Ref. 20. The membrane approximation of the faceted shell proposed in Ref. 35 was improved on in Ref. 37 by using the quadratic strain triangle with the resultant element having a total of 27 degrees-of-freedom. Another higher-order element with 27 degrees-of-freedom, which utilizes the LST to represent the membrane action, and a quartic plate bending element, has been formulated in Ref. 38.

In Ref. 39 a faceted shell element was presented in which subcubic functions of the type in Ref. 20 were used to describe the in-plane displacements. The derivative smoothed bending element,[40] employed to model the slope at midside nodes, exhibits good convergence characteristics. However, the generation of the shape functions for the faceted element is somewhat tedious. In an attempt to incorporate the rotational stiffness about the normal to the shell, Ref. 41 employed a nine-degrees-of-freedom plane stress element with the in-plane rotation as a nodal parameter. This was combined with the bending element of Ref. 20 to yield an 18-degree-of-freedom flat shell element.

The simplest formulation of a faceted shell element is given in Ref. 42, where the CST is combined with the constant moment triangle[19] to yield a 12-degree-of-freedom constant stress shell element. This particular element was originally derived in Ref. 43 through the application of a mixed variational principle. In Ref. 44 a flat shell element was developed using the hybrid stress model by assuming a linear variation of the in-plane forces and a quadratic moment variation. The in-plane displacements are assumed to vary linearly along an edge and cubically normal to it, whereas the transverse displacements possess a cubic variation. In another paper,[45] the membrane approximation was improved on by using second-order polynomials for all interior stress fields. A similar type of element but with linearly varying equilibrium stress fields was presented in Ref. 46. The hybrid stress elements described utilize the in-plane rotation as a nodal parameter. The motivation for this, as is also the case with the displacement-based element presented in Ref. 41, is to prevent a singularity arising in the global stiffness matrix when adjacent faceted elements meeting at a node are coplanar. In Ref. 47 the use of a new family of triangular faceted shell elements has been suggested based on a combination of compatible displacement elements for membrane action and equilibrium stress elements for the bending action.

Recent impetus toward the nonlinear analysis of shell structures has regenerated interest in the relatively simple and economical faceted formulation. The faceted Trump shell element with 18-DOF given in Ref. 48 and formulated through physical lumping ideas was proposed to solve nonlinear plate and shell problems. Another element formulated with nonlinear applications in mind is the triangular element presented in Ref. 49. A constant stress hybrid element was employed to represent the membrane action, while a hybrid stress model utilizing a modified

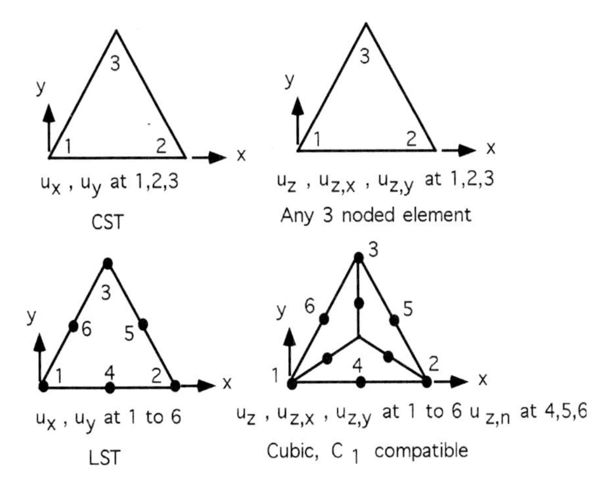

Fig. 4.21 Facet element combinations.

form of Reissner's variational principle was used to derive the bending stiffness. A quadrilateral version of this hybrid flat shell element was also presented in Ref. 49. The three-node triangular discrete Kirchhoff element formulated in Ref. 50 was employed in Ref. 51 to represent the bending action in their flat shell element with the CST modeling the membrane behavior. References 52 and 53 present a detailed account of the development of shell elements. Some such combinations are summarized in Fig. 4.21.

4.8.2 Development of a Facet Triangular Shell Element

One such curved shallow shell element was developed in Ref. 54. This curved element, shown in Fig. 4.22, is assumed to have constant middle surface curvatures $\theta_{x,x}$, $\theta_{y,y}$, and $\theta_{x,y}$, and a quadratic surface in x and y can be used to approximate the middle surface. It is also assumed that the middle surface passes through points 1, 2, 3, and O_m where the tangent plane is parallel to the base triangle. Then one

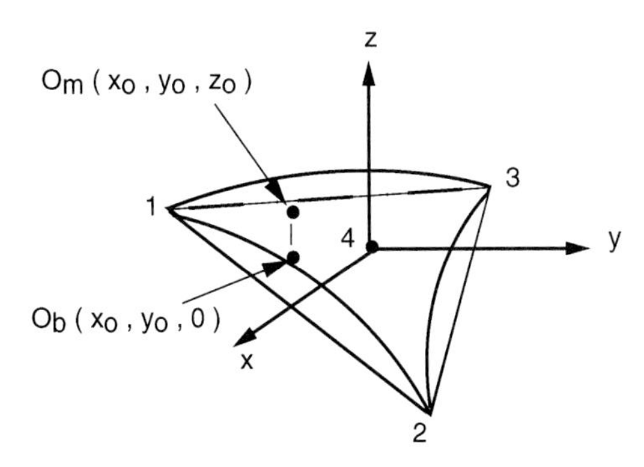

Fig. 4.22 Curved triangular shell element.

may express $z(x, y)$ as

$$z = h_1 x^2 + h_2 y^2 + h_3 xy + h_4 x + h_5 y + h_6 \tag{4.268}$$

and the coefficients h_i can easily be obtained from the assumptions relating to the middle surface, yielding the relationship

$$z = \hat{\boldsymbol{C}}\boldsymbol{C}^{-1}\boldsymbol{z}_0 \tag{4.269}$$

and the corresponding slopes are

$$z_{,x} = \alpha_1(x - x_0) + \alpha_2(y - y_0) \tag{4.270}$$

$$z_{,y} = \alpha_2(x - x_0) + \alpha_3(y - y_0) \tag{4.271}$$

in which

$$\boldsymbol{\alpha} = \boldsymbol{L}\hat{\boldsymbol{C}}^{-1}\boldsymbol{z}_0 \tag{4.272}$$

where

$$\boldsymbol{\alpha} = [\alpha_1 \;\; \alpha_2 \;\; \alpha_3]^T, \quad \hat{\boldsymbol{C}} = [x^2 \;\; y^2 \;\; xy \;\; x \;\; y \;\; 1], \quad \boldsymbol{z}_0 = [0 \;\; 0 \;\; 0 \;\; z_0 \;\; 0 \;\; 0]^T \tag{4.273}$$

$$\boldsymbol{C} = \begin{bmatrix} x_1^2 & y_1^2 & x_1 y_1 & x_1 & y_1 & 1 \\ x_2^2 & y_2^2 & x_2 y_2 & x_2 & y_2 & 1 \\ x_3^2 & y_3^2 & x_3 y_3 & x_3 & y_3 & 1 \\ x_0^2 & y_0^2 & x_0 y_0 & x_0 & y_0 & 1 \\ 2x_0 & 0 & y_0 & 1 & 0 & 0 \\ 0 & 2y_0 & x_0 & 0 & 1 & 0 \end{bmatrix}, \qquad \boldsymbol{L} = \begin{bmatrix} 2 & 0 & 0 & 0 & 0 & 0 \\ 0 & 0 & 1 & 0 & 0 & 0 \\ 0 & 2 & 0 & 0 & 0 & 0 \end{bmatrix} \tag{4.274}$$

Other useful properties of the base triangle are the area given by

$$A = \int_A \mathrm{d}A = \frac{1}{2}(x_{21}y_{31} - x_{31}y_{21}) \tag{4.275}$$

and principal moments of inertia

$$I_x = \int_A y^2 \,\mathrm{d}A = \frac{1}{4}(x_{14}y_{24} - x_{24}y_{14})\left(y_{14}^2 + y_{14}y_{24} + y_{24}^2\right) \tag{4.276}$$

$$I_y = \int_A x^2 \,\mathrm{d}A = \frac{1}{4}(x_{14}y_{24} - x_{24}y_{14})\left(x_{14}^2 + x_{14}x_{24} + x_{24}^2\right) \tag{4.277}$$

$$I_{xy} = \int_A xy \,\mathrm{d}A = \frac{1}{8}(x_{14}y_{24} - x_{24}y_{14})[y_{14}(2x_{14} + x_{24}) + y_{24}(x_{14} + 2x_{24})] \tag{4.278}$$

where $x_{ij} = x_i - x_j$ and $y_{ij} = y_i - y_j$.

The material axes are assumed to be coincident with the (xyz) axes, and the stress–strain relationship can then be expressed as shown next assuming that $\sigma_z = 0$ for thin shells as the first Kirchhoff assumption

$$\boldsymbol{\sigma} = \boldsymbol{D}\boldsymbol{\epsilon} \tag{4.279}$$

$$\boldsymbol{\tau} = \boldsymbol{D}'\boldsymbol{\gamma} \tag{4.280}$$

in which $\boldsymbol{\sigma} = [\sigma_x\ \sigma_y\ \sigma_{xy}]^T$, $\boldsymbol{\tau} = [\sigma_{xz}\ \sigma_{yz}]^T$, $\boldsymbol{\epsilon} = [\epsilon_x\ \epsilon_y\ \gamma_{xy}]^T$, and $\boldsymbol{\gamma} = [\gamma_{xz}\ \gamma_{yz}]^T$. Also, according to the second Kirchhoff assumption, the strain ϵ is assumed to vary linearly across the thickness and $\boldsymbol{\gamma}$ is constant. Since transverse shear strains are not assumed to be zero, the original middle surface normals are no longer normal to the middle surface after deformation. For isotropic material,

$$\boldsymbol{D} = \frac{E}{1-\nu^2}\begin{bmatrix} 1 & \nu & 0 \\ \nu & 1 & 0 \\ 0 & 0 & \dfrac{1-\mu}{2} \end{bmatrix} \tag{4.281}$$

$$\boldsymbol{D}' = G\begin{bmatrix} \dfrac{2}{3} & 0 \\ 0 & \dfrac{2}{3} \end{bmatrix} \tag{4.282}$$

Assuming that the middle surface deflections $u_x, u_y, u_{\theta x}, u_{\theta y}$ vary linearly within the element as

$$u_i = c_j x + c_{j+1} y + c_{j+2} \tag{4.283}$$

with $i = 1, 2, 3, 4$ and $j = 1, 2, \ldots, 12$, this can be expressed in terms of the nodal values as follows:

$$\boldsymbol{U} = \boldsymbol{C}_1\boldsymbol{C}_2\boldsymbol{C}_3 \tag{4.284}$$

in which

$$\boldsymbol{U} = [u_x \quad u_y \quad u_{\theta x} \quad u_{\theta y}], \qquad \boldsymbol{C}_1 = [x \quad y \quad 1]/2A \tag{4.285}$$

$$\boldsymbol{C}_2 = \begin{bmatrix} y_{23} & y_{31} & y_{12} \\ x_{32} & x_{13} & x_{21} \\ r_1 & r_2 & r_3 \end{bmatrix}, \qquad \boldsymbol{C}_3 = \begin{bmatrix} u_{x1} & u_{y1} & u_{\theta x_1} & u_{\theta y_1} \\ u_{x2} & u_{y2} & u_{\theta x_2} & u_{\theta y_2} \\ u_{x3} & u_{y3} & u_{\theta x_3} & u_{\theta y_3} \end{bmatrix} \tag{4.286}$$

and where $r_1 = x_2y_3 - x_3y_2$, $r_2 = x_3y_1 - x_1y_3$, and $r_3 = x_1y_2 - x_2y_1$. Using linear shallow shell theory, the relevant strain–displacement relations may be written as

$$\epsilon_0 = [(u_{x,x} + z_{,x}u_{z,x}) \quad (u_{y,y} + z_{,y}u_{z,y}) \quad (u_{x,y} + u_{y,x} + z_{,x}u_{z,y} + z_{,y}u_{z,x})]^T \tag{4.287}$$

and

$$\epsilon = \epsilon_0 + h\boldsymbol{\chi} \tag{4.288}$$

and for shear strains,

$$\boldsymbol{\gamma} = \begin{bmatrix} u_{z,x} \\ u_{z,y} \end{bmatrix} + \begin{bmatrix} u_{\theta y} \\ -u_{\theta x} \end{bmatrix} \tag{4.289}$$

h being the distance from the middle surface, being positive in the positive z direction in which the changes in curvature are defined as

$$\boldsymbol{\chi} = \begin{Bmatrix} -u_{\theta y,x} \\ u_{\theta x,y} \\ u_{\theta x,x} - u_{\theta y,y} \end{Bmatrix} = -\begin{Bmatrix} u_{z,xx} \\ u_{z,yy} \\ 2u_{z,xy} \end{Bmatrix} \tag{4.290}$$

Using the relations in Eq. (4.284), the preceding relations of Eqs. (4.287) and (4.288) may also be written as

$$\boldsymbol{\epsilon}_0 = \frac{1}{2A}\begin{bmatrix} \boldsymbol{G} & \boldsymbol{H} & z_{,x}\boldsymbol{G} + z_{,y}\boldsymbol{H} \end{bmatrix} \begin{Bmatrix} \boldsymbol{u}_x^e \\ \boldsymbol{u}_y^e \\ \boldsymbol{u}_z^e \end{Bmatrix} \tag{4.291}$$

$$\boldsymbol{\chi} = \frac{1}{2A}[-\boldsymbol{H} \quad \boldsymbol{G}] \begin{Bmatrix} u_{\theta x}^e \\ u_{\theta y}^e \end{Bmatrix} \tag{4.292}$$

in which

$$\boldsymbol{G} = \begin{bmatrix} y_{23} & y_{31} & y_{12} \\ 0 & 0 & 0 \\ x_{32} & x_{13} & x_{21} \end{bmatrix} \tag{4.293}$$

$$\boldsymbol{H} = \begin{bmatrix} 0 & 0 & 0 \\ x_{32} & x_{13} & x_{21} \\ y_{23} & y_{31} & y_{12} \end{bmatrix} \tag{4.294}$$

The expressions for membrane (middle surface stretching) bending and shear strain energies, based on the assumption that strains on planes parallel to the tangent plane of middle surface vary linearly across thickness and that no normal stresses exist across thickness, may be given as

$$U_M = \frac{t}{2}\int_A \boldsymbol{\epsilon}_0^T \boldsymbol{D} \boldsymbol{\epsilon}_0 \, \mathrm{d}A \tag{4.295}$$

$$U_B = \frac{t^3}{24}\int_A \boldsymbol{\chi}^T \boldsymbol{D} \boldsymbol{\chi} \, \mathrm{d}A \tag{4.296}$$

$$U_S = \frac{t}{2}\int_A \boldsymbol{\gamma}^T \boldsymbol{D}' \boldsymbol{\gamma} \, \mathrm{d}A \tag{4.297}$$

t being the shell thickness and total strain energy $U = U_M + U_B + U_S$. Substituting Eqs. (4.270) and (4.271) into Eq. (4.291), and Eqs. (4.291), (4.292), and (4.289) into Eqs. (4.295), (4.296), and (4.297), the strain energies may be expressed after suitable integration as

$$U_M = \tfrac{1}{2}\boldsymbol{u}^T \boldsymbol{K}_M \boldsymbol{u} \tag{4.298}$$

$$U_B = \tfrac{1}{2}\boldsymbol{u}^T \boldsymbol{K}_B \boldsymbol{u} \tag{4.299}$$

$$U_S = \tfrac{1}{2}\boldsymbol{u}^T \boldsymbol{K}_S \boldsymbol{u} \tag{4.300}$$

in which the individual stiffness matrices may be obtained using routine algebraic manipulation. Expressions for the individual stiffness matrices that exclude the membrane–bending coupling are presented for the simplified but commonly used flat element. For this case the membrane and bending stiffness matrices are obtained as

$$\boldsymbol{K}_M^e = \frac{t}{4A}\begin{bmatrix} \boldsymbol{P} & \boldsymbol{R} \\ \text{sym} & \boldsymbol{Q} \end{bmatrix} \tag{4.301}$$

and

$$\boldsymbol{K}_B^e = \frac{t^3}{48A}\begin{bmatrix} \boldsymbol{Q} & -\boldsymbol{R}^T \\ \text{sym} & \boldsymbol{P} \end{bmatrix} \tag{4.302}$$

in which

$$\boldsymbol{P} = \boldsymbol{G}^T \boldsymbol{D} \boldsymbol{G}, \qquad \boldsymbol{Q} = \boldsymbol{H}^T \boldsymbol{D} \boldsymbol{H}, \qquad \boldsymbol{R} = \boldsymbol{G}^T \boldsymbol{D} \boldsymbol{H} \tag{4.303}$$

The transverse shear strain energy is calculated directly from the deflection fields. Denoting u_z', the displacement associated with Kirchhoff behavior, it can be expressed as

$$\begin{bmatrix} u_{\theta_x} \\ u_{\theta_y} \end{bmatrix} = \begin{bmatrix} u'_{z,y} \\ -u'_{z,x} \end{bmatrix} \tag{4.304}$$

The difference between non-Kirchhoff and Kirchhoff transverse displacements is defined as

$$u_z^* = u_z - u_z' \tag{4.305}$$

Then Eq. (4.289) can be written as

$$\boldsymbol{\gamma} = \begin{Bmatrix} \gamma_{xz} \\ \gamma_{yz} \end{Bmatrix} = \begin{Bmatrix} u^*_{z,x} \\ u^*_{z,y} \end{Bmatrix} \tag{4.306}$$

and the shear strain energy obtained from Eq. (4.297). Assuming a consistent distribution of u_z' in the triangular element domain with linear rotation distribution, it is expressed in quadratic form as

$$u_z' = c_1 x^2 + c_2 y^2 + c_3 xy + c_4 x + c_5 y + 1 \tag{4.307}$$

The coefficients c_1–c_5 may be expressed in terms of vertex rotations $u_{\theta x}$ and $u_{\theta y}$ using Eq. (4.304) and by averaging two values of c_3. Further, assuming linear variation of u_z^* within the triangular element, the transverse shear strains can be expressed as shown next, after a series of routine operations:

$$\boldsymbol{\gamma} = \frac{1}{4A}([\boldsymbol{O}][\boldsymbol{I} \quad \tilde{Y} \quad \tilde{X}] + [\boldsymbol{O} \quad \tilde{\boldsymbol{Q}} \quad \boldsymbol{Q}])\tilde{\boldsymbol{u}}^e = \frac{1}{4A}\boldsymbol{R}\tilde{\boldsymbol{u}}^e \tag{4.308}$$

in which

$$\tilde{Y} = \begin{bmatrix} -y_1 & 0 & 0 \\ 0 & -y_2 & 0 \\ 0 & 0 & -y_3 \end{bmatrix}, \quad \tilde{X} = \begin{bmatrix} x_1 & 0 & 0 \\ 0 & x_2 & 0 \\ 0 & 0 & x_3 \end{bmatrix}, \quad \tilde{\boldsymbol{Q}} = \begin{bmatrix} 0 & 0 & 0 \\ -\dfrac{2A}{3} & -\dfrac{2A}{3} & -\dfrac{2A}{3} \end{bmatrix}$$

$$\boldsymbol{Q} = \begin{bmatrix} \dfrac{2A}{3} & \dfrac{2A}{3} & \dfrac{2A}{3} \\ 0 & 0 & 0 \end{bmatrix}, \quad \boldsymbol{I} = \begin{bmatrix} 1 & 0 & 0 \\ 0 & 1 & 0 \\ 0 & 0 & 1 \end{bmatrix}, \quad \tilde{\boldsymbol{u}}^e = \begin{bmatrix} \boldsymbol{u}_z^e \\ \boldsymbol{u}_{\theta_x}^e \\ \boldsymbol{u}_{\theta_y}^e \end{bmatrix}, \quad \boldsymbol{O} = \begin{bmatrix} y_{23} & y_{31} & y_{12} \\ x_{32} & x_{13} & x_{21} \end{bmatrix}$$

Utilizing the formulation of $\boldsymbol{\gamma}$ in Eq. (4.308) neglecting $\tilde{\boldsymbol{Q}}$ and $\boldsymbol{Q}$, and substituting in Eq. (4.306), the transverse shear stiffness matrix is obtained from Eq. (4.297):

$$\boldsymbol{K}_s^e = \frac{t}{4A} \begin{bmatrix} S_{11} & S_{12} & S_{13} \\ & S_{22} & S_{23} \\ \text{sym} & & S_{33} \end{bmatrix} \tag{4.309}$$

in which

$$\begin{aligned} S_{11} &= \boldsymbol{O}^T \boldsymbol{D}' \boldsymbol{O} & S_{21} &= \tfrac{1}{2}\tilde{Y}^T S_{11} & S_{31} &= \tfrac{1}{2}\tilde{X}^T S_{11} \\ S_{22} &= \tfrac{1}{4}\tilde{Y}^T S_{11}\tilde{Y} & S_{32} &= \tfrac{1}{4}\tilde{X}^T S_{11}\tilde{Y} & S_{33} &= \tfrac{1}{4}\tilde{X}^T S_{11}\tilde{X} \end{aligned} \tag{4.310}$$

The total stiffness matrix for the flat triangular shell element may now be defined as

$$\begin{aligned} \boldsymbol{K}^e &= \boldsymbol{K}_M^e + \boldsymbol{K}_B^e + \boldsymbol{K}_S^e \\ &= \frac{t}{4A} \begin{bmatrix} \boldsymbol{P} & \boldsymbol{R} & 0 & 0 & 0 \\ & \boldsymbol{Q} & 0 & 0 & 0 \\ & & S_{11} & S_{21}^T & S_{31}^T \\ \text{sym} & & & \beta\boldsymbol{Q} + S_{22} & -\beta\boldsymbol{R}^T + S_{32}^T \\ & & & & \beta\boldsymbol{P} + S_{33} \end{bmatrix} \end{aligned} \tag{4.311}$$

in which $\beta = t^2/12$. Because all submatrices in Eq. (4.311) have been derived explicitly, it is relatively easy to program the element for the computer. In typical selective reduced integration elements, only the second term in the $\boldsymbol{\gamma}$ formulation is retained in calculations. The mass matrix for the flat-facet triangular shell element is obtained by suitably combining the same for plane- and plate-bending elements.

4.8.3 Flat-Facet Composite Triangular Shell Element

Further extension of the element has been achieved for layered elements made of composite materials.[55] Thus using Eqs. (4.288) and (4.289), the energy expression may be expanded into five terms as follows:

$$
\begin{aligned}
\boldsymbol{U} &= \frac{1}{2}\int \epsilon_0^T \boldsymbol{D}\epsilon_0 \,\mathrm{d}A\,\mathrm{d}z + \frac{1}{2}\int \boldsymbol{X}^T \boldsymbol{D}\boldsymbol{X} z^2 \,\mathrm{d}A\,\mathrm{d}z + \frac{1}{2}\int \boldsymbol{\gamma}^T \boldsymbol{D}'\boldsymbol{\gamma}\,\mathrm{d}A\,\mathrm{d}z \\
&\quad + \frac{1}{2}\int \boldsymbol{X}^T \boldsymbol{D}\epsilon_0 z \,\mathrm{d}A\,\mathrm{d}z + \frac{1}{2}\int \epsilon_0^T \boldsymbol{D}\boldsymbol{X} z \,\mathrm{d}A\,\mathrm{d}z \\
&= U_m + U_B + U_S + U_{A1} + U_{A2}
\end{aligned} \tag{4.312}
$$

Corresponding element matrices are calculated layer by layer by usual differentiation of the energy integral with respect to nodal displacements. Thus integrating through the thickness the membrane energy expression becomes

$$
\begin{aligned}
U_m &= \frac{1}{2}\int \epsilon_0^T \boldsymbol{D}\epsilon_0 \,\mathrm{d}A\,\mathrm{d}z \\
&= \frac{1}{2}\sum_{i=1}^{n}(h_{i+1} - h_i)\int_{Ai} \epsilon_0^T D\epsilon_0 \,\mathrm{d}A
\end{aligned} \tag{4.313}
$$

h_i being the z coordinate of the bottom of layer i, n being the number of layers and from which the stiffness matrix may be obtained as shown next:

$$
\begin{aligned}
\boldsymbol{K}_m^e &= \frac{1}{4A}\sum_{i=1}^{n}(t_{i+1} - t_i)\,[\boldsymbol{G} \quad \boldsymbol{H}]^T \boldsymbol{D}_i\,[\boldsymbol{G} \quad \boldsymbol{H}] \\
&= \frac{1}{4A}\sum_{i=1}^{n}(t_{i+1} - t_i)\begin{bmatrix} \boldsymbol{G}^T\boldsymbol{D}_i\boldsymbol{G} & \boldsymbol{G}^T\boldsymbol{D}_i\boldsymbol{H} \\ \boldsymbol{H}^T\boldsymbol{D}_i\boldsymbol{G} & \boldsymbol{H}^T\boldsymbol{D}_i\boldsymbol{H} \end{bmatrix}
\end{aligned} \tag{4.314}
$$

Similarly,

$$
\begin{aligned}
\boldsymbol{K}_B^e &= \frac{1}{12A}\sum_{i=1}^{n}\left(h_{i+1}^3 - h_i^3\right)[-\boldsymbol{H} \quad \boldsymbol{G}]^T \boldsymbol{D}_i[-\boldsymbol{H} \quad \boldsymbol{G}] \\
&= \frac{1}{12A}\sum_{i=1}^{n}\left(h_{i+1}^3 - h_i^3\right)\begin{bmatrix} \boldsymbol{H}^T\boldsymbol{D}_i\boldsymbol{H} & -\boldsymbol{H}^T\boldsymbol{D}_i\boldsymbol{G} \\ -\boldsymbol{G}^T\boldsymbol{D}_i\boldsymbol{H} & \boldsymbol{G}^T\boldsymbol{D}_i\boldsymbol{G} \end{bmatrix}
\end{aligned} \tag{4.315}
$$

$$
\begin{aligned}
\boldsymbol{K}_S^e &= \frac{1}{16A}\sum_{i=1}^{n}(t_{i+1} - t_i)\boldsymbol{R}^T\boldsymbol{D}_i'\boldsymbol{R} \\
&= \frac{1}{16A}\frac{5}{4}\sum_{i=1}^{n}\left[t_i - \frac{4}{t^2}\left(t_i\bar{z}_i^2 + \frac{t_i^3}{12}\right)\right]
\begin{bmatrix}
4\boldsymbol{S}_{11} & 2\boldsymbol{S}_{11}\tilde{\boldsymbol{Y}} + 2\boldsymbol{S}_{12} & 2\boldsymbol{S}_{11}\tilde{\boldsymbol{X}} + 2\boldsymbol{S}_{13} \\
 & \begin{matrix}\tilde{\boldsymbol{Y}}^T\boldsymbol{S}_{11}\tilde{\boldsymbol{Y}} + \tilde{\boldsymbol{Y}}^T\boldsymbol{S}_{12} \\ +\boldsymbol{S}_{21}\tilde{\boldsymbol{Y}} + \boldsymbol{S}_{22}\end{matrix} & \begin{matrix}\tilde{\boldsymbol{Y}}^T\boldsymbol{S}_{11}\tilde{\boldsymbol{X}} + \tilde{\boldsymbol{Y}}^T\boldsymbol{S}_{13} \\ +\boldsymbol{S}_{21}\tilde{\boldsymbol{X}} + \boldsymbol{S}_{23}\end{matrix} \\
\text{sym} & & \begin{matrix}\tilde{\boldsymbol{X}}^T\boldsymbol{S}_{11}\tilde{\boldsymbol{X}} + \tilde{\boldsymbol{X}}^T\boldsymbol{S}_{13} \\ +\boldsymbol{S}_{31}\tilde{\boldsymbol{X}} + \boldsymbol{S}_{33}\end{matrix}
\end{bmatrix}
\end{aligned} \tag{4.316}
$$

in which

$$
\begin{aligned}
&\boldsymbol{S}_{11}=\boldsymbol{O}^T\boldsymbol{D}_i'\boldsymbol{O} && \boldsymbol{S}_{12}=\boldsymbol{O}^T\boldsymbol{D}_i'\tilde{\boldsymbol{Q}} && \boldsymbol{S}_{13}=\boldsymbol{O}^T\boldsymbol{D}_i'\boldsymbol{Q}\\
&\boldsymbol{S}_{22}=\tilde{\boldsymbol{Q}}^T\boldsymbol{D}_i'\tilde{\boldsymbol{Q}} && \boldsymbol{S}_{23}=\tilde{\boldsymbol{Q}}^T\boldsymbol{D}_i'\boldsymbol{Q} && \boldsymbol{S}_{33}=\boldsymbol{Q}^T\boldsymbol{D}_i'\boldsymbol{Q}\\
&\boldsymbol{S}_{21}=\boldsymbol{S}_{12}^T && \boldsymbol{S}_{31}=\boldsymbol{S}_{13}^T && \boldsymbol{S}_{32}=\boldsymbol{S}_{23}^T
\end{aligned}
\tag{4.317}
$$

and t_i is the layer thickness, t the total thickness, and $\bar{z}_i$ is the distance of the plate middle surface to the center of the layer. This formulation reflects the parabolic distribution of shear stress through the thickness. Also, similarly,

$$
\begin{aligned}
\boldsymbol{K}_{A1}^e &= \frac{1}{8A}\sum_{i=1}^{n}\left(h_{i+1}^2-h_i^2\right)[-\boldsymbol{H}\quad \boldsymbol{G}]^T\boldsymbol{D}_i[\boldsymbol{G}\quad \boldsymbol{H}]\\
&= \frac{1}{8A}\sum_{i=1}^{n}\left(h_{i+1}^2-h_i^2\right)\begin{bmatrix}-\boldsymbol{H}^T\boldsymbol{D}_i\boldsymbol{G} & -\boldsymbol{H}^T\boldsymbol{D}_i\boldsymbol{H}\\ \boldsymbol{G}^T\boldsymbol{D}_i\boldsymbol{G} & \boldsymbol{G}^T\boldsymbol{D}_i\boldsymbol{H}\end{bmatrix}
\end{aligned}
\tag{4.318}
$$

and

$$
\begin{aligned}
\boldsymbol{K}_{A2}^e &= \frac{1}{8A}\sum_{i=1}^{n}\left(h_{i+1}^2-h_i^2\right)[\boldsymbol{G}\quad \boldsymbol{H}]^T\boldsymbol{D}_i[-\boldsymbol{H}\quad \boldsymbol{G}]\\
&= \frac{1}{8A}\sum_{i=1}^{n}\left(h_{i+1}^2-h_i^2\right)\begin{bmatrix}-\boldsymbol{G}^T\boldsymbol{D}_i\boldsymbol{H} & \boldsymbol{G}^T\boldsymbol{D}_i\boldsymbol{G}\\ -\boldsymbol{H}^T\boldsymbol{D}_i\boldsymbol{H} & \boldsymbol{H}^T\boldsymbol{D}_i\boldsymbol{G}\end{bmatrix}
\end{aligned}
\tag{4.319}
$$

in which it may be noted that $\boldsymbol{K}_{A1}=\boldsymbol{K}_{A2}^T$.

The complete stiffness matrix of the composite element may now be appropriately combined as

$$
\begin{aligned}
\boldsymbol{K}^e &= \boldsymbol{K}_M+\boldsymbol{K}_B+\boldsymbol{K}_S+\boldsymbol{K}_{A1}+\boldsymbol{K}_{A2}\\
&= \begin{bmatrix}
\boldsymbol{K}_{M_{11}} & \boldsymbol{K}_{M_{12}} & \cdots & \boldsymbol{K}_{A2_{11}} & \boldsymbol{K}_{A2_{12}}\\
 & \boldsymbol{K}_{M_{22}} & \cdots & \boldsymbol{K}_{A2_{21}} & \boldsymbol{K}_{A2_{22}}\\
 & & \boldsymbol{K}_{S_{11}} & \boldsymbol{K}_{S_{12}} & \boldsymbol{K}_{S_{13}}\\
\text{sym} & & & \boldsymbol{K}_{S_{22}}+\boldsymbol{K}_{B_{11}} & \boldsymbol{K}_{S_{23}}+\boldsymbol{K}_{B_{12}}\\
 & & & & \boldsymbol{K}_{S_{33}}+\boldsymbol{K}_{B_{22}}
\end{bmatrix}
\end{aligned}
\tag{4.320}
$$

The preceding stiffness matrix may be rearranged finally nodewise by suitable interchange of rows and columns for augmentation in any general-purpose code. Stiffness pertaining to $u_{\theta z}^e$ DOF is automatically generated in the global coordinate system when element stiffness matrices are added nodeswise unless there is a 0-DOF constraints associated with the node. It may also be noted that for a single-layer homogeneous flat shell element, $\boldsymbol{K}_{A1}=\boldsymbol{K}_{A2}=\boldsymbol{0}$. This element ensures constant strain states and also rigid-body motion without stretching. It also displays six rigid-body modes. Furthermore in-plane stretching and the slopes are continuous within the element and across interelement boundaries. However, the out-of-plane deformation is not continuous at the element edges. This was not found to be detrimental to solution convergence in numerical studies.

Further improvement in the performance of the preceding element has been achieved by introducing the concept of antisymmetric bending mode. This procedure involves adjustment of the shearing stiffness such that adoption of a cubic bending mode ensures the correct value for the strain energy. Implementation of such a concept is achieved through adjustment in the values of coefficients in the $\boldsymbol{D}'$ matrix. Furthermore, sandwich elements having different material properties and thicknesses are easily incorporated into the preceding element by assuming $\boldsymbol{D}_M$, $\boldsymbol{D}_B$, and $\boldsymbol{D}'_S$ as the relevant material property matrices, t_M, t_B, t_S being corresponding thicknesses. Also a quadrilateral element may be conveniently derived by suitably combining four contiguous triangular elements with the vertex at the center of area of the triangle, as explained previously.

4.9 Numerical Examples

Two example problems are presented herein that verify the solution convergence characteristics.

Example 4.2: Simply Supported Plate

A simply supported plate with unit uniformly distributed loading is shown in Fig. 4.23 ($E = 10^7$, $\nu = 0.3$, thickness $t = 0.1$). Table 4.2 depicts the pattern of solution convergence as a function of mesh size.

Example 4.3: Cylindrical Shell

Figure 4.24 depicts a cantilever cylindrical shell under unit uniform loading ($E = 29.5 \times 10^6$, $\nu = 0.3$, $t = 0.1$, $A = B = 10$, $R = 2.0$). Table 4.3 presents the solution results.

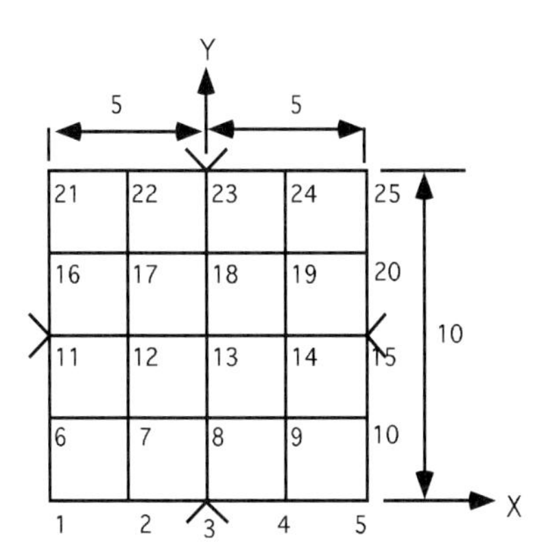

Fig. 4.23 Simply supported plate.

Table 4.2 Solution convergence of the plate problem

Mesh size	Center deflection
4×4	0.04287
8×8	0.04447
16×16	0.04469
32×32	0.04467
Exact value	0.04540

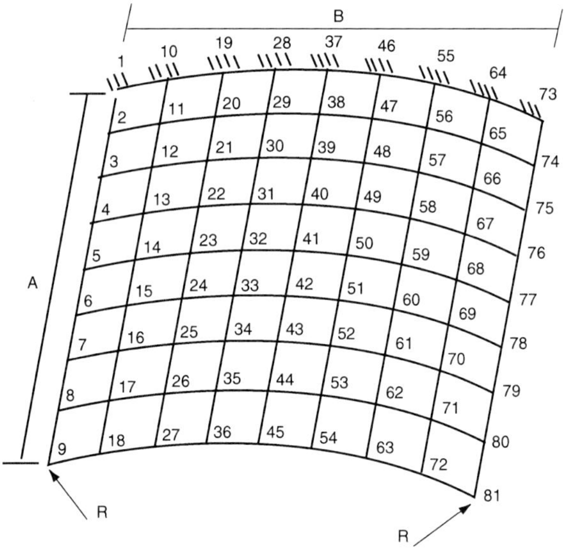

Fig. 4.24 Cantilever cylindrical shell.

Table 4.3 Solution convergence for the cantilever cylindrical shell

Mesh size	Maximum tip deflection
8×8	0.24157×10^{-1}
16×16	0.24283×10^{-1}
32×32	0.24319×10^{-1}

4.10 Concluding Remarks

The finite element approximations for the stiffness and mass matrices for various domains were given in this chapter for a variety of simple and practical cases. For plane, three dimensional, and axisymmetric bodies these elements are relatively simple and are commonly used in practice. For plates and shells a wide variety of elements are available, and so the elements included in the chapter were limited. These elements illustrated the basic concepts used for the various structural types. No mention was made of any of the reduced or selectively integrated elements for plate bending or shells as these are specialty items and can more than likely be found in commercially available software. Once the basic stiffness and mass matrices have been developed it is possible to use them in the static and dynamic analysis required in the multidisciplinary studies of complex structures discussed in this book.

References

[1]Livesley, R. K., *Matrix Methods of Structural Analysis*, Pergamon, New York, 1964.

[2]Przemieniecki, J. C., *Theory of Matrix Structural Analysis*, McGraw–Hill, New York, 1968.

[3]Turner, M. J., Clough, R. W., Martin, H. C., and Topp, L. T., "Stiffness and Deflection Analysis of Complex Structures," *Journal of the Aerospace Sciences*, Vol. 25, No. 9, 1956, pp. 805–823.

[4]Clough, R. W., "The Finite Element Method in Plane Stress Analysis," *Proceedings of the 2nd ASCE Conference on Electronic Computation*, American Society of Civil Engineers, New York, 1960.

[5]Fraeijs de Veubeke, B., "Displacement and Equilibrium Models in the Finite Element Method," *Stress Analysis*, edited by Zienkiewicz and Hollister, Wiley, New York, 1965, Chap. 9.

[6]Argyris, J. H., "Triangular Elements with Linearly Varying Strain for the Matrix Displacement Method," *Journal of the Royal Aeronautical Society*, Vol. 69, 1965, pp. 711–713.

[7]Felippa, C. A., "Refined Finite Element Analysis of Linear and Nonlinear Two-Dimensional Structures," Rept. 66–22, Univ. of California at Berkeley, Berkeley, CA 1966.

[8]Argyris, J. H., "Energy Theorems and Structural Analysis," *Aircraft Engineering*, Vols. 26, 27, Oct. 1954, May 1955.

[9]Clough, R. W., and Rashid, Y. R., "Finite Element Analysis of Axi-Symmetric Solids," *Proceedings of the American Society of Civil Engineers*, Vol. 91, EM. 1, 71, 1965.

[10]Wilson, E. L., "Structural Analysis of Axi-Symmetric Solids," *AIAA Journal* Vol. 3, 1965.

[11]Melosh, R. J., "Structural Analysis of Solids," *J. Str. Div. ASCE*, Vol. 89, No. ST4, 1963, pp. 205–223.

[12]Argyris, J. H., "Matrix Analysis of Three Dimensional Elastic Media," *AIAA Journal*, Vol. 3, No. 1, 1965, pp. 45–51.

[13]Irons, B. M., "Engineering Applications of Numerical Integration in Stiffness Methods," *AIAA Journal*, Vol. 4, No. 11, 1966, pp. 2035–37.

[14]Ergatoudis, J., Irons, B. M., and Zienkiewicz, O. C., "Three Dimensional Stress Analysis of Arch Dams and Their Foundations," *Proceedings of the Symposium on Arch Dams*, Inst. of Civil Engineering, London, 1968, pp. 37–50.

[15]Argyris, J. H., Fried, I., and Scharpf, D. W., "The Tuba Family of Plate Elements for the Matrix Displacement Method," Tech. Note, *Aeronautical Journal*, Vol. 72, 1968, pp. 701–709.

[16]Bell, K. A., "A Refined Triangular Plate Bending Finite Element," *International Journal for Numerical Methods in Engineering*, Vol. 1. No. 1, 1969, pp. 101–122.

[17]Clough, R.W., and Felippa, C. A., "A Refined Quadrilateral Element for Analysis of Plate Bending," *Proceedings of the 2nd Conference on Matrix Methods in Structural Mechanics*, Wright–Patterson AFB, OH, 1968.

[18]Argyris, J. H., "Continua and Discontinua," *Proceedings of the 1st Conference on Matrix Methods in Structural Mechanics*, Wright–Patterson AFB, OH, 1965.

[19]Morley, L. S. D., "On the Constant Moment Plate Bending Element," *Journal of Strain Analysis*, Vol. 6, 1971, pp. 20–24.

[20]Bazeley, G. P., Cheung, Y. K., Irons, B. M., and Zienkiewicz, O. C., "Triangular Elements in Plate Bending-Conforming and Nonconforming Solutions," *Proceedings of the 1st Conference on Matrix Methods in Structural Mechanics*, Wright–Patterson AFB, OH, 1965.

[21]Reissner, E., "On Some Problems in Shell Theory," *Proceedings of the 1st Symposium on Naval Structural Mechanics*, Pergamon, New York, 1969, pp. 74–114.

[22]Koiter, W. T., "A Consistent First Approximation in the General Theory of Thin Elastic Shells," *Theory of Thin Elastic Shells*, edited by W. Koiter, North-Holland, Amsterdam, 1960, pp. 12–23.

[23]Naghdi, P. M., "Foundations of Elastic Shell Theory," *Progress in Solid Mechanics*, edited by A. H. Sneddon, Vol. 4, North-Holland, Amsterdam, 1963, pp. 1–90.

[24]Novozhilov, V. V., *The Theory of Thin Shells*, 2nd ed., Noordhoff, Groningen, 1964.

[25]Kraus, H., *Thin Elastic Shells*, Wiley, New York, 1969.

[26]Ahmad, S., Irons, B. M., and Zienkiewicz, O. C., "Analysis of Thick and Thin Shell Structures by Curved Finite Elements," *International Journal for Numerical Methods in Engineering*, Vol. 2, 1970, pp. 419–51.

[27]Greene, B. E., Strome, D. R., and Weikel, R. C., "Application of the Stiffness Method to the Analysis of Shell Structures," *Proceedings of the Aviation Conference*, American Society of Mechanical Engineers, New York, 1961.

[28]Zienkiewicz, O. C., and Cheung, Y. K., "Finite Element Method of Analysis of Arch Dam Shells and Comparison with Finite Difference Procedures," *Proceedings of the Symposium on Theory of Arch Dams*, Pergamon, New York, 1965.

[29]Gallagher, R. H., Gellatly, R. A., Padlog, J., and Mallet, R. H., "A Discrete Element Procedure for Thin Shell Instability Analysis," *AIAA Journal*, Vol. 5, No. 1, 1967, pp. 138–145.

[30]Johnson, C. P., "The Analysis of Thin Shells by a Finite Element Procedure," Ph.D. Dissertation, Rept. SESM 67–22, Univ. of California, Berkeley, 1967.

[31]Yeh, C., "Large Deflection Dynamic Analysis of Thin Shells Using the Finite Element Method," Ph.D. Thesis, Rept. SESM 70–18, Univ. of California, Berkeley, 1970.

[32]Clough, R. W., and Tocher, J. L., "Analysis of Thin Arch Dams by the Finite Element Method," *Theory of Arch Dams*, edited by J. R. Rydzewski, Pergamon, Oxford, England, U.K., 1965, pp. 107–21.

[33]Petyt, M., "The Application of Finite Element Techniques to Plate and Shell Problems," ISVR Report. 120, Inst. of Sound and Vibration, Univ. of Southampton, England, U.K., 1965.

[34]Argyris, J. H., "Matrix Displacement Analysis of Anisotropic Shells by Triangular Elements," *Journal of the Royal Aeronautical Society*, Vol. 69, 1965, pp. 801–805.

[35]Clough, R. W., and Johnson, C. P., "A Finite Element Approximation for the Analysis of Thin Shells," *International Journal of Solids and Structures*, Vol. 4, 1968, pp. 43–60.

[36]Zienkiewicz, O. C., Parekh, C. J., and King, I. P., "Arch Dam Analysis by a Linear Finite Element Shell Solution Program," *Proceedings of the Symposium on Arch Dams*, edited by T. L. Dennis, Inst. of Civil Engineering, London, 1968, pp. 19–22.

[37]Carr, A. J., "A Refined Finite Element Analysis of Thin Shell Structures Including Dynamic Loadings," Rept. SESM 67-9, Univ. of California, Berkeley, 1967.

[38]Chu, T. C., and Schnobrich, W. C., "Finite Element Analysis of Translational Shells," *Computers and Structures*, Vol. 2, 1972, pp. 197–222.

[39]Razzaque, A., "Finite Element Analysis of Plates and Shells," Ph.D. Dissertation, Dept. of Civil Engineering, Univ. of Wales, Swansea, Wales, U.K., 1972.

[40]Razzaque, A., "Program for Triangular Bending Element with Derivative Smoothing," *International Journal for Numerical Methods in Engineering*, Vol. 5, 1973, pp. 588, 589.

[41]Olsen, M. D., and Beardon, T. W., "The Simple Flat Triangular Shell Element Revisited," *International Journal for Numerical Methods in Engineering*, Vol. 14, 1979, pp. 51–68.

[42]Dawe, D. J., "Shell Analysis Using a Facet Element," *Journal of Strain Analysis*, Vol. 7, 1972, pp. 266–270.

[43]Herrmann, L. R., and Campbell, D. M., "A Finite Element Analysis for Thin Shells," *AIAA Journal*, Vol. 6, 1968, pp. 1842–1847.

[44]Dungar, R., Severn, R. T., and Taylor, P. R., "Vibration of Plate and Shell Structures Using Triangular Finite Elements," *Journal of Strain Analysis*, Vol. 2, 1967, pp. 73–83.

[45]Dungar, R., and Severn, R. T., "Triangular Finite Elements of Variable Thickness and Their Application to Plate and Shell Structures," *Journal of Strain Analysis*, Vol. 4, 1969, pp. 10–20.

[46]Yoshida, Y., "A Hybrid Stress Element for Thin Shell Analysis," *Proceedings of the International Conference on Finite Element Methods in Engineering*, Univ. of New South Wales, Australia, 1974, pp. 271–284.

[47]Sander, G., and Becker, P., "Delinquent Finite Elements for Shell Idealization," *Proceedings of the World Congress on Finite Element Methods in Structural Mechanics*, Vol. 2, Bournemouth, England, U.K., 1975, pp. 1–31.

[48]Argyris, J. H., Dunne, P. C., Malejannakis, G. A., and Schelke, E., "A Simple Triangular Facet Shell Element with Application to Linear and Nonlinear Equilibrium and Elastic Stability Problems," *Computer Methods in Applied Mechanics and Engineering*, Vol. 10, 1977, pp. 371–403.

[49]Horrigmoe, G., and Bergan, P. G., "Non-Linear Analysis of Free-Form Shells," Div. of Structural Mechanics, Univ. of Trondheim, Norway, 1977.

[50]Batoz, J. L., Bathe, K. J., and Ho, L. W., "A Study of Three-Node Triangular Plate Bending Elements," *International Journal for Numerical Methods in Engineering*, Vol. 15, 1980, pp. 1771–1812.

[51]Bathe, K. J., and Ho, L. W., "A Simple and Effective Element for Analysis of General Shell Structures," *Computers and Structures*, Vol. 13, 1987, pp. 673–81.

[52]Meek, J. L., *Computer Methods in Structural Analysis*, E&F Sponn, London, 1991.

[53]Zienkiewicz, O. C., and Taylor, R. L., *The Finite Element Method*, Vols. 1 and 2, 4th ed., McGraw–Hill, London, 1991.

[54]Utku, S., "Stiffness Matrices for Thin Triangular Elements of Nonzero Gaussian Curvature," *AIAA Journal*, Vol. 5, No. 7, 1967, pp. 1659–1667.

[55]Martin, C. W., Lung, S. F., and Gupta, K. K., "A Three-Node C^o Element for Analysis of Laminated Composite Sandwich Shells," NASA TM 4125, June 1989.

5
Spinning Structures

5.1 Introduction

Occurrence of spinning systems is common in aerospace and mechanical engineering, among other disciplines. Relevant structures may be made of a combination of nonspinning and spinning components; the spinning components may rotate around any arbitrary axes with differing speeds of rotation. An accurate free vibration analysis is a vital prerequisite for subsequent dynamic response and aeroservoelastic analysis of such structural systems. Such an analysis is especially complicated for the case of spinning structures. Examples for the relevant structures include helicopters, spacecraft, and rotating machinery.

5.2 Derivation of Equation of Motion

Figure 5.1 depicts the elastic deformation $\boldsymbol{u}_j$ of a point j in a flexible structure rotating at a constant angular velocity Ω in an arbitrary axis having components $\Omega_X, \Omega_Y, \Omega_Z$ along the reference coordinate system axes. Then the position, velocity, and acceleration vectors are defined as

$$\boldsymbol{r} = \boldsymbol{r}_j + \boldsymbol{u}_j \tag{5.1}$$

$$\boldsymbol{v} = \frac{\partial \boldsymbol{r}}{\partial t} + \boldsymbol{\Omega} \times \boldsymbol{r} \tag{5.2}$$

$$\begin{aligned} \boldsymbol{a} &= \frac{\partial \boldsymbol{v}}{\partial t} + \boldsymbol{\Omega} \times \boldsymbol{v} \\ &= \frac{\partial^2 \boldsymbol{r}}{\partial t^2} + \dot{\boldsymbol{\Omega}} \times \boldsymbol{r} + 2\boldsymbol{\Omega} \times \dot{\boldsymbol{r}} + \boldsymbol{\Omega} \times (\boldsymbol{\Omega} \times \boldsymbol{r}) \end{aligned} \tag{5.3}$$

in which the second term in the right-hand side of Eq. (5.3) is zero for steady spin rate and the third and fourth terms relate to Coriolis and centripetal accelerations, respectively; also

$$\boldsymbol{\Omega r} = \begin{bmatrix} 0 & -\Omega_Z & \Omega_Y \\ \Omega_Z & 0 & -\Omega_X \\ -\Omega_Y & \Omega_X & 0 \end{bmatrix} \begin{Bmatrix} r_X \\ r_Y \\ r_Z \end{Bmatrix} \tag{5.4}$$

Further, the first term of Eq. (5.3) relates to effects of in-plane stretching on out-of-plane deformation. In the absence of damping, the governing equations of motion for the entire structure can be derived using the preceding equation in formulating strain and kinetic energies and using Hamilton's principle as

$$\boldsymbol{M}\ddot{\boldsymbol{u}} + 2\boldsymbol{M\Omega}\dot{\boldsymbol{u}} + (\boldsymbol{K}_E + \boldsymbol{M\Omega\Omega})\boldsymbol{u} = -\boldsymbol{M\Omega\Omega r} \tag{5.5}$$

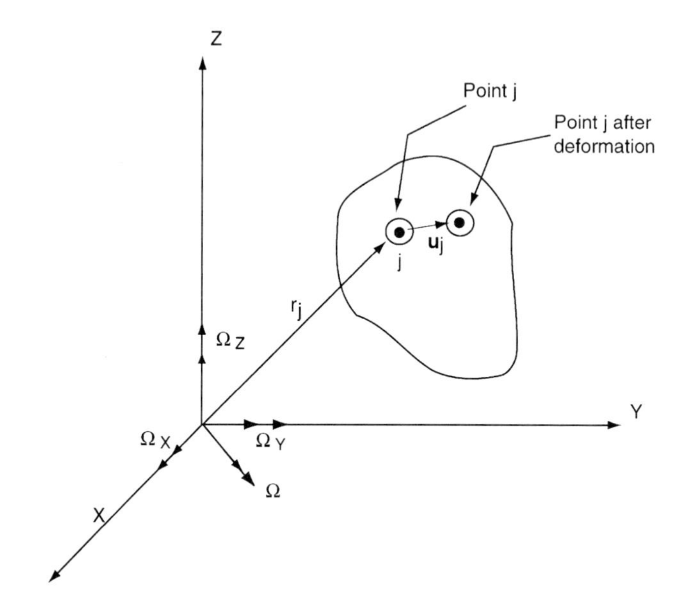

Fig. 5.1 Flexible structure subjected to arbitrary spinning motion.

or

$$\boldsymbol{M}\ddot{\boldsymbol{u}} + \boldsymbol{C}_c\dot{\boldsymbol{u}} + (\boldsymbol{K}_E + \boldsymbol{K}')\boldsymbol{u} = \boldsymbol{f}_c \tag{5.6}$$

with $\boldsymbol{r}$ representing all $\boldsymbol{r}_j$, the term on the right-hand side being the centrifugal force vector.

The analysis starts with computing steady-state deflected position because of constant spin by solving

$$(\boldsymbol{K}_E + \boldsymbol{K}')\boldsymbol{u} = \boldsymbol{f}_c \tag{5.7}$$

and the geometric stiffness matrix $\boldsymbol{K}_G$ is calculated next by taking into account the internal loads introduced by constant spin Ω. The position for all j points may also be updated if significant, and $\boldsymbol{K}_E$ and $\boldsymbol{M}$ matrices recalculated if necessary. The equation of free vibration may then be written as

$$\boldsymbol{M}\ddot{\boldsymbol{u}} + \boldsymbol{C}_c\dot{\boldsymbol{u}} + (\boldsymbol{K}_E + \boldsymbol{K}' + \boldsymbol{K}_G)\boldsymbol{u} = \boldsymbol{0} \tag{5.8}$$

or

$$\boldsymbol{M}\ddot{\boldsymbol{u}} + \boldsymbol{C}_c\dot{\boldsymbol{u}} + \boldsymbol{K}\boldsymbol{u} = \boldsymbol{0} \tag{5.9}$$

An appropriate solution of this equation yields the natural frequencies and mode shapes of the spinning structural system. In the presence of damping, Eq. (5.9) takes the following form:

$$\boldsymbol{M}\ddot{\boldsymbol{u}} + (\boldsymbol{C}_c + \boldsymbol{C}_d)\dot{\boldsymbol{u}} + \boldsymbol{K}(1 + i^*g)\boldsymbol{u} = \boldsymbol{0} \tag{5.10}$$

in which $\boldsymbol{C}_d$ is the viscous damping matrix and g is the structural damping parameter, with i^* being the imaginary member $\sqrt{-1}$.

To be able to perform relevant dynamic analysis, it is first essential to evolve the element centrifugal force and geometric stiffness matrices as well as the Coriolis matrix and the usual elastic stiffness and inertia matrices. To derive such matrices, it is necessary to evolve expressions for nodal centrifugal forces of various finite elements spinning about an arbitrary axis. The generated in-plane element nodal forces are then assembled in the global coordinate system and a subsequent static analysis yields element stresses and forces that in turn are used for the generation of structural geometrical stiffness matrix $\boldsymbol{K}_G^e$. The out-of-plane components of the element centrifugal force vector, on the other hand, are used for the derivation of the related centripetal stiffness matrix $\boldsymbol{K}'^e$. Examples of these matrices are derived next for some typical finite elements.

5.3 Derivation of Nodal Centrifugal Forces

5.3.1 Flat Shell Element

A general formulation for the centrifugal forces at the nodes of a triangular finite element, generated as the result of an arbitrary spin rate vector, is derived next,[1] such results being valid for flat shell, plate, and plane elements. Expressions for in-plane forces are developed first and are followed by similar derivations for out-of-plane components. A simple extension of the formulation for a quadrilateral shell element and similar expressions for line elements are also described in some detail. An earlier work[2] in this connection involved triangular plate bending finite elements spinning about an axis perpendicular to the plane of the rotating plate. Another such effort with isoparametric solid elements is described in Ref. 3.

Figure 5.2 shows a typical flat triangular shell finite element defined by vertices i, j, and k, rotating about an arbitrary axis with a uniform spin rate Ω_R that represents

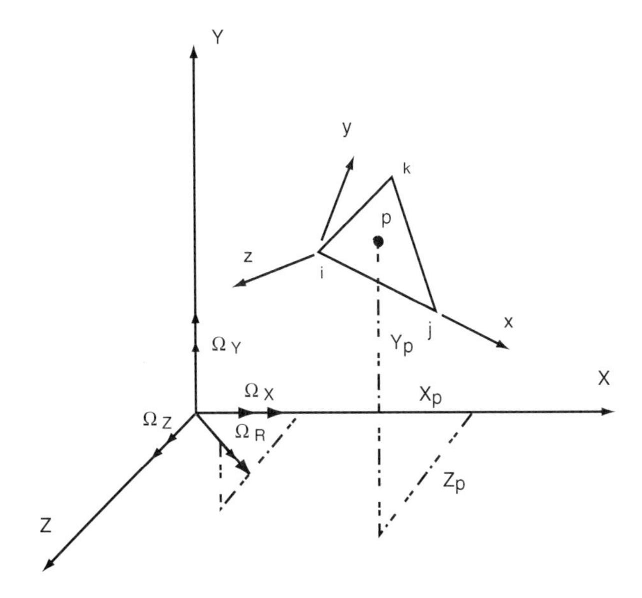

Fig. 5.2 Triangular shell element subjected to arbitrary spin rate.

a resultant spin vector and that can be resolved into components Ω_X, Ω_Y, and Ω_Z along global reference axes X, Y, and Z, respectively. It may be noted in this connection that the position of the element in space and the direction of the spin axis are quite arbitrary. The shape function matrix $\boldsymbol{N}$, pertaining to the in-plane motion of the element, may be expressed as

$$\boldsymbol{u} = \boldsymbol{N}\boldsymbol{u}^e \tag{5.11}$$

where

$$\boldsymbol{N} = \boldsymbol{R}\boldsymbol{Q}^{-1} \tag{5.12}$$

and in which $\boldsymbol{u}$ is the displacement vector of a typical point P within the element, expressed in the local coordinate system (LCS) defined by the x, y, and z axes, $\boldsymbol{u}^e$ is the element nodal displacement vector in the LCS, $\boldsymbol{R}$ is the portion of the shape function matrix having elements that are functions of the local coordinates x and y, and $\boldsymbol{Q}$ is the portion of the shape function matrix with elements expressed in terms of the element nodal coordinate values in the LCS. Also, defining the internal force vector at any point within the element in the LCS as

$$\boldsymbol{f} = [f_x \quad f_y \quad f_z]^T \tag{5.13}$$

the equivalent concentrated nodal forces of a triangular element in the LCS may be defined by

$$\boldsymbol{f}^e = [f_{x1} \quad f_{y1} \quad f_{z1} \quad \cdots \quad f_{x3} \quad f_{y3} \quad f_{z3}]^T \tag{5.14}$$

which may also be expressed in the following form:

$$\boldsymbol{f}^e = \int_v \boldsymbol{N}^T \boldsymbol{f} \, \mathrm{d}v \tag{5.15}$$

or

$$\boldsymbol{f}^e = \int_v \boldsymbol{N}^T \boldsymbol{\lambda} \boldsymbol{p} \, \mathrm{d}v \tag{5.16}$$

where $\boldsymbol{p}$ represents the forces in the global coordinate system at any point within an element, having three components p_X, p_Y, and p_Z in the X, Y, and Z directions, respectively, and $\boldsymbol{\lambda}$ is the element direction cosine matrix:

$$\boldsymbol{\lambda} = \begin{bmatrix} l_x & m_x & n_x \\ l_y & m_y & n_y \\ l_z & m_z & n_z \end{bmatrix} \tag{5.17}$$

where $l_x = \cos(X, x)$, and so on. Furthermore, defining X_i, Y_i, and Z_i as the global coordinates of node i of the triangular element, such coordinates for any point within the element may simply be obtained as

$$\begin{bmatrix} X \\ Y \\ Z \end{bmatrix} = \begin{bmatrix} X_i \\ Y_i \\ Z_i \end{bmatrix} + \boldsymbol{\lambda}^T \begin{bmatrix} x \\ y \\ 0 \end{bmatrix} \tag{5.18}$$

5.3.1.1 In-plane forces. The in-plane element nodal centrifugal forces are derived next for each of the three components of the spin rate Ω_R that may then be combined to yield the final desired expressions. Thus, corresponding to the spin component Ω_X, the appropriate forces are expressed as

$$\boldsymbol{p} = \rho \Omega_X^2 \begin{bmatrix} 0 \\ Y \\ Z \end{bmatrix} \tag{5.19}$$

where ρ is the mass density. Equation (5.19), upon the employment of the relationships expressed by Eqs. (5.17) and (5.18), takes the following form:

$$\boldsymbol{p} = \rho \Omega_X^2 \begin{bmatrix} 0 \\ m_x x + m_y y + Y_i \\ n_x x + n_y y + Z_i \end{bmatrix} \tag{5.20}$$

The desired expression for the element nodal forces is then derived by appropriate substitution of Eqs. (5.12), (5.17), and (5.20) into Eq. (5.16), yielding

$$\boldsymbol{f}_p^e(\Omega_X) = \rho \Omega_X^2 t [\boldsymbol{Q}^{-1}]^T \int_0^{y_k} \int_{x_l}^{x_u} \boldsymbol{R}^T \boldsymbol{\chi} \, \mathrm{d}x \, \mathrm{d}y \tag{5.21}$$

where t is the element thickness, y_k is the y coordinate of node k, $x_l = x_k y / y_k$ is the lower bound of the x coordinate in the integration process, $x_u = x_j - (x_j - x_k) y / y_k$ is the upper bound of the x coordinate, and

$$\boldsymbol{\chi} = \begin{bmatrix} M_x(M_x x + M_y y + Y_i) + N_x(N_x x + N_y y + Z_i) \\ M_y(M_x x + M_y y + Y_i) + N_y(N_x x + N_y y + Z_i) \end{bmatrix} \tag{5.22}$$

In Eq. (5.21) the vector $\boldsymbol{f}_p^e$ contains the in-plane element nodal forces in local x and y directions, being expressed as $[f_{x_i}\, f_{y_i}\, f_{x_j}\, f_{y_j}\, f_{x_k}\, f_{y_k}]^T$.

Similar formulations are derived for forces because of spin rates in the Y and Z directions that are expressed as

$$\boldsymbol{f}_p^e(\Omega_Y) = \rho \Omega_Y^2 t [\boldsymbol{Q}^{-1}]^T \int_0^{y_k} \int_{x_l}^{x_u} \boldsymbol{R}^T \boldsymbol{\gamma} \, \mathrm{d}x \, \mathrm{d}y \tag{5.23}$$

$$\boldsymbol{f}_p^e(\Omega_Z) = \rho \Omega_Z^2 t [\boldsymbol{Q}^{-1}]^T \int_0^{y_k} \int_{x_l}^{x_u} \boldsymbol{R}^T \boldsymbol{\xi} \, \mathrm{d}x \, \mathrm{d}y \tag{5.24}$$

in which

$$\boldsymbol{\gamma} = \begin{bmatrix} L_x(L_x x + L_y y + X_i) + N_x(N_x x + N_y y + Z_i) \\ L_y(L_x x + L_y y + X_i) + N_y(N_x x + N_y y + Z_i) \end{bmatrix} \tag{5.25}$$

$$\boldsymbol{\xi} = \begin{bmatrix} L_x(L_x x + L_y y + X_i) + M_x(M_x x + M_y y + Y_i) \\ L_y(L_x x + L_y y + X_i) + M_y(M_x x + M_y y + Y_i) \end{bmatrix} \tag{5.26}$$

The total element in-plane nodal forces are simply obtained as

$$\boldsymbol{f}_p^e = \boldsymbol{f}_p^e(\Omega_X) + \boldsymbol{f}_p^e(\Omega_Y) + \boldsymbol{f}_p^e(\Omega_Z) \tag{5.27}$$

which may then be transformed into the global coordinate system as follows:

$$\boldsymbol{p}^e = \boldsymbol{\lambda}^T \boldsymbol{f}_p^e \tag{5.28}$$

to be used for a stress analysis of the entire structure for subsequent generation of $\boldsymbol{K}_G$.

5.3.1.2 Out-of-plane forces. The normal components of the element centrifugal force vector are generated in the same manner as the in-plane forces by consideration of an appropriate shape function pertaining to a plate bending element. Thus, at any point within the element, the normal displacement in the LCS may be expressed as

$$u_z = \boldsymbol{dc} \tag{5.29}$$

in which $\boldsymbol{c}$ is a vector of size (9×1) having the coefficients of the displacement function, whereas $\boldsymbol{d}$ is a (1×9) vector with elements expressed as functions of the local coordinates x and y. Such coefficients may be evaluated in terms of all nine nodal displacements and slopes, leading to the matrix equation

$$\boldsymbol{u}^e = \boldsymbol{Cc} \tag{5.30}$$

in which the elements of the (9×9) $\boldsymbol{C}$ matrix are expressed in terms of the local coordinates of the three nodes. Combining Eqs. (5.29) and (5.30), the following relationship is obtained:

$$u_z = \boldsymbol{N}_z \boldsymbol{u}^e \tag{5.31}$$

in which

$$\boldsymbol{N}_z = \boldsymbol{dC}^{-1} \tag{5.32}$$

The nodal centrifugal forces may next be derived for each of the three spin components when the third row of Eq. (5.18) is defined as $Z = Z_i + \boldsymbol{\lambda}^T u_z$ to take into account the effect of out-of-plane deformation. Thus, for the spin component Ω_X, the out-of-plane force vector having nine components is expressed as

$$\boldsymbol{f}_b(\Omega_X) = \rho\Omega_X^2 \int_v \boldsymbol{N}_z^T \left[\alpha + \left(N_z^2 + M_z^2\right) u_z\right] \mathrm{d}v \tag{5.33}$$

in which

$$\alpha = [M_z(M_x x + M_y y + Y_i) + N_z(N_x x + N_y y + Z_i)] \tag{5.34}$$

The first term on the right-hand side of Eq. (5.33) does not depend on displacement, and hence has no effect on the natural frequencies. Thus, retaining only the second term and using Eq. (5.31) the following relationship is obtained:

$$\boldsymbol{f}_b(\Omega_X) = \boldsymbol{K}'^e(\Omega_X)\boldsymbol{u} \tag{5.35}$$

where the element centrifugal force matrix is defined as

$$\boldsymbol{K}'^e(\Omega_X) = \Omega_X^2 \left(N_z^2 + M_z^2\right)\boldsymbol{M} \tag{5.36}$$

where $\boldsymbol{M}$ is the element inertia matrix. Similar expressions are obtained pertaining to the Ω_Y and Ω_Z spin rates:

$$\boldsymbol{K}'^{e}(\Omega_Y) = \Omega_Y^2\left(N_z^2 + L_z^2\right)\boldsymbol{M} \tag{5.37}$$

$$\boldsymbol{K}'^{e}(\Omega_Z) = \Omega_Z^2\left(L_z^2 + M_z^2\right)\boldsymbol{M} \tag{5.38}$$

The complete expression for the matrix is obtained simply as $\boldsymbol{K}'^{e} = \boldsymbol{K}'^{e}(\Omega_X) + \boldsymbol{K}'^{e}(\Omega_Y) + \boldsymbol{K}'^{e}(\Omega_Z)$ that when appropriately combined yields the global centrifugal force matrix $\boldsymbol{K}'$ for the entire structure, as defined in Eq. (5.6).

For quadrilateral flat shell elements, the preceding procedure is used to find the nodal centrifugal forces for each of the four constituent triangular elements typically formed by any two nodes joining a node situated at the center of an area of the quadrilateral. Such forces are next suitably combined to yield the quadrilateral element nodal centrifugal forces in the global coordinate system.

5.3.2 Line Elements

Assuming that the local x axis of the element[1,4,5] is defined by joining node 1 (i) to node 2 (j) (see Fig. 5.3), the coordinates of any point in the element may be obtained as before:

$$\begin{bmatrix} X \\ Y \\ Z \end{bmatrix} = \begin{bmatrix} X_i \\ Y_i \\ Z_i \end{bmatrix} + \boldsymbol{\lambda}^T \begin{bmatrix} x \\ u_y \\ u_z \end{bmatrix} \tag{5.39}$$

in which u_y and u_z define deformations of any point on the element in the local y and z directions, respectively. Proceeding in the same manner as with the triangular flat

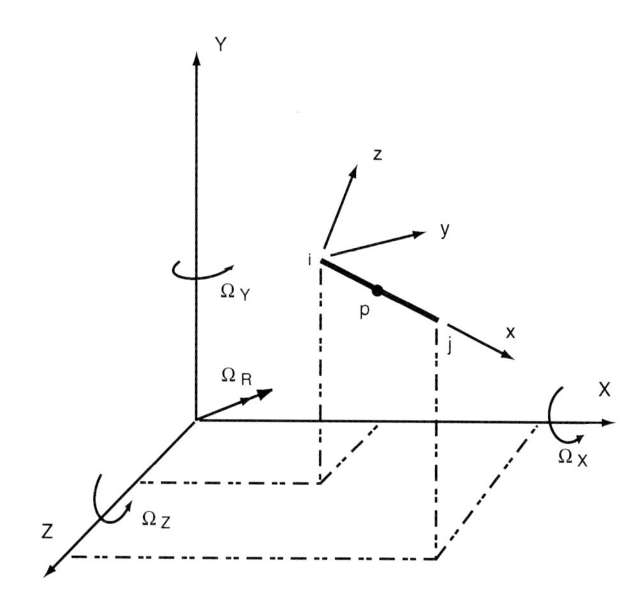

Fig. 5.3 Line element subjected to arbitrary spin rate.

shell element, the force vector defined by Eq. (5.13) at a point within the element may be derived as

$$f(\Omega_x) = \rho\Omega_x^2[\alpha_1 \quad \alpha_2 \quad \alpha_3]^T \tag{5.40}$$

with

$$\alpha_1 = M_x(M_x x + YI) + N_x(N_x x + ZI)$$
$$\alpha_2 = M_y(M_x x + YI) + N_y(N_x x + ZI)$$
$$\alpha_3 = M_z(M_x x + YI) + N_z(N_x x + ZI)$$
$$+ M_z(M_y u_y + M_z u_z) + N_z(N_y u_y + N_z u_z)$$

from which the element nodal forces are obtained by employing Eq. (5.15) when appropriate shape functions are used to compute the element in-plane forces for the determination of the $\boldsymbol{K}_G$ matrix as well as to determine the effect of the out-of-plane forces in the shape of the $\boldsymbol{K}'$ matrices. Similar expressions for the force vector because of the other spin rates are

$$f(\Omega_Y) = \rho\Omega_Y^2[\beta_1 \quad \beta_2 \quad \beta_3]^T \tag{5.41}$$

$$f(\Omega_Z) = \rho\Omega_Z^2[\gamma_1 \quad \gamma_2 \quad \gamma_3]^T \tag{5.42}$$

in which

$$\beta_1 = L_x(L_x x + XI) + N_x(N_x x + ZI)$$
$$\beta_2 = L_y(L_x x + XI) + N_y(N_x x + ZI)$$
$$\beta_3 = L_z(L_x x + XI) + N_z(N_x x + ZI)$$
$$+ L_z(L_y u_y + L_z u_z) + N_z(N_y u_y + N_z u_z)$$
$$\gamma_1 = L_x(L_x x + XI) + M_x(M_x x + YI)$$
$$\gamma_2 = L_y(L_x x + XI) + M_y(M_x x + YI)$$
$$\gamma_3 = L_z(L_x x + XI) + M_z(M_x x + YI)$$
$$+ L_z(L_y u_y + L_z u_z) + M_z(M_y u_y + M_z u_z)$$

For line elements with equivalent nodal lumped masses, the centrifugal forces at a typical ith node expressed in the global coordinate system are obtained as follows:

$$\begin{aligned} p_X^i &= m\Omega_Y^2 X_i + m\Omega_Z^2 X_i \\ p_Y^i &= m\Omega_X^2 Y_i + m\Omega_Z^2 Y_i \\ p_Z^i &= m\Omega_X^2 Z_i + m\Omega_Y^2 Z_i \end{aligned} \tag{5.43}$$

in which m is the appropriate lumped mass at the node under consideration.

Once the element nodal centrifugal force vectors have been derived for the various elements by the foregoing procedures, the element in-plane stresses and axial forces for shell and line elements, respectively, are next obtained by solving $\boldsymbol{K u} = \boldsymbol{p}$ where the vector $\boldsymbol{p}$ is derived by combining the forces $\boldsymbol{f}_P$ for all structural elements. The matrices $\boldsymbol{K}_G^e$ are then derived by standard procedures, which are then suitably combined to yield the $\boldsymbol{K}_G$ matrix for the entire structure in the global coordinate system.[6]

5.4 Derivation of Element Matrices

Following the derivation of Eq. (5.8), the equation of free vibration of an element may be written as

$$\boldsymbol{M}^e \ddot{\boldsymbol{u}} + \boldsymbol{C}_c^e \dot{\boldsymbol{u}} + \left(\boldsymbol{K}^e + \boldsymbol{K}'^e + \boldsymbol{K}_G^e\right)\boldsymbol{u} = \boldsymbol{0} \tag{5.44}$$

in which the inertia and elastic stiffness matrices $\boldsymbol{M}^e$ and $\boldsymbol{K}^e$ have been derived earlier for various finite elements. The other matrices are derived as follows using the LCS.

5.4.1 *Coriolis Acceleration Matrix*

This matrix for any structural element is expressed as

$$\boldsymbol{C}_c^e = 2\boldsymbol{M}^e \boldsymbol{\Omega}^e \tag{5.45}$$

and may be easily derived for any element as the element inertia matrix is calculated by employing standard formulation.[6]

5.4.2 *Centripetal Acceleration Matrix*

Expression for the matrix is as follows:

$$\boldsymbol{K}'^e = \boldsymbol{M}^e \boldsymbol{\Omega}^e \boldsymbol{\Omega}^e \tag{5.46}$$

which may again be computed routinely for any element.

5.4.3 *Geometric Stiffness Matrix*

Development of this matrix[6] uses nonlinear total strain displacement relation in an elastic continuum:

$$\epsilon_x = \frac{\partial u_x}{\partial x} + \frac{1}{2}\left[\left(\frac{\partial u_x}{\partial x}\right)^2 + \left(\frac{\partial u_y}{\partial x}\right)^2 + \left(\frac{\partial u_z}{\partial x}\right)^2\right]$$

$$\epsilon_y = \frac{\partial u_y}{\partial y} + \frac{1}{2}\left[\left(\frac{\partial u_x}{\partial y}\right)^2 + \left(\frac{\partial u_y}{\partial y}\right)^2 + \left(\frac{\partial u_z}{\partial y}\right)^2\right]$$

$$\epsilon_z = \frac{\partial u_z}{\partial z} + \frac{1}{2}\left[\left(\frac{\partial u_x}{\partial z}\right)^2 + \left(\frac{\partial u_y}{\partial z}\right)^2 + \left(\frac{\partial u_z}{\partial z}\right)^2\right]$$

$$\epsilon_{xy} = \frac{\partial u_y}{\partial x} + \frac{\partial u_x}{\partial y} + \frac{\partial u_x}{\partial x}\frac{\partial u_x}{\partial y} + \frac{\partial u_y}{\partial x}\frac{\partial u_y}{\partial y} + \frac{\partial u_z}{\partial x}\frac{\partial u_z}{\partial y}$$
$$\epsilon_{yz} = \frac{\partial u_z}{\partial y} + \frac{\partial u_y}{\partial z} + \frac{\partial u_x}{\partial y}\frac{\partial u_x}{\partial z} + \frac{\partial u_y}{\partial y}\frac{\partial u_y}{\partial z} + \frac{\partial u_z}{\partial y}\frac{\partial u_z}{\partial z}$$
$$\epsilon_{zx} = \frac{\partial u_x}{\partial z} + \frac{\partial u_z}{\partial x} + \frac{\partial u_x}{\partial z}\frac{\partial u_x}{\partial x} + \frac{\partial u_y}{\partial z}\frac{\partial u_y}{\partial x} + \frac{\partial u_z}{\partial z}\frac{\partial u_z}{\partial x} \quad (5.47)$$

which may also be written as

$$\epsilon = \epsilon_l + \tilde{\epsilon} \quad (5.48)$$

where ϵ_l is the linear strain, whereas $\tilde{\epsilon}$ is the vector of the nonlinear strains proportional to the squares of displacements. The displacements u_x, u_y, u_z may be expressed in terms of a set of discrete values as

$$\boldsymbol{u}_i = \boldsymbol{N}_i \boldsymbol{u}^e \qquad i = x, y, z \quad (5.49)$$

where $\boldsymbol{N}$ is the usual shape function matrix and $\boldsymbol{u} = [u_1, u_2, \ldots, u_n]^T$. In the context of a finite element, $\boldsymbol{u}$ contains the element nodal displacement values. Introducing

$$\boldsymbol{B}_{ij} = \frac{\partial \boldsymbol{N}_i}{\partial j} \qquad i, j = x, y, z \quad (5.50)$$

the strains are given by

$$\epsilon = \begin{bmatrix} \boldsymbol{B}_{xx} \\ \boldsymbol{B}_{yy} \\ \boldsymbol{B}_{zz} \\ \boldsymbol{B}_{yx} + \boldsymbol{B}_{xy} \\ \boldsymbol{B}_{zy} + \boldsymbol{B}_{yz} \\ \boldsymbol{B}_{xz} + \boldsymbol{B}_{zx} \end{bmatrix} \boldsymbol{u} + \begin{bmatrix} \frac{\boldsymbol{B}_{xx}}{\sqrt{2}} \\ \frac{\boldsymbol{B}_{xy}}{\sqrt{2}} \\ \frac{\boldsymbol{B}_{xz}}{\sqrt{2}} \\ \boldsymbol{B}_{xx} \\ \boldsymbol{B}_{xy} \\ \boldsymbol{B}_{xz} \end{bmatrix} \boldsymbol{u} * \begin{bmatrix} \frac{\boldsymbol{B}_{xx}}{\sqrt{2}} \\ \frac{\boldsymbol{B}_{xy}}{\sqrt{2}} \\ \frac{\boldsymbol{B}_{xz}}{\sqrt{2}} \\ \boldsymbol{B}_{xy} \\ \boldsymbol{B}_{xz} \\ \boldsymbol{B}_{xx} \end{bmatrix} \boldsymbol{u} + \begin{bmatrix} \frac{\boldsymbol{B}_{yx}}{\sqrt{2}} \\ \frac{\boldsymbol{B}_{yy}}{\sqrt{2}} \\ \frac{\boldsymbol{B}_{yz}}{\sqrt{2}} \\ \boldsymbol{B}_{yx} \\ \boldsymbol{B}_{yy} \\ \boldsymbol{B}_{yz} \end{bmatrix} \boldsymbol{u} * \begin{bmatrix} \frac{\boldsymbol{B}_{yx}}{\sqrt{2}} \\ \frac{\boldsymbol{B}_{yy}}{\sqrt{2}} \\ \frac{\boldsymbol{B}_{yz}}{\sqrt{2}} \\ \boldsymbol{B}_{yy} \\ \boldsymbol{B}_{yz} \\ \boldsymbol{B}_{yx} \end{bmatrix}$$

$$\boldsymbol{u} + \begin{bmatrix} \frac{\boldsymbol{B}_{zx}}{\sqrt{2}} \\ \frac{\boldsymbol{B}_{zy}}{\sqrt{2}} \\ \frac{\boldsymbol{B}_{zz}}{\sqrt{2}} \\ \boldsymbol{B}_{zx} \\ \boldsymbol{B}_{zy} \\ \boldsymbol{B}_{zz} \end{bmatrix} \boldsymbol{u} * \begin{bmatrix} \frac{\boldsymbol{B}_{zx}}{\sqrt{2}} \\ \frac{\boldsymbol{B}_{zy}}{\sqrt{2}} \\ \frac{\boldsymbol{B}_{zz}}{\sqrt{2}} \\ \boldsymbol{B}_{zy} \\ \boldsymbol{B}_{zz} \\ \boldsymbol{B}_{zx} \end{bmatrix} \boldsymbol{u} \quad (5.51)$$

in which the asterisk symbol represents element-by-element multiplication.

The strain–displacement relations may also be written as

$$\begin{aligned}\epsilon &= \hat{\boldsymbol{B}}\boldsymbol{u} + \sum_i (\boldsymbol{B}_{i1}\boldsymbol{u}) * (\boldsymbol{B}_{i2}\boldsymbol{u}) \qquad i = x, y, z \\ &= \hat{\epsilon} + \tilde{\epsilon} \end{aligned} \tag{5.52}$$

The expression for strain energy may be derived from

$$U = \frac{1}{2}\int_V \epsilon^T \sigma \, \mathrm{d}V \tag{5.53}$$

in which

$$\sigma = \boldsymbol{D}\epsilon \tag{5.54}$$

and the linear stress

$$\hat{\boldsymbol{\sigma}} = \boldsymbol{D}\hat{\epsilon} \tag{5.55}$$

The final expression for the strain energy is obtained as

$$U = \frac{1}{2}\int_V \left(\boldsymbol{u}^T \hat{\boldsymbol{B}}^T \boldsymbol{D}\hat{\boldsymbol{B}}\boldsymbol{u} + 2\sum_{i=x,y,z} \hat{\boldsymbol{\sigma}}^T (\boldsymbol{B}_{i1}\boldsymbol{u}) * (\boldsymbol{B}_{i2}\boldsymbol{u}) \right) \mathrm{d}V \tag{5.56}$$

Performing the matrix differentiation

$$\begin{aligned}\frac{\partial U}{\partial \boldsymbol{u}} &= \int_V \hat{\boldsymbol{B}}^T \boldsymbol{D}\hat{\boldsymbol{B}} \, \mathrm{d}V \, \boldsymbol{u} + \int_V \sum_{i=x,y,z} \left(\boldsymbol{B}_{i1}^T \hat{\boldsymbol{\sigma}}^D \boldsymbol{B}_{i2} + \boldsymbol{B}_{i2}^T \hat{\boldsymbol{\sigma}}^D \boldsymbol{B}_{i1} \right) \mathrm{d}V \, \boldsymbol{u} \\ &= \left(\boldsymbol{K}_E^e + \boldsymbol{K}_G^e \right) \boldsymbol{u} \end{aligned} \tag{5.57}$$

in which $\boldsymbol{K}_E^e$ is the standard element elastic stiffness matrix, and $\boldsymbol{K}_G^e$, the element geometric stiffness matrix, is derived in terms of a general expression as

$$\boldsymbol{K}_G = \int_V \sum_{i=x,y,z} \left(\boldsymbol{B}_{i1}^T \hat{\boldsymbol{\sigma}}^D \boldsymbol{B}_{i2} + \boldsymbol{B}_{i2}^T \hat{\boldsymbol{\sigma}}^D \boldsymbol{B}_{i1} \right) \mathrm{d}V \tag{5.58}$$

applicable to any typical finite element; $\hat{\boldsymbol{\sigma}}^D$ is a diagonal matrix containing scalars of the $\hat{\boldsymbol{\sigma}}$ vector. Introducing $\boldsymbol{b}_1 = [\boldsymbol{b}_{x1}\,\boldsymbol{b}_{y1}\,\boldsymbol{b}_{z1}]^T$ and $\boldsymbol{b}_2 = [\boldsymbol{b}_{x2}\,\boldsymbol{b}_{y2}\,\boldsymbol{b}_{z2}]^T$ and noting that in most practical applications only one out of the three nonlinear terms is retained in the strain–displacement relation, the other two terms being of high order of magnitude, Eq. (5.58) reduces to the following form:

$$\boldsymbol{K}_G = \int_v \left(\boldsymbol{B}_1^T \hat{\boldsymbol{\sigma}}^D \boldsymbol{B}_2 + \boldsymbol{B}_2^T \hat{\boldsymbol{\sigma}}^D \boldsymbol{B}_1 \right) \mathrm{d}V \tag{5.59}$$

This expression can be effectively employed for routine computation of $\boldsymbol{K}_G$ for any finite element. Typical derivations are given next.

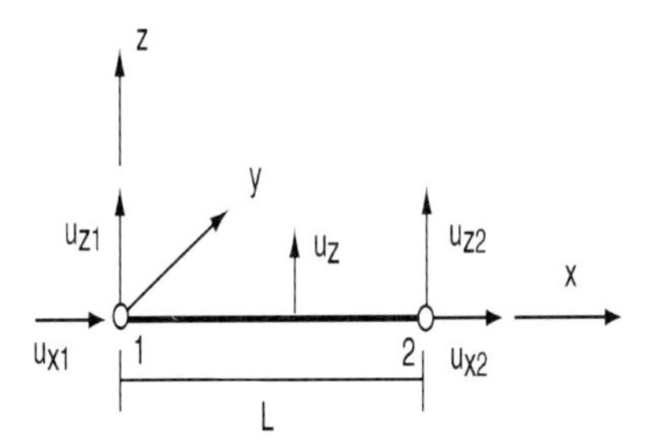

Fig. 5.4 Bar element.

5.4.3.1 Bar element. The strain–displacement relationship for the bar in Fig. 5.4 may be derived from Eqs. (5.47) and (5.51):

$$\begin{aligned}\epsilon_x &= \frac{\partial u_x}{\partial x} + \frac{1}{2}\left(\frac{\partial u_z}{\partial x}\right)^2 \\ &= \hat{\boldsymbol{b}}\boldsymbol{u} + \left(\frac{1}{\sqrt{2}}\boldsymbol{b}_{zx}\boldsymbol{u}\right) * \left(\frac{1}{\sqrt{2}}\boldsymbol{b}_{zx}\boldsymbol{u}\right)\end{aligned} \tag{5.60}$$

Lateral displacement u_z is expressed in terms of nodal displacements as

$$u_z = \boldsymbol{N}_z\boldsymbol{u} \tag{5.61}$$

where $\boldsymbol{N}_z = [0\ \ (1 - x/l)\ \ 0\ \ x/l]$ and $\boldsymbol{u} = [u_{x_1}\ u_{z_1}\ u_{x_2}\ u_{z_2}]^T$. Then

$$\boldsymbol{B}_{zx} = \frac{\partial \boldsymbol{N}_z}{\partial x} = \frac{1}{l}[0 \quad -1 \quad 0 \quad 1] \tag{5.62}$$

and

$$\begin{aligned}\boldsymbol{B}_1 = \boldsymbol{B}_2 &= \frac{1}{\sqrt{2}}\boldsymbol{B}_{zx} \\ &= \frac{1}{\sqrt{2}l}[0 \quad -1 \quad 0 \quad 1]\end{aligned} \tag{5.63}$$

Also $\hat{\boldsymbol{\sigma}}^D = F/A$, F being the axial force in the bar and $\boldsymbol{A}$ the cross-sectional area. Then

$$\begin{aligned}\boldsymbol{K}_G^e &= \int_v 2\boldsymbol{B}_1^T\hat{\boldsymbol{\sigma}}^D\boldsymbol{B}_1\,\mathrm{d}v \\ &= \frac{F}{l}\begin{bmatrix} 0 & 0 & 0 & 0 \\ 0 & 1 & 0 & -1 \\ 0 & 0 & 0 & 0 \\ 0 & -1 & 0 & 1 \end{bmatrix}\end{aligned} \tag{5.64}$$

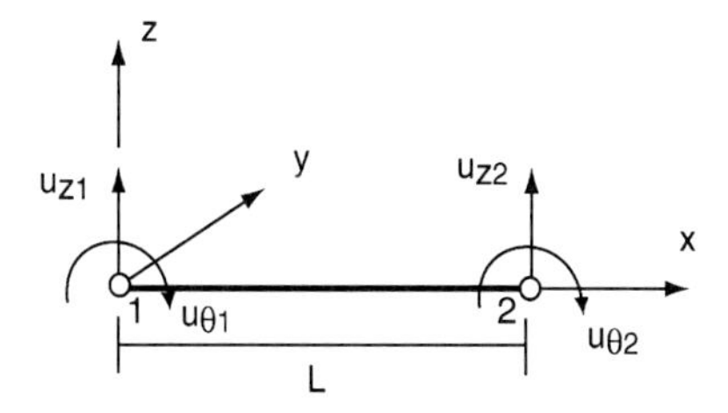

Fig. 5.5 Beam element.

5.4.3.2 Beam element. Neglecting the effects of shear deformation, the transverse deformation in the beam element in Fig. 5.5 is expressed as

$$\begin{aligned} u_z &= \boldsymbol{N}_z \boldsymbol{u}^e = [N_1\ N_2\ N_3\ N_4]^T \boldsymbol{u}^e \\ &= \left[\frac{1}{L^3}(L^3 - 3x^2L + 2x^3)\ \ \frac{1}{L^2}(xL^2 - 2x^2L + x^3)\right. \\ &\quad \left.\frac{1}{L^3}(3x^2L - 2x^3)\ \ \frac{1}{L^2}(-x^2L + x^3)\right]\boldsymbol{u}^e \end{aligned} \tag{5.65}$$

and hence

$$\boldsymbol{B}_{zx} = \frac{\partial \boldsymbol{N}_z}{\partial x} = \left[\left(-\frac{6x}{L^2}\frac{6x^2}{L^3}\right)\left(1 - \frac{4x}{L} + \frac{3x^2}{L^2}\right)\left(\frac{6x}{L^2} - \frac{6x^2}{L^3}\right)\left(-\frac{2x}{L} + \frac{3x^2}{L^2}\right)\right] \tag{5.66}$$

Assuming usual beam bending theory and noting that

$$\boldsymbol{B}_1 = \boldsymbol{B}_2 = \frac{1}{\sqrt{2}}\boldsymbol{B}_{zx}$$

the expression for the geometric stiffness matrix is obtained from Eq. (5.59) as

$$\boldsymbol{K}_G^e = \begin{bmatrix} 36 & 3l & -36 & 3l \\ & 4l^2 & -3l & -l^2 \\ & & 36 & -3l \\ \text{sym} & & & 4l^2 \end{bmatrix} \tag{5.67}$$

The geometric stiffness matrix for a three-dimensional beam element may be obtained by simply combining the bar and beam elements, resulting in the following matrix, using relevant shape functions and strain–displacement relationships derived earlier:

$$
\boldsymbol{K}_G^e = \frac{F}{l}\begin{bmatrix}
0 & 0 & 0 & 0 & 0 & 0 & 0 & 0 & 0 & 0 & 0 & 0 \\
0 & \frac{6}{5} & 0 & 0 & 0 & \frac{l}{10} & 0 & -\frac{6}{5} & 0 & 0 & 0 & \frac{l}{10} \\
0 & 0 & \frac{6}{5} & 0 & \frac{l}{10} & 0 & 0 & 0 & -\frac{6}{5} & 0 & \frac{l}{10} & 0 \\
0 & 0 & 0 & 0 & 0 & 0 & 0 & 0 & 0 & 0 & 0 & 0 \\
0 & 0 & \frac{l}{10} & 0 & \frac{2l^2}{15} & 0 & 0 & 0 & -\frac{l}{10} & 0 & -\frac{l^2}{30} & 0 \\
0 & \frac{l}{10} & 0 & 0 & 0 & \frac{2l^2}{15} & 0 & -\frac{l}{10} & 0 & 0 & 0 & -\frac{l^2}{30} \\
0 & 0 & 0 & 0 & 0 & 0 & 0 & 0 & 0 & 0 & 0 & 0 \\
0 & -\frac{6}{5} & 0 & 0 & 0 & -\frac{l}{10} & 0 & \frac{6}{5} & 0 & 0 & 0 & -\frac{l}{10} \\
0 & 0 & -\frac{6}{5} & 0 & -\frac{l}{10} & 0 & 0 & 0 & \frac{6}{5} & 0 & -\frac{l}{10} & 0 \\
0 & 0 & 0 & 0 & 0 & 0 & 0 & 0 & 0 & 0 & 0 & 0 \\
0 & 0 & \frac{l}{10} & 0 & -\frac{l^2}{30} & 0 & 0 & 0 & -\frac{l}{10} & 0 & \frac{2l^2}{15} & 0 \\
0 & \frac{l}{10} & 0 & 0 & 0 & -\frac{l^2}{30} & 0 & -\frac{l}{10} & 0 & 0 & 0 & \frac{2l^2}{15}
\end{bmatrix} \tag{5.68}
$$

5.4.3.3 Triangular shell and plate element. Assuming that the plate in Fig. 5.6 is subjected to membrane stresses only, an expression for $\boldsymbol{K}_G^e$ can be derived by adopting the standard procedure. Thus the transverse displacement

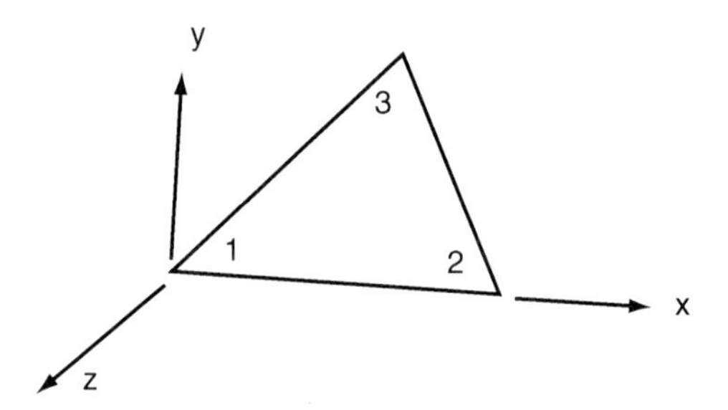

Fig. 5.6 Triangular plate element.

may be expressed as

$$u_z = (1/2A)[(y_{32}(x - x_2) - x_{32}(y - y_2)) \ (-y_{31}(x - x_3) + x_{31}(y - y_3)) \ (y_{21}(x - x_1) - x_{21}(y - y_1))]\boldsymbol{u} \tag{5.69}$$

in which

$$\boldsymbol{u} = [u_{z1} \quad u_{z2} \quad u_{z3}]^T \tag{5.70}$$

$$x_{ij} = x_i - x_j; \qquad y_{ij} = y_i - y_j \tag{5.71}$$

and A is the area of the triangle. The matrices $\boldsymbol{B}_{zx}$ and $\boldsymbol{B}_{zy}$ may then be obtained from Eq. (5.69) and the $\boldsymbol{K}_G^e$ matrix obtained from Eq. (5.59) as

$$\boldsymbol{K}_G^e = \hat{\sigma}_{oxx}\frac{t}{4A}\begin{bmatrix} y_{32}^2 & -y_{32}y_{31} & y_{32}y_{21} \\ & y_{31}^2 & -y_{31}y_{21} \\ \text{sym} & & y_{21}^2 \end{bmatrix} + \hat{\sigma}_{oyy}\frac{t}{4A}\begin{bmatrix} x_{32}^2 & -x_{32}x_{31} & x_{32}x_{21} \\ & x_{31}^2 & -x_{31}x_{21} \\ \text{sym} & & x_{21}^2 \end{bmatrix} + \hat{\sigma}_{oxy}\frac{t}{4A}\begin{bmatrix} -2x_{32}y_{32} & y_{32}x_{31} + x_{32}y_{31} & -(y_{32}x_{21} + x_{32}y_{21}) \\ & -2y_{31}x_{31} & y_{32}x_{21} + x_{31}y_{21} \\ \text{sym} & & -2y_{21}x_{21} \end{bmatrix} \tag{5.72}$$

where t is the thickness of the element and $\hat{\sigma}_{oxx}$, $\hat{\sigma}_{oyy}$, and $\hat{\sigma}_{oxy}$ are the stresses in the middle plane. In other element types the $\boldsymbol{K}_G^e$ matrix may be derived readily by using the standard formulation of Eq. (5.59). Alternative procedures for derivation of these matrices are summarized in Refs. 7–10.

5.5 Numerical Examples

A large variety of relevant example problems may be solved by using the NASA STARS[11] (Structural Analysis Routines) program that incorporates the procedures developed in this chapter. Two such representative problems and their solutions are presented next, followed by a third, helicopter example.

Example 5.1: Spinning Cantilever Beam

Figure 5.7 depicts a cantilever beam spinning along the Y axis. The beam is idealized by a number of finite elements having the following basic structural properties: Young's modulus $E = 30 \times 10^6$, cross sectional area $A = 1.0$, moment of inertia about Y axis $I_Y = 1/12$, moment of inertia about Z axis $I_Z = 1/24$, $L = 60$, mass per unit length $= 0.1666$. Table 5.1 presents the results of free vibration analysis for the beam for varying the number of elements and also spin rates.

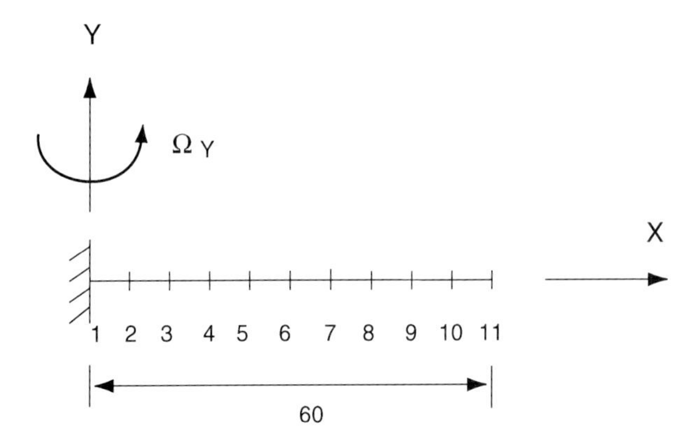

Fig. 5.7 Spinning cantilever beam.

Example 5.2: Spinning Plate

This example problem relates to a square cantilever plate rotating with a uniform spin rate about an arbitrary axis. Figure 5.8 depicts the plate with a 6×6 finite element mesh, the edge along the X axis being clamped and the structure having the following basic structural data: Young's modulus $E = 10^7$, thickness $t = 0.1$, side length $l = 10$, Poisson's ratio $\nu = 0.3$, mass density $\rho = 0.259 \times 10^{-3}$. A free vibration analysis of the nonspinning structure was initially performed, yielding a first natural frequency (ω_1) value of 215.59 rad/s. Subsequent analyses of the structure were performed for the case of $\Omega_Z = 100$ rad/s as well as for a resulting spin vector Ω_R having components $\Omega_X = \Omega_Y = \Omega_Z = 100/\sqrt{3}$ (57.735 rad/s). Numerical results for these analyses are presented in Table 5.2.

Example 5.3: Helicopter

A coupled helicopter rotor–fuselage system freely floating in space, as shown in Fig. 5.9 is chosen as the next example. Associated varying stiffness and mass

Table 5.1 Natural frequencies of a spinning cantilever beam

	Natural frequency, rad/s					
	Spin rate $\Omega_Y = 0$ Number of elements		Spin rate $\Omega_Y = 0.1$ Hz Number of elements		Spin rate $\Omega_Y = 0.2$ Hz Number of elements	
Mode number	5	10	5	10	5	10
1	2.6747	2.6747	2.7642	2.7620	3.0161	3.0081
2	3.7826	3.7824	3.7928	3.7925	3.8231	3.8224
3	16.7680	16.7600	16.8486	16.8370	17.0893	17.0670
4	23.7090	23.6970	23.7570	23.7430	23.8781	23.8781
5	47.0830	46.9260	47.1625	47.0020	47.3999	47.2290
6	66.5560	66.3333	66.6080	66.3840	66.7650	66.5360

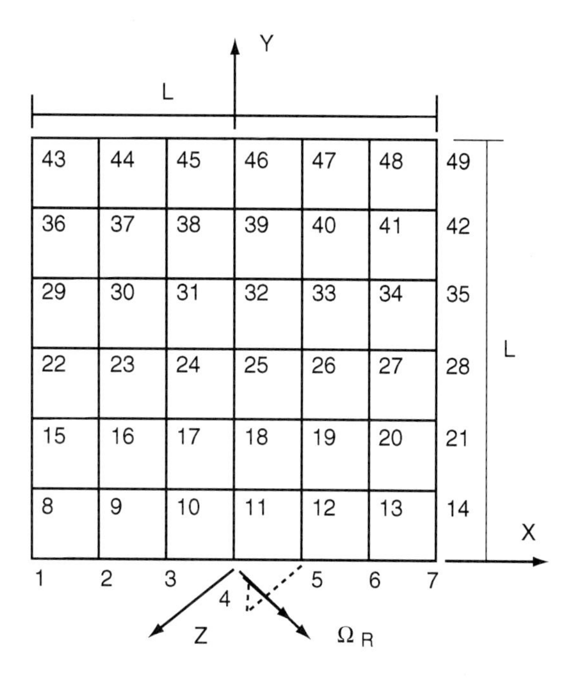

Fig. 5.8 Cantilever plate spinning along a specified axis.

distributions are suitably approximated for the discrete-element modeling of the structure. Free vibration analysis was performed with the rotor spinning at 10 rad/s ($\Omega_Y = 10$). The results are presented in Table 5.3 along with the results for the corresponding nonspinning case. Associated mode shapes, which correspond to the nonspinning rotors, are shown in Fig. 5.10.

5.6 Concluding Remarks

A generalized numerical formulation has been presented for the effective evaluation of nodal centrifugal forces in various finite elements due to any arbitrary spin

Table 5.2 Spinning cantilever plate: natural frequencies for various spin rates

Mode	$\Omega_R = 0$	$\Omega_R = \Omega_Z = 100$ rad/s	$\Omega_R = 100$ rad/s $\Omega_X = \Omega_Y = \Omega_Z = 57.735$ rad/s
1	215.59	243.89	159.29
2	518.78	538.33	503.35
3	1292.52	1316.00	1296.60
4	1620.84	1633.60	1620.40
5	1854.17	1872.60	1859.20
6	3151.41	3161.30	3156.20

Table 5.3 Natural frequencies of a helicopter

Mode number	Natural frequencies ω (rad/s) for spin rates $\Omega_Y = 0$	$\Omega_Y = 10$	Mode shape
1–6	0.0000	0.0000	Rigid body
7	4.6374	10.4720	Rotor 1st symmetric bending
8	4.6374	10.4720	Rotor 1st antisymmetric bending
9	5.0407	11.6660	Fuselage 1st symmetric bending
10	5.0407	11.6660	Rotor 2nd antisymmetric bending
11	22.1274	22.1940	Rotor 2nd symmetric bending
12	22.1274	22.1940	Fuselage 1st antisymmetric bending

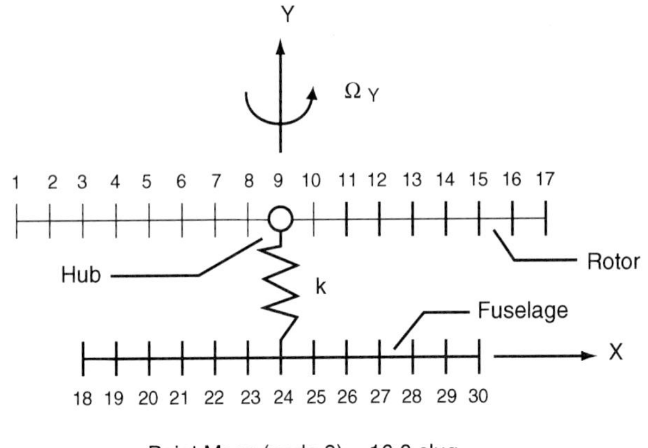

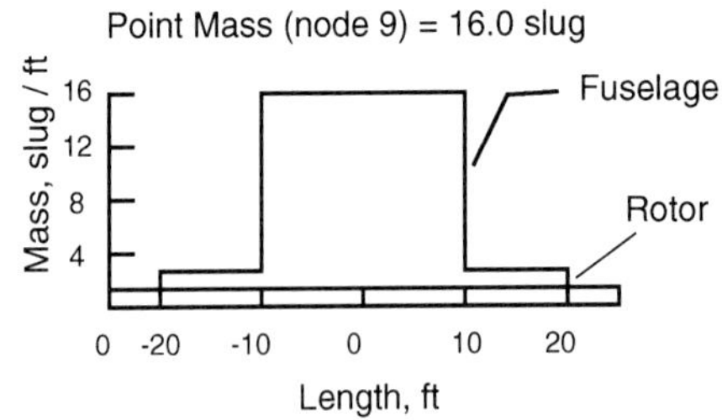

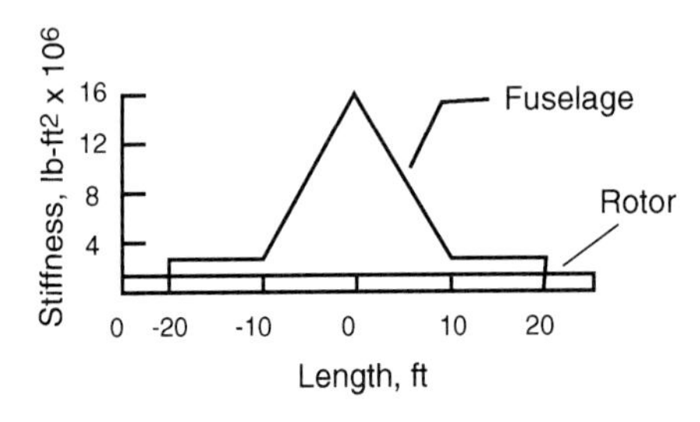

Fig. 5.9 Coupled helicopter rotor–fuselage system.

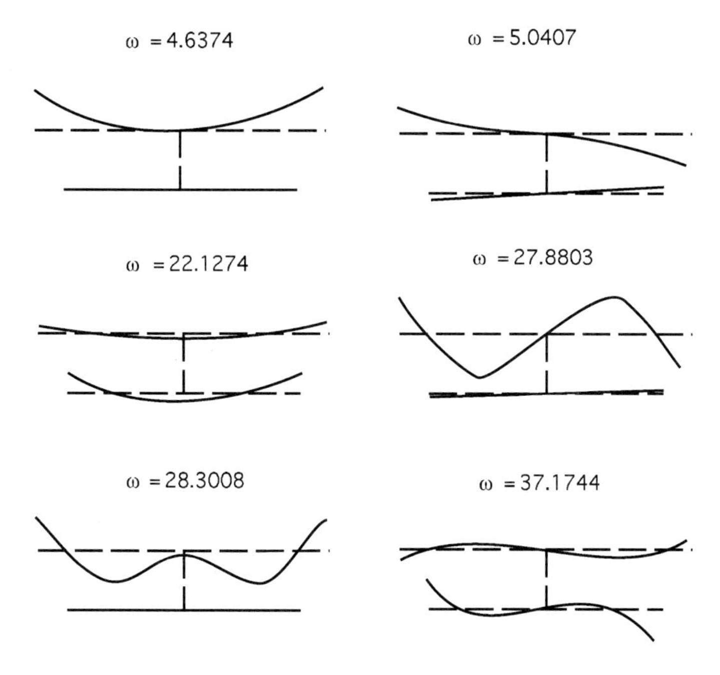

Fig. 5.10 Helicopter mode shapes.

rate. This in turn enables the derivation of element geometric stiffness and centrifugal force matrices $\boldsymbol{K}_G$ and $\boldsymbol{K}'$, respectively, which are vital for the free vibration analysis of spinning structures. A recently developed general purpose finite element computer program, STARS,[11] incorporates the current techniques described herein. A number of example problems are also presented that demonstrate the effectiveness of the current analysis program.

References

[1]Gupta, K. K., "Formulation of Numerical Procedures for Dynamic Analysis of Spinning Structures," *International Journal for Numerical Methods in Engineering*, Vol. 23, 1986, pp. 2347–2357.

[2]Dokainish, M. A., and Rawtani, S., "Vibration Analysis of Rotating Cantilever Plates," *International Journal for Numerical Methods in Engineering*, Vol. 3, 1971, pp. 233–248.

[3]Bossak, M. A. J., and Zienkiewicz, O. C. Z., "Free Vibration of Initially Stressed Solids with Particular Reference to Centrifugal-Force Effects in Rotating Machinery," *Journal of Strain Analysis*, Vol. 8, No. 4, 1973, pp. 245–252.

[4]Leung, A. Y. T., and Fung, T. C., "Spinning Finite Elements," *Journal of Sound and Vibration*, Vol. 125, No. 3, 1988, pp. 523–537.

[5]Wittrick, W. H., and Williams, F. W., "On the Free Vibration Analysis of Spinning Structures by Using Discrete or Distributed Mass Models," *Journal of Sound and Vibration*, Vol. 82, 1982, pp. 1–19.

[6]Przemieniecki, J. S., "Discrete Element Methods for Stability Analysis of Complex

Structures," *Proceedings of the Symposium on Structural Stability and Optimization*, Loughborough Univ., England, U.K., 1967.

[7]Martin, H. C., "On the Derivation of Stiffness Matrices for the Analysis of Large Deflection and Stability Problems," *Proceedings of the Conference on Matrix Methods in Structural Mechanics*, Wright–Patterson AFB, OH, Oct. 1965; AFFDL TR 66-80, 1966.

[8]Turner, M. J., Dill, E. H., Martin, H. C., and Melosh, R. J., "Large Deflections of Structures Subjected to Heating and External Loads," *Journal of the Aeronautical Sciences*, Vol. 27, 1960, pp. 97–102.

[9]Argyris, J. H., "Continua and Discontinua," *Proceedings of the Conference on Matrix Methods in Structural Mechanics*, Wright–Patterson AFB, OH, Oct. 1965; AFFDL TR 66-80, 1966.

[10]Gallagher, R. H., and Padlog, J., "Discrete Element Approach to Structural Instability Analysis," *AIAA Journal*, Vol. 1, 1963, pp. 1537–1539.

[11]Gupta, K. K., "STARS—An Integrated, Multidisciplinary, Finite Element, Structural, Fluids, Aeroelastic and Aeroservoelastic Analysis Computer Program," NASA TM 4795, 1997.

6
Dynamic Element Method

6.1 Introduction

In the usual finite element procedure, the continuous displacement field within an element is defined in terms of its unknown nodal displacement parameters by the fundamental relationship

$$u = Nu^e \tag{6.1}$$

in which the matrix N represents the shape functions. Such a relationship, however, is strictly valid only for static loading since, for the general dynamic problem, N is not unique; it is then a function of the entire time history of the nodal displacements. In the special case of free vibration involving harmonic motion, Eq. (6.1) is valid when N is a function of the instantaneous nodal displacements and also of the frequency of such motion. The resulting stiffness and mass matrices are then obtained as functions of the frequency of the harmonic motion.[1,2] Thus, as an example, the equation of motion for a simple bar element in Fig. 6.1 has the following form:

$$c^2 \frac{\partial^2 u_x}{\partial x^2} - \ddot{u}_x = 0 \tag{6.2}$$

in which $c^2 = E/\rho$. The solution of Eq. (6.2) may be expressed as

$$\begin{aligned} u_x &= N(\omega)u^e \\ &= N(\omega)q^e e^{i\omega t} \\ &= g(x)e^{i\omega t} \end{aligned} \tag{6.3}$$

where $u^e = [u_{x1}\ u_{x2}]^T$ and $q^e = [q_{x1}\ q_{x2}]^T$. Substituting Eq. (6.3) into Eq. (6.2) yields

$$\frac{d^2 g}{dx^2} + \frac{\omega^2}{c^2} g = 0 \tag{6.4}$$

The solution to Eq. (6.4) is

$$g = A \sin \frac{\omega x}{c} + B \cos \frac{\omega x}{c} \tag{6.5}$$

Using boundary conditions

$$u_{x1} = q_1 e^{i\omega t}, \qquad u_{x2} = q_2 e^{i\omega t} \tag{6.6}$$

the coefficients A and B are evaluated appropriately yielding the following

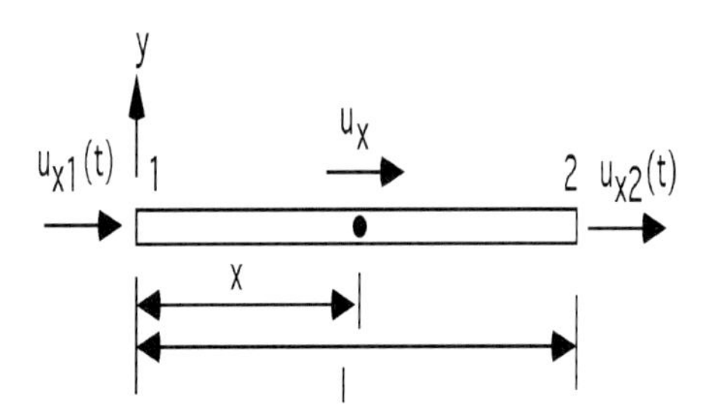

Fig. 6.1 Bar element with time-varying nodal displacements.

expression for the shape function:

$$\boldsymbol{N} = \left[\left(\cos \frac{\omega x}{c} - \cot \frac{\omega l}{c} \sin \frac{\omega x}{c} \right) \csc \frac{\omega l}{c} \sin \frac{\omega x}{c} \right] \tag{6.7}$$

The strain–displacement relations may then be obtained as

$$\epsilon = \frac{\mathrm{d}u_x}{\mathrm{d}x} = \boldsymbol{B}\boldsymbol{u}^e$$

resulting in

$$\boldsymbol{B} = \left[-\frac{\omega}{c} \left(\sin \frac{\omega x}{c} + \cot \frac{\omega l}{c} \cos \frac{\omega x}{c} \right) \frac{\omega}{c} \left(\csc \frac{\omega l}{c} \cos \frac{\omega x}{c} \right) \right] \tag{6.8}$$

Element stiffness and mass matrices are then obtained as follows, where A is the cross sectional area of the bar:

$$\begin{aligned} \boldsymbol{K}^e(\omega) &= \int_V \boldsymbol{B}^T E \boldsymbol{B} \, \mathrm{d}V \\ &= \frac{AE}{2l} \frac{\omega l}{c} \csc \frac{\omega l}{c} \begin{bmatrix} \left(\frac{\omega l}{c} \csc \frac{\omega l}{c} + \cos \frac{\omega l}{c} \right) & -\left(1 + \frac{\omega l}{c} \cot \frac{\omega l}{c} \right) \\ -\left(1 + \frac{\omega l}{c} \cot \frac{\omega l}{c} \right) & \left(\frac{\omega l}{c} \csc \frac{\omega l}{c} + \cos \frac{\omega l}{c} \right) \end{bmatrix} \end{aligned} \tag{6.9}$$

$$\begin{aligned} \boldsymbol{M}^e(\omega) &= \int_V \rho \boldsymbol{N}^T \boldsymbol{N} \, \mathrm{d}V \\ &= \frac{\rho A l}{2} \frac{c}{\omega l} \csc \frac{\omega l}{c} \begin{bmatrix} \left(\frac{\omega l}{c} \csc \frac{\omega l}{c} - \cos \frac{\omega l}{c} \right) & \left(1 - \frac{\omega l}{c} \cot \frac{\omega l}{c} \right) \\ \left(1 - \frac{\omega l}{c} \cot \frac{\omega l}{c} \right) & \left(\frac{\omega l}{c} \csc \frac{\omega l}{c} - \cos \frac{\omega l}{c} \right) \end{bmatrix} \end{aligned} \tag{6.10}$$

Extraction of roots and vectors from initially unknown frequency-dependent matrices, as just shown, is rather uneconomical, being prohibitive for most practical problems. The matrices $\boldsymbol{N}$ and $\boldsymbol{B}$ are thus expressed in ascending powers of the natural frequency ω,

$$\boldsymbol{N} = \sum_{r=0}^{\infty} \omega^r \boldsymbol{N}_r \tag{6.11}$$

$$\boldsymbol{B} = \sum_{r=0}^{\infty} \omega^r \boldsymbol{B}_r \tag{6.12}$$

resulting in the following expressions for stiffness and mass matrices:

$$\boldsymbol{K}^e = \boldsymbol{K}_0^e + \omega^2 \boldsymbol{K}_2^e + \omega^4 \boldsymbol{K}_4^e + \cdots \tag{6.13}$$

$$\boldsymbol{M}^e = \boldsymbol{M}_0^e + \omega^2 \boldsymbol{M}_2^e + \omega^4 \boldsymbol{M}_4^e + \cdots \tag{6.14}$$

The equation of free vibration takes the following form after assembly of the element matrices:

$$\left[\boldsymbol{K}_0 - \omega^2(\boldsymbol{M}_0 - \boldsymbol{K}_2) - \omega^4(\boldsymbol{M}_2 - \boldsymbol{K}_4) - \cdots\right]\boldsymbol{q} = 0 \tag{6.15}$$

where $\boldsymbol{q}$ is the amplitude of the nodal deformation, and $\boldsymbol{M}_0$, $\boldsymbol{K}_0$ are the static mass and stiffness matrices, whereas the other higher-order terms constitute the dynamic corrections. Such elements incorporating the higher-order terms are referred to as finite dynamic elements (FDE) and the related solution technique as the dynamic element method (DEM).[3] In general, the first three terms only are retained in Eq. (6.15) for subsequent analysis, resulting in a quadratic matrix equation of the form

$$(\boldsymbol{A} - \lambda \boldsymbol{B} - \lambda^2 \boldsymbol{C})\boldsymbol{q} = 0 \tag{6.16}$$

where $\lambda = \omega^2$, the solution of which yields the desired roots and vectors. Derivations of some such FDEs are given next.

6.2 Bar Element

Using Eq. (6.11) the displacement function may be written as

$$\begin{aligned} u_x &= \boldsymbol{N}\boldsymbol{u}^e \\ &= \sum_{r=0}^{\infty} \omega^r \boldsymbol{N}_r \boldsymbol{q}^e e^{i\omega t} \end{aligned} \tag{6.17}$$

which is substituted in Eq. (6.2) to yield the equation of motion

$$\left(c^2 \sum_{r=0}^{\infty} \omega^r \frac{\partial^2 \boldsymbol{N}_r}{\partial x^2} + \omega^2 \sum_{r=0}^{\infty} \omega^r \boldsymbol{N}_r\right) \boldsymbol{q}^e e^{i\omega t} = 0 \tag{6.18}$$

and equating to zero the coefficients of the same powers of ω yields the following equations:

$$\frac{\partial^2 \boldsymbol{N}_0}{\partial x^2} = 0 \tag{6.19}$$

$$\frac{\partial^2 \boldsymbol{N}_1}{\partial x^2} = 0 \tag{6.20}$$

$$c^2 \frac{\partial^2 \boldsymbol{N}_2}{\partial x^2} = -\boldsymbol{N}_0 \tag{6.21}$$

$$c^2 \frac{\partial^2 \boldsymbol{N}_3}{\partial x^2} = -\boldsymbol{N}_1 \tag{6.22}$$

The solution for Eq. (6.19) is expressed as

$$\boldsymbol{N}_0 \boldsymbol{u}^e = c_1 + c_2 x \tag{6.23}$$

and the unknown coefficients are determined to satisfy the boundary conditions $u_x = u_{x1}$ at $x = 0$ and $u_x = u_{x2}$ at $x = l$, yielding

$$\boldsymbol{N}_0 = \left[1 - \frac{x}{l} \ \frac{x}{l}\right] \tag{6.24}$$

The remaining matrices must all vanish at $x = 0$ and l, and they are evaluated similarly to yield

$$\boldsymbol{N}_1 = 0 \tag{6.25}$$

$$\boldsymbol{N}_2 = \frac{\rho l^2}{6E}\left[\left(\frac{2x}{l} - \frac{3x^2}{l^2} + \frac{x^3}{l^3}\right)\left(x - \frac{x^3}{l^3}\right)\right] \tag{6.26}$$

$$\boldsymbol{N}_3 = 0 \tag{6.27}$$

Thus the shape function matrix is taken as

$$\boldsymbol{N} = \boldsymbol{N}_0 + \omega^2 \boldsymbol{N}_2 \tag{6.28}$$

The strain–displacement matrices are calculated from

$$\begin{aligned} \epsilon &= \frac{\partial u_x}{\partial x} \\ &= \sum_{r=0}^{\infty} \omega^r \frac{\partial \boldsymbol{N}_r}{\partial x} \boldsymbol{q} e^{i\omega t} \\ &= \sum_{r=0}^{\infty} \omega^r \boldsymbol{B}_r \boldsymbol{q} e^{i\omega t} \end{aligned} \tag{6.29}$$

where

$$\boldsymbol{B}_0 = \frac{\mathrm{d}\boldsymbol{N}_0}{\mathrm{d}x} = \frac{1}{l}[-1 \quad 1] \tag{6.30}$$

$$\boldsymbol{B}_1 = \frac{\mathrm{d}\boldsymbol{N}_1}{\mathrm{d}x} = 0 \tag{6.31}$$

$$\boldsymbol{B}_2 = \frac{\mathrm{d}\boldsymbol{N}_2}{\mathrm{d}x} = \frac{\rho l}{6E}\left[\left(2 - \frac{6x}{l} + \frac{3x^2}{l^2}\right)\left(1 - \frac{3x^2}{l^2}\right)\right] \tag{6.32}$$

and then

$$\boldsymbol{B} = \boldsymbol{B}_0 + \omega^2 \boldsymbol{B}_2 \tag{6.33}$$

Substituting Eq. (6.33) into the standard formulation for the stiffness matrix,

$$\boldsymbol{K}^e = \int_V \boldsymbol{B}^T \boldsymbol{D} \boldsymbol{B} \, \mathrm{d}V \tag{6.34}$$

one obtains

$$\boldsymbol{K}_0^e = \frac{AE}{l} \begin{bmatrix} 1 & -1 \\ -1 & 1 \end{bmatrix} \tag{6.35}$$

$$\boldsymbol{K}_4^e = \rho^2 \frac{Al^3}{45E} \begin{bmatrix} 1 & \frac{7}{8} \\ \frac{7}{8} & 1 \end{bmatrix} \tag{6.36}$$

Similarly the inertia matrices are obtained by using Eq. (6.28) for the mass matrix,

$$\boldsymbol{M}^e = \int_V \rho \boldsymbol{N}^T \boldsymbol{N} \, \mathrm{d}V \tag{6.37}$$

resulting in

$$\boldsymbol{M}_0^e = \frac{\rho A l}{6} \begin{bmatrix} 2 & 1 \\ 1 & 2 \end{bmatrix} \tag{6.38}$$

and

$$\boldsymbol{M}_2^e = \frac{2\rho^2 A l^3}{45E} \begin{bmatrix} 1 & \frac{7}{8} \\ \frac{7}{8} & 1 \end{bmatrix} \tag{6.39}$$

noting that $\boldsymbol{K}_0^e$, $\boldsymbol{M}_0^e$ pertain to the usual FEM, whereas $\boldsymbol{M}_2^e$, $\boldsymbol{K}_4^e$ are the dynamic correction terms.

6.3 Beam Element

Figure 6.2 shows a typical beam element with four nodal degrees of freedom. The equation of motion in the y direction is given by

$$c^4 \frac{\partial^4 u_y}{\partial x^4} + \ddot{u}_y = 0 \tag{6.40}$$

in which $c^4 = EI_z/\rho A$, I_z being the beam cross-sectional moment of inertia. Expressing displacements in ascending powers of ω (Ref. 2)

$$u_x = \left(\sum_{r=0}^{\infty} \omega^r \boldsymbol{N}_{rx} \right) \boldsymbol{q} e^{i\omega t} \tag{6.41}$$

$$u_y = \left(\sum_{r=0}^{\infty} \omega^r \boldsymbol{N}_{ry} \right) \boldsymbol{q}^{i\omega t} \tag{6.42}$$

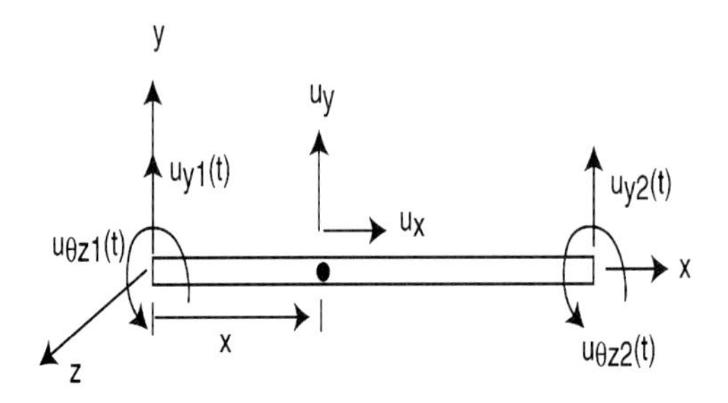

Fig. 6.2 Beam elements.

Noting that $u_x = (-\partial u_y/\partial x)y$, in accordance with engineering bending theory and continuing as in the preceding section,

$$u_x = \boldsymbol{N}_{0x} + \omega^2 \boldsymbol{N}_{2x} \tag{6.43}$$

$$u_y = \boldsymbol{N}_{0y} + \omega^2 \boldsymbol{N}_{2y} \tag{6.44}$$

which may be derived by following the standard procedure as previously.

The dynamic correction matrices are given next, noting that $\boldsymbol{K}_0^e$ and $\boldsymbol{M}_0^e$ have been derived in an earlier section:

$$\boldsymbol{K}_4^e = (\rho A l)^2 \frac{l^3}{EI} \begin{bmatrix} 0.364872 & 0.076616l & 0.329571 & -0.072193l \\ & 0.016262l^2 & 0.072193l & -0.015704l^2 \\ & \text{sym} & 0.364872 & -0.076616l \\ & & & 0.016262l^2 \end{bmatrix} \tag{6.45}$$

$$\boldsymbol{M}_2^e = (\rho A l)^2 \frac{l^3}{EI} \begin{bmatrix} 0.729746 & 0.153233l & 0.659142 & -0.144386l \\ & 0.032525l^2 & 0.144386l & -0.031408l^2 \\ & \text{sym} & 0.729746 & -0.0153233l \\ & & & 0.0325248l^2 \end{bmatrix}$$
$$+ (\rho A l)^2 \frac{l^3}{EI} \left(\frac{r}{l}\right)^2 \begin{bmatrix} 0.317460 & 0.793651l & -0.317460 & -0.595238l \\ & 0.317460l^2 & 0.595238l & -0.277998l^2 \\ & \text{sym} & 0.317460 & -0.793651l \\ & & & 0.317460l^2 \end{bmatrix} \tag{6.46}$$

In Eq. (6.46), r is the radius of gyration of the beam cross section. Similar expressions may also be derived for motion involving the z direction. These matrices may then be suitably combined along with the bar and torsional elements for the general 12-DOF line element.

6.4 Rectangular Prestressed Membrane Element

The first derivation of an FDE pertaining to a continuum relates to a thin rectangular membrane element[3] subjected to a uniform stress along the four edges in

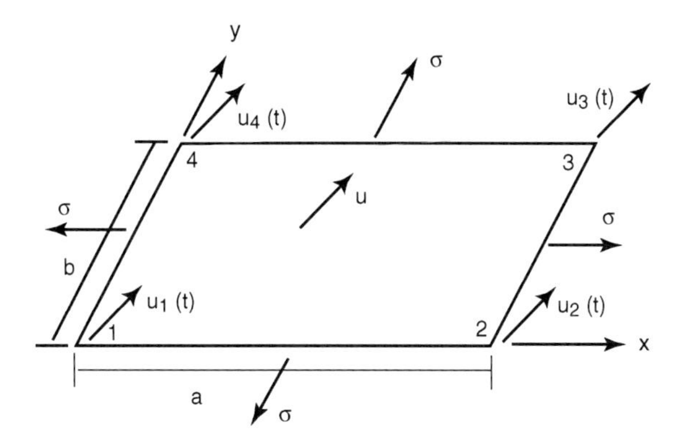

Fig. 6.3 Membrane rectangular element.

Fig. 6.3. The associated differential equation of motion is expressed as

$$\frac{\partial^2 u}{\partial x^2} + \frac{\partial^2 u}{\partial y^2} = \frac{\rho}{\sigma h}\frac{\partial^2 u}{\partial t^2} \tag{6.47}$$

in which ρ is the mass of the membrane per unit area, h is the thickness, and σh is the uniform tensile force per unit length assumed to remain unchanged during free vibration. Due to the symmetry and uniform nature of the stretching forces, in-plane shear stresses are zero. Expressing the solution of Eq. (6.47) is ascending powers of ω:

$$\begin{aligned} \boldsymbol{u} &= \boldsymbol{N}\boldsymbol{u}^e \\ &= \left(\sum_{r=0}^{\infty} \omega^r \boldsymbol{N}_r\right) \boldsymbol{q} e^{i\omega t} \end{aligned} \tag{6.48}$$

Substituting Eq. (6.48) into Eq. (6.47), the equation of motion becomes

$$\sum_{r=0}^{\infty} \omega^r \frac{\partial^2 \boldsymbol{N}_r}{\partial x^2} \boldsymbol{q} e^{i\omega t} + \sum_{r=0}^{\infty} \omega^r \frac{\partial^2 \boldsymbol{N}_r}{\partial y^2} \boldsymbol{q} e^{i\omega t} + \frac{\rho}{\sigma h}\omega^2 \sum_{r=0}^{\infty} \omega^r \boldsymbol{N}_r \boldsymbol{q} e^{i\omega t} = 0 \tag{6.49}$$

The coefficients of the same powers of ω in the preceding equation are next equated to zero, yielding the following equations:

$$\frac{\partial^2 \boldsymbol{N}_0}{\partial x^2} + \frac{\partial^2 \boldsymbol{N}_0}{\partial y^2} = 0 \tag{6.50}$$

$$\frac{\partial^2 \boldsymbol{N}_1}{\partial x^2} + \frac{\partial^2 \boldsymbol{N}_1}{\partial y^2} = 0 \tag{6.51}$$

$$\frac{\partial^2 \boldsymbol{N}_2}{\partial x^2} + \frac{\partial^2 \boldsymbol{N}_2}{\partial y^2} + \frac{\rho}{\sigma h}\boldsymbol{N}_0 = 0 \tag{6.52}$$

where only the first three terms of the series form of Eq. (6.48) have been retained in the analysis. Solution of the preceding equations is obtained by using $\boldsymbol{N}_0$ to satisfy the boundary conditions: $u = u_1(x = 0,\ y = 0)$, $u = u_2(x = a,\ y = 0)$, $u = u_3(x = a,\ y = b)$, and $u = u_4(x = 0,\ y = b)$ while observing that $\boldsymbol{N}_1$ and $\boldsymbol{N}_2$ must all vanish at the boundaries.

The solution of homogeneous Eq. (6.50) is taken in a series form

$$\boldsymbol{N}_0\boldsymbol{u}^e = c_1 + c_2 x + c_3 y + c_4 xy \tag{6.53}$$

where the coefficients c_1, c_2, c_3, c_4 are evaluated by satisfying the appropriate boundary conditions. A similar solution is adopted for the differential Eq. (6.51). The solution of Eq. (6.52) is assumed in the form

$$\boldsymbol{N}_2\boldsymbol{u}^e = \hat{c}_1 + \hat{c}_2 x + \hat{c}_3 y + \hat{c}_4 xy - \frac{\rho}{\sigma h}\left[\frac{c_1}{4}x^2 + \frac{c_1}{4}y^2 + \frac{c_2}{12}x^3 + \frac{c_2}{4}xy^2 + \frac{c_3}{4}x^2 y + \frac{c_3}{12}y^3 + \frac{c_4}{12}x^3 y + \frac{c_4}{12}xy^3\right] \tag{6.54}$$

in which the coefficients $\hat{c}_1, \hat{c}_2, \hat{c}_3, \hat{c}_4$ of the complementary function are obtained from the boundary conditions, and where it is noted that the coefficients appearing in the particular integral are those derived from Eq. (6.53). A number of algebraic operations finally yield the solution as follows:

$$\boldsymbol{N}_0 = \left[\left(1 - \frac{x}{a} - \frac{y}{b} + \frac{xy}{ab}\right)\left(\frac{x}{a} - \frac{xy}{ab}\right)\frac{xy}{ab}\left(\frac{y}{b} - \frac{xy}{ab}\right)\right] \tag{6.55}$$

$$\boldsymbol{N}_1 = 0 \tag{6.56}$$

$$\boldsymbol{N}_2^T = \frac{\rho}{\sigma h} \times \begin{bmatrix} \left(\frac{ax}{6} + \frac{by}{6} - \frac{x^2}{4} - \frac{y^2}{4} - \frac{bxy}{6a} - \frac{axy}{6b} + \frac{x^3}{12a} + \frac{y^3}{12b} + \frac{xy^2}{4a} + \frac{x^2 y}{4b} - \frac{xy^3}{12ab} - \frac{x^3 y}{12ab}\right) \\ \left(\frac{ax}{12} + \frac{bxy}{6a} - \frac{x^3}{12a} - \frac{xy^2}{4a} - \frac{axy}{12b} + \frac{xy^3}{12ab} + \frac{x^3 y}{12ab}\right) \\ \left(\frac{axy}{12b} + \frac{bxy}{12a} - \frac{xy^3}{12ab} - \frac{x^3 y}{12ab}\right) \\ \left(\frac{by}{12} + \frac{axy}{6b} - \frac{bxy}{12a} - \frac{y^3}{12b} - \frac{x^2 y}{4b} + \frac{xy^3}{12ab} + \frac{x^3 y}{12ab}\right) \end{bmatrix} \tag{6.57}$$

and the matrix $\boldsymbol{N}$ is defined as

$$\boldsymbol{N} = \boldsymbol{N}_0 + \omega^2 \boldsymbol{N}_2 \tag{6.58}$$

The strain–displacement equations of the membrane may be determined next from such relationships for the general two-dimensional problem. Thus, assuming that the strain energy caused by in-plane displacement is negligible compared to that caused by lateral displacement u, the strain–displacement relation is given as

$$\epsilon_x = \frac{1}{2}\left(\frac{\partial u}{\partial x}\right)^2, \qquad \epsilon_y = \frac{1}{2}\left(\frac{\partial u}{\partial y}\right)^2 \tag{6.59}$$

noting that the shear strain is zero because of the uniform tensile load.

The preceding relationship may then be utilized to develop the stiffness and mass matrices of the dynamic element. Thus the increase in potential (strain) energy is given by

$$\Delta\tilde{U} = \int_{y=0}^{b}\int_{x=0}^{a} \sigma h\left[\frac{1}{2}\left(\frac{\partial u}{\partial x}\right)^2 + \frac{1}{2}\left(\frac{\partial u}{\partial y}\right)^2\right] \mathrm{d}x\,\mathrm{d}y \tag{6.60}$$

which may also be expanded as

$$\begin{aligned}\Delta\tilde{U} &= \frac{\sigma h}{2}\iint_A \left[\left(\frac{\partial \boldsymbol{N}_0\boldsymbol{u}^e}{\partial x}\right)^2 + 2\omega^2\left(\frac{\partial \boldsymbol{N}_0\boldsymbol{u}^e}{\partial x}\frac{\partial \boldsymbol{N}_2\boldsymbol{u}^e}{\partial x}\right) + \omega^4\left(\frac{\partial \boldsymbol{N}_2\boldsymbol{u}^e}{\partial x}\right)^2\right]\mathrm{d}x\,\mathrm{d}y \\ &\quad + \frac{\sigma h}{2}\iint_A \left[\left(\frac{\partial \boldsymbol{N}_0\boldsymbol{u}^e}{\partial y}\right)^2 + 2\omega^2\left(\frac{\partial \boldsymbol{N}_0\boldsymbol{u}^e}{\partial y}\frac{\partial \boldsymbol{N}_2\boldsymbol{u}^e}{\partial y}\right) + \omega^4\left(\frac{\partial \boldsymbol{N}_2\boldsymbol{u}^e}{\partial y}\right)^2\right]\mathrm{d}x\,\mathrm{d}y\end{aligned} \tag{6.61}$$

The corresponding kinetic energy of the transversely vibrating membrane is expressed as

$$\tilde{T} = \iint_A \frac{1}{2}\rho\dot{u}^2\,\mathrm{d}x\,\mathrm{d}y \tag{6.62}$$

which takes the form

$$\begin{aligned}\tilde{T} &= \omega^2\rho\iint_A \frac{1}{2}[(\boldsymbol{N}_0 + \omega^2\boldsymbol{N}_2)\boldsymbol{u}^e]^2\,\mathrm{d}x\,\mathrm{d}y \\ &= \omega^2\frac{\rho}{2}\iint_A [(\boldsymbol{N}_0\boldsymbol{u}^e)^2 + 2\omega^2(\boldsymbol{N}_0\boldsymbol{u}^e)(\boldsymbol{N}_2\boldsymbol{u}^e) + \cdots]\,\mathrm{d}x\,\mathrm{d}y\end{aligned} \tag{6.63}$$

The two energies defined by Eqs. (6.61) and (6.63) may be equated by using Hamilton's principle. Expressing the strain and kinetic energies as

$$\Delta\tilde{U} = \tfrac{1}{2}\boldsymbol{u}^{eT}(\boldsymbol{K}_{0x}^e + \omega^2\boldsymbol{K}_{2x}^e + \omega^4\boldsymbol{K}_{4x}^e)\boldsymbol{u}^e + \tfrac{1}{2}\boldsymbol{u}^{eT}(\boldsymbol{K}_{0y}^e + \omega^2\boldsymbol{K}_{2y}^e + \omega^4\boldsymbol{K}_{4y}^e)\boldsymbol{u}^e \tag{6.64}$$

$$\tilde{T} = \omega^2\tfrac{1}{2}\boldsymbol{u}^{eT}(\boldsymbol{M}_0^e + \omega^2\boldsymbol{M}_2^e)\boldsymbol{u}^e \tag{6.65}$$

the equation of free vibration of the membrane element is obtained as

$$\left(\left(\boldsymbol{K}_{0x}^e + \boldsymbol{K}_{0y}^e\right) - \omega^2\left[\boldsymbol{M}_0^e - \left(\boldsymbol{K}_{2x}^e + \boldsymbol{K}_{2y}^e\right)\right] - \omega^4\left[\boldsymbol{M}_2^e - \left(\boldsymbol{K}_{4x}^e + \boldsymbol{K}_{4y}^e\right)\right]\right)\boldsymbol{q} = \boldsymbol{0} \tag{6.66}$$

When combined for the entire structure, this expression yields the quadratic matrix equation of the form of Eqs. (6.15) and (6.16). In the usual free vibration analysis, only the static element stiffness and mass matrices $\boldsymbol{K}_0$ and $\boldsymbol{M}_0$ are retained for the eigenproblem solution.

The expressions for the various stiffness matrices are obtained by comparing Eq. (6.64) with Eq. (6.61). Although the integral of Eq. (6.61) may be evaluated by numerical means, they have been solved algebraically to present the stiffness matrices in explicit form. These matrices are

$$\boldsymbol{K}_{0x}^e = \sigma h \begin{bmatrix} \frac{b}{3a} & -\frac{b}{3a} & -\frac{b}{6a} & \frac{b}{6a} \\ & \frac{b}{3a} & \frac{b}{6a} & -\frac{b}{6a} \\ & & \frac{b}{3a} & -\frac{b}{3a} \\ \text{sym} & & & \frac{b}{3a} \end{bmatrix} \tag{6.67}$$

$$\boldsymbol{K}_{2x}^e = \rho\frac{b^3}{a} \begin{bmatrix} \frac{1}{45} & -\frac{1}{45} & -\frac{7}{360} & \frac{7}{360} \\ & \frac{1}{45} & \frac{7}{360} & -\frac{7}{360} \\ & & \frac{1}{45} & -\frac{1}{45} \\ \text{sym} & & & \frac{1}{45} \end{bmatrix} \tag{6.68}$$

$$\boldsymbol{K}_{4x}^e = \frac{\rho^2}{\sigma h} \times \begin{bmatrix} \left(\frac{a^3b}{540} + \frac{b^5}{a}\frac{1}{1890}\right) & \left(\frac{7a^3b}{4320} - \frac{b^5}{a}\frac{1}{1890}\right) & \left(\frac{7a^3b}{8640} - \frac{b^5}{a}\frac{31}{60480}\right) & \left(\frac{a^3b}{1080} + \frac{b^5}{a}\frac{31}{60480}\right) \\ & \left(\frac{a^3b}{540} + \frac{b^5}{a}\frac{1}{1890}\right) & \left(\frac{a^3b}{1080} + \frac{b^5}{a}\frac{31}{60480}\right) & \left(\frac{7a^3b}{8640} - \frac{b^5}{a}\frac{31}{60480}\right) \\ \text{sym} & & \left(\frac{a^3b}{540} + \frac{b^5}{a}\frac{1}{1890}\right) & \left(\frac{7a^3b}{4320} - \frac{b^5}{a}\frac{1}{1890}\right) \\ & & & \left(\frac{a^3b}{540} + \frac{b^5}{a}\frac{1}{1890}\right) \end{bmatrix} \tag{6.69}$$

The corresponding matrices $\boldsymbol{K}_{0y}$, $\boldsymbol{K}_{2y}$, and $\boldsymbol{K}_{4y}$ pertaining to the y direction may be obtained from $\boldsymbol{K}_{0x}$, $\boldsymbol{K}_{2x}$, and $\boldsymbol{K}_{4x}$ by simply interchanging a and b.

Similarly the mass matrices are obtained by comparing Eq. (6.65) with Eq. (6.63), yielding the following expressions:

$$M_0^e = \rho ab \begin{bmatrix} \frac{1}{9} & \frac{1}{18} & \frac{1}{36} & \frac{1}{18} \\ & \frac{1}{9} & \frac{1}{18} & \frac{1}{36} \\ & & \frac{1}{9} & \frac{1}{18} \\ \text{sym} & & & \frac{1}{9} \end{bmatrix} \tag{6.70}$$

$$M_2^e = \frac{\rho^2}{\sigma h} \frac{ab}{2160} \begin{bmatrix} 16(a^2+b^2) & (14a^2+8b^2) & 7(a^2+b^2) & (8a^2+14b^2) \\ & 16(a^2+b^2) & (8a^2+14b^2) & 7(a^2+b^2) \\ & & 16(a^2+b^2) & (14a^2+8b^2) \\ \text{sym} & & & 16(a^2+b^2) \end{bmatrix} \tag{6.71}$$

Equations (6.67–6.71) thus provide all of the expressions for the stiffness and mass matrices of the membrane dynamic element. These matrices are combined by the usual process to yield the stiffness and mass matrices of an entire structure, which are then used to form the quadratic eigenvalue equation in the form of Eq. (6.16).

6.5 Plane Triangular Element

The differential equations of motion of a plane continuum undergoing free vibration are

$$\frac{\partial^2 u_x}{\partial x^2} + \frac{\partial^2 u_x}{\partial y^2} + \frac{1}{(1-2\nu)} \frac{\partial}{\partial x}\left(\frac{\partial u_x}{\partial x} + \frac{\partial u_y}{\partial y}\right) = 2(1+\nu)\frac{\rho}{E}\frac{\partial^2 u_x}{\partial t^2} \tag{6.72}$$

$$\frac{\partial^2 u_y}{\partial x^2} + \frac{\partial^2 u_y}{\partial y^2} + \frac{1}{(1-2\nu)} \frac{\partial}{\partial y}\left(\frac{\partial u_x}{\partial x} + \frac{\partial u_y}{\partial y}\right) = 2(1+\nu)\frac{\rho}{E}\frac{\partial^2 u_y}{\partial t^2} \tag{6.73}$$

where u_x, u_y are in-plane deformations, and ρ, E, and ν are the element mass per unit volume, Young's modulus, and Poisson's ratio, respectively. The plane triangular element under consideration is shown in Fig. 6.4; each node has two degrees of freedom. Both local (x, y) and global (X, Y) coordinate systems are also shown in the illustration. Solution of Eqs. (6.72) and (6.73) pertaining to the element may be expressed in series form as follows:

$$u_x = \boldsymbol{N}\boldsymbol{u}^e = \sum_{r=0}^{\infty} \omega^r \boldsymbol{N}_{rx}\boldsymbol{u}^e = \sum_{r=0}^{\infty} \omega^r \boldsymbol{N}_{rx}\boldsymbol{q}e^{i\omega t} \tag{6.74}$$

and similarly

$$u_y = \sum_{r=0}^{\infty} \omega^r \boldsymbol{N}_{ry}\boldsymbol{q}e^{i\omega t} \tag{6.75}$$

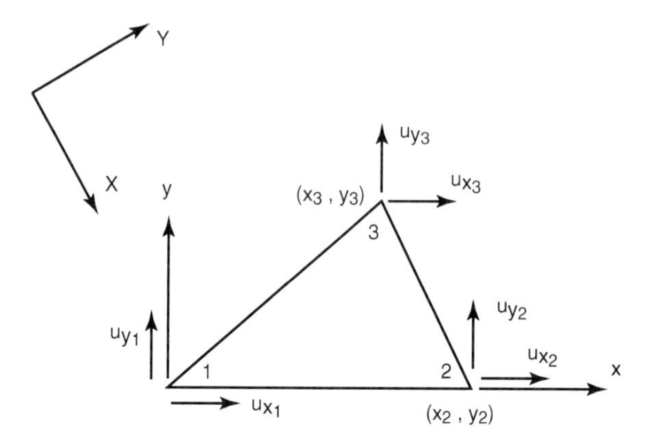

Fig. 6.4 Plane triangular element.

These expressions are next substituted in Eqs. (6.72) and (6.73) to yield the differential equations of motion in infinite series form. For the present case,[4] only the first three terms of the series are retained for subsequent analysis. The resulting differential equations of motion, e.g., in the x direction, are given next:

$$\frac{\partial^2 \boldsymbol{N}_{0x}}{\partial x^2} + \frac{\partial^2 \boldsymbol{N}_{0x}}{\partial y^2} + \alpha_2 \left(\frac{\partial^2 \boldsymbol{N}_{0x}}{\partial x^2} + \frac{\partial \boldsymbol{N}_{0y}}{\partial x \partial y} \right) = 0 \tag{6.76}$$

$$\frac{\partial^2 \boldsymbol{N}_{1x}}{\partial x^2} + \frac{\partial^2 \boldsymbol{N}_{1x}}{\partial y^2} + \alpha_2 \left(\frac{\partial^2 \boldsymbol{N}_{1x}}{\partial x^2} + \frac{\partial \boldsymbol{N}_{1y}}{\partial x \partial y} \right) = 0 \tag{6.77}$$

$$\alpha_1 \frac{\partial^2 \boldsymbol{N}_{2x}}{\partial x^2} + \frac{\partial^2 \boldsymbol{N}_{2x}}{\partial y^2} + \alpha_2 \frac{\partial^2 \boldsymbol{N}_{2y}}{\partial x \partial y} = -\beta \boldsymbol{N}_{0x} \tag{6.78}$$

where

$$\alpha_1 = \frac{2(1-\nu)}{(1-2\nu)}, \qquad \alpha_2 = \frac{1}{(1-2\nu)}, \qquad \text{and} \qquad \beta = 2\frac{\rho}{E}(1+\nu)$$

Similar expressions in the y direction may be obtained when the x and y suffixes are interchanged. The solutions of the differential equations are secured when $\boldsymbol{N}_0$ is allowed to satisfy appropriate boundary conditions, and noting that $\boldsymbol{N}_1$ and $\boldsymbol{N}_2$ must vanish at the boundaries. Thus, the expressions for $\boldsymbol{N}_{0x}$ and $\boldsymbol{N}_{0y}$ that satisfy differential Eq. (6.76) and its counterpart are taken as follows:

$$\boldsymbol{N}_{0x} = c_1 + c_2 x + c_3 y \tag{6.79}$$

$$\boldsymbol{N}_{0y} = c_4 + c_5 x + c_6 y \tag{6.80}$$

and the coefficients c_1–c_6 are evaluated from the boundary conditions

$$u_x = u_{x1}, \qquad u_y = u_{y1} \qquad \text{at} \qquad x = 0, \quad y = 0$$

$$u_x = u_{x2}, \qquad u_y = u_{y2} \qquad \text{at} \qquad x = x_2, \quad y = 0$$

$$u_x = u_{x3}, \qquad u_y = u_{y3} \qquad \text{at} \qquad x = x_3, \quad y = y_3$$

Expressions for $\boldsymbol{N}_{1x}$ and $\boldsymbol{N}_{1y}$ are chosen in a form similar to Eqs. (6.79) and

(6.80), respectively; after the satisfaction of appropriate boundary conditions, they are determined as $\boldsymbol{N}_{1x} = 0$, $\boldsymbol{N}_{1y} = 0$. The relevant solutions of Eq. (6.78) and its counterpart are assumed as follows[4]:

$$\boldsymbol{N}_{2x} = \hat{c}_1 + \hat{c}_2 x + \hat{c}_3 y - \beta\left[\frac{c_1}{4}\frac{x^2}{\alpha_1} + \frac{c_1}{4}y^2 + \frac{c_2}{12}\frac{x^3}{\alpha_1} + \frac{c_2}{4}xy^2 + \frac{c_3}{4}\frac{x^2 y}{\alpha_1} + \frac{c_3 y^3}{12}\right] \tag{6.81}$$

$$\boldsymbol{N}_{2y} = \hat{c}_4 + \hat{c}_5 x + \hat{c}_6 y - \beta\left[\frac{c_4}{4}\frac{y^2}{\alpha_1} + \frac{c_4}{4}x^2 + \frac{c_5}{12}x^3 + \frac{c^5}{4}\frac{xy^2}{\alpha_1} + \frac{c_6}{4}x^2 y + \frac{c_6}{12}\frac{y^3}{\alpha_1}\right] \tag{6.82}$$

The coefficients c_1–c_6 are computed earlier from the solution of Eqs. (6.79) and (6.80). The coefficients $\hat{c}_1$–$\hat{c}_6$ pertaining to the complementary functions are determined by satisfaction of the appropriate boundary conditions. After some algebraic manipulation, the final forms of the shape functions are determined in terms of the element nodal coordinates expressed in the local coordinate system:

$$\boldsymbol{N}_{0x} = \boldsymbol{N}_{0y} = \left[\left(\frac{x_3}{x_2}\frac{y}{y_3} - \frac{y}{y_3} - \frac{x}{x_2} + 1\right)\left(\frac{x}{x_2} - \frac{x_3}{x_2}\frac{y}{y_3}\right)\frac{y}{y_3}\right] \tag{6.83}$$

$$\boldsymbol{N}_{2x}^{T} = \beta\begin{bmatrix} \left(-\frac{x_3 y y_3}{6x_2} + \frac{y y_3}{6} - \frac{x_3 y^3}{12 x_2 y_3} + \frac{y^3}{12 y_3} + \frac{x_3^3 y}{6\alpha_1 x_2 y_3} - \frac{x_2 x_3 y}{6\alpha_1 y_3} \right.\\ \left. - \frac{x^2 x_3 y}{4\alpha_1 x_2 y_3} + \frac{x^2 y}{4\alpha_1 y_3} + \frac{x y^2}{4 x_2} - \frac{y^2}{4} + \frac{x x_2}{6\alpha_1} + \frac{x^3}{12\alpha_1 x_2} - \frac{x^2}{4\alpha_1}\right) \\ \left(\frac{x_3 y y_3}{6x_2} + \frac{x_3 y^3}{12 x_2 y_3} - \frac{x_3^3 y}{6\alpha_1 x_2 y_3} - \frac{x_2 x_3 y}{12\alpha_1 y_3} + \frac{x^2 x_3 y}{4\alpha_1 x_2 y_3}\right.\\ \left. - \frac{x y^2}{4 x_2} + \frac{x x_2}{12\alpha_1} - \frac{x^3}{12\alpha_1 x_2}\right) \\ \left(\frac{y y_3}{12} - \frac{y^3}{12 y_3} + \frac{x_3^2 y}{4\alpha_1 y_3} - \frac{x^2 y}{4\alpha_1 y_3}\right) \end{bmatrix} \tag{6.84}$$

$$\boldsymbol{N}_{2y}^{T} = \beta\begin{bmatrix} \left(-\frac{x_3 y y_3}{6\alpha_1 x_2} + \frac{y y_3}{6\alpha_1} - \frac{x_3 y^3}{12\alpha_1 x_2 y_3} + \frac{y^3}{12\alpha_1 y_3} + \frac{x_3^3 y}{6 x_2 y_3} - \frac{x_2 x_3 y}{6 y_3} \right.\\ \left. - \frac{x^2 x_3 y}{4 x_2 y_3} + \frac{x^2 y}{4 y_3} + \frac{x y^2}{4\alpha_1 x_2} - \frac{y^2}{4\alpha_1} + \frac{x x_2}{6} + \frac{x^3}{12 x_2} - \frac{x^2}{4}\right) \\ \left(\frac{x_3 y y_3}{6\alpha_1 x_2} + \frac{x_3 y^3}{12\alpha_1 x_2 y_3} - \frac{x_3^3 y}{6 x_2 y_3} - \frac{x_2 x_3 y}{12 y_3} + \frac{x^2 x_3 y}{4 x_2 y_3} - \frac{x y^2}{4\alpha_1 x_2}\right.\\ \left. + \frac{x x_2}{12} - \frac{x^3}{12 x_2}\right) \\ \left(\frac{y y_3}{12\alpha_1} - \frac{y^3}{12\alpha_1 y_3} + \frac{x_3^2 y}{4 y_3} - \frac{x^2 y}{4 y_3}\right) \end{bmatrix} \tag{6.85}$$

Shape function vectors are thus defined as

$$\boldsymbol{N}_x = \boldsymbol{N}_{0x} + \omega^2 \boldsymbol{N}_{2x}, \qquad \boldsymbol{N}_y = \boldsymbol{N}_{0y} + \omega^2 \boldsymbol{N}_{2y} \tag{6.86}$$

in which each of the scalars of these vectors is coupled to the appropriate nodal degrees-of-freedom. The associated matrix that defines strain–displacement relationship,

$$\epsilon = \boldsymbol{B}\boldsymbol{u}^e \tag{6.87}$$

is assumed to be in the form

$$\boldsymbol{B} = \boldsymbol{B}_0 + \omega^2 \boldsymbol{B}_2 \tag{6.88}$$

each individual matrix being obtained from the following known formulation:

$$\epsilon_x = \frac{\partial u_x}{\partial x} = \frac{\partial}{\partial x}\left(\boldsymbol{N}_{0x} + \omega^2 \boldsymbol{N}_{2x}\right)\boldsymbol{u}^e = \boldsymbol{B}_{0xx}\boldsymbol{u}^e + \omega^2 \boldsymbol{B}_{2xx}\boldsymbol{u}^e \tag{6.89}$$

$$\epsilon_y = \frac{\partial u_y}{\partial y} = \frac{\partial}{\partial y}\left(\boldsymbol{N}_{0y} + \omega^2 \boldsymbol{N}_{2y}\right)\boldsymbol{u}^e = \boldsymbol{B}_{0yy}\boldsymbol{u}^e + \omega^2 \boldsymbol{B}_{2yy}\boldsymbol{u}^e \tag{6.90}$$

$$\begin{aligned}\gamma_{xy} &= \frac{\partial u_x}{\partial y} + \frac{\partial u_y}{\partial x} \\ &= \left(\frac{\partial \boldsymbol{N}_{0x}}{\partial y} + \frac{\partial \boldsymbol{N}_{0y}}{\partial x}\right)\boldsymbol{u}^e + \omega^2\left(\frac{\partial \boldsymbol{N}_{2x}}{\partial y} + \frac{\partial \boldsymbol{N}_{2y}}{\partial x}\right)\boldsymbol{u}^e \\ &= \boldsymbol{B}_{0xy}\boldsymbol{u}^e + \omega^2 \boldsymbol{B}_{2xy}\boldsymbol{u}^e \end{aligned} \tag{6.91}$$

Once the various $\boldsymbol{N}$ and $\boldsymbol{B}$ matrices are generated, the element inertia and stiffness matrices may be developed by standard procedures. Thus, the inertia matrices are obtained from the following relations:

$$\begin{aligned}\boldsymbol{M}_x^e &= \rho_v \int_V \boldsymbol{N}_x^T \boldsymbol{N}_x \, \mathrm{d}V \\ &= \rho \int_A \boldsymbol{N}_{0x}^T \boldsymbol{N}_{0x} \, \mathrm{d}x \, \mathrm{d}y + \omega^2 \left[\rho \int_A \boldsymbol{N}_{0x}^T \boldsymbol{N}_{2x} \, \mathrm{d}x \, \mathrm{d}y + \rho \int_A \boldsymbol{N}_{2x}^T \boldsymbol{N}_{0x} \, \mathrm{d}x \, \mathrm{d}y\right] \\ &= \boldsymbol{M}_{0x}^e + \omega^2 \boldsymbol{M}_{2x}^e \end{aligned} \tag{6.92}$$

$$\begin{aligned}\boldsymbol{M}_y^e &= \rho_v \int_V \boldsymbol{N}_y^T \boldsymbol{N}_y \, \mathrm{d}V \\ &= \rho \int_A \boldsymbol{N}_{0y}^T \boldsymbol{N}_{0y} \, \mathrm{d}x \, \mathrm{d}y + \omega^2 \left[\rho \int_A \boldsymbol{N}_{0y}^T \boldsymbol{N}_{2y} \, \mathrm{d}x \, \mathrm{d}y + \rho \int_A \boldsymbol{N}_{2y}^T \boldsymbol{N}_{0y} \, \mathrm{d}x \, \mathrm{d}y\right] \\ &= \boldsymbol{M}_{0y}^e + \omega^2 \boldsymbol{M}_{2y}^e \end{aligned} \tag{6.93}$$

when $\boldsymbol{M}_0^e$ and $\boldsymbol{M}_2^e$ are finally obtained by appropriate combination of $\boldsymbol{M}_{0x}^e$, $\boldsymbol{M}_{0y}^e$ and $\boldsymbol{M}_{2x}^e$, $\boldsymbol{M}_{2y}^e$, respectively, and ρ_v is the mass per unit volume, ρ the mass per unit area of the element, and A the area. Similarly, the stiffness matrices are derived as

$$\boldsymbol{K}^e = \int_V \boldsymbol{B}^T \boldsymbol{D} \boldsymbol{B} \, \mathrm{d}V \tag{6.94}$$

$$K_0^e = \int_V B_0^T D B_0 \, \mathrm{d}V \tag{6.95}$$

$$K_4^e = \int_V B_2^T D B_2 \, \mathrm{d}V \tag{6.96}$$

in which $\boldsymbol{D}$ is the stress–strain matrix pertaining to two-dimensional elasticity. Derivations of the various $\boldsymbol{N}$, $\boldsymbol{B}$, $\boldsymbol{M}^e$, and $\boldsymbol{K}^e$ matrices involve a rather large number of algebraic manipulations, which may by conveniently achieved by numerical integration. The elements of these matrices are finally determined in the form of simple algebraic expressions; a suitable instruction is used next to program them automatically in FORTRAN.

Once the element matrices are generated, they are appropriately combined to form the global matrices pertaining to a structure to yield the quadratic matrix eigenvalue problem base on the first three terms of Eq. (6.15).

6.6 Shell Element

A composite shell dynamic finite element[5,6] has been developed on the basis of the flat shell element described earlier and is suitable for laminated and sandwich structures as well as a single layer shell. Figure 6.5 depicts such an element. The displacement components within the element may be written as

$$\begin{aligned} u_x(x, y, z, t) &= u_x^0(x, y, t) + z\psi_x(x, y, t) \\ u_y(x, y, z, t) &= u_y^0(x, y, t) + z\psi_y(x, y, t) \\ u_z(x, y, z, t) &= u_z(x, y, t) \end{aligned} \tag{6.97}$$

in which u_x^0 and u_y^0 are the displacement values of u_x and u_y at the middle surface, ψ_x and ψ_y being rotation of the normal. As previously, the shape functions are

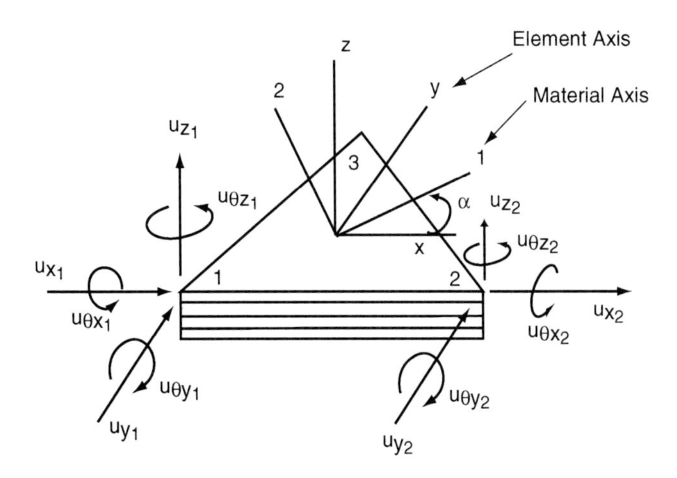

Fig. 6.5 Layered shell element.

expanded in ascending power of the frequency

$$\begin{aligned} \boldsymbol{u} &= [u_x \; u_y \; u_z]^T = \boldsymbol{N}\boldsymbol{u}^e \\ &= \left(\boldsymbol{N}_0 + \omega^2 \boldsymbol{N}_2 + \cdots\right)\boldsymbol{u}^e \end{aligned} \tag{6.98}$$

in which $\boldsymbol{N}_0$ is the conventional finite element interpolation function and the function $\boldsymbol{N}_2$ is chosen to satisfy exactly the dynamic differential equations. They should also vanish on the element boundaries to satisfy continuity conditions.

In the Mindlin plate theory, the normals to the middle surface of the plate are assumed to remain straight lines but not necessarily remain normal. For an anisotropic layered Mindlin plate, the shape functions must satisfy five partial differential equations, whereas for an isotropic Mindlin plate only three such equations need to be satisfied. The interpolation functions are assumed as

$$u_x^0 = \left[\boldsymbol{N}_{0x}\boldsymbol{u}_x^e + \boldsymbol{N}_{2x}\hat{\boldsymbol{u}}^e\right] \tag{6.99}$$

$$u_y^0 = \left[\boldsymbol{N}_{0y}\boldsymbol{u}_y^e + \boldsymbol{N}_{2y}\hat{\boldsymbol{u}}^e\right] \tag{6.100}$$

$$\psi_x = u_{\theta y} = \left[\boldsymbol{N}_{0\theta y}\boldsymbol{u}_{\theta y}^e + \boldsymbol{N}_{2\theta y}\hat{\boldsymbol{u}}^e\right] \tag{6.101}$$

$$-\psi_y = u_{\theta x} = \left[\boldsymbol{N}_{0\theta x}\boldsymbol{u}_{\theta x}^e + \boldsymbol{N}_{2\theta y}\hat{\boldsymbol{u}}^e\right] \tag{6.102}$$

$$u_z = \left[\boldsymbol{N}_{0z} + \boldsymbol{N}_{2z}\right]\tilde{\boldsymbol{u}}^e \tag{6.103}$$

in which $\hat{\boldsymbol{u}}^e = [\boldsymbol{u}_x^e \; \boldsymbol{u}_y^e \; \boldsymbol{u}_{\theta x}^e \; \boldsymbol{u}_{\theta y}^e]^T$ and $\tilde{\boldsymbol{u}}^e = [\boldsymbol{u}_z^e \; \boldsymbol{u}_{\theta x}^e \; \boldsymbol{u}_{\theta y}^e]^T$. A significant effort[6] in this connection used dynamic shape functions that satisfied the five partial differential equations exactly and vanished at nodal points, but not everywhere on the element boundaries, without effecting solution convergence.

The higher-order matrices are derived as previously. Thus the mass matrix is generated as

$$\boldsymbol{M}_2 = \int_V \rho\left(\boldsymbol{N}_0^T \boldsymbol{N}_2 + \boldsymbol{N}_2^T \boldsymbol{N}_0\right) \mathrm{d}V \tag{6.104}$$

where both $\boldsymbol{N}_0$ and $\boldsymbol{N}_2$ are (3×15) matrices and in which

$$\boldsymbol{N}_0 = \begin{bmatrix} \boldsymbol{N}_{0x} + z\boldsymbol{N}_{0\theta y} \\ \boldsymbol{N}_{0y} - z\boldsymbol{N}_{0\theta x} \\ \boldsymbol{N}_{0z} \end{bmatrix} \tag{6.105}$$

$$\boldsymbol{N}_2 = \begin{bmatrix} \boldsymbol{N}_{2x} + z\boldsymbol{N}_{2\theta y} \\ \boldsymbol{N}_{2y} - z\boldsymbol{N}_{2\theta x} \\ \boldsymbol{N}_{2z} \end{bmatrix} \tag{6.106}$$

and

$$\begin{aligned} \boldsymbol{u} &= [u_x \; u_y \; u_z]^T \\ &= \left[\boldsymbol{N}_0 + \omega^2 \boldsymbol{N}_2\right]\left[\boldsymbol{u}_x^e \; \boldsymbol{u}_y^e \; \boldsymbol{u}_z^e \; \boldsymbol{u}_{\theta x}^e \; \boldsymbol{u}_{\theta y}^e\right]^T \end{aligned} \tag{6.107}$$

The $\boldsymbol{K}_4$ matrix also can be generated similarly. Complete details of the shape functions and element matrices are given in Ref. 6.

6.7 Numerical Examples

A number of example problems are presented next that illustrate the use of a variety of dynamic elements; these solutions are also compared with the usual finite element procedure.

Example 6.1: Prestressed Membrane

A simply supported square membrane is first considered with the following data in parametric form: tension per unit area $(\sigma) = 1$, mass per unit area $(\rho) = 1$, side length $= 5$, and thickness $= 1$. An increasing number of square elements are used to discretize the membrane, and the results of such computations are depicted in Table 6.1.[3] Figure 6.6 presents the pattern of root convergence as a function of mesh size.

Table 6.1 Natural frequencies of a square membrane[a]

	Eigenvalues, rad/s				
Mesh size	ω_1	$\omega_2 = \omega_3$	ω_4	$\omega_5 = \omega_6$	$\omega_7 = \omega_8$
2 × 2	0.979838 0.936101				
3 × 3	0.929516 0.900712	1.609969 1.573654	2.078463 1.995600		
4 × 4	0.911555 0.892881	1.528224 1.478296	1.959581 1.872221	2.342161 2.364136	2.643908 2.598400
5 × 5	0.903256 0.890449	1.484406 1.443888	1.894968 1.822838	2.247806 2.191767	2.537759 2.443643
6 × 6	0.898759 0.889529	1.460097 1.428606	1.858942 1.801452	2.173449 2.110863	2.459268 2.367514
7 × 7	0.896052 0.889112	1.445403 1.420714	1.837215 1.791163	2.124928 2.069099	2.408652 2.328801
8 × 8	0.894297 0.888884	1.435878 1.416174	1.823010 1.785756	2.092707 2.045165	2.375164 2.307410
9 × 9	0.893095 0.888781	1.429360 1.413343	1.813270 1.782710	2.070461 2.030297	2.352076 2.294721
10 × 10	0.892236 0.888706	1.424705 1.411466	1.806476 1.780960	2.054522 2.02471	2.335543 2.286750
11 × 11	0.891600 0.888664	1.421267 1.410159	1.801403 1.779835	2.042733 2.013653	2.323317 2.281497
13 × 12	0.891117 0.888640	1.418655 1.409216	1.797519 1.779025	2.033775 2.008735	2.314029 2.277896
14 × 14	0.890444 0.888612	1.415014 1.407972	1.792002 1.778193	2.021307 2.002265	2.301100 2.273489
Exact results	0.888577	1.404963	1.777153	1.986918	2.265435

[a]FEM values in upper rows, DEM values in lower rows.

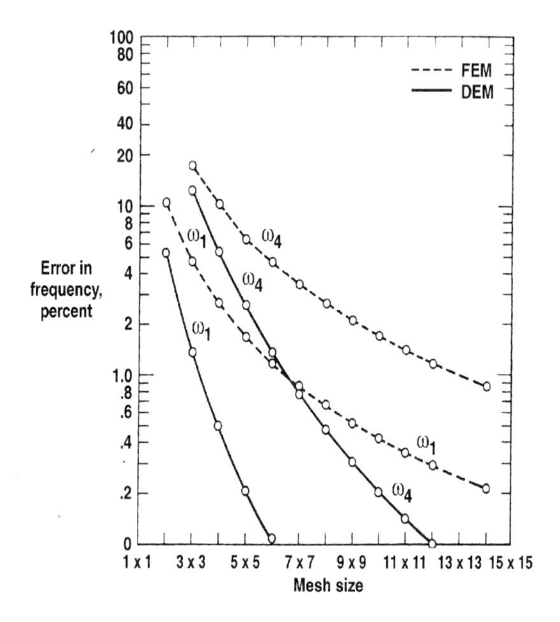

Fig. 6.6 Convergence characteristics of frequencies of a square membrane by the finite static and dynamic element methods.

Example 6.2: Cantilever Plate

A square cantilever plate undergoing in-plane vibration is shown in Fig. 6.7. The plate is discretized by six-noded plane triangular elements[8] employing both FEM and DEM formulations to yield the first few natural frequencies and associated modes; relevant structural data employed for the analyses are Young's modulus $E = 10 \times 10^6$, mass density $\rho = 0.259 \times 10^{-3}$, side length $l = 10$, and thickness

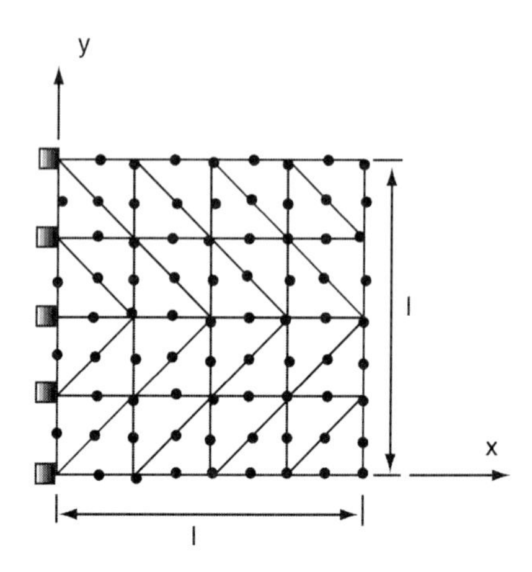

Fig. 6.7 Square cantilever plate with a (4 × 4) mesh.

Table 6.2 Natural frequencies of a square cantilever plate[a]

Number of elements	Natural frequency parameter, $\lambda = \omega/\sqrt{E/\rho} \times 10$					
	λ_1	λ_2	λ_3	λ_4	λ_5	λ_6
2	0.0753	0.1669	0.2005	0.3289		
	0.0713	0.1601	0.1897	0.3169		
8	0.0693	0.1615	0.1895	0.2990	0.3251	0.3418
	0.0668	0.1585	0.1825	0.2873	0.3151	0.3302
32	0.0666	0.1597	0.1801	0.2847	0.3125	0.3230
	0.0661	0.1581	0.1780	0.2812	0.3056	0.3230
Exact	0.0658	0.1579	0.1769	0.2796	0.3033	0.3214

[a]FEM values in upper rows and DEM values in lower rows.

$t = 0.1$. Analyses were performed for varying mesh sizes, and Table 6.2 presents details of the solution results. Figure 6.8 depicts percentage solution errors in graphical form for two typical modes.

Example 6.3: Composite Plates and Shells

A swept, tapered cantilever plate[6] with the shape resembling a wing is shown in Fig. 6.9. The plate has 0.2-in.-thick graphite–epoxy faces and a 3.6-in.-thick

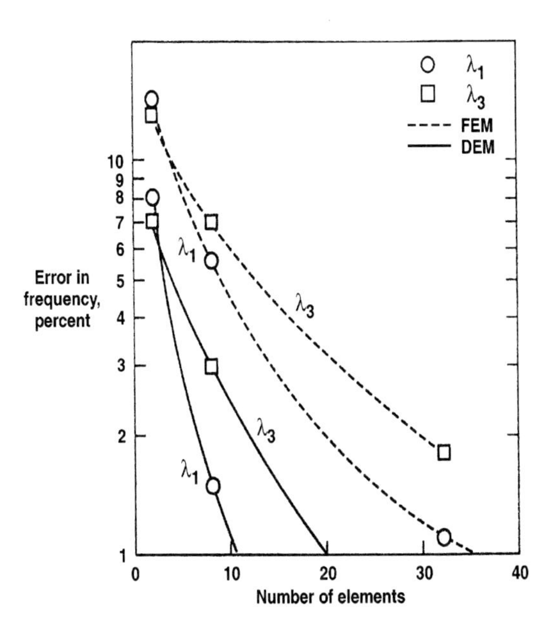

Fig. 6.8 Root convergence characteristics for plane six-noded finite and dynamic elements.

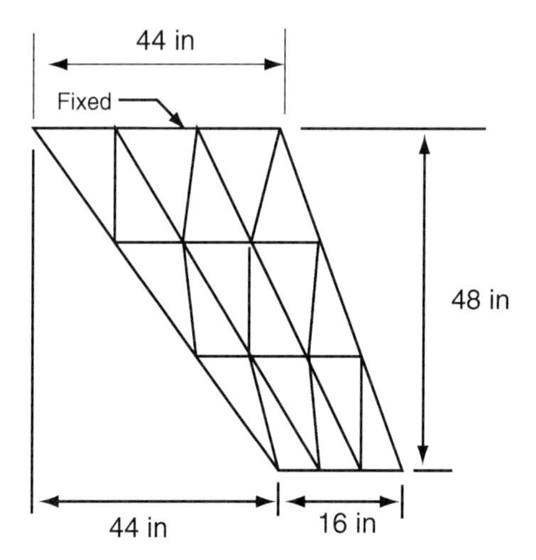

Fig. 6.9 Swept, tapered sandwich plate.

honeycomb core. The chord varies from 44 in. at the root (fixed end) to 16 in. at the tip; the span is 4 ft. The first 12 frequencies were calculated using meshes for which no frequency has an error greater than 3%. The cost parameter nm^2 for the meshes required to achieve this accuracy using various relevant elements is depicted in Table 6.3; n is the order and m is the semibandwidth of the stiffness matrix.

6.8 Concluding Remarks

Related work for tapered beams,[7] higher-order plane elements,[8] and solid elements[9] is published in the open literature. Other similar efforts in this connection are published in Refs. 10 and 11. Relevant example problems provide a comparison of solution convergence characteristics of the FEM and DEM analyses. A symbolic manipulation program MACSYMA[12] can be effectively used for all algebraic manipulations involved in the development of element matrices. The STARS program[13] was used for the solution of problems presented in this section.

Table 6.3 Natural frequencies of a tapered sandwich plate[a]

Element	Mesh	Cost, nm^2	Cost rating
3-node FEM	12 × 12	5.69×10^5	2.93
3-node DEM	9 × 9	1.94×10^6	1
8-node FEM	6 × 6	11.43×10^6	5.88

[a]FEM values in upper rows and DEM values in lower rows.

References

[1]Przemieniecki, J. S., "Quadratic Matrix Equations for Determining Vibration Modes and Frequencies of Continuous Elastic Systems," *Proceedings of the Conference on Matrix Methods in Structural Mechanics*, Wright–Patterson AFB, OH, 1965, pp. 779–802; AFFDL TR 66-80, 1966.

[2]Przemieniecki, J. S., *Theory of Matrix Structural Analysis*, McGraw–Hill, New York, 1968.

[3]Gupta, K. K., "On a Finite Dynamic Element Method for Free Vibration Analysis of Structures," *Computer Methods in Applied Mechanics and Engineering*, Vol. 9, 1976, pp. 105–120.

[4]Gupta, K. K., "Finite Dynamic Element Formulation for a Plane Triangular Element," *International Journal for Numerical Methods in Engineering*, Vol. 14, 1984, pp. 1407–1414.

[5]Martin, C. W., Lung, S. F., and Gupta, K. K., "A Three-Node C^o Element for Analysis of Laminated Composite Sandwich Shells," NASA TM 4125, June 1989.

[6]Martin, C. W., and Lung, S. F., "A Finite Dynamic Element for Laminated Composite Plates and Shells," *Computers & Structures*, Vol. 40, No. 5, 1991, pp. 1249–1259.

[7]Gupta, A. K., "Frequency Dependent Matrices for Tapered Beams," *Journal of Structural Engineering*, Vol. 112, 1986, pp. 85–103.

[8]Gupta, K. K., Lawson, C. L., and Ahmadi, A. R., "On Development of a Finite Dynamic Element and Solution of Associated Eigenproblem by a Block Lanczos Procedure," *International Journal of Numerical Methods in Engineering*, Vol. 33, 1992, pp. 1611–1623.

[9]Gupta, K. K., "Development of a Solid Hexahedron Finite Dynamic Element," *International Journal of Numerical Methods in Engineering*, Vol. 20, 1984, pp. 2143–2150.

[10]Fricker, A. J., "A New Approach to the Dynamic Analysis of Structures using Fixed Frequency Dynamic Stiffness Matrices," *International Journal of Numerical Methods in Engineering*, Vol. 19, 1983, pp. 1111–1129.

[11]Voss, H., "A New Justification of Finite Dynamic Element Methods," *Numerical Mathematics*, Vol. 83, 1987, pp. 222–242.

[12]Bogen, R., *MACSYMA Reference Manual*, The Mathlab Group, MIT, Cambridge, MA, 1975.

[13]Gupta, K. K., "STARS—An Integrated, Multidisciplinary, Finite-Element, Structural, Fluids, Aeroelastic and Aeroservoelastic Analysis Computer Program," NASA TM 4795, 1997.

7
Generation of System Matrices

7.1 Introduction

In Chapters 4–6, all structural characteristic matrices, such as stiffness and inertia matrices, have been conveniently expressed in individual element, attached local coordinate systems (LCS). Usually an entire structure is defined in terms of a single global coordinate system (GCS). Also, for many practical structures, because of ease of data input, each substructure is defined in terms of the coordinate system unique to that substructure, which may conveniently be referred to as the local–global coordinate system (LGCS). The relevant element characteristic matrices expressed in any specific coordinate system, either in LCS or LGCS, are then required to be transformed in the GCS before they can be assembled for the entire structure.[1,2] Once assembled, the various displacement boundary conditions (DBC) need to be incorporated into such matrices before the relevant solution process may be started.

Typically, an assembled matrix tends to be highly banded because of the very nature of FEM discretization. Incorporation of DBCs have a tendency to increase the bandwidth. Because the solution effort of these matrices is usually a function of $(N \times M^2)$, N being the order and M the semibandwidth, it is essential to evolve efficient bandwidth minimization techniques to keep the bandwidth to a possible minimum. These features are addressed in this chapter.

7.2 Coordinate Systems and Transformations

For a line element in Fig. 7.1, the local x axis is defined by the straight line joining the two end nodes and the y axis is chosen along the direction of a principal moment of inertia axis lying in the plane of the cross section. The local z axis is then the axis perpendicular to the local x-y plane, and is in the direction of the other principal axes of the cross section.

In the case of either a triangular or a quadrilateral planar element (Fig. 7.2), the x axis is again defined as the line joining nodes 1 and 2. Lines joining nodes 1 and 3 for the triangular element and nodes 1 and 4 for the quadrilateral element define the in-plane vector for the element, and their cross product with the respective x axis vector defines the local z axis. Finally a cross product of the x axes and z axes yields the respective y axis. An identical procedure is adopted for three-dimensional solid tetrahedron and hexahedron elements given in Fig. 7.3.

For many practical problems it is necessary to use a number of LGCSs for the ease of defining nodal definitions such as coordinates. Thus Fig. 7.4 provides a simple depiction of a spacecraft with a number of LGCS. It is sometimes useful to obtain the global coordinates of any point defined in say the Ith LGCS, which

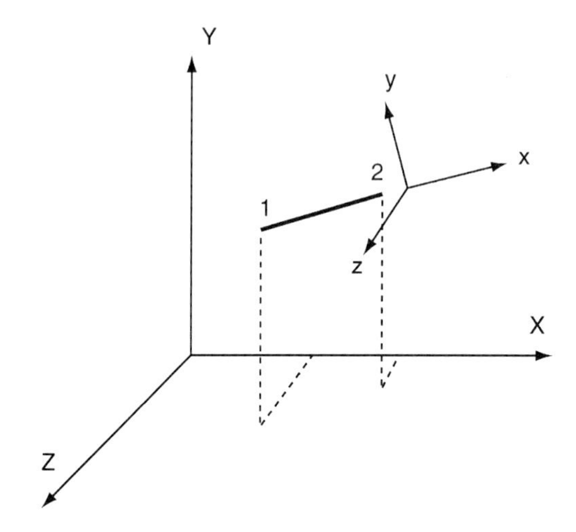

Fig. 7.1 Line element—coordinate systems.

may be obtained as

$$\boldsymbol{C} = [XI \quad YI \quad ZI]^T + \boldsymbol{\Lambda}_I^T [x \quad y \quad z]_I^T$$

in which XI, YI, ZI are global coordinates of the origin of the Ith LGCS, and x, y, z are the coordinates of the point expressed in the Ith LGCS; $\boldsymbol{\Lambda}_I$ is the direction cosine matrix of the Ith LGCS with respect to the GCS.

7.2.1 Other Coordinate Systems

For many practical problems, particularly with geometry displaying some degree of cyclic symmetry, it is useful to express relevant nodal data in cylindrical (Fig. 7.5) and also spherical (Fig. 7.6) coordinate systems. Thus a point s in the cylindrical

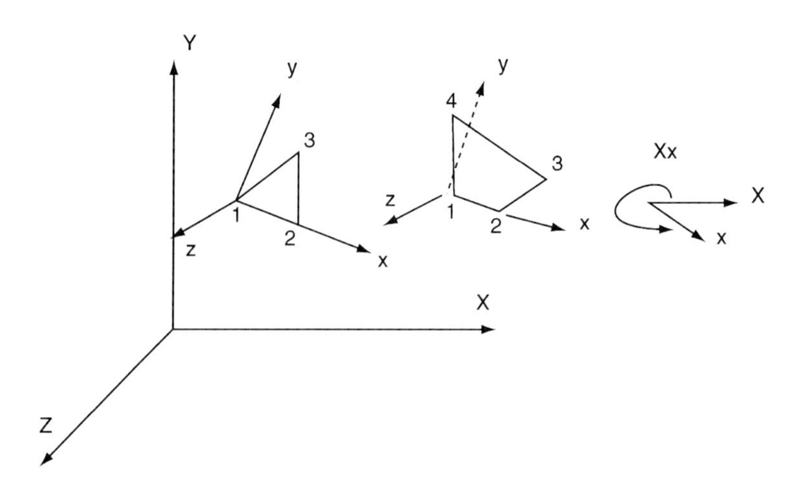

Fig. 7.2 Planar element—coordinate systems.

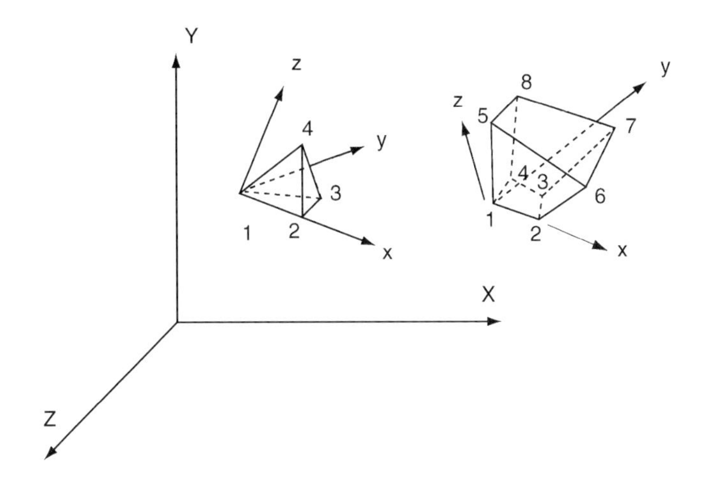

Fig. 7.3 Solid element—coordinate systems.

coordinate system is defined by the radius r and angle θ measured from the x^i axis defined in the ith LGCS and its coordinates in the rectangular coordinate system are obtained as

$$x_s^i = r\cos\theta, \qquad y_s^i = r\sin\theta, \qquad z_s^i = z$$

Similarly in a spherical coordinate system (r, θ, ϕ), the relevant coordinates are given by

$$x_s^i = r\cos\theta\sin\phi, \qquad y_s^i = r\sin\theta\sin\phi, \qquad z_s^i = r\cos\phi$$

All such coordinates expressed in LGCS are next transformed into GCS by the standard procedure described earlier.

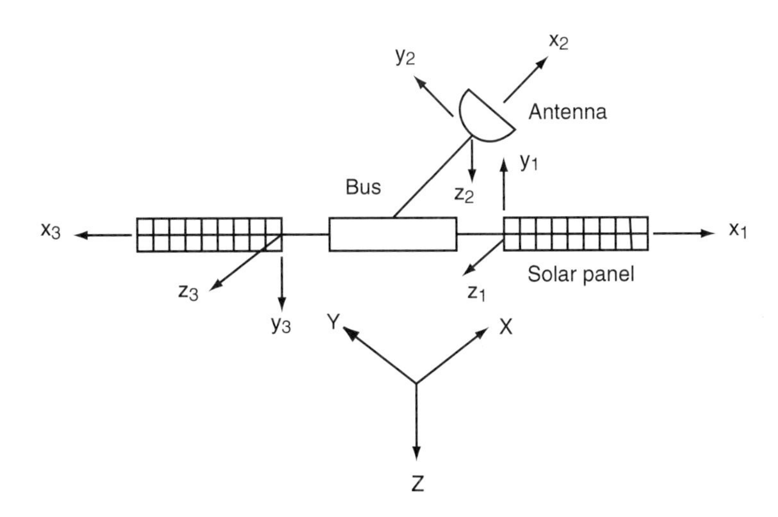

Fig. 7.4 Structural system with multiple LGCS.

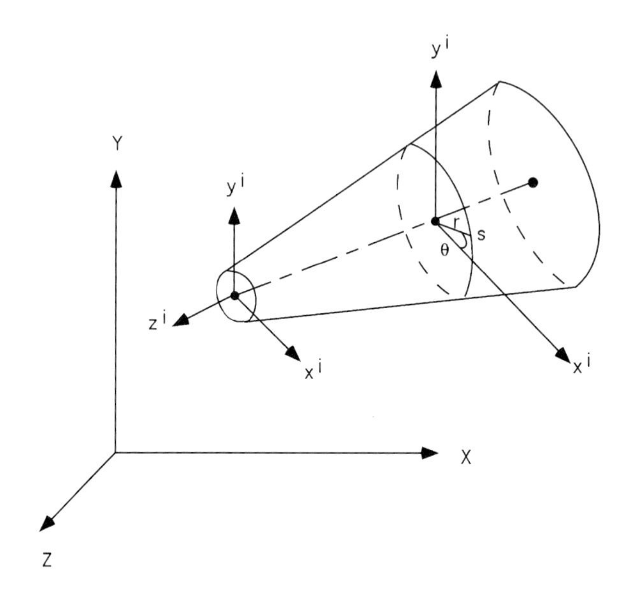

Fig. 7.5 Cylindrical coordinate system.

7.2.2 *Coordinate Transformation*

The direction cosine matrix for the general three-dimensional case is defined as

$$\boldsymbol{\lambda} = \begin{bmatrix} l_x & m_x & n_x \\ l_y & m_y & n_y \\ l_z & m_z & n_z \end{bmatrix} \tag{7.1}$$

in which, e.g., l_x is the cosine of the angle measured from global X to local x axis

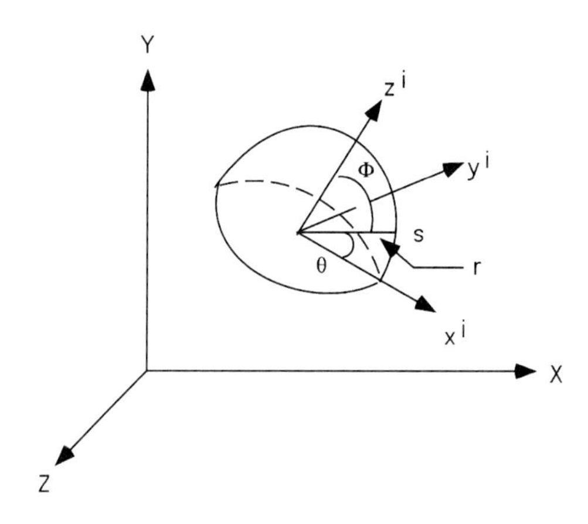

Fig. 7.6 Spherical coordinate system.

traversing in the counterclockwise direction with similar definition for the other angles in Eq. (7.1). Then the following relationships can easily be derived that define the relationship of variables between LCS and GCS:

$$\begin{array}{llll} \text{Displacement} & \boldsymbol{u}^e = \boldsymbol{\Lambda} \boldsymbol{q}^e & \text{or} & \boldsymbol{q}^e = \boldsymbol{\Lambda}^T \boldsymbol{u}^e \\ \text{Load} & \boldsymbol{f}^e = \boldsymbol{\Lambda} \boldsymbol{p}^e & \text{or} & \boldsymbol{p}^e = \boldsymbol{\Lambda}^T \boldsymbol{f}^e \end{array} \tag{7.2}$$

$$\begin{array}{ll} \text{Element stiffness} & \tilde{\boldsymbol{K}}^e = \boldsymbol{\Lambda}^T \boldsymbol{K}^e \boldsymbol{\Lambda} \\ \text{Element inertia} & \tilde{\boldsymbol{M}}^e = \boldsymbol{\Lambda}^T \boldsymbol{M}^e \boldsymbol{\Lambda} \end{array} \tag{7.3}$$

in which $\boldsymbol{u}^e$ and $\boldsymbol{q}^e$ are element nodal displacement in LCS and GCS, respectively. Similarly, $\boldsymbol{f}^e$ and $\boldsymbol{p}^e$ are loads in LCS and GCS; $\tilde{\boldsymbol{K}}^e$ and $\tilde{\boldsymbol{M}}^e$ are in the GCS and in which

$$\boldsymbol{\Lambda} = \begin{bmatrix} \lambda_1 & & & \\ & \lambda_2 & & \boldsymbol{0} \\ & & \ddots & \\ \boldsymbol{0} & & & \\ & & & \lambda_{NE2} \end{bmatrix} \tag{7.4}$$

is the direction cosine matrix for each element, the order of which is equal to the total number of element degrees of freedom ($NE2 = 2 \times$ number of nodes in the element).

7.3 Matrix Assembly

Once the element matrices have been obtained in the GCS, they may be assembled appropriately to yield the matrices for the entire structure, written symbolically as

$$\boldsymbol{K} = \sum \tilde{\boldsymbol{K}}^e \tag{7.5}$$

$$\boldsymbol{M} = \sum \tilde{\boldsymbol{M}}^e \tag{7.6}$$

and so on for the relevant stiffness and inertia matrices. This is achieved easily by adopting a systematic assembly process in which each element of the typical $\tilde{\boldsymbol{K}}^e$ matrix, with its row location identified as the nodal degree of freedom whereas the column location is indicative of a similar identification of the coupled node, is inserted in an exactly similar location of the assembled matrix. Thus for the simple bar in Fig. 7.7, any constituent matrix such as the stiffness can be represented as

$$\boldsymbol{K}^{e(i)} = \begin{bmatrix} K_{11}^{(i)} & K_{12}^{(i)} \\ K_{21}^{(i)} & K_{22}^{(i)} \end{bmatrix} \tag{7.7}$$

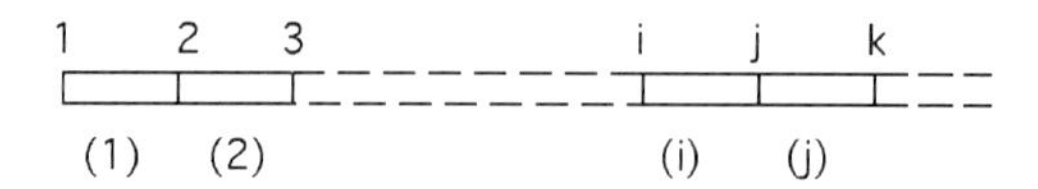

Fig. 7.7 Simple bar.

whose elements can simply be inserted in the global assembled matrix as shown:

$$
K = \begin{bmatrix} K_{11}^{(1)} & K_{12}^{(1)} & & & & \\ K_{21}^{(1)} & K_{22}^{(1)} + K_{11}^{(2)} & & & & \\ & & \ddots & & & \\ & & & K_{22}^{(i-1)} + K_{11}^{(i)} & K_{12}^{(i)} & \\ & & & K_{21}^{(i)} & K_{22}^{(i)} + K_{11}^{(i+1)} & \\ & & & & & \ddots \end{bmatrix} \tag{7.8}
$$

in which it can be observed that $K_{11}^{(i)}$ in the i, i location and $K_{12}^{(i)}$ in the i, j location are added to exactly the same locations of the global stiffness matrix. This observation is valid for any combination of one-, two-, and three-dimensional elements, and the process is easily automated in computer software. It may also be noted that for any such finite element idealization the assembled matrix tends to be symmetric and highly banded, particularly if care is taken to number the nodes in an optimized fashion.

7.4 Imposition of Deflection Boundary Conditions

The general algebraic expression for a prescribed nodal displacement boundary condition may be expressed as

$$
\begin{aligned} u_{i,j} &= \sum a_{m,n} u_{m,n} \\ &= a_{i,j} u_{i,j} + a_{k,l} u_{k,l} + \cdots \end{aligned} \tag{7.9}
$$

in which $u_{i,j}$ is the dependent displacement to be set, whereas $u_{k,l}, u_{m,n}, \ldots,$ are independent displacement variables; the subscripts i, k are node numbers, whereas $j, l, \ldots,$ are the associated degrees of freedom. Data input for processing by any computer code takes the following form:

$i \; j \; i \; j \; 0 \rightarrow$ Zero displacement boundary condition (ZDBC)
$i \; j \; i \; j \; a_{i,j} \rightarrow$ Finite displacement boundary condition (FDBC)
$i \; j \; k \; l \; a_{k,l} \rightarrow$ Interdependent displacement boundary condition (IDBC)

in which the IDBC can have any number of admissible independent nodal degrees of freedom to which the dependent node may be related. Implementation of these boundary conditions in the system matrices [Eq. (7.5)] is achieved as follows, in which NDOF is the number of active degrees of freedom per node.

ZDBC: To implement $u_{i,j} = 0$ located in the rth row, where $r = i * NDOF + j$, set rth row/column $= 0$ for $\boldsymbol{K}$ (also $\boldsymbol{K}_G$, $\boldsymbol{K}'$, $\boldsymbol{K}_2$, $\boldsymbol{K}_4$), $\boldsymbol{M}$ (also $\boldsymbol{M}_2$), and $\boldsymbol{C}$ (includes $\boldsymbol{C}_C, \boldsymbol{C}_D$) matrices.

Place 1 on diagonal of $\boldsymbol{K}$, $\boldsymbol{K}_2$, $\boldsymbol{K}_4$ matrices.

Set $p_r = 0$, set external load to zero at rth row for static problems.

FDBC: To implement $u_{i,j} = a_{i,j}$, a finite value, set rth row/column $= 0$ for all $\boldsymbol{K}$, $\boldsymbol{M}$, and $\boldsymbol{C}$ matrices.

Place 1 on diagonal of $\boldsymbol{K}$, $\boldsymbol{K}_2$, and $\boldsymbol{K}_4$ matrices.

For static problems, set the right-hand side $= \boldsymbol{p} - a_{i,j} \times$ column r of $\boldsymbol{K}$. Set $\boldsymbol{p}_r = a_{i,j}$.

IDBC: To implement $u_{i,j} = a_{k,l} u_{k,l}$, set $r = i \times NDOF + j$, $s = k \times NDOF + l$. Perform the following steps:

1) Multiply the rth column by $a_{k,l}$ and add to the sth column of all $\boldsymbol{K}$, $\boldsymbol{M}$, and $\boldsymbol{C}$ matrices.

2) Multiply the rth row by $a_{k,l}$ and add to the sth row of all $\boldsymbol{K}$, $\boldsymbol{M}$, and $\boldsymbol{C}$ matrices.

3) For static problems, multiply p_r by $a_{k,l}$ and add to p_s.

4) Set the rth column/row of all $\boldsymbol{K}$, $\boldsymbol{M}$, and $\boldsymbol{C}$ matrices to zero.

5) Put 1 on diagonal for $\boldsymbol{K}$, $\boldsymbol{K}_2$, $\boldsymbol{K}_4$ matrices.

6) Set $p_r = 0$ for the static case.

7) The solution yields $u_r = 0$; then set $u_r = a_{k,l} u_s$.

A dependent degree of freedom may be connected to a number of independent degrees of freedom as

$$\begin{array}{ccccc} i & j & k & l & a_{k,l} \\ i & j & k_1 & l_1 & a_{k_1,l_1} \\ \vdots & \vdots & \vdots & \vdots & \vdots \end{array}$$

but may not appear subsequently in columns 3 and 4 as an independent degree of freedom. However, independent degrees of freedom appearing in columns 3 and 4 may subsequently be dependent degrees of freedom. It may be noted that the preceding row/column interchanges for the implementation of the IDBC will result in some increase of the bandwidth. For this reason, an efficient bandwidth minimization routine should be designed to take this phenomenon into consideration.

For many practical problems it may be advantageous to express the DBCs in LGCSs. A standard coordinate transformation is then needed to convert the resulting IDBC relationships in GCS. Thus, the following procedure may be adopted to convert ZDBC relationships expressed in an LGCS (Fig. 7.8) into IDBCs in which

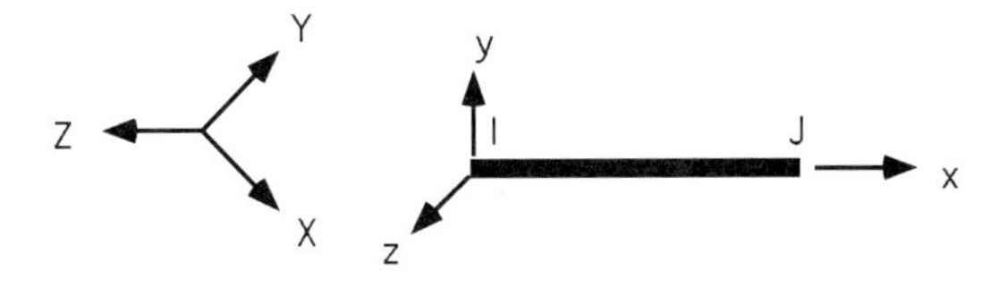

Fig. 7.8 ZDBC in LGCS.

a typical Jth node has zero deformation in the ith degree of freedom. Then for the Jth node

$$u^{(J)} = \lambda q^{(J)} \tag{7.10}$$

and also for the ith row

$$\begin{aligned} u_i^{(J)} &= \sum_{j=1}^{6} \lambda_{i,j} q_j^{(J)} \\ &= \lambda_{i,1} q_X^{(J)} + \lambda_{i,2} q_Y^{(J)} + \lambda_{i,3} q_Z^{(J)} + \lambda_{i,4} q_{\theta X}^{(J)} + \lambda_{i,5} q_{\theta Y}^{(J)} + \lambda_{i,6} q_{\theta Z}^{(J)} \\ &= 0 \end{aligned} \tag{7.11}$$

which also can be written as follows, assuming perhaps the first occurrence of nonzero $\lambda_{i,j}$ denoted as $\lambda_{i,j'}$:

$$q_{j'}^{(J)} = -\frac{1}{\lambda_{i,j'}} \sum \lambda_{i,j} q_j \tag{7.12}$$

The relationship of Eq. (7.12) can be written in IDBC format as

$$\begin{matrix} J & j' & J & 1 & \alpha_{i,1} \\ J & j' & J & 2 & \alpha_{i,2} \\ \vdots & \vdots & \vdots & \vdots & \vdots \end{matrix}$$

in which $\alpha_{i,j} = -\lambda_{i,j}/\lambda_{i,j'}$. Thus every ZDBC in an LGCS is translated into up to five IDBCs. Similar formulations are adopted for IDBCs initially expressed in LGCS.

7.5 Matrix Bandwidth Minimization

The bandwidth of a matrix is defined as the maximum difference between any two connected nodes multiplied by the relevant number of nodal degrees of freedom. A number of efficient techniques are described in the literature[3–7] that describes procedures to minimize matrix bandwidth. An effective minimization algorithm reduces the bandwidth of the stiffness, inertia, and all other relevant system matrices to a minimum by reordering input nodal numbers, taking into consideration first-order as well as second-order nodal connectivity.[7] The renumbering process starts with a node with minimum such connectivity and then computing the second-order connectivity numbers for each of the nodes connected to the starting node and numbering the later sequentially in order of diminishing connectivity. The procedure is repeated with each of the renumbered nodes

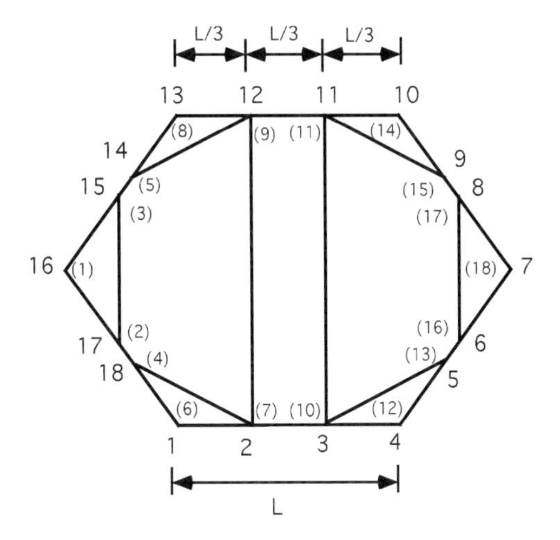

Fig. 7.9 Bandwidth minimization scheme (revised nodal numbering in parenthesis).

until all of the nodes with original numbering are renumbered to yield minimum connectivity.

With reference to Fig. 7.9, the existing nodal numbering may be modified to minimize the bandwidth of associated matrices. Therefore, any node with minimum first-order connectivity may be chosen as the starting node. Accordingly, any one of nodes 1, 4, 7, 10, 13, and 16, all of which have a minimum first-order nodal connectivity of two, may be selected as the first node to start the nodal numbering scheme. However, nodes 1, 4, 10, and 13 possess a higher second-order connectivity condition than do nodes 7 and 16. For example, nodes connected to node 1 (namely nodes 2 and 18) are, in turn, connected to a total of seven nodes, whereas such a connectivity number for either node 7 or 16 happens to be only six. As such, either node 7 or 16 may be chosen as the starting node for the renumbering scheme. A revised nodal numbering that minimizes matrix bandwidth is shown in parentheses in Fig. 7.9. Figure 7.10 provides a depiction of the arrangement of matrix elements prior to and after minimization. The minimization scheme should also take into consideration the presence of nodal IDBCCs, if any.

Jennings[8] describes and recommends the use of profile information(the local bandwidth method) as an improvement on the band method. Four papers on bandwidth or profile minimization were mentioned in the 1970 IMA Oxford Conference on Large Sparse Sets of Linear Equations.[9] One of these methods, the Cuthill–McKee(CM) method,[10] has subsequently achieved wide use and provided a basis on which improved algorithms have been built. The CM method aims at achieving a small bandwidth. George[11] noticed that reversing the ordering produced by the CM method generally produces a smaller profile while always leaving the bandwidth unchanged, giving rise to the reverse CM (RCM) method. Jennings[12] explains why the RCM has this property. Further improvements on the CM and RCM were made by Gibbs, Poole, and Stockmeyer.[13] A FORTRAN implementation of this algorithm is given by Lewis.[14] This is incorporated in the STARS[7] program.

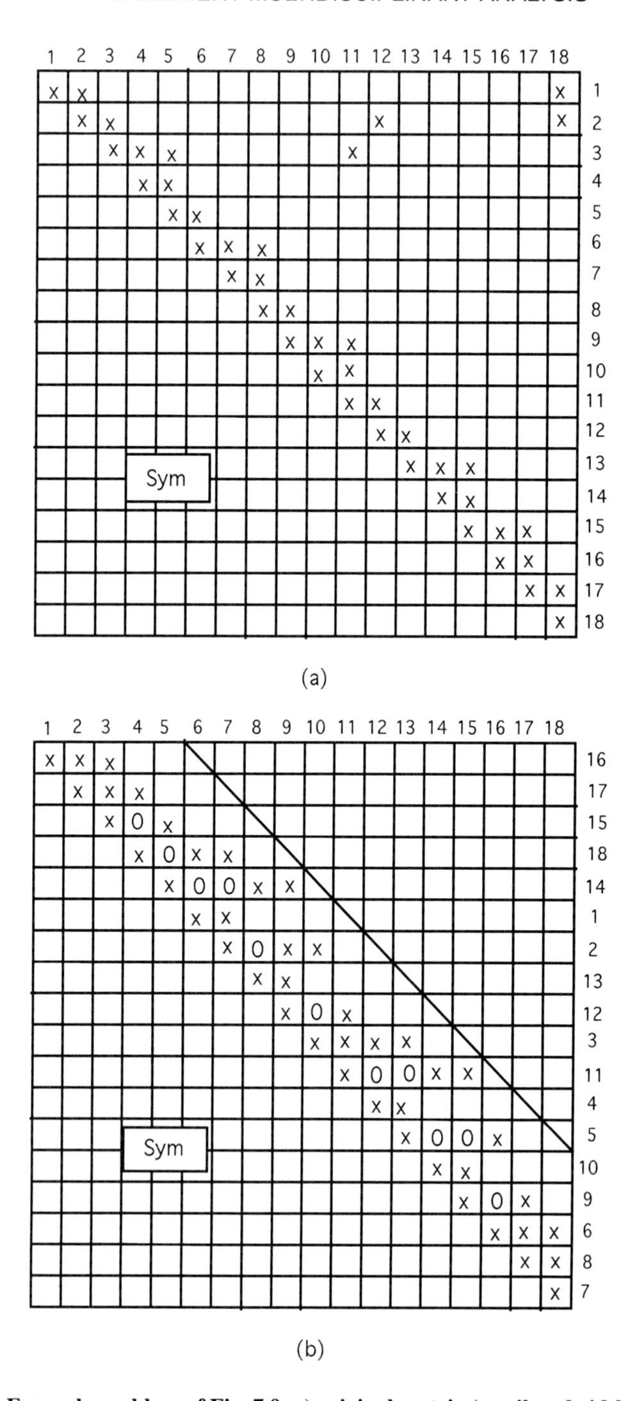

Fig. 7.10 Example problem of Fig. 7.9: a) original matrix (semibandwidth = 18) and b) final matrix (semibandwidth = 5).

7.6 Sparse Matrix Storage Schemes

Because the bandwidth minimization scheme yields matrices in which all nonzero elements reside within the band, considerable solution economy is achieved by storing only such data. Thus, typically a rectangular $(N \times M)$ array, containing elements in the upper symmetric half of the band, is stored for subsequent analysis. However, for many large, complex practical problems, the actual nonzero elements inside the band may constitute only 10–20% or so of the total number of elements. Then further considerable savings in data storage and solution effort are achieved by avoiding storing most of the zero elements within the band.

7.6.1 Sparse Matrix Format for Sequential Access

This format uses an integer array and a real array to store the upper triangular part of a symmetric or skew-symmetric matrix. This arrangement is convenient when one-way sequential access to the matrix suffices, as for matrix vector multiplication and for computing the Cholesky factorization of a matrix. Although this format conceptually uses just two, possibly large arrays, these arrays can be easily partitioned into segments relating to separate rows or groups of rows for storage on external files. These two arrays have the following form:

$$\text{Integer array} : -I_r, I_s, I_e, I_s, I_e, \ldots, -I_r, I_s, I_e, I_s, I_e, \ldots, -(N+1)$$

$$\text{Real array} : V_1, V_2, \ldots$$

I_r is a row number; the negative sign signifies the start of a new row. Row numbers must occur in increasing order. If a row has no nonzero elements it can be omitted. For a matrix of order N, the end of the integer array is signaled by the entry $-(N+1)$.

Each successive pair of positive integers I_s, I_e gives the starting and ending matrix column numbers of the next $(I_e - I_s + 1)$ elements of the V array. Only columns on and to the right of the diagonal position are recorded for each row. The array V contains all of the nonzero elements of the matrix and possibly also a small percentage of the zero values.

7.6.2 Sparse Matrix Format for Rowwise and Columnwise Access

This format was developed particularly to support the needs of the IDBC computation (Sec. 7.4), where both rowwise and columnwise access are needed. A matrix is partitioned into (6×6) blocks. Assuming, for convenience of description, the dimension N of the matrix is a multiple of 6, define $NB = N/6$. Then the blocked matrix has NB block rows and NB block columns and each block can be identified by a block index pair (I, J), meaning the block is at the intersection of the Ith block row and the Jth block column. The matrix will be stored with rowwise indexing in a pair of external direct access files, e.g., **`fdata`** and **`findex`**.

In the file **`fdata`**, each record holds 36 real values, i.e., one block of the matrix. The records are numbered beginning with one. Blocks may be scattered arbitrarily among the records of this file. The indexing for finding a particular block in this file is contained in the file `findex`. In the file `findex`, the Ith record contains indexing information for the Ith block row. This consists of a sequence of pairs of integers, e.g., (J,K), and is terminated by a pair of zeros (0,0). The index pair (J,K) in record I means the block at position (I,J) in the matrix is stored in record K of file `fdata`. Within each record of `findex` the column indices, i.e., the J values, will be strictly increasing. Only blocks that contain some nonzero elements and are in the upper triangle of the matrix are stored.

When columnwise access is needed, a columnwise index can be constructed by appropriate processing of the rowwise index contained in `findex`. Use of this approach was found to very significantly reduce the execution time of the IDBC computation in cases in which the matrix was too large to permit storage of all of its nonzero elements in memory simultaneously.

7.7 Concluding Remarks

This chapter covered some practical aspects of matrix structural analysis. Thus, details were provided on various coordinate systems for data input that proves to be useful in practice. Sparse matrix storage schemes were described, showing savings in both computer storage and solution time. Also provided were details of various matrix bandwidth minimization techniques, including one that takes into account second order connectivity, which is essential for the economical solution of large order problems.

References

[1]Livesley, R. K., *Matrix Methods of Structural Analysis*, Pergamon, New York, 1964.

[2]Przemieniecki, J. S., *Theory of Matrix Structural Analysis*, McGraw–Hill, New York, 1968.

[3]Alway, G. G., and Martin, D. W., "An Algorithm for Reducing the Bandwidth of a Matrix of Symmetrical Configuration," *Computer Journal*, Vol. 8, 1965, pp. 264–272.

[4]Akyuz, F. A., and Utku, S., "An Automatic Node-Relabeling Scheme for Bandwidth Minimization of Stiffness Matrices," *AIAA Journal*, Vol. 6, No. 4, 1968, pp. 728–730.

[5]Grooms, H. R., "Algorithm for Matrix Bandwidth Reduction," *American Society of Civil Engineers, Journal of Structural Division*, Vol. 98, STI, 1972, pp. 203–214.

[6]Collins, R. J., "Bandwidth Reduction by Automatic Renumbering," *International Journal of Numerical Methods in Engineering*, Vol. 6, 1973, pp. 345–356.

[7]Gupta, K. K., *STARS—An Integrated Multidisciplinary, Finite-Element, Structural, Fluids, Aeroelastic and Aeroservoelastic Analysis Computer Program*, NASA TM 4795, 1997; also NASA RP 1129; 1984.

[8]Jennings, A., "A Compact Storage Scheme for the Solution of Symmetric Linear Simultaneous Equations" *Computer Journal*, Vol. 9, 1966, pp. 281–285.

[9]Reid, J. K., Large Sparse Sets of Linear Equations," *Proceedings of Oxford IMA Conference*, April 1970, Academic, New York, 1971.

[10]Cuthill, E., and Mckee, J., "Reducing the Bandwidth of Sparse Symmetric Matrices," *Proceedings of the 24th ACM National Conference*, San Francisco, 1969, pp. 157–172.

[11]George. A., and Liu, J. W-H., "Computer Solution of Large Sparse Positive Definite Systems," *SIAM Journal Numerical Analysis*, Vol. 13, 1976, pp. 236–250.

[12]Jennings, A., *Matrix Computation for Engineers and Scientists*, Wiley, New York, 1997.

[13]Gibbs, N. E., Poole, W. G., Jr., and Stockmeyer, P. K., "An Algorithm for Reducing the Bandwidth and Profile of a Sparse Matrix," *SIAM Journal of Numerical Analysis*, Vol. 13, 1976, pp. 236–250.

[14]Lewis, J. G., "Implementation of the Gibbs–Poole–Stockmeyer and Gibbs–King Algorithms," *ACM Transactions of Mathematical Software*, Vol. 8, 1982, pp. 180–194.

MINIMUM FILL IN L

Sparse L

MINIMUM DEGREE

DOMAIN DECOMPOSITION

SYMMETRIC PERMUTATION OF A

MANAGE FACTORIZATION

PERFORMS ONLY ESSENTIAL OPERATIONS

MULTIFRONTAL METHOD

8
Solution of System Equations

8.1 Introduction

Effective simulation of the pattern of behavior of discrete structural systems requires the solution of equilibrium, eigenvalue, or propagation problems, depending on the type of external excitation. Under static loading, it is necessary to solve the equilibrium problem involving the solution of a set of simultaneous algebraic equations. For stability and free vibration problems, the relevant analysis involves the eigenvalue problem solution of a pair of mostly symmetric matrices. A vital preliminary for the computation of the response of a structure because of time-dependent loading is the solution of the eigenvalue problem of matrices, with distinctive characteristics, if the modal superposition method is used for the calculations. Alternatively, a suitable step-by-step integration technique may be adopted for the dynamic problem. This involves the repeated solution of a set of simultaneous algebraic equations. In general, though, this latter technique is reserved for nonlinear problems because for linear problems the modal superposition method proves to be more efficient.

For efficient analysis of static and dynamic problems, it is vital to generate economical solution schemes for a set of algebraic simultaneous equations. Section 8.2 provides details of such conventional analysis schemes followed by more advanced techniques pertaining to sparse matrices, in Sec. 8.3.

8.2 Formulation and Solution of System Equation

These problems are characterized by the matrix formulation

$$\boldsymbol{K}\boldsymbol{q} = \boldsymbol{p} \tag{8.1}$$

in which the stiffness matrix $\boldsymbol{K}$ is usually symmetric and positive definite in nature; $\boldsymbol{p}$ is the external load vector consisting of mechanical and thermal forces, and $\boldsymbol{q}$ is the nodal displacement vector to be determined by solving Eq. (8.1). Because, for most practical problems $\boldsymbol{K}$ is sparse and can be formulated to be highly banded, an efficient analysis scheme exploits this sparsity to the fullest extent to yield an accurate and efficient solution.

8.2.1 Solution of Equilibrium Equations—A Direct Method

The simplest of all methods for solving the linear equations in Eq. (8.1) is the Gauss elimination method. Since, for practical problems, the stiffness matrix $\boldsymbol{K}$ is positive definite, it has the property that all pivots in the elimination method chosen down the diagonal are nonzero as well as positive. The status of a matrix is defined as follows:

$$\left.\begin{array}{l}\text{positive definite}\\ \text{negative definite}\\ \text{nonnegative definite}\\ \text{nonpositive definite}\end{array}\right\} \text{according as } \boldsymbol{q}^T\,\boldsymbol{K}\boldsymbol{q} \left\{\begin{array}{l} >0\\ <0\\ \geq 0\\ \leq 0\end{array}\right.$$

for all real nonzero vectors $\boldsymbol{q}$. A solution of the static equation may be obtained by the Cholesky method in which

$$\boldsymbol{K} = \hat{\boldsymbol{L}}\boldsymbol{D}\hat{\boldsymbol{L}}^T \tag{8.2}$$

where $\hat{\boldsymbol{L}}$ is a unit lower triangular matrix derived from multipliers in the elimination process and $\boldsymbol{D}$ is a diagonal matrix whose elements are the pivots in the elimination process. Equation (8.2) may also be written as

$$\boldsymbol{K} = \boldsymbol{L}\boldsymbol{L}^T \tag{8.3}$$

which is known as the Cholesky factorization, the lower triangular matrix now having diagonals that are the square roots of the elements of the $\boldsymbol{D}$ matrix. It is also noted that the decomposition retains the original bandwidth as shown in Fig. 8.1, where N is the order and M the semibandwidth of $\boldsymbol{K}$, including the diagonal element. Thus only the symmetric half of the matrix that is eventually transformed into the $\boldsymbol{L}$ matrix needs to be stored.

Solution of the Eq. (8.1) may be obtained in two simple steps. Rewriting Eq. (8.1) as

$$\boldsymbol{K}\boldsymbol{q} = \boldsymbol{L}\boldsymbol{L}^T\boldsymbol{q} = \boldsymbol{p} \tag{8.4}$$

the solution steps are 1) obtain $\boldsymbol{y}$ by simple forward substitution

$$\boldsymbol{L}\boldsymbol{y} = \boldsymbol{p} \tag{8.5}$$

and 2) solve by backward substitution

$$\boldsymbol{L}^T\boldsymbol{q} = \boldsymbol{y} \tag{8.6}$$

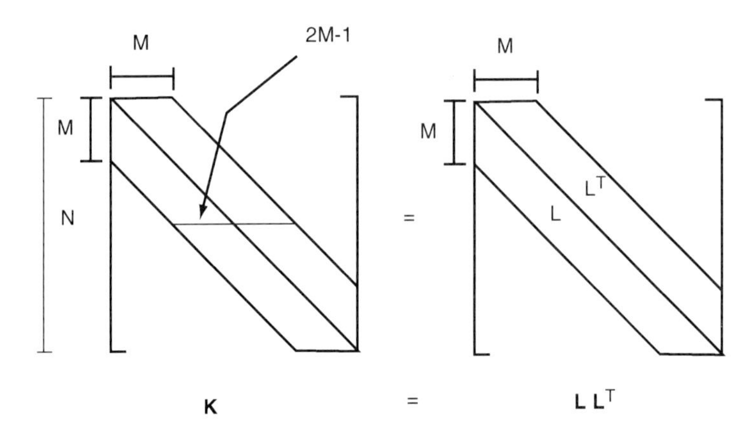

Fig. 8.1 Matrix decomposition.

yielding the required solution $\boldsymbol{q}$, the deformation vector. It may be noted that matrix bandedness is retained throughout the solution process. This procedure can be easily automated for multiple sets of loading on the right-hand side while requiring only one Cholesky decomposition of the $\boldsymbol{K}$ matrix. A simple example illustrates the solution process. Suppose the solution of the problem

$$\begin{bmatrix} 4.0 & 5.0 & 0.0 \\ 5.0 & 6.5 & 1.5 \\ 0.0 & 1.5 & 10.0 \end{bmatrix} \begin{bmatrix} q_1 \\ q_2 \\ q_3 \end{bmatrix} = \begin{bmatrix} 2.0 \\ 3.5 \\ 7.0 \end{bmatrix}$$

is sought by the present method. Then routine elimination of subdiagonal elements is given:

$$\boldsymbol{K} \Longrightarrow {}^{5/4}\begin{bmatrix} 4.0 & 5.0 & 0.0 \\ 5.0 & 6.5 & 1.5 \\ 0.0 & 1.5 & 10.0 \end{bmatrix} \Longrightarrow 1.5/0.25 \begin{bmatrix} 4.0 & 5.0 & 0.0 \\ 0.0 & 0.25 & 1.5 \\ 0.0 & 1.5 & 10.0 \end{bmatrix}$$

$$\Longrightarrow \begin{bmatrix} 4.0 & 5.0 & 0.0 \\ 0.0 & 0.25 & 1.5 \\ 0.0 & 0.0 & 1.0 \end{bmatrix} = \boldsymbol{U} = \begin{bmatrix} 4.0 & 0.0 & 0.0 \\ 0.0 & 0.25 & 0.0 \\ 0.0 & 0.0 & 1.0 \end{bmatrix} \begin{bmatrix} 1.0 & 1.25 & 0.0 \\ 0.0 & 1.0 & 6.0 \\ 0.0 & 0.0 & 1.0 \end{bmatrix} = \boldsymbol{D}\boldsymbol{L}'^T \tag{8.7}$$

The row multiplication factors are shown along the row under consideration. Thus row 1 is multiplied by $K_{1,2}/K_{1,1}$ and subtracted from row 2 to eliminate $K_{2,1}$ and proceeding in the same way to eliminate the entire column; the pivot is moved down the diagonal and the process repeated until all subdiagonal elements are eliminated. The matrix $\boldsymbol{L}^T$ is obtained by dividing rows of $\boldsymbol{U}$ by the diagonals, being a unit upper triangle. Equation (8.7) may simply be formulated in the format of Eq. (8.3) in which the columns of $\boldsymbol{L}'$ are simply multiplied by the square root of the respective diagonal elements of $\boldsymbol{D}$, storing the pivots in the elimination process. Thus

$$\boldsymbol{K} = \boldsymbol{L}\boldsymbol{L}^T = \left(\boldsymbol{L}'\boldsymbol{D}^{\frac{1}{2}}\right)\left(\boldsymbol{D}^{\frac{1}{2}}\boldsymbol{L}'\right)$$
$$= \begin{bmatrix} 2.0 & 0.0 & 0.0 \\ 2.5 & 0.5 & 0.0 \\ 0.0 & 3.0 & 1.0 \end{bmatrix} \begin{bmatrix} 2.0 & 2.5 & 0.0 \\ 0.0 & 0.5 & 3.0 \\ 0.0 & 0.0 & 1.0 \end{bmatrix} \tag{8.8}$$

A solution of the problem may then be obtained using Eqs. (8.5) and (8.6), respectively.

Step 1: $\boldsymbol{L}\boldsymbol{y} = \boldsymbol{p}$

$$\begin{bmatrix} 2.0 & 0.0 & 0.0 \\ 2.5 & 0.5 & 0.0 \\ 0.0 & 3.0 & 1.0 \end{bmatrix} \begin{bmatrix} y_1 \\ y_2 \\ y_3 \end{bmatrix} = \begin{bmatrix} 2.0 \\ 3.5 \\ 7.0 \end{bmatrix}$$

yielding $\boldsymbol{y} = [1\ 2\ 1]^T$.

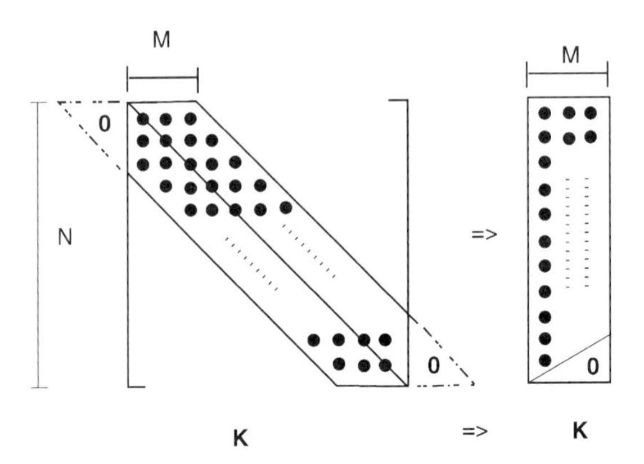

Fig. 8.2 Storage of upper symmetric part of K.

Step 2: $L^T q = y$

$$\begin{bmatrix} 2.0 & 2.5 & 0.0 \\ 0.0 & 0.5 & 3.0 \\ 0.0 & 0.0 & 1.0 \end{bmatrix} \begin{bmatrix} q_1 \\ q_2 \\ q_3 \end{bmatrix} = \begin{bmatrix} 1.0 \\ 2.0 \\ 1.0 \end{bmatrix}$$

yielding $q = [3 - 2\ 1]^T$.

Because of symmetry, only elements of the L matrix within the band need to be stored for subsequent use in the solution of the static equations obtained from Eqs. (8.5) and (8.6). In practice L can overwrite K, stored as an $(N \times M)$ array. Assuming that only the upper symmetric banded part of the stiffness matrix K is stored in an $(N \times M)$ array (Fig. 8.2), a banded upper triangular matrix $U = L^T$ can be generated and stored in K as follows.

```
c      ==============================================================
       subroutine book_chol(n, m, mdim, amat)
c    Cholesky factorization of band matrix.
c    Diag and upper triangle packed into amat(1:M,1:n)
c      --------------------------------------------------------------
       implicit none
       integer incr, ipiv, irow, j
       integer m, n, mdim, num
       double precision amat(mdim, n), fac
c      --------------------------------------------------------------
       do ipiv = 1, n
          if(amat(1,ipiv) .le. 0.0d0) then
             write(*,'(/a/a,i10/a,g18.10)')
     *        ' ERROR stop in Book_Chol. Nonpositive diagonal element',
     *        '                             at position ',ipiv,
     *        '       Element is ',amat(1,ipiv)
             stop ' Nonpositive diagonal in Book_Chol.'
          endif !(amat(...)...)
          amat(1,ipiv) = sqrt(amat(1,ipiv))
          fac = 1.0d0 / amat(1,ipiv)
          num = min(m-1, n-ipiv)
```

```
         do j = 2, 1+num
            amat(j,ipiv) = fac * amat(j,ipiv)
         enddo ! j
         do incr = 1, num
            irow = ipiv+incr
            do j = 1, 1+num-incr
               amat(j,irow) = amat(j,irow) -
     *                        amat(1+incr,ipiv) * amat(j+incr,ipiv)
            enddo ! j
         enddo ! incr
      enddo ! ipiv
      end ! subroutine book_chol
c     ================================================================
```

A simple transposition of L^T yields L, which can be stored in the same space for band forward substitution to obtain y from $Ly = p$ as shown next, in which the algorithm overwrites p with the solution.

```
c     ================================================================
      subroutine book_solvl(n, m, mdim, amat, vec)
c     Given lower triangular matrix L packed in amat(),
c     and vector z stored in vec(),
c     solve the system  L * y = z  for y and store y in vec().
c     ----------------------------------------------------------------
      implicit none
      integer incr, ipiv
      integer m, n, mdim
      double precision amat(mdim, n), vec(n)
c     ----------------------------------------------------------------
      do ipiv = 1,n
         vec(ipiv) = vec(ipiv) / amat(1,ipiv)
         do incr = 1, min(m-1, n-ipiv)
            vec(ipiv+incr) = vec(ipiv+incr) -
     *                       vec(ipiv) * amat(1+incr,ipiv)
         enddo ! incr
      enddo ! ipiv
      end ! subroutine book_solvl
c     ================================================================
```

The unknown solution vector q is obtained from the relationship $L^T q = y$ by back substitution as follows (in which the solution overwrites p).

```
c     ================================================================
      subroutine book_solvlt(n, m, mdim, amat, vec)
c     Given lower triangular matrix L packed in amat(),
c     and vector y stored in vec(),
c     solve the system  L**t * x = y  for x and store x in vec().
c     ----------------------------------------------------------------
      implicit none
      integer incr, ipiv
      integer m, n, mdim
      double precision amat(mdim, n), sum, vec(n)
c     ----------------------------------------------------------------
```

```
do ipiv = n, 1, -1
   sum = 0.0d0
   do incr = 1, min(m-1, n-ipiv)
      sum = sum + amat(1+incr,ipiv) * vec(ipiv+incr)
   enddo ! incr
   vec(ipiv) = (vec(ipiv) - sum) / amat(1,ipiv)
enddo ! ipiv
end ! subroutine book_solvlt
```

The solution effort for the preceding procedures involves NM^2, NM, and NM major floating point operations (flops) for the Cholesky factorization, forward, and backward substitution, respectively.

8.2.2 Solution Accuracy and Precision

For any typical symmetric positive definite matrix $\boldsymbol{K}$, the uncertainties in the solution of related equations [Eq. (8.1)] may be caused by two sources of error: first, rounding errors introduced by the solution process and second, inexact data input. It is useful to obtain an estimate of the maximum error in the solution; rounding errors in general can be attributed to 1) the conversion of data stored in floating point form into the binary form used by the computer, and 2) rounding off of numbers to the nearest binary place during the entire computational process in the computer.

The effect of these errors is dependent on the type of the problem and almost invariably is related to the magnitude of the coefficients of the inverse matrix. Typically, for example, the difference in solution of $\boldsymbol{Kq} = \boldsymbol{p}$ and $\boldsymbol{Kq} = \boldsymbol{p} + \delta\boldsymbol{p}$ is simply $\boldsymbol{K}^{-1}\delta\boldsymbol{p}$. These errors may become very pronounced if the matrix is ill-conditioned with respect to inversion when small changes in $\boldsymbol{p}$ will effect large perturbation in $\boldsymbol{q}$. In the extreme case $\boldsymbol{K}$ will be nearly singular and this is made manifest by the steady decrease in absolute value of the size of the pivots during the decomposition process.

It is possible to obtain some quantitative measure of the degree of ill-conditioning of the matrix $\boldsymbol{K}$ by using known results of the algebra of norms; the norm of a vector $\boldsymbol{q}$ is defined by the expression

$$\begin{aligned}\|\boldsymbol{q}\|_p &= \left(\sum |\boldsymbol{q}|^p\right)^{1/p} \\ &= \left(|\boldsymbol{q}_1|^p + |\boldsymbol{q}_2|^p + \cdots + |\boldsymbol{q}_n|^p\right)^{1/p}; \qquad p \geq 1 \end{aligned} \tag{8.9}$$

whereas the Frobenius norm, also known as the Euclidean matrix norm, for a typical matrix $\boldsymbol{K}$ is defined as

$$\|\boldsymbol{K}\|_F = \left(\sum_{i=1}^{n}\sum_{j=1}^{n} |\boldsymbol{K}_{ij}|^2\right)^{\frac{1}{2}} \tag{8.10}$$

Also, the spectral norm of the matrix is defined as

$$\|\boldsymbol{K}\|_s = \max_{\|\boldsymbol{x}\|_2 = 1} \|\boldsymbol{Kx}\|_2$$

Assuming that the computed inverse is the exact inverse of a perturbed matrix $\boldsymbol{K} + \boldsymbol{E}$ of the original matrix $\boldsymbol{K}$, the expression for the relative error for the computed inverse is derived as

$$\frac{\|(\boldsymbol{K} + \boldsymbol{E})^{-1} - \boldsymbol{K}^{-1}\|_s}{\|\boldsymbol{K}^{-1}\|_s} = \frac{\kappa\epsilon}{1 - \kappa\epsilon}, \qquad \kappa\epsilon \leq 1 \tag{8.11}$$

and

$$\kappa = \|\boldsymbol{K}\|_s \|\boldsymbol{K}^{-1}\|_s, \qquad \epsilon = \|\boldsymbol{E}\|_s / \|\boldsymbol{K}\|_s \tag{8.12}$$

where $\boldsymbol{x}$ is a vector of unit norm and ϵ relates the perturbation $\boldsymbol{E}$ to the original $\boldsymbol{K}$. The error is proportional to the spectral condition number κ, which is the ratio of the numerically largest root to the smallest defined by $\|\boldsymbol{K}\|_s$ and $1/\|\boldsymbol{K}^{-1}\|_s$, respectively; while the relative error has $\|\boldsymbol{K}^{-1}\|_s$ as the factor, the absolute error is factored with the square of this possible large quantity. The larger the value of κ, the more ill-conditioned is the matrix $\boldsymbol{K}$. Typically ill-conditioning occurs in a matrix if the rows of $\boldsymbol{K}$ are almost linearly dependent and specifically if two rows are very similar. For ill-conditioned matrices the loss in the solution accuracy may be recovered by adapting the residual correction strategy. Since initially the computed solution will tend to be erroneous because of possible ill-conditioning of $\boldsymbol{K}$, a correction in the applied load value $\Delta \boldsymbol{p}$ is computed and the correction in the solution $\Delta \boldsymbol{q}$ computed from $\Delta \boldsymbol{p}$, which is added to the computed solution. This process, repeated until adequate solution convergence has been achieved, can be expressed as

$$\begin{aligned}
\boldsymbol{q}^{(1)} &= (\boldsymbol{L}\boldsymbol{L}^T)^{-1}\boldsymbol{p} \\
\Delta\boldsymbol{p}^{(1)} &= \boldsymbol{p} - \boldsymbol{K}\boldsymbol{q}^{(1)} \\
\boldsymbol{q}^{(2)} &= \boldsymbol{q}^{(1)} + (\boldsymbol{L}\boldsymbol{L}^T)^{-1}\Delta\boldsymbol{p}^{(1)} \qquad \text{using computed } \boldsymbol{L} \\
&\vdots \\
\Delta\boldsymbol{p}^{(s)} &= \boldsymbol{p} - \boldsymbol{K}\boldsymbol{q}^{(s)}
\end{aligned} \tag{8.13}$$

$$\boldsymbol{q}^{(s+1)} = \boldsymbol{q}^{(s)} + (\boldsymbol{L}\boldsymbol{L}^T)^{-1}\Delta\boldsymbol{p}^{(s)} \tag{8.14}$$

In general only a small number of steps need to be performed to achieve the required accuracy. The procedure is quite efficient because $\boldsymbol{L}$ has already been computed.

8.2.3 Iterative Methods

For very large banded, sparse matrix, iterative solution methods may be used effectively. The procedure starts with an arbitrary initial solution approximation and successively improves it until the required precision is obtained. In problems for which convergence is known to be rapid, the solution is obtained with much less work than the direct method. Also, for large sparse matrices the iterative procedure is computationally less laborious and requires relatively less storage. For ill-conditioned problems, however, the solution convergence may become slow.

The two simplest iteration schemes are Jacobi and Gauss–Seidel iteration. The latter is more efficient because it uses the most recently computed information derived from previous rows. This procedure for a typical ith row and $(r+1)$th iteration involving a banded matrix $\boldsymbol{K}$ (Fig. 8.1) is written as

$$\boldsymbol{K}_{ii}q_i^{(r+1)} = \boldsymbol{p}_i - \sum_{j=i+1}^{\hat{j}} \boldsymbol{K}_{ij}\boldsymbol{q}_j^{(r)} - \sum_{j=\tilde{j}}^{i-1} \boldsymbol{K}_{ij}\boldsymbol{q}_j^{(r+1)} \tag{8.15}$$

in which

$$\hat{j} = \min(M + i - 1, N)$$
$$\tilde{j} = \max(1, i - M + 1) \tag{8.16}$$

and where, for $i = 1$, the last term in Eq. (8.15) is nonexistent, and for $i = N$, the next to last term is nonexistent.

The element $q_i^{(r+1)}$ is simply computed by dividing the right-hand side by $\boldsymbol{K}_{ii}$. The process is continued involving all rows, and the solution step $(r+1)$ is repeated until adequate convergence is achieved. In the simpler, less efficient Jacobi method, $q_j^{(r+1)}$ is replaced by $q_j^{(r)}$, resulting in a slower convergence rate.

Although rather attractive because of its inherent simplicity, the procedure may prove to be prohibitively slow. This can be rectified by adopting the method of successive overrelaxation that involves modification of the Gauss–Seidel step as follows:

$$\boldsymbol{K}_{ii}\boldsymbol{q}_i^{(r+1)} = \phi\left(\boldsymbol{p}_i - \sum_{j=i+1}^{\hat{j}} \boldsymbol{K}_{ij}^{(r)}\boldsymbol{q}_j^{(r)} - \sum_{j=\tilde{j}}^{i-1} \boldsymbol{K}_{ij}\boldsymbol{q}_j^{(r+1)}\right) + (1 - \phi)\boldsymbol{q}_i^{(r)} \tag{8.17}$$

An optimum value of the overrelaxation factor ϕ can be estimated following established guidelines. Considerable improvement in the solution convergence is achieved with the successive overrelaxation procedure. Other iterative methods studied and used extensively in recent years include the preconditioned conjugate gradient and the generalized minimal residual techniques.[3]

It is somewhat difficult to assess the amount of work involved in any one of these iterative methods. A number of factors including the ability to find good accelerating factors, form of the matrix, and also the rate of convergence contribute to the overall performance of the adopted procedure. Each cycle of iteration typically involves NM multiplications, and for large sparse matrices such a procedure will in general be faster than a direct method and furthermore will involve much less storage. However, for very ill-conditioned matrices, the process may have a very poor rate of convergence or even fail.

8.3 Sparse Cholesky Factorization

8.3.1 Introduction

The computational algorithms described earlier for static problems and in the next chapter for eigenvalue problems require computation of the Cholesky factorization of certain symmetric positive-definite (SPD) matrices. These matrices are typically very sparse, that is, a relatively small percentage of their elements are nonzero. If these matrices are also of very large order, the execution time to compute the Cholesky factorization, and subsequent operations of matrix–vector multiplication or solving systems using the Cholesky factor matrix, will generally be a dominant part of the overall execution time. Thus there is strong motivation to organize the factorization process and the representation of the Cholesky factor to take as much advantage of the sparsity as possible.

Let $\boldsymbol{A}$ denote the SPD matrix of interest, and let $\boldsymbol{L}$ denote its (lower triangular) Cholesky factor matrix, i.e., $\boldsymbol{A} = \boldsymbol{L}\boldsymbol{L}^T$. Even if $\boldsymbol{A}$ is sparse, the matrix $\boldsymbol{L}$ may or may not have significant sparsity, but this can be controlled, within some range, by the ordering of the rows and columns of $\boldsymbol{A}$. More specifically, except for very low probability instances of exact numerical cancelation, the matrix $\boldsymbol{L}$ will at least have nonzero values in all of the positions in which the lower triangular part of $\boldsymbol{A}$ has nonzero values, and will generally have nonzero elements besides these. These additional nonzero elements are commonly referred to as *fill*.

To avoid unnecessarily long execution times, it is important 1) to determine and apply a symmetric permutation of $\boldsymbol{A}$ that will cause the matrix $\boldsymbol{L}$ to be as sparse as feasible, i.e., to have a minimal amount of fill, and 2) to manage the factorization process so that, as far as possible, the only floating point arithmetic operations executed are those essential to the computation of the nonzero elements of $\boldsymbol{L}$. The purpose of this section is to introduce the reader to two classes of methods that have been found to be very effective in determining a permutation to minimize fill. Minimum degree methods are described in Sec. 8.3.7 and domain decomposition methods in Sec. 8.3.8; also, the multifrontal method described in Sec. 8.3.9 is used to efficiently limit the floating point computation to the essential operations. The combination of these techniques provides dramatic reduction of computation time and storage for large problems, in comparison with the usual band or profile methods.

Sections 8.3.2–8.3.6 provide essential preliminary material, and Sec. 8.3.10 reports on a large computation showing the remarkable efficiency of these methods, which focus on minimizing fill and avoiding unnecessary floating point operations.

8.3.2 Essential Arithmetic of Cholesky Factorization

Let $\boldsymbol{A}$ denote the SPD matrix of interest. The Cholesky factorization process computes a lower triangular matrix $\boldsymbol{L}$ such that

$$\boldsymbol{A} = \boldsymbol{L}\boldsymbol{L}^T \tag{8.18}$$

Because $\boldsymbol{A}$ is symmetric and $\boldsymbol{L}$ is lower triangular, it suffices to give attention to elements of $\boldsymbol{A}$ and $\boldsymbol{L}$ on and below the diagonal, i.e., elements in positions (i, j) with $i \geq j$.

From Eq.(8.18), the elements of $\boldsymbol{A}$ are related to elements of $\boldsymbol{L}$ by

$$a_{i,j} = \sum_{k=1}^{j} l_{i,k} l_{j,k} \qquad \text{for } i \geq j \tag{8.19}$$

This equation will ultimately be used to compute $l_{i,j}$, which occurs as a factor in the last term of the sum. To this end, bring the last term in the preceding sum to the left side, writing

$$l_{i,j}l_{j,j} = a_{i,j} - \sum_{k=1}^{j-1} l_{i,k}l_{j,k} \qquad \text{for } i \geq j \tag{8.20}$$

When $j = 1$ the summation in Eq.(8.20) will be empty and thus is to be treated as having the value zero. Introduce the symbol $\hat{a}_{i,j}$ to stand for the right side of Eq.(8.20):

$$\hat{a}_{i,j} = a_{i,j} - \sum_{k=1}^{j-1} l_{i,k}l_{j,k} \qquad \text{for } i \geq j \tag{8.21}$$

Then Eq.(8.20) can be written as

$$l_{i,j}l_{j,j} = \hat{a}_{i,j} \qquad \text{for } i \geq j \tag{8.22}$$

From Eq.(8.22), the diagonal elements of $\boldsymbol{L}$ can be computed as

$$l_{j,j} = \text{sqrt}(\hat{a}_{j,j}) \tag{8.23}$$

and the subdiagonal elements as

$$l_{i,j} = \hat{a}_{i,j}/l_{j,j} \qquad \text{for } i > j \tag{8.24}$$

The notation in Eq.(8.21) is intended to convey the idea that $\hat{a}_{i,j}$ can be regarded as an "update" of $a_{i,j}$. In fact it would be common in computer implementations for $a_{i,j}$ to be replaced by $\hat{a}_{i,j}$ in memory when the computation of Eq.(8.21) is completed and subsequently for $\hat{a}_{i,j}$ to be replaced by $l_{i,j}$ when Eq.(8.23) or (8.24) is computed. The whole Cholesky factorization process can be regarded as the computation of the quantities $\hat{a}_{i,j}$ of Eq.(8.21) followed by Eq.(8.23) when $i = j$ and Eq.(8.24) when $i > j$. There is a great deal of freedom available in the ordering of these operations. It is only necessary that quantities be computed before they are needed in another formula.

8.3.3 Dense Factorization Algorithms

A *dense* factorization algorithm is one that treats all elements of the matrix as potentially being nonzero, that is, no sparsity is assumed. It is useful, before studying the sparse algorithm in Sec. 8.3.9, to have an understanding of how the operations in Eqs.(8.21), (8.23), and (8.24) can be ordered for dense factorization. Two versions of dense Cholesky algorithms will be described.

It is assumed the diagonal and subdiagonal elements of a SPD matrix $\boldsymbol{A}$ of order n are initially in a working array `w`. On termination the lower triangular Cholesky factor matrix $\boldsymbol{L}$ will have replaced $\boldsymbol{A}$ in `w`. It is assumed the `do` statements test at the top of the loop so that in a case of `do j = j1, j2`, with `j1 > j2` the body of the loop will be skipped entirely. The doubly subscripted array `w` is used for clarity in the exposition of the algorithms. In an actual implementation one would probably use some form of packed storage.

The column-oriented right-looking Cholesky factorization algorithm is depicted as

```
do j = 1, n
   w(j,j) = sqrt(w(j,j))
   do i = j+1, n
      w(i,j) = w(i,j)/w(j,j)
  enddo

  do k = j+1, n
    do i = k, n
       w(i,k) = w(i,k) - w(i,j) x w(k,j)
    enddo
  enddo
enddo
```

The column-oriented left-looking Cholesky factorization algorithm as depicted as

```
do j = 1,  n
  do i = j, n
    do k = 1, j-1
       w(i,j) = w(i,j) - w(j,k) x w(i,k)
    enddo
  enddo

  w(j,j) = sqrt(w(j,j))
  do i = j+1, n
     w(i,j) = w(i,j)/w(j,j)
  enddo
enddo
```

The two preceding algorithms are both called *column-oriented* because at each step of the outer loop they each complete computation of one column of the $\boldsymbol{L}$ matrix. (Row-oriented algorithms are equally possible.) Each of these two algorithms contains two blocks of code as the body of the outer loop. The block of four lines starting with the `sqrt` line may be regarded as the completion block because it completes computation of column `j`. This block implements Eqs.(8.23) and (8.24). The other block of five lines, consisting of two nested `do` loops, may be called the update block as it computes and propagates update values from completed columns to unfinished columns. This update block computes and accumulates terms in Eq.(8.21).

During the `jth` time through the outer loop, the *right*-looking algorithm does the completion block first, completing column `j` of $\boldsymbol{L}$, and then does the update block to propagate updates from column `j` to columns to the *right* of column `j`. In contrast, the *left*-looking algorithm does the update block first, propagating updates from the completed columns to the *left* of column `j` to column `j`, and then does the completion block, completing column `j` of $\boldsymbol{L}$.

The efficiency of a linear algebra algorithm generally depends strongly on the efficiency of its innermost loop. Ideally the innermost loop should not have any conditional statements or indirect addressing or separate index incrementation statements, and the sequence of memory locations referenced as the loop progresses should be contiguous. In the right-looking algorithm, the inner loop references `w(i,k)`and `w(i,j)` for increasing values of `i`, so the array `w` should be stored by columns. In the left-looking algorithm the inner loop references `w(j,k)` and `w(i,k)` for increasing values of `k`, so the array `w` should be stored by rows.

As a very small example of application of the right-looking dense algorithm consider the SPD matrix:

$$A = \begin{bmatrix} 100 & 120 & 140 \\ 120 & 225 & 276 \\ 140 & 276 & 404 \end{bmatrix}$$

Move the diagonal and lower triangle into a working array:

$$w = \begin{bmatrix} 100 & & \\ 120 & 225 & \\ 140 & 276 & 404 \end{bmatrix}$$

Complete computation of column 1 of L by replacing the (1,1) element by its square root and dividing the rest of the first column by this value:

$$\begin{bmatrix} 10 & & \\ 12 & 225 & \\ 14 & 276 & 404 \end{bmatrix}$$

Propagate updates from column 1 to columns 2 and 3 by subtracting 12 times $\begin{bmatrix} 12 \\ 14 \end{bmatrix}$ from the second column and subtracting 14×14 from the third column:

$$\begin{bmatrix} 10 & & \\ 12 & 81 & \\ 14 & 108 & 208 \end{bmatrix}$$

Complete computation of column 2 of L by replacing the (2,2) element with its square root and dividing the rest of the second column by this value:

$$\begin{bmatrix} 10 & & \\ 12 & 9 & \\ 14 & 12 & 208 \end{bmatrix}$$

Propagate an update from column 2 to column 3 by subtracting 12×12 from the (3,3) element, getting 64, and then complete computation of column 3 of L by replacing the (3,3) element with its square root:

$$L = \begin{bmatrix} 10 & & \\ 12 & 9 & \\ 14 & 12 & 8 \end{bmatrix}$$

8.3.4 Determining Fill

In the Cholesky factor matrix L, *filled* elements are those that are nonzero at locations that contained zeroes in the A matrix. It is useful to be able to determine the amount of fill that will occur for a given matrix A without actually carrying out the floating point arithmetic of the Cholesky factorization. A method of doing this can be derived from Eqs.(8.21), (8.23), and (8.24).

Eqs.(8.23) and 8.24) show that an element $l_{i,j}$ will be nonzero if and only if the corresponding quantity $\hat{a}_{i,j}$ is nonzero, and so the problem becomes one of determining which values $\hat{a}_{i,j}$ will be nonzero. Assuming the possibility of numerical cancelation is not to be considered, Eq.(8.21) implies $\hat{a}_{i,j}$ will be nonzero if any of the summands in the equation are nonzero, i.e., if $a_{i,j} \neq 0$ or if both $l_{i,k}$ and $l_{j,k}$ are nonzero for some $k < j$. This operation shows that to determine the nonzero status of all elements of $\boldsymbol{L}$ in some column j, the nonzero status of all elements of $\boldsymbol{L}$ in all columns to the left of column j must be known already. This process leads to the following symbolic algorithm to model the fill produced in Cholesky factorization.

Let C be an $n \times n$ array of single characters, initialized to all blanks.
Set the diagonal elements of C to “X”.
For each element $a_{i,j}$ below the diagonal that is nonzero set $C(i, j) =$ “X”.
For $k = 1$ to n
  In column k of C, for each pair of entries below the diagonal
  that are nonblank, say $C(j, k)$ and $C(i, k)$ with $j < i$,
  test $C(i, j)$, and if it is blank set it to “f”.
End for

On termination of this algorithm the array C will display the nonzero structure of the Cholesky factor $\boldsymbol{L}$ with “X” indicating locations at which there was a nonzero in $\boldsymbol{A}$ and “f” indicating locations that are nonzero because of fill. Displays constructed using this algorithm will be shown in the following sections.

This analysis algorithm can be interpreted directly in terms of the Cholesky factor matrix $\boldsymbol{L}$, noting that if column k of $\boldsymbol{L}$ has m nonzero values below the diagonal, each of the $m(m-1)/2$ pairs of these nonzero values makes a nonzero additive contribution to an element in a column to the right of column k, and so this will be a fill position in $\boldsymbol{L}$ if it was not already nonzero in $\boldsymbol{A}$. Specifically, if $k < j < i$, and $\boldsymbol{L}(j, k)$ and $\boldsymbol{L}(i, k)$ are nonzero, then $\boldsymbol{L}(i, j)$ will be nonzero.

8.3.5 Adjacency Graph

The adjacency graph of a SPD matrix **A**, of order n, is a set of n nodes (also called *vertices*) numbered from 1 to n, and a set of *edges*, such that an edge is present connecting nodes i and j, $(i \neq j)$, if and only if the element $a_{i,j}$ (and thus also $a_{j,i}$) in $\boldsymbol{A}$ is nonzero. The number of edges connecting to a node i is called the *degree* of node i.

8.3.6 Band and Profile

Because $\boldsymbol{A}$ is SPD there is no need for any row and column interchanges to preserve numerical stability during Cholesky factorization, and therefore no such interchanges will be considered. With these assumptions, no fill will be introduced in a row of $\boldsymbol{L}$ to the left of the position of the first nonzero element of $\boldsymbol{A}$ in that row.

Let $\alpha(i)$ denote the column index of the first nonzero element in row i of $\boldsymbol{A}$. As was just noted, there will be no fill in row i of $\boldsymbol{L}$ to the left of column $\alpha(i)$. On the other hand, if all rows of $\boldsymbol{A}$ except the first have at least one nonzero to the left of the diagonal, that is, $\alpha(i) < i$ for all $i > 1$, then all elements in row i of $\boldsymbol{L}$ from column $\alpha(i)$ through the diagonal will be nonzero.

Define

$$\omega(i) = i - \alpha(i)$$
$$\beta = \max\{\omega(i), i = 1, \ldots, n)\}$$
$$\pi = \sum \omega(i)$$

The quantity β is called the *semi bandwidth* of $\boldsymbol{A}$ and π is the *profile* of $\boldsymbol{A}$. The semi bandwidth β also is the largest value of $|i - j|$ among elements $a_{i,j}$ that are nonzero. A *band* method is one in which storage of elements of $\boldsymbol{A}$ and arithmetic operations on elements of $\boldsymbol{A}$ are limited to elements that are at most β positions to the left of the diagonal, i.e., within the band. Similarly in a *profile* (or skyline, envelope or locally variable bandwidth) method, one only stores and operates upon elements of each row i that are in columns $\alpha(i)$ through i, i.e., within the profile.

The Cuthill–McKee (CM) method[4] appeared in 1969 as a method of finding a symmetric permutation of a symmetric matrix to achieve an approximately minimal semi bandwidth. It was later noted[5,19] that the Reverse Cuthill–McKee (RCM) method achieved an approximately minimal profile while producing the same semi bandwidth as the CM method. Further improvements on the CM and RCM methods were made during the 1970s. An efficient and conveniently usable FORTRAN subroutine, GPSKCA, for bandwidth and/or profile minimization was authored by Lewis.[7] This code, based particularly on Refs. 6 and 20, uses carefully designed data structures to accomplish efficient analysis and manipulation of the adjacency graph. The STARS software uses GPSKCA as well as the SPOOLES library, which will be mentioned in Sec. 8.3.9.

Figure 8.3 shows a graph consisting of 18 nodes and 26 edges. Let this represent the structure of a system of 18 equations in which each equation involves one of the nodes and all of the other nodes connected to that node by an edge in the graph. To construct a matrix $\boldsymbol{A}$ representing this system of equations one must assign the numbers from 1 to 18 to the nodes. Then, a pair of symmetrically located elements, $a_{i,j}$ and $a_{j,i}$, will be nonzero if and only if nodes i and j are connected by an edge. (Diagonal elements, $a_{i,i}$ all will be assumed nonzero for any numbering.)

Figure 8.4 shows a simple sequential numbering of the graph that might be used just as a convenience for input of data to a computer program. This would not be expected to be an advantageous numbering for the factorization process. Figure 8.5

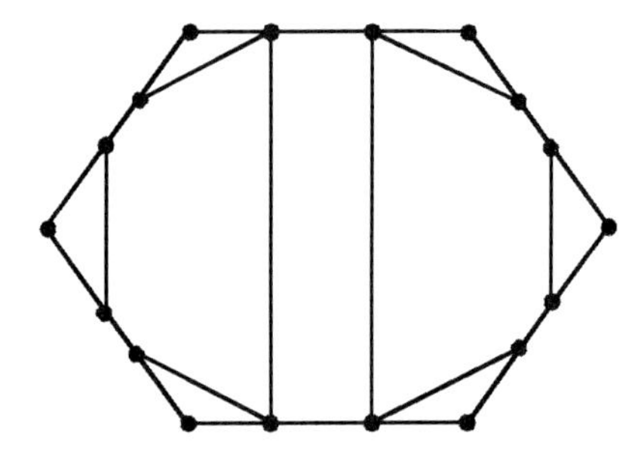

Fig. 8.3 Graph with 18 nodes and 26 edges.

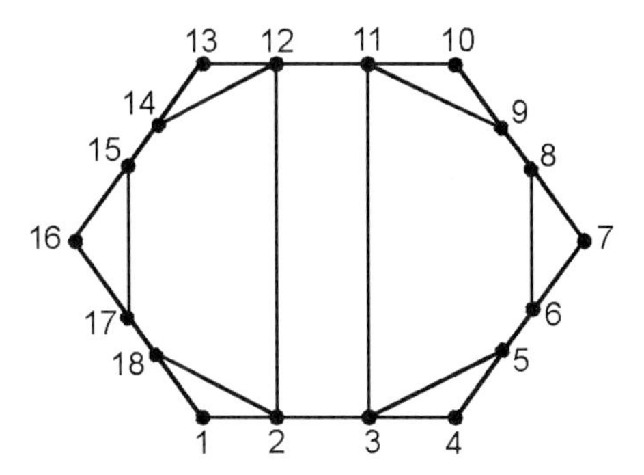

Fig. 8.4 Sequential numbering.

shows the nonzero structure of the matrix resulting from this numbering. The diagonal elements are all assumed to be nonzero, so use is made of those locations in the displays to show the row and column numbers. The X's denote nonzero values of ***A*** to the left of the diagonal, and these locations will remain nonzero in the Cholesky factor L. There are 26 of these, regardless of the numbering, because these X's simply correspond to the 26 edges in Fig. 8.3. The f's indicate locations that are zero in ***A*** but become nonzero fill in ***L***. The number of f's is the fill of ***L***. The fill depends strongly on the numbering.

Using subroutine GPSKCA to reduce the semi bandwidth β gives the numbering shown in Fig. 8.6, and the nonzero structure of Fig. 8.7. Using subroutine GPSKCA to reduce the profile π gives the numbering shown in Fig. 8.8 and the nonzero structure of Fig. 8.9.

```
 1   1
 2   X   2
 3       X   3
 4           X   4
 5           X   X   5
 6                   X   6
 7                       X   7
 8                       X   X   8
 9                               X   9
10                                   X   10
11           X   f   f   f   f   f   X   X   11
12       X   f   f   f   f   f   f   f   f   X   12
13                                               X   13
14                                               X   X   14
15                                                       X   15
16                                                           X   16
17                                                           X   X   17
18   X   X   f   f   f   f   f   f   f   f   f   f   f   f   f   f   x   18
```

Fig. 8.5 Sequential numbering: $\beta = 17$, $\pi = 53$, and fill = 27.

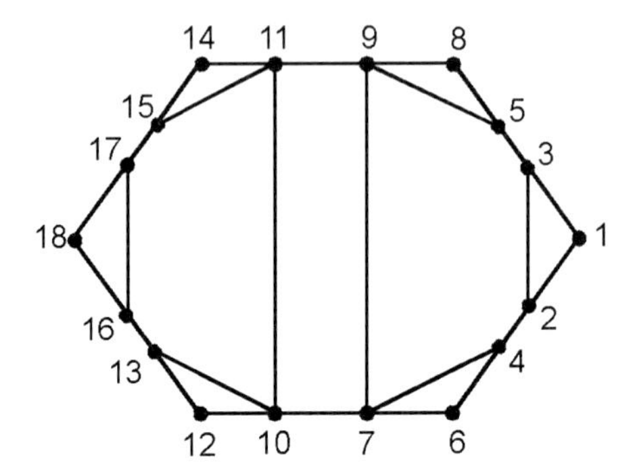

Fig. 8.6 Minimize semi bandwidth β using GPSKCA.

```
 1   1
 2   X  2
 3   X  X  3
 4      X  f  4
 5         X  f  5
 6            X  f  6
 7            X  f  X  7
 8               X  f  f  8
 9               X  f  X  X  9
10                     X  f  f 10
11                           X  X 11
12                              X  f 12
13                              X  f  X 13
14                                 X  f  f 14
15                                 X  f  f  X 15
16                                       X  f  f 16
17                                             X  X 17
18                                                X  X 18
```

Fig. 8.7 Minimize semi bandwidth β using GPSKCA: $\beta = 4$, $\pi = 43$, and fill $= 17$.

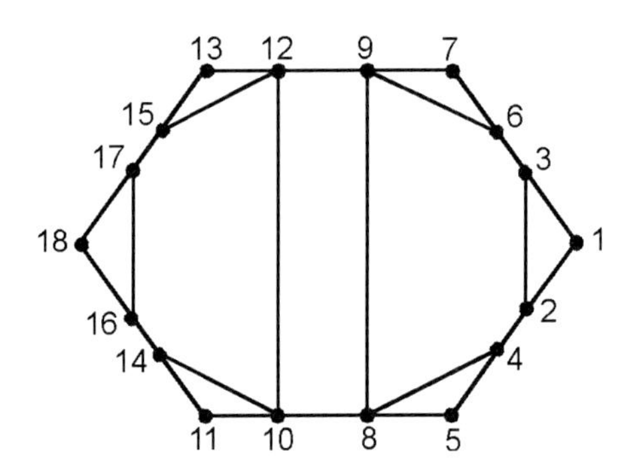

Fig. 8.8 Minimize profile π using GPSKCA.

```
 1   1
 2   X  2
 3   X  X  3
 4      X  f  4
 5            X  5
 6         X  f  f  6
 7                  X  7
 8            X  X  f  f  8
 9                  X  X  X  9
10                        X  f 10
11                              X 11
12                           X  X  f 12
13                                    X 13
14                              X  X  f  f 14
15                                    X  X  f 15
16                                          X  f 16
17                                             X  X 17
18                                                X  X 18
```

Fig. 8.9 Minimize profile π using GPSKCA: $\beta = 4$, $\pi = 37$, and fill $= 11$.

8.3.7 Minimum Degree

From the discussion at the end of the Sec. 8.3.4, if there are m off-diagonal nonzero elements in the first column of $\boldsymbol{A}$, the $m(m-1)/2$ products of pairs of these numbers will be added into following columns in the course of the factorization. These numbers produce fill to the extent that they are added into locations that were formerly zero.

With this in mind, one approach to reducing fill is to permute to the first column (and row) a column (and row) that has the least number of off-diagonal nonzero values. In terms of the adjacency graph of $\boldsymbol{A}$, this procedure amounts to selecting a node of minimal degree in the graph. After that, one can determine by operations on the adjacency graph of the matrix the nonzero structure the remaining $(n-1)$-order matrix will have following the elimination step using the first row and column. This is done by removing the node (and edges connected to it) associated with the row and column eliminated, and introducing edges, if not already present, connecting each pair of nodes that were previously connected by an edge to the removed node. This rule for adding edges in a graph can be regarded as just a different implementation of the algorithm for determining fill given in Sec. 8.3.4.

This process can be repeated to find the next permutation using the altered graph of the remaining $(n-1)$–order matrix, and so on, until a complete set of $n-1$ permutations has been determined.

This minimum degree procedure does not necessarily produce a permutation giving minimum fill. However, the idea has proved to be quite effective, so that with refinements, versions of minimum degree procedures are still found to be very useful in a number of current sparse matrix packages. The introduction of this general idea is attributed to Ref. 12, where it was applied to unsymmetric matrices, which is a bit more involved than the SPD case previously described. It was later described for SPD matrices in Ref. 13. For a 1989 survey of the evolution

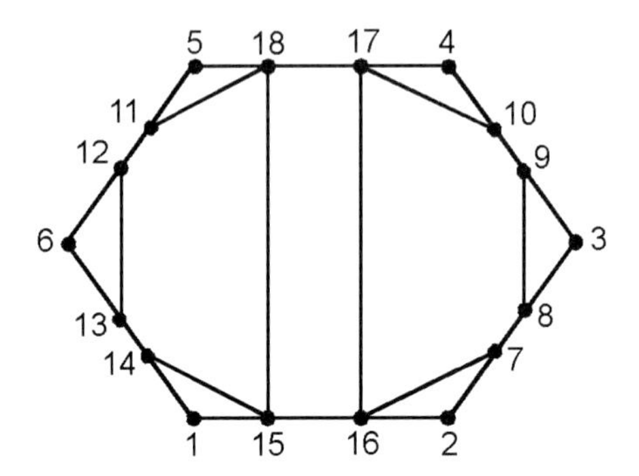

Fig. 8.10 Minimum degree numbering: fill = 7.

of minimum degree methods, see Ref. 14. More recent variants are described in Refs. 15 and 16.

Figure 8.3 will be used again as an example, this time for a minimum degree numbering. The initial sequential numbering of Fig. 8.4 will be assumed in order to have a means of referencing particular nodes and edges. When necessary for clarity this initial numbering will be called the old numbering. The nodes of minimum degree in Fig. 8.4 are seen to be nodes 1, 4, 7, 10, 13, and 16, each having degree 2. Node 1 can be removed, along with the edges connecting it to nodes 18 and 2. This process does not require inserting a new edge because the edge connecting 18 and 2 is already present. Similarly nodes 4, 7, 10, 13, and 16 can be removed and no new edges are required. New numbers 1, 2, 3, 4, 5, and 6 will be assigned respectively to the nodes whose old numbers were 1, 4, 7, 10, 13, and 16. See Fig. 8.10 for the new numbering.

After removing the six cited nodes from Fig. 8.2, there will be eight nodes with minimal degree 2. These are the nodes with old numbers 5, 6, 8, 9, 14, 15, 17, and 18. Removing node 5 requires inserting an edge connecting nodes 6 and 3. Next, removing node 6 requires inserting an edge from 8 to 3. Removing node 8 requires inserting an edge from 9 to 3, but then removing node 9 will not require a new edge because the edge connecting nodes 11 and 3 is already present. Summarizing: nodes 5, 6, 8, and 9 can be removed at a cost of three new edges (implying fill at locations, (6,3), (8,3) and (9,3) in L) and these nodes will be assigned the new numbers 7, 8, 9, and 10, respectively.

Remove nodes 14, 15, 17, and 18 similarly at a cost of three new edges and assign these nodes the new numbers 11, 12, 13, and 14, respectively.

Just the four nodes are new remaining whose old numbers are 2, 3, 11, and 12, each at this stage having degree 2. Any one of these nodes can be removed at a cost of one new edge, and the remaining three can be removed in any order, requiring no further new edges. For definiteness, remove nodes 2, 3, 11, and 12 in that order at a cost of just one new edge from 12 to 3, and assign new numbers 15, 16, 17, and 18, respectively.

The total number of new edges inserted and thus the total fill in the corresponding $\boldsymbol{L}$ matrix is 7. The graph with this numbering is shown in Fig. 8.10, and the nonzero structure of the corresponding $\boldsymbol{A}$ and $\boldsymbol{L}$ matrices is shown in Fig. 8.11.

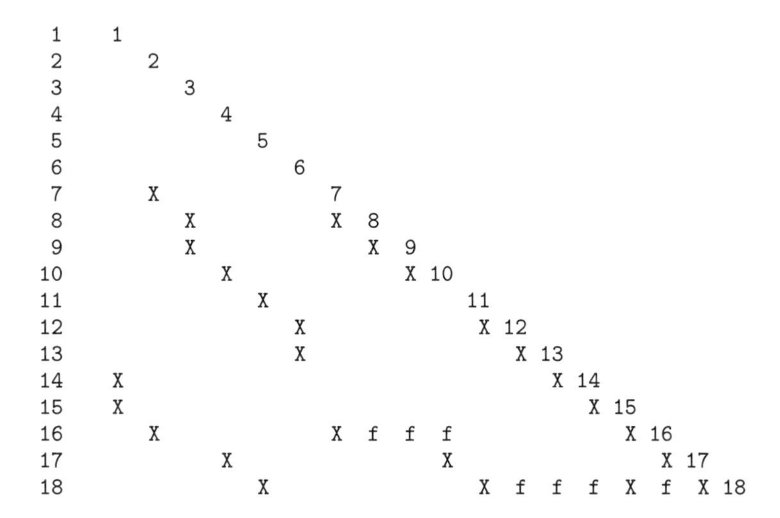

Fig. 8.11 Minimum degree numbering: fill = 7.

8.3.8 Domain Decomposition

Within domain decomposition are grouped methods that classify the nodes of the adjacency graph of the matrix $\boldsymbol{A}$ into two or more connected *domains* and one or more *separators*. Each node is assigned to exactly one domain or separator. Domains are defined so that nodes in one domain are not linked by an edge to nodes in any other domain. Roughly speaking, domains are separated from each other, and separators are buffer regions between the domains.

Different methods have been developed for classifying the variables of a problem into domains and separators. Reference 8 presents *one-way dissection* and *nested dissection*, and these methods are also described in Ref. 9. The *multisection* method is given in Ref. 17, where comparisons are made with the fill-reducing performance of implementations of a minimum degree method and a nested dissection method.

In a nested dissection method one first divides the nodes of the adjacency graph into two connected domains and one separator, aiming to have the separator contain relatively few nodes and having the domains contain nearly equal numbers of nodes and being isolated from each other. This process is applied recursively to each domain, replacing each domain with a new separator and two new domains, until a point is reached at which the domains are below a specified size.

Node numbers are also assigned recursively. Suppose the initial dissection produces a separator S_1 with s_1 nodes and domains D_{1a} and D_{1b} with d_{1a} and d_{1b} nodes, respectively. For a problem of order n, one would assign the largest s_1 numbers, i.e., the numbers $n-s_1+1$ through n, to the nodes of S_1. The smallest d_{1a} numbers would be allocated to D_{1a}, but not yet assigned to individual nodes. Similarly, the remaining d_{1b} numbers would be allocated to D_{1b}, but not yet assigned to individual nodes. When the dissection is applied recursively, say to D_{1a}, the largest of the numbers allocated to D_{1a} would be assigned to the nodes of the new separator, and the smaller numbers, in two contiguous blocks, would be allocated to the two new domains.

As an example, consider application of two stages of nested dissection to the graph of Fig. 8.3. For convenience refer immediately to the final configuration in Fig. 8.12. The four nodes enclosed by the large dotted loop constitute the first separator, S_1. This leaves a domain D_{1a} of seven nodes to the left of the separator and a second domain D_{1b}, also of seven nodes, on the right. The largest four numbers, 15, 16, 17, and 18 are assigned immediately to the nodes of the separator. The numbers 1–7 are reserved for the domain D_{1a} and the numbers 8–14 are reserved for the domain D_{1b}.

As a second stage of dissection, the domain D_{1a} on the left is divided into a single node separator S_2, shown enclosed in the small dotted oval on the left, and two domains, D_{2a} and D_{2b}, below and above S_2. The largest number reserved for the domain D_{1a}, which is 7, is assigned to the separator. The numbers 1 and 2 are assigned to the domain D_{2a} and the numbers 3–6 are assigned to the domain D_{2b}. Assignment of the available numbers to individual nodes in these two small domains could be done in various ways. A minimum degree assignment is one possibility.

Similarly, the domain D_{1b}, consisting of the seven nodes on the right, is dissected into a one-node separator, given the number 14, and two small domains, given the numbers 8 and 9 and 10–13, respectively.

The final assignment of numbers is shown in Fig. 8.12, and the nonzero structure of the corresponding $\boldsymbol{A}$ and $\boldsymbol{L}$ matrices is shown in Fig. 8.13. The total fill is 7.

In Ref. 8. it is proved that nested dissection is optimal in the sense of the order of magnitude of the number of nonzero values in $\boldsymbol{L}$ and the number of arithmetic operations needed to produce $\boldsymbol{L}$ for a problem arising from a two-dimensional $m \times m$ regular grid. In Ref. 17, the relative performance of nested dissection and other methods is reported for a variety of two-and three-dimensional grids.

The one-way dissection and multisection methods set out from the beginning to produce a specified number of connected domains isolated from each other by one or more separators. The higher numbers would be allocated to the separators, and the lower numbers, in contiguous blocks, would be assigned to the domains. Assignment of numbers from the allocations for a domain or separator to individual nodes would depend on details of the different methods.

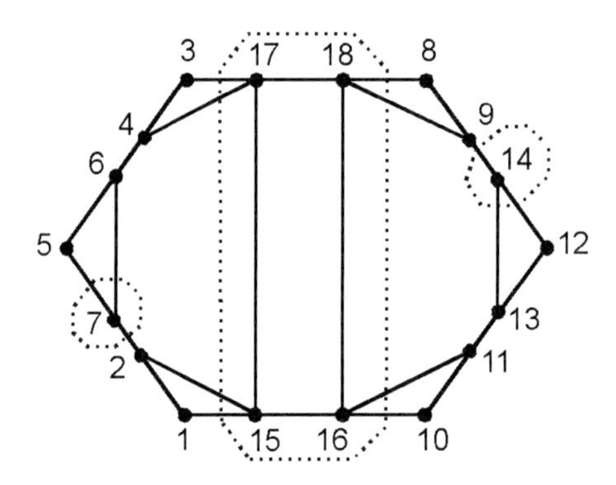

Fig. 8.12 Numbering by nested dissection: fill = 7.

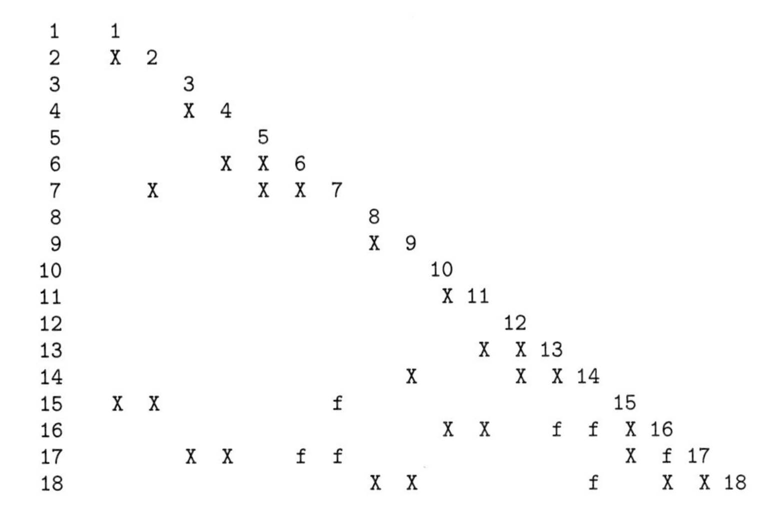

Fig. 8.13 Numbering by nested dissection: fill = 7.

8.3.9 Management of the Factorization Process

Assuming one has permuted the rows and columns of $\boldsymbol{A}$ using a fill-reducing method, and has determined by appropriate manipulation of the adjacency graph where fill will occur, a primary goal in organizing the factorization procedure is to avoid, as much as possible, bringing elements of $\boldsymbol{L}$ into the computation that are known to be zero. Considerable flexibility is available in sequencing of computation needed for Eqs. (8.21), (8.23), and (8.24) and the accumulation of the partial sums needed in Eq. (8.21).

The order 18 example will be used to illustrate how this computation can be organized using a column-oriented forward-looking multifrontal method as described in Ref. 10. The name *frontal* and basic concepts of this approach apparently originated with Ref. 21. The authors of Ref. 9 have been very active contributors to the development of multifrontal methods. Either of the low-fill numbering methods would be appropriate to use, that is, minimum degree method or nested dissection. The former will be used, referencing Fig. 8.11.

The first step is to construct the elimination tree shown in Fig. 8.14 that shows the dependency relations between the columns of $\boldsymbol{L}$, in the sense of showing for each column, j, which columns to the left of column j must be computed before column j can be computed. Each node of this tree represents a column of $\boldsymbol{L}$, symbolized by its column number followed by the row numbers of its subdiagonal nonzero values in parenthesis. For example column 2 is symbolized by "2(7,16)."

A *tree* is a graph with no cycles. Graph theory uses both arboreal and genealogical terminology in referring to relations in a tree. A tree is called a rooted tree if one node, in this example, node 18, is distinguished as the root. This imposes a partial ordering on the nodes based on distance along a path to the root. For two neighboring nodes, say p and q, if p is closer to the root than q, it is called the parent of q, and q is called the child of p. All nodes farther from the root than

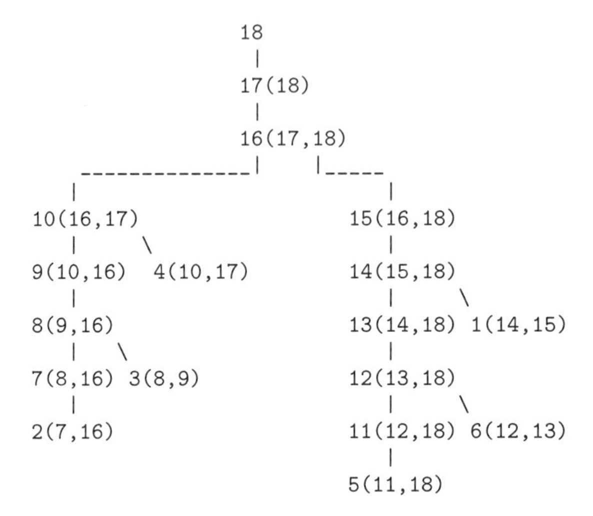

Fig. 8.14 Elimination tree based on Fig. 8.11.

p along any path through p are descendants of p. The root, and all other nodes closer to the root than p along the (unique) path from p to the root are ancestors of p. A node that has no children is called a leaf.

The tree in Fig. 8.14 was built by starting with node 18 as the root. Then for $j = 18, 17, \ldots, 2$, each column of $\boldsymbol{L}$, if any, whose representation is $k(j, \ldots)$ is inserted into the tree as a child of node j.

The significance of an elimination tree for the Cholesky factorization problem is that computation can begin with any leaf node, or simultaneously with any number of leaf nodes in a multiprocessor implementation, and proceed up the tree, i.e., toward the root. Once the nonzero elements of all columns that appear as children of any particular node j have been computed, the nonzero values of column j can be computed.

Before illustrating this computation, there is one more renumbering that can be applied. There is a bit more efficiency attainable, depending on details of an implementation, by having the numbering in the elimination tree be sequential from a child to its parent, except where this is not possible because of the branching in the tree. This renumbering does not alter the amount of fill. The tree will be renumbered by changing the numbers

$$1, 2, 4, 5, 6, 7, 8, 9, 10, 11, 12, 13$$

to

$$13, 1, 6, 8, 10, 2, 4, 5, 7, 9, 11, 12$$

This renumbering changes the tree of Fig. 8.14 to Fig. 8.15. The adjacency graph of the matrix $\mathbf{A}$ with this renumbering is shown in Fig. 8.16, and the nonzero structure of $\boldsymbol{A}$ and $\boldsymbol{L}$ in Fig. 8.17.

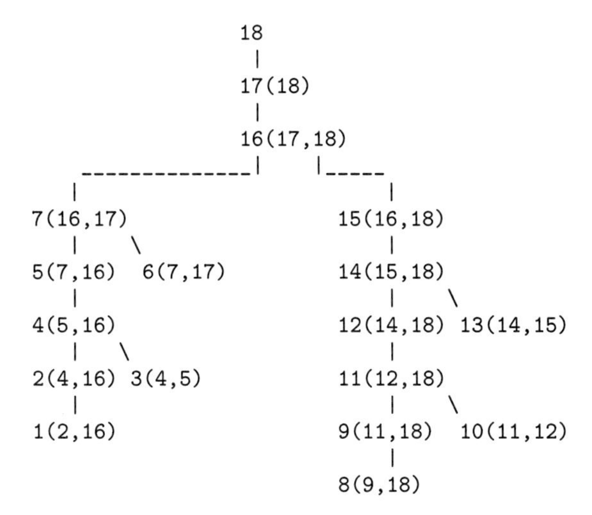

Fig. 8.15 Elimination tree renumbered from Fig. 8.14.

To provide an illustration of a multifrontal method of organizing Cholesky factorization, a sample SPD matrix $\boldsymbol{A}$ of order 18 is defined in Fig. 8.18, having the nonzero structure indicated by the X's in Fig. 8.17, by assigning the values $21, 22, \ldots, 38$ to the diagonal positions and the values $0.1, 0.2, \ldots, 2.6$ to the subdiagonal nonzero positions and to the symmetric super-diagonal positions.

Associated with each node j in Fig. 8.15 there will be a frontal matrix $\boldsymbol{F}_j$. Each frontal matrix $\boldsymbol{F}_j$ will be a symmetric matrix (often not positive-definite) of order equal to the number of nonzero values in column j of $\boldsymbol{L}$. For example, column 5 of $\boldsymbol{L}$ (see Fig. 8.17) has nonzero values in rows 5, 7, and 16, and so the frontal matrix $\boldsymbol{F}_5$ will be of order 3, and its first, second and third rows and columns will contain quantities that eventually contribute to the final values in rows and columns 5, 7,

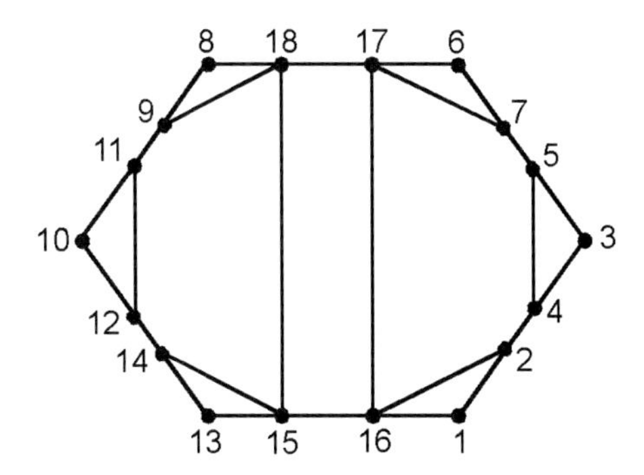

Fig. 8.16 Adjacency graph. (Minimum degree with renumbering to improve the elimination tree.)

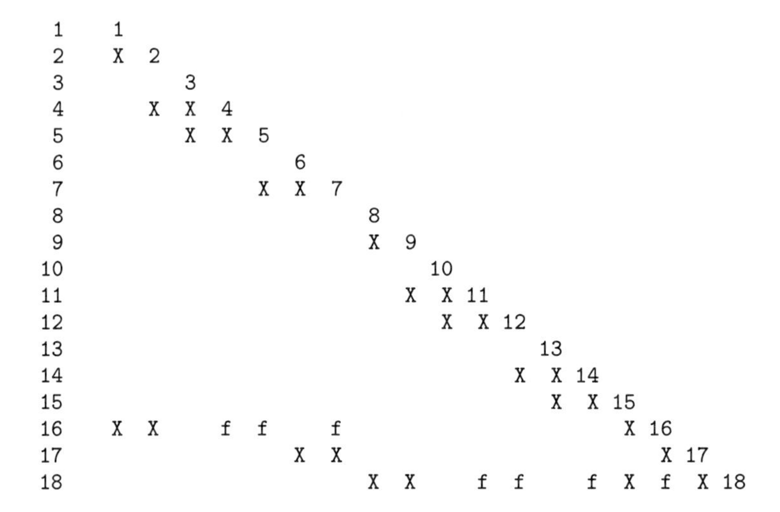

Fig. 8.17 Nonzero structure of A and L after minimum degree and renumbering to improve the elimination tree: fill = 7.

and 16 of $\boldsymbol{L}$. The index values 5, 7, and 16 will be called alias indices for rows and columns 1, 2 and 3 of $\boldsymbol{F}_5$. Because each $\boldsymbol{F}_j$ is symmetric, only the diagonal and lower triangle of each $\boldsymbol{F}_j$ will be stored and operated upon.

Frontal matrices come into existence and go out of existence during the factorization process as the locus of the process flows up from the leaves to the root of the elimination tree. It is not necessary to have memory space for all of the frontal matrices simultaneously. In our example, space for two or three frontal matrices existing simultaneously would be adequate, depending on details of an implementation.

(Symmetric)

1	21																	
2	0.1	22																
3	0	0	23															
4	0	0.2	0.3	24														
5	0	0	0.4	0.5	25													
6	0	0	0	0	0	26												
7	0	0	0	0	0.6	0.7	27											
8	0	0	0	0	0	0	0	28										
9	0	0	0	0	0	0	0	0.8	29									
10	0	0	0	0	0	0	0	0	0	30								
11	0	0	0	0	0	0	0	0	0.9	1.0	31							
12	0	0	0	0	0	0	0	0	0	1.1	1.2	32						
13	0	0	0	0	0	0	0	0	0	0	0	0	33					
14	0	0	0	0	0	0	0	0	0	0	0	1.3	1.4	34				
15	0	0	0	0	0	0	0	0	0	0	0	0	1.5	1.6	35			
16	1.7	1.8	0	0	0	0	0	0	0	0	0	0	0	0	1.9	36		
17	0	0	0	0	0	20	2.1	0	0	0	0	0	0	0	0	2.2	37	
18	0	0	0	0	0	0	0	2.3	2.4	0	0	0	0	0	2.5	0	2.6	38

Fig. 8.18 Sample SPD matrix A.

The input to frontal matrix $\boldsymbol{F}_j$ consists of the nonzero values of column j of A and update matrices from frontal matrices that are children of node j in the elimination graph. The output of $\boldsymbol{F}_j$ are the nonzero values of column j of $\boldsymbol{L}$ and an update matrix $\boldsymbol{U}_j$ to be passed up to the parent of node j.

To describe the processing of one frontal matrix, consider $\boldsymbol{F}_7$ (see Fig. 8.15). Assume $\boldsymbol{F}_5$ and $\boldsymbol{F}_6$ already have been processed and they produced the update matrices $\boldsymbol{U}_5$ and $\boldsymbol{U}_6$, respectively, with $\boldsymbol{U}_5$ being an order 2 matrix, with alias indices 7 and 16, and $\boldsymbol{U}_6$ being an order 2 matrix with alias indices 7 and 17.

Because column 7 of $\boldsymbol{L}$ has nonzero values in rows 7, 16, and 17, the frontal matrix $\boldsymbol{F}_7$ will be an order 3 triangular matrix with alias indices 7, 16, and 17. Initially set this matrix to zero. Copy the nonzero elements of column 7 of A (on and below the diagonal) to $\boldsymbol{F}_7$, placing them according to the alias indices. Thus the nonzero element $a_{7,7}$ will go to position (1,1) of $\boldsymbol{F}_7$, and $a_{17,7}$ will go to position (3,1) of $\boldsymbol{F}_7$.

Add update matrix $\boldsymbol{U}_5$ into $\boldsymbol{F}_7$, adding elements of like alias indices. Here $\boldsymbol{F}_7$ is of order 3 with alias indices 7, 16, and 17, whereas $\boldsymbol{U}_5$ is of order 2 with alias indices 7 and 16.

Add update matrix $\boldsymbol{U}_6$ into $\boldsymbol{F}_7$, adding elements of like alias indices.

Do one step of column-oriented forward-looking Cholesky factorization on the frontal matrix $\boldsymbol{F}_7$ of order 3. Here is this algorithm for the case of a frontal matrix F of order nf:

```
F(1,1) = sqrt( F(1,1))
do i = 2, nf
   F(i,1) = F(i,1) / F(1,1)
enddo

do j = 2, nf
   do i = j, nf
      F(i,j) = F(i,j) - F(i,1) x F(j,1)
   enddo
enddo
```

Note that the preceding algorithm implements the square root and quotients needed for Eqs.(8.23), and (8.24) and the accumulation of partial sums of products needed for Eq.(8.21). The operation of adding update matrices into frontal matrices accomplishes combining of partial sums to eventually form the complete sums for Eq.(8.21).

After this partial Cholesky factorization of $\boldsymbol{F}_7$, the first column of $\boldsymbol{F}_7$ contains the nonzero elements of column 7 of $\boldsymbol{L}$, and the remaining two rows and columns of $\boldsymbol{F}_7$ contain the update matrix $\boldsymbol{U}_7$, which in this case will be an order 2 matrix with alias indices 16 and 17.

Considering the computer memory required by this algorithm on a uniprocessor system, the input is the nonzero values of $\boldsymbol{A}$, one column on and below the diagonal at a time. Only one frontal matrix needs to exist at a time, so it suffices to provide space to accommodate the largest frontal matrix. For the update matrices, space must be provided for the largest set of update matrices that need to be held at any point in time for future use. In our example, this would be one or two update matrices, depending on details of the implementation. Output consists of the nonzero values of $\boldsymbol{L}$, one column at a time.

Considering computational efficiency, note that the only elements of $\boldsymbol{L}$ that are computed are the nonzero values. All of the update matrices are dense triangular matrices, and so no zeros are being passed from one frontal matrix to the next. Most of the arithmetic operations, including all of the multiplications and divisions, occur in the preceding algorithm, which uses sequential direct addressing, as contrasted with sparse matrix indirect addressing, and thus can potentially benefit significantly from cache memory because of the locality of the operands.

Following are the steps of applying this algorithm to the present example. Alias indices are shown to the left of each row.

```
Initialize Front Matrix, No.   1, Size =   3
   1   21.000
   2  0.10000        0.0000
  16   1.7000        0.0000        0.0000

          Partially Factored Front Matrix, No.   1
            1    4.5826
            2  0.21822E-01  -0.47619E-03
           16  0.37097      -0.80952E-02 -0.13762

Initialize Front Matrix, No.   2, Size =   3
   2   22.000
   4  0.20000        0.0000
  16   1.8000        0.0000        0.0000

        Front Matrix, No.   2 after adding updates from front 1
           2   22.000
           4  0.20000        0.0000
          16   1.7919        0.0000       -0.13762

          Partially Factored Front Matrix, No.   2
            2    4.6904
            4  0.42641E-01  -0.18182E-02
           16  0.38204      -0.16290E-01 -0.28357

Initialize Front Matrix, No.   3, Size =   3
   3   23.000
   4  0.30000        0.0000
   5  0.40000        0.0000        0.0000

          Partially Factored Front Matrix, No.   3
            3    4.7958
            4  0.62554E-01  -0.39130E-02
            5  0.83406E-01  -0.52174E-02 -0.69565E-02
```

```
Initialize Front Matrix, No.   4, Size =   3
  4   24.000
  5  0.50000        0.0000
 16   0.0000        0.0000         0.0000

    Front Matrix, No.   4 after adding updates from fronts 2
      and 3
      4   23.994
      5  0.49478     -0.69565E-02
     16 -0.16290E-01  0.0000      -0.28357

        Partially Factored Front Matrix, No.   4
          4   4.8984
          5  0.10101     -0.17159E-01
         16 -0.33257E-02  0.33592E-03 -0.28358

Initialize Front Matrix, No.   5, Size =   3
  5   25.000
  7  0.60000        0.0000
 16   0.0000        0.0000         0.0000

    Front Matrix, No.   5 after adding updates from front 4
      5   24.983
      7  0.60000        0.0000
     16  0.33592E-03    0.0000      -0.28358

        Partially Factored Front Matrix, No.   5
          5   4.9983
          7  0.12004     -0.14410E-01
         16  0.67207E-04 -0.80677E-05 -0.28358

Initialize Front Matrix, No.   6, Size =   3
  6   26.000
  7  0.70000        0.0000
 17   2.0000        0.0000         0.0000

        Partially Factored Front Matrix, No.   6
          6   5.0990
          7  0.13728     -0.18846E-01
         17  0.39223     -0.53846E-01 -0.15385

Initialize Front Matrix, No.   7, Size =   3
  7   27.000
 16   0.0000        0.0000
 17   2.1000        0.0000         0.0000

    Front Matrix, No.   7 after adding updates from fronts 5
      and 6
```

```
       7   26.967
      16 -0.80677E-05 -0.28358
      17   2.0462        0.0000      -0.15385

          Partially Factored Front Matrix, No.   7
           7   5.1930
          16 -0.15536E-05 -0.28358
          17  0.39403       0.61215E-06 -0.30910

Initialize Front Matrix, No.   8, Size =   3
  8   28.000
  9  0.80000       0.0000
 18   2.3000       0.0000        0.0000

          Partially Factored Front Matrix, No.   8
           8   5.2915
           9  0.15119     -0.22857E-01
          18  0.43466     -0.65714E-01 -0.18893

Initialize Front Matrix, No.   9, Size =   3
  9   29.000
 11  0.90000       0.0000
 18   2.4000       0.0000        0.0000

     Front Matrix, No.   9 after adding updates from front 8
       9   28.1177
      11  0.90000       0.0000
      18   2.3343       0.0000      -0.18893

          Partially Factored Front Matrix, No.   9
           9   5.3830
          11  0.16719     -0.27953E-01
          18  0.43364     -0.72500E-01 -0.37697

Initialize Front Matrix, No.  10, Size =   3
 10   30.000
 11   1.0000       0.0000
 12   1.1000       0.0000        0.0000

          Partially Factored Front Matrix, No.  10
          10   5.4772
          11  0.18257     -0.33333E-01
          12  0.20083     -0.36667E-01 -0.40333E-01

Initialize Front Matrix, No.  11, Size =   3
 11   31.000
 12   1.2000       0.0000
 18   0.0000       0.0000        0.0000
```

```
    Front Matrix, No.  11 after adding updates from fronts 9
      and 10
     11   30.939
     12   1.1633      -0.40333E-01
     18 -0.72500E-01   0.0000      -0.37697


        Partially Factored Front Matrix, No.  11
         11   5.5623
         12  0.20915     -0.84076E-01
         18 -0.13034E-01  0.27261E-02 -0.37714

Initialize Front Matrix, No.  12, Size =   3
 12   32.000
 14   1.3000        0.0000
 18   0.0000        0.0000        0.0000

    Front Matrix, No.  12 after adding updates from front 11
     12   31.916
     14   1.3000        0.0000
     18  0.27261E-02   0.0000      -0.37714

        Partially Factored Front Matrix, No.  12
         12   5.6494
         14  0.23011     -0.52952E-01
         18  0.48255E-03 -0.11104E-03 -0.37714

Initialize Front Matrix, No.  13, Size =   3
 13   33.000
 14   1.4000        0.0000
 15   1.5000        0.0000        0.0000

        Partially Factored Front Matrix, No.  13
         13   5.7446
         14  0.24371     -0.59394E-01
         15  0.26112     -0.63636E-01 -0.68182E-01

Initialize Front Matrix, No.  14, Size =   3
 14   34.000
 15   1.6000        0.0000
 18   0.0000        0.0000        0.0000

    Front Matrix, No.  14 after adding updates from fronts
      12 and 13
     14   33.888
     15   1.5364      -0.68182E-01
     18 -0.11104E-03   0.0000      -0.37714
```

```
          Partially Factored Front Matrix, No.  14
           14   5.8213
           15  0.26392      -0.13784
           18 -0.19075E-04  0.50342E-05 -0.37714

 Initialize Front Matrix, No.  15, Size =   3
  15   35.000
  16   1.9000        0.0000
  18   2.5000        0.0000        0.0000

      Front Matrix, No.  15 after adding updates from front 14
       15   34.862
       16   1.9000        0.0000
       18   2.5000        0.0000      -0.37714

          Partially Factored Front Matrix, No.  15
           15   5.9044
           16  0.32179      -0.10355
           18  0.42341      -0.13625      -0.55642

 Initialize Front Matrix, No.  16, Size =   3
  16   36.000
  17   2.2000        0.0000
  18   0.0000        0.0000        0.0000

      Front Matrix, No.  16 after adding updates from fronts 7
        and 15
       16   35.613
       17   2.2000     -0.30910
       18 -0.13625       0.0000      -0.55642

          Partially Factored Front Matrix, No.  16
           16   5.9677
           17  0.36865      -0.44501
           18 -0.22832E-01  0.84170E-02 -0.55694

 Initialize Front Matrix, No.  17, Size =   2
  17   37.000
  18   2.6000        0.0000

      Front Matrix, No.  17 after adding updates from front 16
       17   36.555
       18   2.6084     -0.55694

          Partially Factored Front Matrix, No.  17
           17   6.0461
           18  0.43142      -0.74307
```

```
Initialize Front Matrix, No.  18, Size =   1
 18   38.000

     Front Matrix, No.  18 after adding updates from front 17
      18   37.257

          Partially Factored Front Matrix, No.  18
           18   6.1038
```

In larger, more realistic problems it will commonly be the case that some sets of adjacent columns of $\boldsymbol{L}$ will have a dense triangle on the diagonal and identical locations of nonzero values below the triangle. These columns can be treated together in one frontal matrix. When adjacent columns are grouped together for this or any other reason, the group is called a supernode in the tree that guides the factorization. Our example has a rather trivial example of this grouping with columns 16, 17, and 18.

If the columns making up a supernode satisfy the stated conditions then the frontal matrix for the first column of the supernode will be of adequate dimension and have suitable alias indices to handle all of the columns of the supernode, thus avoiding some data movement that would occur if the columns were treated in separate frontal matrices, and not requiring any increase in the number of floating point operations . Even if a set of adjacent columns only partly satisfy the stated conditions, it may be worthwhile to group them together, paying a penalty by doing some arithmetic operations on zero elements in exchange for avoiding some data movement.

Many refinements and variations have been developed around the basic idea of organizing the factorization of a sparse matrix into operations on a number of small dense matrices that for the most part avoid doing arithmetic on zero values. The paper[11] surveys and reports on tests of seven unsymmetric sparse matrix software packages of this type that were individually the subjects of papers published in the 1997 – 2002 time period. Although the focus of Ref. 11 is unsymmetric matrices, the text and the 45 references contain much valuable information applicable to SPD matrices. At least three of the packages surveyed have options to give special treatment to SPD matrices. One of the packages treated in Ref. 11, SPOOLES,[18] not only provides for special treatment of SPD matrices, but has versions for serial, multithreaded, and multiprocessor usage. The SPOOLES software library is used in the STARS package.

8.3.10 Numerical Example

A representative, practical, large and complex problem is presented that provides numerical estimates of the relative efficiencies of various sparse matrix schemes.

Example 8.1: HyperX Stack Flight Vehicle

A practical example of a large, complex structure is chosen to demonstrate the superior performance of the multifrontal factorization procedure over the conventional band minimization schemes. The HyperX stack structural finite element model is shown in Fig. 8.19. It is characterized by 20,018 nodes and 22,410

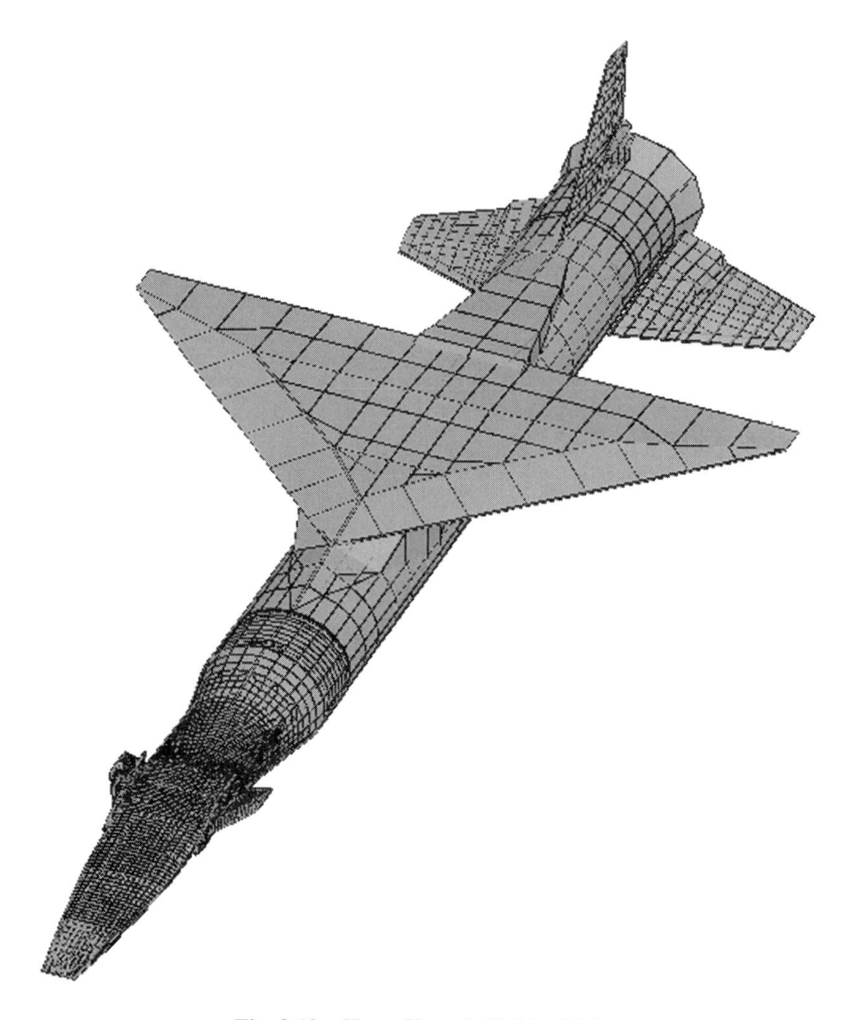

Fig. 8.19 HyperX stack flight vehicle.

elements. The order of the associated matrices is 120,108. Free vibration analysis of the structure was performed for the first 20 modes and frequencies using the usual band minimization (method 1), and the procedure that uses minimum fill techniques in conjunction with the multifrontal solution method (method 2). The relevant results are summarized in Table 8.1; the analyses were performed using a 1.7 GHZ PC machine.

8.4 Concluding Remarks

A number of sparse matrix storage and decomposition techniques have been described in this chapter. Techniques that ensure minimum storage requirements include the minimum fill procedures. As the numerical example indicates, a combination of the minimum fill method along with a multifrontal decomposition

Table 8.1 Solution comparison

Item	Method 1	Method 2
Storage(***L*** matrix, sparse)	1650 Mbyte	133 Mbyte
Solution time (cpu)	9 h, 38 min	29 min, 48s

procedure ensures the most economical solution in terms of both storage and CPU time. Both of these two important parameters are reduced by more than an order of magnitude. Therefore, for practical problems of very large magnitude, these techniques prove to be both economical and useful.

References

[1]Wilkinson, J.H., *The Algebraic Eigenvalue Problem*, Clarendon Press, Oxford, England, United Kingdom, 1965.

[2]Fox, L., *An Introduction to Numerical Linear Algebra*, Oxford Univ. Press, Oxford, England, United Kingdom, 1964.

[3]Barrett, R., et. al., *Templates for the Solution of Linear Systems: Building Blocks for Iterative Methods*, Society of Industrial and Applied Mathematics, 1994.

[4]Cuthill, E. and McKee, J., "Reducing the Bandwidth of Sparse Symmetric Matrices," *Proceedings of the 24th National Conference, Association of Computing Machinery*, ACM, pp. 157–172, 1969.

[5]Jennings, A., Matrix Computation for Engineers and Scientists, Wiley, New York, 1977.

[6]Gibbs, N. E., Poole, W. G., Jr., and Stockmeyer, P. K., An Algorithm for Reducing the Bandwidth and Profile of a Sparse Matrix, *SIAM Journal of Numerical Analysis*, Vol. 13, 1976, pp. 236–250.

[7]Lewis, J. G., "Implementation of the Gibbs–Poole–Stockmeyer and Gibbs–King Algorithms," *ACM Transactions on Mathematical Software*, Vol. 8, 1982, pp. 180–194.

[8]George, A., and Liu, J. W-H., *Computer Solution of Large Sparse Positive Definite Systems*, Prentice-Hall, Englewood Cliffs, NJ, 1981.

[9]Duff, I. S., Erisman, A. M., and Reid, J. K., Direct Methods for Sparse Matrices, Oxford Univ. Press, Oxford, England, United Kingdom, 1986.

[10]Liu, J. W-H., "The Multifrontal Method for Sparse Matrix Solution: Theory and Practice," *SIAM Review*, Vol. 34, 1992, pp. 82–109.

[11]Gupta, A., "Recent Advances in Direct Methods for Solving Unsymmetric Sparse Systems of Linear Equations," *ACM Transactions on Mathematical Software*, Vol. 28, No. 3, 2002, pp. 301–324.

[12]Markowitz, H. M., "The elimination Form of the Inverse and Its Application to Linear Programming," *Management Science*, Vol. 3, 1957, pp. 255–269.

[13]Tinney, W. F., and Walker, J. W., "Direct Solutions of Sparse Network Equations by Optimally Ordered Triangular Factorization," *Journal of the Proceedings of the IEEE*, Vol. 55, 1967, pp. 1801–1809.

[14]George, J., and Liu, J. W-H., "The Evolution of the Minimum Degree Ordering Algorithm," *SIAM Review*, Vol. 31, 1989, pp. 1–19.

[15]Amestoy, P. R., Davis, T. A., and Duff, I. S., "An Approximate Minimum Degree Ordering Algorithm," *SIAM Journal of Matrix Analysis and Applications*, Vol. 17, No. 4, 1996, pp. 886–905.

[16]Davis, T. A., Gilbert, J. R., Larimore, S. I., and Ng, E. G.-Y., "A Column Approximate Minimum Degree Ordering Algorithm," Tech. Rept. TR-00-005, Computer and Information Sciences Dept., Univ. of Florida, Gainsville, FL, 2000.

[17]Ashcraft,C., and Liu,J,W.-H., Robust Ordering of Sparse Matrices Using Multisection, *SIAM J. Matrix Anal. Appl.*, Vol. 19, No. 3, 1998, pp. 816–832.

[18]Ashcraft, C., and Grimes, R. G., "SPOOLES: An Object-Oriented Sparse Matrix Library," *Proceedings of the Ninth SIAM Conference on Parallel Processing for Scientific Computing*, 1999.

[19]George, A., Computer Implementation of the Finite Element Method, Tech. Rept. STAN-CS-208, Stanford Univ., 1971.

[20]Gibbs, N. E., "A Hybrid Profile Reduction Algorithm," *ACM Trans. Math. Softw.*, Vol. 2, No. 4, 1976, 378–387.

[21]Irons, B. M., "A Frontal Solution Program for Finite-Element Analysis," *Int. J. Numer. Meth. Engng.*, Vol. 2, 1970, 5–23.

[22]Golub, G.H., and Van Loan, C.F., *Matrix Computations*, Third Edition, Johns Hopkins Univ. Press, Baltimore, MD, 1996.

9
Eigenvalue Problems

9.1 Introduction

Some commonly occurring eigenvalue problems encountered in practice are as follows:

- Free vibration of nonspinning structures.
- Free vibration of spinning structures.
- Free vibration formulation based on dynamic element formulation.
- Stability of structural systems.

Their solution methods are described here in detail. At least three different procedures are relevant to their solution and these are classified as 1) single vector iteration, 2) progressive simultaneous iteration, and 3) the Lanczos method. Each technique has relative advantages over the others. The matrix formulation pertaining to the various eigenvalue problems may be conveniently derived from their respective dynamic equations. Each problem and its solution is discussed in the following sections.

9.2 Free Vibration Analysis of Undamped Nonspinning Structures

The matrix equation of motion for a typical problem is given by a set of linear second-order differential equations

$$\boldsymbol{K}\boldsymbol{q} + \boldsymbol{M}\ddot{\boldsymbol{q}} = \boldsymbol{0} \tag{9.1}$$

whose solution has the form

$$\boldsymbol{q} = \boldsymbol{\phi} e^{i\omega t} \tag{9.2}$$

in which $\boldsymbol{\phi}$ is the amplitude of the displacements $\boldsymbol{q}$ at time zero, ω is the frequency of oscillations, and t is the time variable, as usual. Substituting Eq. (9.2) in Eq. (9.1), and canceling the common factor $e^{i\omega t}$, yields the standard structural eigenvalue problem,

$$(\boldsymbol{K} - \omega^2 \boldsymbol{M})\boldsymbol{\phi} = \boldsymbol{0} \tag{9.3}$$

or alternatively,

$$(\boldsymbol{K} - \lambda \boldsymbol{M})\boldsymbol{\phi} = \boldsymbol{0}, \qquad \lambda = \omega^2 \tag{9.4}$$

where λ is an eigenvalue and $\boldsymbol{\phi}$ the corresponding eigenvector. Solution of a wide range of eigenspectrum involving the number of eigenvalues p can be written in the form as

$$\boldsymbol{K\Phi} - \boldsymbol{M\Phi\Lambda} = \boldsymbol{0} \tag{9.5}$$

In Eq. (9.5) $\boldsymbol{\Phi}$ is an $(n \times p)$ matrix containing columns of p eigenvectors and $\boldsymbol{\Lambda}$ is a diagonal matrix of eigenvalues. Since Eq. (9.4) forms a set of homogeneous algebraic equations, its matrix form is rank deficient by one and thus the vectors are determined within a multiplication constant. These vectors are also orthogonal with respect to matrices $\boldsymbol{K}$ and $\boldsymbol{M}$. These conditions are expressed by

$$\boldsymbol{\phi}_r^T \boldsymbol{M} \boldsymbol{\phi}_s = \boldsymbol{0}, \qquad \boldsymbol{\phi}_r^T \boldsymbol{K} \boldsymbol{\phi}_s = \boldsymbol{0}, \qquad \text{for } r \neq s \tag{9.6}$$

and

$$\boldsymbol{\phi}_r^T \boldsymbol{M} \boldsymbol{\phi}_r = M_r, \qquad \boldsymbol{\phi}_r^T \boldsymbol{K} \boldsymbol{\phi}_r = \omega_r^2 M_r \qquad \text{for } r = s \tag{9.7}$$

where M_r is a constant. Also,

$$\boldsymbol{\Phi}^T \boldsymbol{M} \boldsymbol{\Phi} = \hat{\boldsymbol{M}} \qquad \boldsymbol{\Phi}^T \boldsymbol{K} \boldsymbol{\Phi} = \boldsymbol{\Lambda} \hat{\boldsymbol{M}} = \hat{\boldsymbol{K}} \tag{9.8}$$

where $\hat{\boldsymbol{M}}$ and $\hat{\boldsymbol{K}}$ are diagonal matrices. If the eigenvectors are mass orthonormalized, i.e.,

$$\boldsymbol{\psi}_r = \boldsymbol{\phi}_r / \sqrt{M_r} \tag{9.9}$$

then the following results are obtained:

$$\boldsymbol{\Psi}^T \boldsymbol{M} \boldsymbol{\Psi} = \boldsymbol{I} \tag{9.10}$$

$$\boldsymbol{\Psi}^T \boldsymbol{K} \boldsymbol{\Psi} = \boldsymbol{\Lambda} \tag{9.11}$$

with

$$\boldsymbol{\Lambda} = \begin{bmatrix} \omega_1^2 & & & \\ & \omega_2^2 & & \\ & & \ddots & \\ & & & \omega_p^2 \end{bmatrix}, \qquad \text{and} \qquad \boldsymbol{\Psi} = [\boldsymbol{\psi}_1 \boldsymbol{\psi}_2 \ldots \boldsymbol{\psi}_p] \tag{9.12}$$

and $\boldsymbol{I}$ is the identity matrix.

Various techniques suitable for the efficient eigenproblem solution involving large sparse matrix equations [Eq. (9.3)] that are characteristic of complex practical problems are described next.

9.2.1 Single Vector Iteration

An inverse iteration technique[1] combined with a repeated bisection strategy and a Sturm sequence count may be conveniently employed to obtain the first few roots and vectors of the associated eigenvalue problem,

$$(\boldsymbol{K} - \lambda \boldsymbol{M})\boldsymbol{\phi} = \boldsymbol{0} \tag{9.13}$$

within specified bounds $(\lambda^\ell, \lambda^u)$, λ being ω^2. It is well known that the Sturm sequence property is valid for a pair of symmetric matrices $\boldsymbol{K}$ and $\boldsymbol{M}$ in which at least one of the two matrices is also positive definite or nonnegative definite. The latter being characterized by zero-valued roots signifying the presence of rigid-body modes. For any specified value of the root λ^s, the number of negative diagonals or the changes in the signs of the leading principal minors of the triangularized

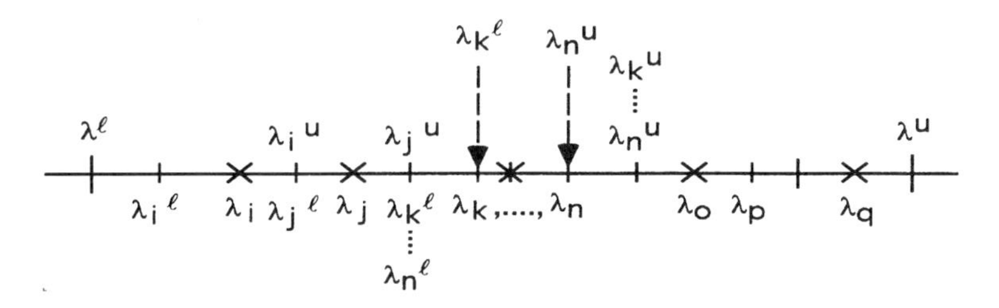

Fig. 9.1 Convergence scheme for SS/II solution scheme.

$(\boldsymbol{K} - \lambda^s \boldsymbol{M})$ is equal to the number of roots of the system having algebraic value less than λ^s. It is also known that an inverse iteration procedure converges to the root nearest the starting root iteration value. A solution strategy has the following salient features:

1) During repeated bisection, change upper and lower bounds of each root under consideration to reduce the intervals containing each root.

2) During the bisection process, while monitoring bounds of a group of roots, if the bound remains unchanged while upper and lower bounds each change at least once, then inverse iteration is performed from both ends to converge to a set of repeated roots.

3) During repeated bisection, also implement inverse iteration to converge to the nearest root.

A combined Sturm sequence and inverse iteration (SS/II) procedure[2,3] is shown in Fig. 9.1. Assuming the lowest number of roots (NR) are required to be computed within a bound $(\lambda^\ell, \lambda^u)$, then during a typical bisection step, if λ^ℓ and λ^u are the current lower and upper bounds associated with NL and NU roots, respectively, the following solution steps are adopted:

Step 1: Calculate $\lambda^m = (\lambda^u + \lambda^l)/2$, middle point of the bounds.

Step 2: Triangularize $\boldsymbol{K} - \lambda^m \boldsymbol{M}$ and compute NM, the number of roots of the system having algebraic values lower than λ^m, employing the Sturm sequence method.

Step 3: If NM < NL or NM >NU, then go to step 8.

Step 4: Perform inverse iteration: the typical $(i + 1)$th iteration is

$$[\boldsymbol{K} - \lambda^m \boldsymbol{M}]\boldsymbol{\phi}_{i+1} = N_{i+1}\boldsymbol{M}\boldsymbol{\phi}_i \tag{9.14}$$

N_{i+1} being a normalizing factor.

Step 5: Check possible root convergence by comparing norms of vectors. If

$$(\|\boldsymbol{\phi}_{i+1}\| - \|\boldsymbol{\phi}_i\|)/\|\boldsymbol{\phi}_i\| \leq \text{EPS} \tag{9.15}$$

then perform computation in step 6; EPS is a user-specified accuracy parameter. Otherwise, repeat step 4.

Step 6: If adequate convergence has been achieved, compute the converged eigenvalue by employing the Rayleigh quotient:

$$\lambda_{i+1} = \frac{\boldsymbol{\phi}_{i+1}^T \boldsymbol{K} \boldsymbol{\phi}_{i+1}}{\boldsymbol{\phi}_{i+1}^T \boldsymbol{M} \boldsymbol{\phi}_{i+1}} \tag{9.16}$$

Step 7: From the knowledge of values of λ^m, NM, NL, and NU, check if the converged root is a desired one, in which case store λ_{i+1} and $\boldsymbol{\phi}_{i+1}$, respectively.

Step 8: Adjust bounds of all of the roots under consideration, as appropriate, and reset values of NL and NU pertaining to the next root to be determined in the continuing bisection scheme. Repeat steps 1 through 8 until all roots are determined.

This procedure enables computation of any number of roots within a specified bound. The majority of the computational effort revolves around NR number of triangularizations of the left-hand side of Eq. (9.14), i.e., ($NR \times N \times M^2$) multiplications. Also, for certain values of λ^s, $\boldsymbol{K} - \lambda^s \boldsymbol{M}$ may not be positive definite. In such a case partial pivoting[2] may have to be performed during the triangularization procedure that in turn will increase the bandwidth, resulting in increased computational effort. Alternatively, a slight shift in the value of λ^m usually proves to be adequate to avoid the conditioning problem.

9.2.2 Progressive Simultaneous Iteration

In this procedure a set of vectors is used for simultaneous determination of a number of roots and associated eigenvectors. The original simultaneous iteration (SI) method developed for the determination of the first few roots and vectors is described in Ref. 4 and was further elaborated in Ref. 5. The progressive simultaneous iteration (PSI) method, described herein, is a novel simultaneous iteration procedure with significantly accelerated convergence characteristics.[6]

The associated numerical algorithm is next described for the most general case when a desired NR number of roots in the vicinity of a specified shift value λ^s and associated modes are the usual requirement. With $\lambda^s = 0$, the algorithm computes the first few desired roots only. Thus the original eigenvalue program defined by Eq. (9.4) is written in the shifted form:

$$(\boldsymbol{K} - \lambda^s \boldsymbol{M} - (\lambda - \lambda^s)\boldsymbol{M})\boldsymbol{\phi} = \boldsymbol{0} \tag{9.17}$$

or

$$(\tilde{\boldsymbol{K}} - \tilde{\lambda}\boldsymbol{M})\boldsymbol{\phi} = \boldsymbol{0} \tag{9.18}$$

with $\tilde{\lambda} = \lambda - \lambda^s$, and the solution is sought for NR number of roots in the vicinity of λ^s as shown in Fig. 9.2. As a special case for $\lambda_s = 0$ the solution is sought for the first NR. The following solution steps describe the PSI algorithm.

Step 1: Form a set of NRT (number of random trial) vectors, NRT > NR:

$$\hat{\boldsymbol{\Phi}}_1 = \begin{bmatrix} \hat{\boldsymbol{\phi}}_1^1 & \hat{\boldsymbol{\phi}}_1^2 & \hat{\boldsymbol{\phi}}_1^3 & \cdots & \hat{\boldsymbol{\phi}}_1^{\mathrm{NRT}} \end{bmatrix}$$

Step 2: Perform Cholesky factorization on the $\tilde{\boldsymbol{K}}$ matrix, initially only $\tilde{\boldsymbol{K}} = \boldsymbol{LDL}^T$, where $\boldsymbol{L}$ and $\boldsymbol{D}$ are a unit lower triangular and a diagonal matrix, respectively.

Step 3: At a typical ith iteration step, solve $\tilde{\boldsymbol{K}}\boldsymbol{\Phi}_{i+1} = \boldsymbol{M}\hat{\boldsymbol{\Phi}}_i$ progressively, involving only the unconverged roots, by the usual back substitution method.

Step 4: Estimate the magnitude of the jth eigenvalue λ^j, progressively, on unconverged roots only, as

$$\lambda_{i+1}^j = \|\phi_i^j\| / \|\phi_{i+1}^j\|$$

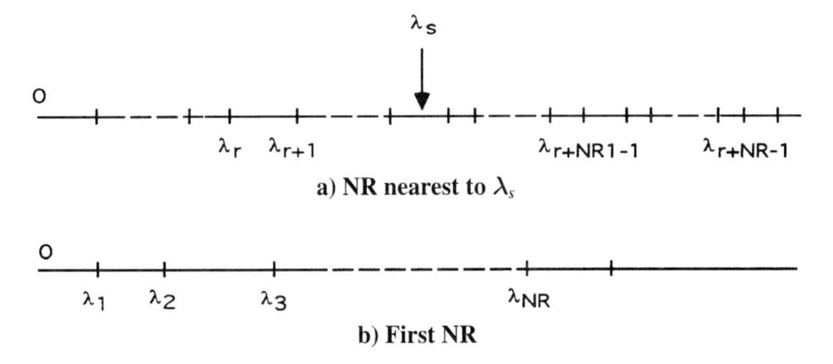

Fig. 9.2 Distribution of desired roots.

If this is not the first iteration, check for convergence using the test

$$\left(\lambda_{i+1}^{j} - \lambda_{i}^{j}\right) / \lambda_{i+1}^{j} \leq \text{EPSN}$$

Let NR1 be the number of leading consecutive roots that have converged; EPSN is the specified norm tolerance factor. If NR1 = NR, go to step 7.

Step 5: Perform progressive mass orthonormalization on unconverged (NRT – NR1) vectors only:

$$\tilde{\boldsymbol{\phi}}_{i+1}^{k} = \boldsymbol{\phi}_{i+1}^{k} - \sum_{l=1}^{\text{NR1}} [(\boldsymbol{\phi}^{l})^{T} \boldsymbol{M} \boldsymbol{\phi}_{i+1}^{k}] \boldsymbol{\phi}^{l} - \sum_{l=\text{NR1}+1}^{k-1} [(\hat{\boldsymbol{\phi}}_{i+1}^{l})^{T} \boldsymbol{M} \boldsymbol{\phi}_{i+1}^{k}] \hat{\boldsymbol{\phi}}_{i+1}^{l}$$

$$\hat{\boldsymbol{\phi}}_{i+1}^{k} = \tilde{\boldsymbol{\phi}}_{i+1}^{k} / [(\tilde{\boldsymbol{\phi}}_{i+1}^{k})^{T} \boldsymbol{M} \tilde{\boldsymbol{\phi}}_{i+1}^{k}]^{\frac{1}{2}}, \qquad k = \text{NR1} + 1, \text{NRT}$$

Step 6: Continue iteration, go to step 3; $\hat{\boldsymbol{\Phi}}_{i+1}$ being made of $\hat{\boldsymbol{\phi}}_{i+1}^{k}$ vectors:

$$\hat{\boldsymbol{\Phi}}_{i+1} = \begin{bmatrix} \hat{\boldsymbol{\phi}}_{i+1}^{\text{NR1}+1} & \hat{\boldsymbol{\phi}}_{i+1}^{\text{NR1}+2} & \cdots & \hat{\boldsymbol{\phi}}_{i+1}^{\text{NRT}-1} \end{bmatrix}$$

Step 7: Compute roots by Rayleigh–Ritz method, employing the progressive procedure involving only the unconverged roots by solving the further reduced eigenvalue problem

$$\hat{\boldsymbol{K}}_{i+1} \boldsymbol{X}_{i+1} = \hat{\boldsymbol{M}}_{i+1} \boldsymbol{X}_{i+1} \hat{\boldsymbol{\Lambda}}$$

$\hat{\lambda}$ being an approximation of $\tilde{\lambda}$ and in which

$$\hat{\boldsymbol{K}}_{i+1} = \hat{\boldsymbol{\Phi}}_{i+1}^{T} \tilde{\boldsymbol{K}} \hat{\boldsymbol{\Phi}}_{i+1}, \qquad \hat{\boldsymbol{M}}_{i+1} = \hat{\boldsymbol{\Phi}}_{i+1}^{T} \boldsymbol{M} \hat{\boldsymbol{\Phi}}_{i+1}$$

noting that $\hat{\boldsymbol{M}}_{i+1} = \boldsymbol{I}$ because of the mass orthonormalization effected in step 5.

Step 8: Perform a root convergence test for each root under consideration:

$$|(\hat{\lambda}_{i+1}^{j} - \hat{\lambda}_{i}^{j})| / |\hat{\lambda}_{i+1}^{j}| \leq \text{EPS}$$

EPS being a specified root convergence factor. Update NR1 to be the number of consecutive leading roots that have converged.

Step 9: Recalculate eigenvectors $\tilde{\boldsymbol{\Phi}}_{i+1} = \hat{\boldsymbol{\Phi}}_{i+1}\boldsymbol{X}_{i+1}$; set $\hat{\boldsymbol{\Phi}}_{i+1} = \tilde{\boldsymbol{\Phi}}_{i+1}$, noting that $\boldsymbol{X}_{i+1}$ must be orthonormalized, so that $\boldsymbol{X}_{i+1}^T\boldsymbol{X}_{i+1} = \boldsymbol{I}$.

Step 10: If NR1 $<$ NR, then go to step 3 and perform steps 3, 5, 7, 8, and 9 only. Otherwise, go to step 11.

Step 11: End of analysis.

The preceding solution scheme results in progressive saving in the solution process involving four key areas of operations: 1) back substitution (step 3), 2) mass orthonormalization (step 5), 3) two Rayleigh triple matrix multiplications (step 7), and 4) Rayleigh vector computation (step 9).

This scheme results in a very efficient eigenproblem solution procedure when compared to other similar existing techniques. For subsequent references, step 3–6 involved in checking root norm convergence will be termed as phase 1 operation. The Rayleigh–Ritz root convergence is checked in phase 2 of this formulation and involves steps 3, 5, and 7–10, respectively. At the conclusion of the progressive solution process in phase 1, phase 2 analysis starts with the full set of trial vectors, subsequent steps being progressive in nature. It may be noted that for a zero shift value, i.e., $\lambda^s = 0$, the procedure will yield the first few desired roots and corresponding vectors.

Usually the vector norm convergence tolerance factor EPSN pertaining to phase 1 of the solution process is set to a value between 0.03 and 0.07, typically at 0.05. The Rayleigh root convergence factor EPS for the remaining phase 2 effort is set to 0.00001. Since the last one or two roots of the required series is always determined to a slightly lower accuracy, no matter which technique is used for the eigenproblem solution, a somewhat lower value for EPS, to be termed as EPSR, is used for their determination. This factor is set to a value between 0.002 and 0.00005, yielding an overall accurate and efficient solution.

Solution efficiency is dictated by the choice of EPSR and NRT, the number of trial vectors adapted for the analysis. For a general algorithm these values are found to be optimum as EPSR $=$ 0.0002 and NRT $=$ Max(1.1 $\times$ NR, NR $+$ 4) when, usually, solution norm convergence is achieved in a finite number of steps in phase 1 followed by a few more steps in phase 2, and this procedure will be termed the PSI method.

In phase 1 of this procedure, at any particular step, a vector norm check is conducted sequentially from the lower to the higher eigenvalues and only the first consecutive leading roots that pass the test are accepted as converged ones. Also, at each step, an approximate value of each root is computed as $\lambda_{i+1}^j = \|\boldsymbol{\phi}_{i+1}^j\| / \|\boldsymbol{\phi}_i^j\|$ and the vectors are reorganized in sequential order of increasing root values in the $\boldsymbol{\Phi}_{i+1}$ matrix. This is advisable, at least initially, since at the beginning of the scheme, trial vectors were generated at random and in general it results in faster solution convergence. In phase 2, the roots are automatically computed by the Rayleigh–Ritz method and the vectors are thus reorganized accordingly. Further, the relevant coding fully exploits matrix sparsity, and only nonzero elements within the bandwidth of the $\boldsymbol{M}$ and $\boldsymbol{K}$ matrices are stored and operated on thereby effecting significant savings in computational effort in steps 5 and 7, respectively. The usual SI procedure involves steps 3 and 7–10 without progressive features.

9.2.3 Lanczos Method

To apply this procedure[7] it is first necessary to recast the standard eigenvalue problem [Eq. (9.4)]

$$(\boldsymbol{K} - \lambda \boldsymbol{M})\boldsymbol{\phi} = \boldsymbol{0}$$

in terms of a single matrix. This is easily achieved by performing a Cholesky decomposition on the stiffness matrix $\boldsymbol{K}$, which is usually positive definite, as $\boldsymbol{K} = \boldsymbol{L}\boldsymbol{L}^T$ and pre- and postmultiplying both matrices by $\boldsymbol{L}^{-1}$ and $\boldsymbol{L}^{-T}$, respectively, yielding

$$(\boldsymbol{L}^{-1}\boldsymbol{L}\boldsymbol{L}^T\boldsymbol{L}^{-T} - \lambda \boldsymbol{L}^{-1}\boldsymbol{M}\boldsymbol{L}^{-T})\boldsymbol{\phi} = \boldsymbol{0}$$

or

$$\left(\boldsymbol{A} - \frac{1}{\lambda}\boldsymbol{I}\right)\boldsymbol{\phi} = \boldsymbol{0}$$

or

$$(\boldsymbol{A} - \gamma \boldsymbol{I})\boldsymbol{\phi} = \boldsymbol{0} \tag{9.19}$$

in which $\gamma = 1/\lambda$.

The block Lanczos algorithm is most popular in practice because it is capable of computing multiple roots and also it proves to be faster than the usual procedures that employ a single vector or a multivector SI procedure. This scheme is described as follows.

Step 1: Generate $\boldsymbol{R}_0$, a set of $(N \times \text{NB})$ random vectors; N is the order of $\boldsymbol{A}$ and NB the block size, fixed usually between 2 and 4.

Step 2: For $j = 1, 2, 3, \ldots$

Step 3: Perform standard QR factorization $\boldsymbol{R}_{j-1} = \boldsymbol{Q}_j\boldsymbol{B}_{j-1}$, in which $\boldsymbol{Q}_j$ is orthonormalized $\boldsymbol{R}_{j-1}$ of dimension $(N\times \text{NB})$ and $\boldsymbol{B}_{j-1}$ is the $(\text{NB} \times \text{NB})$ upper triangular matrix defining the relationship between $\boldsymbol{R}_{j-1}$ and $\boldsymbol{Q}_j$.

Step 4: Form $\boldsymbol{R}_j = \boldsymbol{A}\boldsymbol{Q}_j$ for $j = 1$ and $\boldsymbol{R}_j = \boldsymbol{A}\boldsymbol{Q}_j - \boldsymbol{Q}_{j-1}\boldsymbol{B}_{j-1}^T$ for $j >= 2$, noting that the second term on the right-hand side is valid for $j \geq 2$.

Step 5: Compute $\boldsymbol{A}_j = \boldsymbol{Q}_j^T\boldsymbol{R}_j$, which is a $(\text{NB} \times \text{NB})$ matrix.

Step 6: Compute new residual $\boldsymbol{R}_j = \boldsymbol{R}_j - \boldsymbol{Q}_j\boldsymbol{A}_j$.

Step 7: Form the block tridiagonal matrix

$$\boldsymbol{T}_j = \begin{bmatrix} \boldsymbol{A}_1 & \boldsymbol{B}_1^T & & & \\ \boldsymbol{B}_1 & \boldsymbol{A}_2 & \boldsymbol{B}_2 & & \\ & & & \ddots & \\ & & & & \boldsymbol{B}_{j-1}^T \\ & & & \boldsymbol{B}_{j-1} & \boldsymbol{A}_j \end{bmatrix}$$

which is of order $(j \times \text{NB}, j \times \text{NB})$.

Step 8: Compute eigenvalues of $\boldsymbol{T}_j$, and check convergence. If convergence is achieved, go to the next step; otherwise, go to step 2.

Step 9: Compute eigenvectors of $\boldsymbol{T}_j$, $\boldsymbol{Y}$ and transform the same to obtain such vectors for the original matrix: $\boldsymbol{\Phi} = [\boldsymbol{Q}_1\boldsymbol{Q}_2\boldsymbol{Q}_3, \ldots, \boldsymbol{Q}_j][\boldsymbol{Y}_1\boldsymbol{Y}_2 \cdots \boldsymbol{Y}_j]$, which is of the dimension $(N, \mathrm{NB} \times j)$.

In theory $\boldsymbol{R}_j$ computed in step 6 is orthogonal to $\boldsymbol{Q}_1, \boldsymbol{Q}_2, \ldots, \boldsymbol{Q}_j$. However, in practice this orthogonality is less well achieved as the algorithm proceeds. Thus a practical solution algorithm must take into account this degradation in orthogonality. One possible approach of dealing with this is the adoption of a technique such as selective orthogonalization.[8]

9.2.4 Numerical Examples

The first example relates to a free–free aircraft structure. Analyses are performed using various eigenproblem solution techniques to afford a clear comparison of relative solution efficiencies. The second example is for a cantilever plate, and analysis results are presented for varying mesh sizes to establish a pattern of solution convergence.

Example 9.1: Generic Hypersonic Vehicle

Figure 9.3 depicts the various views of the generic hypersonic vehicle (GHV) with relevant structural dimensions. This vehicle was generated in support of the National Aerospace Plane project and the associated finite element numerical model of the vehicle has the following details: number of elements = 4990, number

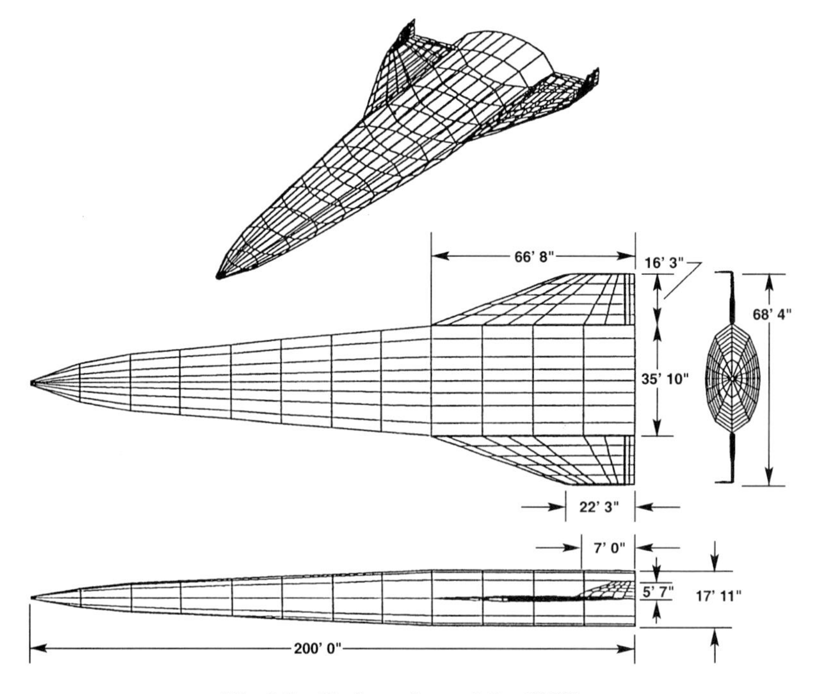

Fig. 9.3 Various views of the GHV.

Table 9.1 First 20 natural frequencies of the GHV

Mode	Natural frequencies, ω, rad/s		
	PSI	Lanczos	SI
1–6	0.0	0.0	0.0
7	19.2357	19.2357	19.2357
8	24.0874	24.0874	24.0874
9	24.2588	24.2588	24.2588
10	33.5716	33.5716	33.5716
11	39.5108	39.5108	39.5108
12	39.6883	39.6883	39.6883
13	41.6569	41.6569	41.6569
14	42.0390	42.0390	42.0390
15	44.8386	44.8386	44.8386
16	56.0925	56.0923	56.0923
17	56.7742	56.7740	56.7740
18	62.4139	62.4095	62.4139
19	62.7898	62.7993	62.7899
20	62.7932	62.8283	62.7931
	Number of iterations		
Phase 1	9	—	0
Phase 2	12	—	21
Solution time for CPU, s	145	215	290

of nodes = 2812, number of degrees of freedom = 16,872, and semibandwidth (minimized) = 1062.

A series of vibration analyses was performed to afford a comparison of their relative efficiencies. Table 9.1 presents the results involving the computation of the first 20 roots and associated vectors. Also some typical mode shapes pertaining to the first few roots are shown in Fig. 9.4. The essential parameters for the PSI solution were set as EPSN = 0.05, EPSR = 0.0002, and EPS = 0.00001.

An analysis was performed employing the current PSI technique to determine a few intermediate roots around a given shift value and the associated vectors. The results, pertaining to 30 roots, are then compared with the standard SI procedure as shown in Table 9.2.

Example 9.2: Cantilever Plate

A square cantilever plate was analyzed to yield the natural frequencies and associated mode shapes. Figure 9.5 shows the plate with a (4×4) finite element mesh; the bottom edge along the x axis is clamped. Analyses were performed for varying mesh sizes to study convergence characteristics. Data parameters for the problem are Young's modulus $E = 10 \times 10^6$, side length $L = 10$, plate thickness $t = 0.1$, Poisson's ratio $\nu = 0.3$, and mass density $\rho = 0.259 \times 10^{-3}$. Table 9.3 shows the first few natural frequencies of the plate as a function of mesh sizes.

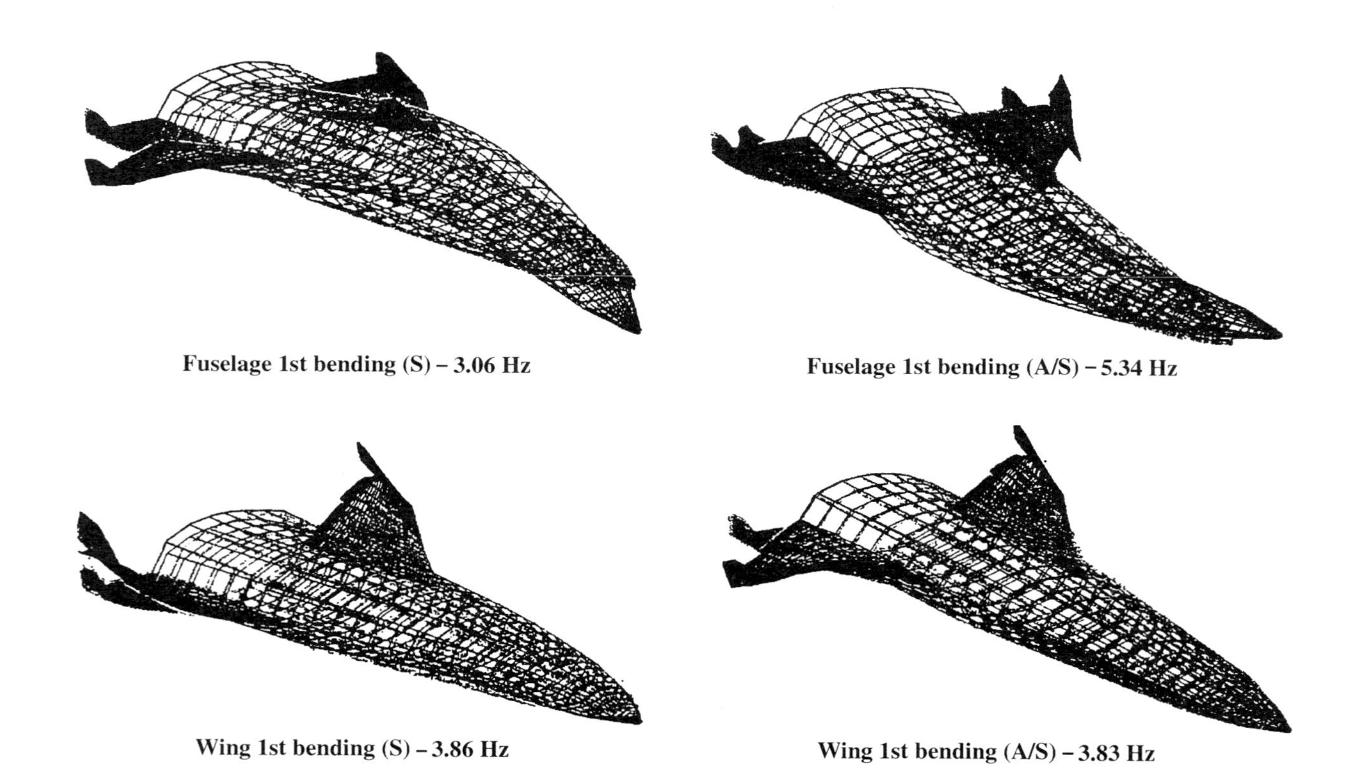

Fig. 9.4 Typical mode shapes of the GHV.

Table 9.2 Intermediate natural frequencies of the generic hypersonic vehicle (shift = 65.0)

	Natural frequencies, rad/s	
Mode	PSI	SI
10	33.5720	33.5716
11	39.5109	39.5108
12	39.6885	39.6883
13	41.6570	41.6570
14	42.0391	42.0391
15	44.8386	44.8386
16	56.0923	56.0923
17	56.7740	56.7740
18	62.4057	62.4057
19	62.7897	62.7897
20	62.7932	62.7932
21	62.8170	62.8170
22	63.5561	63.5561
23	67.0402	67.0401
24	69.7192	69.7192
25	70.2888	70.2888
26	72.6345	72.6344
27	73.6461	73.6461
28	73.7215	73.7215
29	77.6818	77.6816
30	77.6818	77.6816
31	78.6454	78.6454
32	78.6853	78.6853
33	79.7320	79.7320
34	79.7448	79.7448
35	81.6240	81.6240
36	81.6288	81.6288
37	84.9109	84.9111
38	87.1963	87.1963
39	88.6063	88.6063
	Number of iterations	
Phase 1	16	0
Phase 2	21	36
Solution time for CPU, s	314	733

The PSI algorithm presented herein adopts a progressive solution strategy, employing only the unconverged roots and vectors at any stage of the solution process. This procedure enables accurate computation of the first few roots and vectors as well as intermediate ones only, without having to compute any other. Numerical examples of practical problems characterized by large matrices and bandwidth

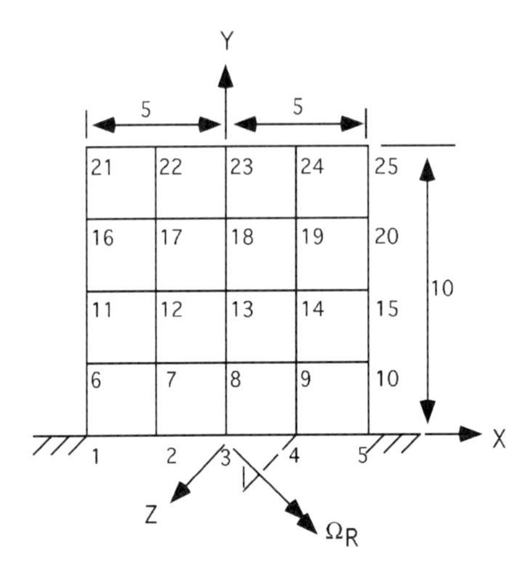

Fig. 9.5 Square cantilever plate.

have been presented. Results of an analysis performed by the current PSI, the usual SI, and the now widely used Lanczos procedure are also presented and compared in some detail. Comparisons indicate considerable advantage of the PSI technique, being vastly superior to the usual SI procedure and significantly more efficient than the Lanczos method.

For free-free structures, such as an aircraft, the associated eigenvalue problem is characterized by up to six rigid-body (zero frequencies) modes. It is then necessary to employ an eigenvalue shift, $(\hat{\boldsymbol{K}} - \hat{\lambda}\hat{\boldsymbol{M}})\boldsymbol{y} = \boldsymbol{0}$, in which $\hat{\boldsymbol{K}} = \boldsymbol{K} + \hat{\boldsymbol{M}}$ and $\hat{\boldsymbol{M}} = F\boldsymbol{M}$, where the multiplication factor F is a suitable function of relative norm values of $\boldsymbol{K}$ and $\boldsymbol{M}$ matrices and also machine accuracy; λ is calculated from the relationship $\lambda = \hat{\lambda} + F$.

9.3 Free Vibration Analysis of Spinning Structures

A large variety of practical structures, e.g., spacecraft and satellites, are spin stabilized, whereas others such as turbines and helicopters possess rotating parts

Table 9.3 Natural frequencies of a square cantilever plate

	Natural frequencies ω, rad/s					
Mesh size	ω_1	ω_2	ω_3	ω_4	ω_5	ω_6
4×4	214.0167	506.6163	1248.3956	1538.2863	1765.5332	2889.1137
6×6	215.5942	518.7804	1292.5258	1620.8405	1854.1748	3151.2927
8×8	216.1619	523.4059	1308.3716	1652.7015	1885.7732	3246.2215
10×10	216.4302	525.7005	1315.8046	1668.1704	1900.7870	3290.4513
12×12	216.5782	527.0247	1319.8822	1676.8343	1906.1761	3315.1575
20×20	216.7980	529.1494	1325.9069	1689.9493	1922.0397	3352.5300

arising out of functional requirements. The associated matrix equation of motion for an undamped structure discretized by the finite element method and the structure or portions thereof subjected to a uniform spin rate Ω may be written as

$$\boldsymbol{K}\boldsymbol{q} + \boldsymbol{C}\dot{\boldsymbol{q}} + \boldsymbol{M}\ddot{\boldsymbol{q}} = \boldsymbol{0} \tag{9.20}$$

in which $\boldsymbol{K} = \boldsymbol{K}_E + \boldsymbol{K}' + \boldsymbol{K}_G$ is the stiffness matrix, $\boldsymbol{C} = \boldsymbol{C}_c$ is the Coriolis force matrix, and $\boldsymbol{M}$ is the inertia matrix. The terms $\boldsymbol{K}_E$, $\boldsymbol{K}'$, and $\boldsymbol{K}_G$ are the elastic stiffness, the centrifugal force, and the geometric stiffness matrices, respectively; the latter two matrices being functions of Ω^2. For small vibrations the $\boldsymbol{K}$ and $\boldsymbol{M}$ matrices are real, symmetric, and positive definite, whereas the Coriolis matrix $\boldsymbol{C}_c$ is skew symmetric being also a function of Ω. All of these matrices are sparse in nature being highly banded for most practical problems and an associated eigenproblem solution of Eq. (9.20) takes the following quadratic form:

$$(\boldsymbol{K} + p\boldsymbol{C} + p^2\boldsymbol{M})\boldsymbol{\phi} = \boldsymbol{0} \tag{9.21}$$

by substituting $\boldsymbol{q} = \boldsymbol{\phi}e^{pt}$, in which p is purely imaginary, such roots and associated complex vectors occurring in conjugate pairs.

A convenient approach to the solution of the present eigenvalue problem involves, first, resolving the same into a set of first-order ordinary differential equations resulting in a linear eigenvalue problem:

$$\begin{bmatrix} \boldsymbol{M} & \boldsymbol{0} \\ \boldsymbol{0} & \boldsymbol{K} \end{bmatrix} \begin{Bmatrix} \dot{\boldsymbol{q}} \\ \boldsymbol{q} \end{Bmatrix} + \begin{bmatrix} \boldsymbol{0} & -\boldsymbol{M} \\ \boldsymbol{M} & \boldsymbol{C} \end{bmatrix} \begin{Bmatrix} \ddot{\boldsymbol{q}} \\ \dot{\boldsymbol{q}} \end{Bmatrix} = 0 \tag{9.22}$$

or

$$\boldsymbol{E}\boldsymbol{y} + \boldsymbol{G}\dot{\boldsymbol{y}} = \boldsymbol{0} \tag{9.23}$$

in which

$$\boldsymbol{y} = \begin{bmatrix} \dot{\boldsymbol{q}} \\ \boldsymbol{q} \end{bmatrix} \tag{9.24}$$

Both matrices $\boldsymbol{E}$ and $\boldsymbol{G}$ are of the order $(2 \times N)$. The solution of Eq. (9.22) may be taken as $\boldsymbol{y} = \boldsymbol{\phi}e^{\omega t}$, yielding

$$(\boldsymbol{E} + \omega\boldsymbol{G})\boldsymbol{\phi} = \boldsymbol{0} \tag{9.25}$$

The eigenvalues are purely imaginary, whereas the vectors occur in complex conjugate pairs. Equation (9.25) may also be further rearranged as

$$(\boldsymbol{E} - \lambda\boldsymbol{F})\boldsymbol{\phi} = \boldsymbol{0} \tag{9.26}$$

in which $\boldsymbol{F} = i^*\boldsymbol{G}$ and can be written in its expanded form as

$$\left(\begin{bmatrix} \boldsymbol{M} & \boldsymbol{0} \\ \boldsymbol{0} & \boldsymbol{K} \end{bmatrix} - \lambda \begin{bmatrix} \boldsymbol{0} & -i^*\boldsymbol{M} \\ i^*\boldsymbol{M} & i^*\boldsymbol{C} \end{bmatrix} \right) \begin{bmatrix} \dot{\boldsymbol{\psi}} \\ \boldsymbol{\psi} \end{bmatrix} = \boldsymbol{0} \tag{9.27}$$

where $\boldsymbol{F}$ is a pure imaginary Hermitian matrix, $\boldsymbol{\phi} = [\dot{\boldsymbol{\psi}}\ \boldsymbol{\psi}]^T$, the roots $\lambda = i^*\omega$ are real, and i^* is the imaginary number $\sqrt{-1}$. The roots of Eq. (9.27) occur as pairs of real numbers $\lambda_1, -\lambda_1, \ldots, \lambda_N, -\lambda_N$ and corresponding eigenvectors as complex conjugate pairs. The eigenvalues of the original system defined by Eq. (9.25) may then be simply obtained as λ/i^*, noting that eigenvectors are the same for both cases.

9.3.1 Single Vector Iteration

It is well known that the Sturm sequence property[9] is valid for the current eigenproblem defined by Eq. (9.27). This property is exploited along with the bisection process as before, although also implementing inverse iterations at the same time to converge to the roots nearest to the bisection points.[10] Assume that the eigenvalue problem is concerned with the computation of the first NR roots, lying within specified bounds PU and PL having NU and NL corresponding roots and associated vectors, respectively. During a typical bisection step, if AL, AU are the current lower and upper root bounds, respectively, the following solution steps are adopted.

Step 1: Calculate H = (AL | AU)/2.

Step 2: Triangularize $\boldsymbol{K} - i^*\mathrm{H}\boldsymbol{C} - \mathrm{H}^2\boldsymbol{M}$. Compute NH, the number of roots of the system having algebraic values lower than H, employing a Sturm sequence method that involves counting the number of changes in the signs of the leading principal minors.

Step 3: If NH < NL or NH > NU, then go to step 6.

Step 4A: Perform inverse iterations, the typical $(i+1)$th step involves operations as follows. Iterate on

$$(\boldsymbol{E} - \mathrm{H}\boldsymbol{F})\boldsymbol{\phi}_{i+1} = N_{i+1}\boldsymbol{F}\boldsymbol{\phi}_i \tag{9.28}$$

where N_{i+1} has a suitable normalizing factor, and Eq. (9.28) has the detailed form:

$$\begin{bmatrix} \boldsymbol{M} & i^*\mathrm{H}\boldsymbol{M} \\ -i^*\mathrm{H}\boldsymbol{M} & \boldsymbol{K} - i^*\mathrm{H}\boldsymbol{C} \end{bmatrix} \begin{bmatrix} \dot{\boldsymbol{\psi}}_{i+1} \\ \boldsymbol{\psi}_{i+1} \end{bmatrix} = N_{i+1} \begin{bmatrix} -i^*\boldsymbol{M}\boldsymbol{\psi}_i \\ i^*\boldsymbol{M}\dot{\boldsymbol{\psi}}_i + i^*\boldsymbol{C}\boldsymbol{\psi}_i \end{bmatrix} \tag{9.29}$$

Noting that $\dot{\boldsymbol{\psi}} = \lambda\boldsymbol{\psi}$ and substituting $\dot{\boldsymbol{\psi}}_{i+1} = \mathrm{H}\boldsymbol{\psi}_{i+1}$, the lower half of Eq. (9.29) can be written as

$$[\boldsymbol{K} - i^*\mathrm{H}\boldsymbol{C} - \mathrm{H}^2\boldsymbol{M}]\boldsymbol{\psi}_{i+1} = N_{i+1}[\mathrm{H}\boldsymbol{M}\boldsymbol{\psi}_i + i^*\boldsymbol{M}\dot{\boldsymbol{\psi}}_i + i^*\boldsymbol{C}\boldsymbol{\psi}_i] \tag{9.30}$$

and the upper half as

$$[\boldsymbol{M}\dot{\boldsymbol{\psi}}_{i+1} + i^*\mathrm{H}\boldsymbol{M}\boldsymbol{\psi}_{i+1}] = -N_{i+1}i^*\boldsymbol{M}\boldsymbol{\psi}_i \tag{9.31}$$

Step 4B: Check possible root convergence by comparing norms of vectors. If

$$(\|\boldsymbol{\phi}_{i+1}\| - \|\boldsymbol{\phi}_i\|)/\|\boldsymbol{\phi}_i\| \leq \mathrm{EPS} \tag{9.32}$$

then perform the computations in step 5; EPS is a user-specified accuracy parameter. Otherwise repeat step 4A.

Step 5A: If adequate convergence is achieved in step 4, compute the converged eigenvalue by employing the Rayleigh quotient:

$$(\lambda_{i+1})^2 = (\bar{\boldsymbol{\phi}}_{i+1})^T\boldsymbol{E}\boldsymbol{\phi}_{i+1}/(\bar{\boldsymbol{\phi}}_{i+1})^T\boldsymbol{F}^*\boldsymbol{\phi}_{i+1} \tag{9.33}$$

in which

$$(\bar{\boldsymbol{\phi}}_{i+1})^T\boldsymbol{F}\boldsymbol{\phi}_{i+1} = -i^*(\bar{\dot{\boldsymbol{\psi}}}_{i+1})^T\boldsymbol{M}\boldsymbol{\psi}_{i+1} + i^*(\bar{\boldsymbol{\psi}}_{i+1})^T\boldsymbol{M}\dot{\boldsymbol{\psi}}_{i+1} + i^*(\bar{\boldsymbol{\psi}}_{i+1})^T\boldsymbol{C}\boldsymbol{\psi}_{i+1} \tag{9.34}$$

$$(\bar{\boldsymbol{\phi}}_{i+1})^T\boldsymbol{E}\boldsymbol{\phi}_{i+1} = (\bar{\dot{\boldsymbol{\psi}}}_{i+1})^T\boldsymbol{M}\dot{\boldsymbol{\psi}}_{i+1} + (\bar{\boldsymbol{\psi}}_{i+1})^T\boldsymbol{K}\boldsymbol{\psi}_{i+1} \tag{9.35}$$

Step 5B: From the knowledge of values of H, NH, AL, and AU and other associated parameters, check if the converged root is a required one, in which case store λ_{i+1} and $\boldsymbol{\phi}_{i+1}$, respectively.

Step 6: Adjust the bounds of all roots under consideration appropriately, and set values of AL and AU pertaining to the next root under consideration, in the continuing bisection scheme; repeat steps 1–6.

The preceding procedure enables computation of roots lying within any specified bounds.

9.3.2 PSI

Equations (9.25) and (9.26) may be conveniently used in conjunction with a multiple set of vectors to yield the first few roots and vectors and the solution steps are given next.

Step 1: Start with a $2 \times \text{NRT}(\text{NRT} > \text{NR})$ set of randomly generated real trial vectors of order $2N$,

$$[{}^1\hat{\boldsymbol{x}} \quad {}^1\hat{\boldsymbol{y}} \quad {}^2\hat{\boldsymbol{x}} \quad {}^2\hat{\boldsymbol{y}} \quad \cdots \quad {}^{\text{NRT}}\hat{\boldsymbol{x}} \quad {}^{\text{NRT}}\hat{\boldsymbol{y}}] \tag{9.36}$$

in which

$${}^j\hat{\boldsymbol{\phi}} = {}^j\hat{\boldsymbol{x}} + i^{*j}\hat{\boldsymbol{y}} \tag{9.37}$$

Step 2: Perform Cholesky factorization of the $\boldsymbol{E}$ matrix, initially only,

$$\boldsymbol{E} = \boldsymbol{LDL}^T \tag{9.38}$$

$\boldsymbol{L}$ and $\boldsymbol{D}$ being a unit lower triangular and a diagonal matrix, respectively. Form

$$\hat{\boldsymbol{X}}_1 = \left[{}^1\hat{\boldsymbol{x}}_1 \quad {}^2\hat{\boldsymbol{x}}_1 \quad \cdots \quad {}^{\text{NRT}}\hat{\boldsymbol{x}}_1\right] \tag{9.39}$$

Step 3: Solve progressively, involving only unconverged roots,

$$\boldsymbol{EY}_{i+1} = \boldsymbol{G}\hat{\boldsymbol{X}}_i \tag{9.40}$$

and

$$\boldsymbol{EX}_{i+1} = \boldsymbol{GY}_{i+1} \tag{9.41}$$

by the usual back substitution process, which involves the following operations for Eq. (9.40)

$$\boldsymbol{KY}_{i+1}^{(l)} = \boldsymbol{M}\hat{\boldsymbol{X}}_i^{(u)} + \boldsymbol{C}\hat{\boldsymbol{X}}_i^{(l)} \qquad \text{yielding} \qquad \boldsymbol{Y}_{i+1}^{(l)} \tag{9.42}$$

$$\boldsymbol{MY}_{i+1}^{(u)} = -\boldsymbol{M}\hat{\boldsymbol{X}}_i^{(l)} \qquad \text{or} \qquad \boldsymbol{Y}_{i+1}^{(u)} = -\hat{\boldsymbol{X}}_i^{(l)} \qquad \text{yielding} \qquad \boldsymbol{Y}_{i+1}^{(u)} \tag{9.43}$$

and similarly elaborating Eq. (9.41)

$$\boldsymbol{KX}_{i+1}^{(l)} = \boldsymbol{MY}_{i+1}^{(u)} + \boldsymbol{CY}_{i+1}^{(l)} \qquad \text{yielding} \qquad \boldsymbol{X}_{i+1}^{(l)} \tag{9.44}$$

$$\boldsymbol{MX}_{i+1}^{(u)} = -\boldsymbol{MY}_{i+1}^{(l)} \qquad \text{yielding} \qquad \boldsymbol{X}_{i+1}^{(u)} \tag{9.45}$$

in which the superscripts u and l refer to the upper and lower half of a vector, respectively.

Step 4: Perform a vector norm check progressively on unconverged vectors only; compute approximate estimate of magnitudes of typical roots

$$ {}^j\lambda_{i+1} = \left\| {}^j\boldsymbol{x}_i \right\| / \left\| {}^j\boldsymbol{x}_{i+1} \right\| \tag{9.46} $$

and check if $({}^j\lambda_{i+1} - {}^j\lambda_i)/{}^j\lambda_{i+1} \leq \text{EPSN}$, a prescribed convergence parameter. Suppose that first NR1 roots have converged. If $\text{NR1} = \text{NR}$, go to step 7.

Step 5: Orthonormalize each ${}^j\boldsymbol{p}_{i+1}$ vector with respect to the $\boldsymbol{E}$ matrix progressively for vectors associated with unconverged $(\text{NR1} - \text{NR})$ eigenvalues only, yielding $\hat{\boldsymbol{p}}$

$$ {}^j\tilde{\boldsymbol{p}}_{i+1} = {}^j\boldsymbol{p}_{i+1} - \sum_{l=1}^{2\times\text{NR1}} \left[({}^l\boldsymbol{p})^T \boldsymbol{E}\,{}^j\boldsymbol{p}_{i+1}\right]{}^l\boldsymbol{p} - \sum_{l=2\times\text{NR1}+1}^{j-1} \left[({}^l\hat{\boldsymbol{p}}_{i+1})^T \boldsymbol{E}\,{}^j\boldsymbol{p}_{i+1}\right]{}^l\hat{\boldsymbol{p}}_{i+1} $$

$$ {}^j\hat{\boldsymbol{p}}_{i+1} = {}^j\tilde{\boldsymbol{p}}_{i+1} / \left[\left({}^j\tilde{\boldsymbol{p}}_{i+1}\right)^T \boldsymbol{E}\,{}^j\tilde{\boldsymbol{p}}_{i+1}\right]^{\frac{1}{2}}, \qquad j = 2\times\text{NR1}+1, 2\times\text{NRT} $$

in which typically ${}^1\boldsymbol{p} = {}^1\boldsymbol{x}$, ${}^2\boldsymbol{p} = {}^1\boldsymbol{y}$, ${}^3\boldsymbol{p} = {}^2\boldsymbol{x}$, ${}^4\boldsymbol{p} = {}^2\boldsymbol{y}$, and so on, and similarly set ${}^1\hat{\boldsymbol{x}} = {}^1\hat{\boldsymbol{p}}$, ${}^1\hat{\boldsymbol{y}} = {}^2\hat{\boldsymbol{p}}$, ${}^2\hat{\boldsymbol{x}} = {}^3\hat{\boldsymbol{p}}$, ${}^2\hat{\boldsymbol{y}} = {}^4\hat{\boldsymbol{p}}$, and so on.

Step 6: Continue iteration, go to step 3 in which

$$ \hat{\boldsymbol{X}}_{i+1} = \left[{}^{\text{NR1}+1}\hat{\boldsymbol{x}}_{i+1} \quad {}^{\text{NR1}+2}\hat{\boldsymbol{x}}_{i+1} \quad \cdots \quad {}^{\text{NRT}}\hat{\boldsymbol{x}}_{i+1}\right] $$

Step 7: Employing the Rayleigh–Ritz method and the progressive procedure involving only the unconverged roots, solve the further reduced eigenvalue problem

$$ \hat{\boldsymbol{E}}_{i+1}\boldsymbol{Q}_{i+1} = \hat{\boldsymbol{F}}_{i+1}\boldsymbol{Q}_{i+1}\hat{\boldsymbol{\Lambda}} $$

in which

$$ \hat{\boldsymbol{E}}_{i+1} = \bar{\hat{\boldsymbol{\Phi}}}_{i+1}^T \boldsymbol{E}\hat{\boldsymbol{\Phi}}_{i+1}, \qquad \hat{\boldsymbol{F}}_{i+1} = \bar{\hat{\boldsymbol{\Phi}}}_{i+1}^T \boldsymbol{F}\hat{\boldsymbol{\Phi}}_{i+1} $$

with

$$ \hat{\boldsymbol{\Phi}}_{i+1} = \left[{}^{\text{NR1}+1}\hat{\boldsymbol{\phi}}_{i+1} \quad {}^{\text{NR1}+2}\hat{\boldsymbol{\phi}}_{i+1} \quad \cdots \quad {}^{\text{NRT}}\hat{\boldsymbol{\phi}}_{i+1}\right] \tag{9.47} $$

$\hat{\lambda}$ being an approximation of λ.

Step 8: Perform a root convergence test on each unconverged root under consideration, in which EPS is the root convergence factor:

$$ \left|{}^j\hat{\lambda}_{i+1} - \hat{\lambda}_i\right| / \left|{}^j\lambda_{i+1}\right| \leq \text{EPS} $$

Step 9: Recalculate eigenvectors

$$ \tilde{\boldsymbol{\Phi}}_{i+1} = \hat{\boldsymbol{\Phi}}_{i+1}\boldsymbol{Q}_{i+1} \tag{9.48} $$

and relabel $\tilde{\boldsymbol{\Phi}}_{i+1}$ as $\hat{\boldsymbol{\Phi}}_{i+1}$ for possible use in step 3 involving $\hat{\boldsymbol{X}}_{i+1}$ components of $\hat{\boldsymbol{\Phi}}_{i+1}$.

Step 10: Go to step 3 and perform steps 3, 5, and 7–9 only if the number of converged roots $\text{NR1} < \text{NR}$. Otherwise, go to step 11.

Step 11: End of analysis.

This analysis procedure possesses solution characteristics similar to that described for the nonspinning case.

9.3.3 Block Lanczos Method and Algorithm

To be able to develop the block Lanczos algorithm it is necessary to perform a reduction procedure on the original formulation $(\boldsymbol{E} - \lambda \boldsymbol{F})\boldsymbol{\phi} = \boldsymbol{0}$ in Eq. (9.26) to yield a single matrix. Thus a Cholesky decomposition of matrix $\boldsymbol{E}$ that is symmetric and positive definite is performed as

$$\boldsymbol{E} = \boldsymbol{L}_E \boldsymbol{L}_E^T \tag{9.49}$$

in which

$$\boldsymbol{L}_E = \begin{bmatrix} \boldsymbol{L}_M & 0 \\ 0 & \boldsymbol{L}_K \end{bmatrix} \tag{9.50}$$

where $\boldsymbol{L}_K, \boldsymbol{L}_M$ are the lower triangular forms of the $\boldsymbol{K}$ and $\boldsymbol{M}$ matrices, respectively; the distributed inertia matrix is assumed to be positive definite and similar in form to the stiffness matrix. Then by substituting Eq. (9.49) into Eq. (9.26) and pre- and postmultiplying the individual matrices by $\boldsymbol{L}_E^{-1}$ and $\boldsymbol{L}_E^{-T}$, respectively, it follows that

$$\left[\boldsymbol{L}_E^{-1}\left(\boldsymbol{L}_E \boldsymbol{L}_E^T\right)\boldsymbol{L}_E^{-T} - \lambda \boldsymbol{L}_E^{-1}\boldsymbol{F}\boldsymbol{L}_E^{-T}\right]\boldsymbol{\phi} = \boldsymbol{0} \tag{9.51}$$

which can be written in the form

$$(\boldsymbol{A} - \gamma \boldsymbol{I})\boldsymbol{\phi} = \boldsymbol{0} \tag{9.52}$$

where $\gamma = 1/\lambda$, ω being expressed as $1/i^*\gamma$. To obtain a detailed expression for matrix $\boldsymbol{A}$, the multiplications of relevant matrices in Eq. (9.51) are performed as

$$\boldsymbol{L}_E^{-1}\boldsymbol{F} = \begin{bmatrix} \boldsymbol{L}_M^{-1} & 0 \\ 0 & \boldsymbol{L}_K^{-1} \end{bmatrix} \begin{bmatrix} 0 & -i^*\boldsymbol{M} \\ i^*\boldsymbol{M} & i^*\boldsymbol{C} \end{bmatrix} = \begin{bmatrix} 0 & -i^*\boldsymbol{L}_M^{-1}\boldsymbol{M} \\ i^*\boldsymbol{L}_K^{-1}\boldsymbol{M} & i^*\boldsymbol{L}_K^{-1}\boldsymbol{C} \end{bmatrix} \tag{9.53}$$

and similarly

$$\boldsymbol{L}_E^{-1}\boldsymbol{F}\boldsymbol{L}_E^{-T} = \begin{bmatrix} 0 & -i^*\boldsymbol{L}_M^T\boldsymbol{L}_K^{-T} \\ i^*\boldsymbol{L}_K^{-1}\boldsymbol{M}\boldsymbol{L}_M^{-T} & i^*\boldsymbol{L}_K^{-1}\boldsymbol{C}\boldsymbol{L}_K^{-T} \end{bmatrix} \tag{9.54}$$

which on substitution of $\boldsymbol{M} = \boldsymbol{L}_M \boldsymbol{L}_M^T$ yields the final expression

$$\boldsymbol{A} = \begin{bmatrix} 0 & -i^*\boldsymbol{L}_M^T\boldsymbol{L}_K^{-T} \\ i^*\boldsymbol{L}_K^{-1}\boldsymbol{L}_M & i^*\boldsymbol{L}_K^{-1}\boldsymbol{C}\boldsymbol{L}_K^{-T} \end{bmatrix} \tag{9.55}$$

noting that matrix $\boldsymbol{A}$ is a pure imaginary Hermitian matrix. In the following developments it is assumed the $\boldsymbol{A}$ is of order $n = 2N$, whereas the half bandwidth of the $\boldsymbol{K}$, $\boldsymbol{M}$, and $\boldsymbol{C}$ matrices is denoted by M, as before.

The Lanczos procedure for eigenvalue problems associated with a real symmetric matrix has been successfully developed earlier.[11,12] The relative merit of such a procedure has been described in detail in Ref. 13. A description of the relative merits of the block version of the Lanczos algorithm over the corresponding conventional nonblock procedure is also presented in detail in Ref. 11. Further, a conventional Lanczos algorithm for gyroscopic systems is presented in Ref. 14. To develop the block Lanczos procedure for a gyroscopic system characterized by a pure imaginary Hermitian matrix $\boldsymbol{A}$ using a block size k, the following definitions are firstly accorded:

$$\boldsymbol{Q}_j = n \times k \text{ complex matrix with orthonormal columns, i.e., } \bar{\boldsymbol{Q}}_j^T \boldsymbol{Q}_j = \boldsymbol{I}$$

$$\boldsymbol{P}_j, \boldsymbol{R}_j = n \times k \text{ complex matrices}$$

$$\boldsymbol{A}_j = k \times k \text{ Hermitian matrix}$$

$$\boldsymbol{B}_j = k \times k \text{ upper Hessenberg complex matrix}$$

$$\boldsymbol{T}_j = jk \times jk \text{ block Hermitian matrix, blocks of } (k \times k)$$

It can be proved that, for the eigenproblem solution of the matrix $\boldsymbol{A}$, a unitary matrix $\boldsymbol{J}_p$ of order $(p \times p)$ can be utilized to relate the complex matrices occurring in the Lanczos algorithm to corresponding real matrices. For any even, positive integer p, the matrix $\boldsymbol{J}_p$ denotes a $(p \times p)$ unitary block diagonal matrix having $p/2$ replications of $\boldsymbol{J}_2$ on the diagonal, which in turn is defined as

$$\boldsymbol{J}_2 = \frac{\sqrt{2}}{2} \begin{bmatrix} 1 & 1 \\ i^* & -i^* \end{bmatrix}$$

Thus if $\boldsymbol{S}$ is an $(n \times n)$ unitary matrix of column of eigenvectors of $\boldsymbol{A}$ occurring in complex conjugate pairs, then the matrix $\hat{\boldsymbol{S}} = \boldsymbol{S}\bar{\boldsymbol{J}}_n^T$ is a real orthogonal matrix of order n. This procedure may then be recast in terms of the real matrices, so that all numerical operations can be carried out in terms of real numbers, thereby effecting savings in solution time and storage requirements.

To implement the block Lanczos algorithm, the following substitutions are made to the procedure:

$$\boldsymbol{A} = i^* \hat{\boldsymbol{A}} \text{ where } \hat{\boldsymbol{A}} \text{ is of order } n \times n \text{ being real skew symmetric}$$

$$\boldsymbol{Q}_j = \hat{\boldsymbol{Q}}_j \boldsymbol{J}_k \text{ where } \hat{\boldsymbol{Q}}_j \text{ is an } n \times k \text{ real matrix with orthonormal columns}$$

$$\hat{\boldsymbol{P}}_j = i^* \hat{\boldsymbol{P}}_j \boldsymbol{J}_k \text{ where } \hat{\boldsymbol{P}}_j \text{ is an } n \times k \text{ real matrix}$$

$$\boldsymbol{A}_j = i^* \bar{\boldsymbol{J}}_k^T \hat{\boldsymbol{A}}_j \boldsymbol{J}_k \text{ where } \hat{\boldsymbol{A}}_j \text{ is a } k \times k \text{ real skew-symmetric matrix}$$

$$\boldsymbol{R}_j = i^* \hat{\boldsymbol{R}}_j \boldsymbol{J}_k \text{ where } \hat{\boldsymbol{R}}_j \text{ is a real } n \times k \text{ matrix}$$

$$\boldsymbol{B}_j = i^* \bar{\boldsymbol{J}}_k \hat{\boldsymbol{B}}_j \boldsymbol{J}_k \text{ where } \hat{\boldsymbol{B}}_j \text{ is a } k \times k \text{ real upper triangular matrix}$$

$$\boldsymbol{T}_j = i^* \bar{\boldsymbol{J}}_{jk} \hat{\boldsymbol{T}}_j \boldsymbol{J}_{jk} \text{ where } \hat{\boldsymbol{T}}_j \text{ is a } (jk \times jk) \text{ real skew-symmetric matrix}$$

It may be noted that if $\hat{\boldsymbol{B}}_j$ is upper triangular, $\boldsymbol{B}_j$ will generally have nonzero elements immediately below the diagonal at specific locations, which, however, is of no consequence for the present procedure. The essential properties for the technique are that $\boldsymbol{R}_j = \boldsymbol{Q}_{j+1}\boldsymbol{B}_j$ and $\bar{\boldsymbol{Q}}_{j+1}^T \boldsymbol{Q}_{j+1} = \boldsymbol{I}$.

With the preceding substitutions, the Lanczos algorithm can be formulated as follows. Let $\hat{\boldsymbol{Q}}_1$ be an arbitrary $(n \times k)$ real matrix with orthonormal columns, which is achieved by first forming columns of computer-generated random numbers and then orthonormalizing the columns by a standard procedure. Then for $j = 1, 2, \ldots,$ the following computational steps are performed.[15]

Step 1: The first Lanczos computational step

$$\begin{aligned} \boldsymbol{P}_j &= \boldsymbol{A}\boldsymbol{Q}_j && \text{for} \quad j = 1 \\ &= \boldsymbol{A}\boldsymbol{Q}_j - \boldsymbol{Q}_{j-1}\boldsymbol{B}_{j-1}^T && \text{for} \quad j > 1 \end{aligned}$$

on substitution yields

$$\begin{aligned} i^*\hat{\boldsymbol{P}}_j\boldsymbol{J}_k &= i^*\hat{\boldsymbol{A}}\hat{\boldsymbol{Q}}_j\boldsymbol{J}_k && \text{for} \quad j = 1 \\ &= i^*\hat{\boldsymbol{A}}\hat{\boldsymbol{Q}}_j\boldsymbol{J}_k - (\hat{\boldsymbol{Q}}_{j-1}\boldsymbol{J}_k)\left(-i^*\bar{\boldsymbol{J}}_k^T\hat{\boldsymbol{B}}_{j-1}^T\boldsymbol{J}_k\right) && \text{for} \quad j > 1 \end{aligned}$$

or

$$\begin{aligned} \hat{\boldsymbol{P}}_j &= \hat{\boldsymbol{A}}\hat{\boldsymbol{Q}}_j && \text{for} \quad j = 1 \\ &= \hat{\boldsymbol{A}}\hat{\boldsymbol{Q}}_j + \hat{\boldsymbol{Q}}_{j-1}\hat{\boldsymbol{B}}_{j-1}^T && \text{for} \quad j > 1 \end{aligned}$$

Step 2: This step involves computing $\boldsymbol{A}_j = \bar{\boldsymbol{Q}}_j^T\boldsymbol{P}_j$, which on substitution gives

$$i^*\bar{\boldsymbol{J}}_k^T\hat{\boldsymbol{A}}_j\boldsymbol{J}_k = \left(\bar{\boldsymbol{J}}_k^T\hat{\boldsymbol{Q}}_j^T\right)(i^*\hat{\boldsymbol{P}}_j\boldsymbol{J}_k)$$

or

$$\hat{\boldsymbol{A}}_j = \hat{\boldsymbol{Q}}_j^T\hat{\boldsymbol{P}}_j$$

Step 3: To compute $\boldsymbol{R}_j = \boldsymbol{P}_j - \boldsymbol{Q}_j\boldsymbol{A}_j$ and the relevant substitution yields

$$i^*\hat{\boldsymbol{R}}_j\boldsymbol{J}_k = i^*\hat{\boldsymbol{P}}_j\boldsymbol{J}_k - (\hat{\boldsymbol{Q}}_j\boldsymbol{J}_k)\left(i^*\bar{\boldsymbol{J}}_k^T\hat{\boldsymbol{A}}_j\boldsymbol{J}_k\right)$$

or

$$\hat{\boldsymbol{R}}_j = \hat{\boldsymbol{P}}_j - \hat{\boldsymbol{Q}}_j\hat{\boldsymbol{A}}_j$$

Step 4: As QR factors of $\hat{\boldsymbol{R}}_j$, the matrices $\hat{\boldsymbol{Q}}_{j+1}$ and $\hat{\boldsymbol{B}}_j$ are computed next, satisfying $\hat{\boldsymbol{R}}_j = \hat{\boldsymbol{Q}}_{j+1}\hat{\boldsymbol{B}}_j$ and $\hat{\boldsymbol{Q}}_{j+1}^T\hat{\boldsymbol{Q}}_{j+1} = \boldsymbol{I}$, $\hat{\boldsymbol{B}}_j$ being upper triangular. The standard Givens, Householder, or Gramm Schmidt technique may be used for this purpose. The previously defined substitutions then yield $\boldsymbol{R}_j = \boldsymbol{Q}_{j+1}\boldsymbol{B}_j$ and $\bar{\boldsymbol{Q}}_{j+1}^T\boldsymbol{Q}_{j+1} = \boldsymbol{I}$, $\boldsymbol{B}_j$ being upper Hessenberg.

Step 5: To form the block matrix $\hat{\boldsymbol{T}}_j$ of order jk

$$\hat{\boldsymbol{T}}_j = \begin{bmatrix} \hat{\boldsymbol{A}}_1 & -\hat{\boldsymbol{B}}_1^T & & \\ \hat{\boldsymbol{B}}_1 & \hat{\boldsymbol{A}}_2 & \ddots & \\ & \ddots & \ddots & -\hat{\boldsymbol{B}}_{j-1}^T \\ & & \hat{\boldsymbol{B}}_{j-1} & \hat{\boldsymbol{A}}_j \end{bmatrix}$$

which yields the eigenvalues and vectors of the system as the jth stage approximation. It may be noted that the $i^*\hat{\boldsymbol{T}}_j$ is complex Hermitian and pure imaginary, and the matrices $\boldsymbol{T}_j$ and $i^*\hat{\boldsymbol{T}}_j$ have the same eigenvalues, being real and occurring in pairs having opposite signs.

Step 6: To perform a convergence test utilizing the vectors computed in step 5 and also the $\hat{\boldsymbol{B}}_j$ matrix derived in step 4, and if the analysis is to be continued repeating the preceding steps, then a selective orthogonalization process must be carried out for the $\boldsymbol{Q}_{j+1}$ matrix. In this procedure the columns of $\boldsymbol{Q}_{j+1}$ are orthogonalized relative to some of the current Ritz vectors.[11,12]

The eigenvectors associated with step 5 may be computed as follows. Thus if α and $-\alpha$ are a pair of real roots of $i^*\hat{\boldsymbol{T}}_j$, the related eigenvectors will occur as complex conjugate pairs, denoted by $\hat{\boldsymbol{v}}$ and $\bar{\hat{\boldsymbol{v}}}$. The corresponding vectors for $\hat{\boldsymbol{T}}_j$ will then be defined as $\boldsymbol{v} = \bar{\boldsymbol{J}}^T\hat{\boldsymbol{v}}$ and $\boldsymbol{w} = \bar{\boldsymbol{J}}^T\bar{\hat{\boldsymbol{v}}}^T$, noting that $\boldsymbol{v}$ and $\boldsymbol{w}$ will not be mutually conjugate. The numbers α and $-\alpha$ are Ritz values for $\boldsymbol{A}$, the corresponding Ritz vectors being $\boldsymbol{\phi} = \boldsymbol{u}_j\boldsymbol{v} = \boldsymbol{u}_j\bar{\boldsymbol{J}}^T\hat{\boldsymbol{v}} = \hat{\boldsymbol{u}}_j\hat{\boldsymbol{v}}$ and $\boldsymbol{z} = \boldsymbol{u}_j\boldsymbol{w} = \boldsymbol{u}_j\bar{\boldsymbol{J}}^T\bar{\hat{\boldsymbol{v}}}^T$, in which

$$
\begin{aligned}
\hat{\boldsymbol{u}}_j = \boldsymbol{u}_j\bar{\boldsymbol{J}}^T &= [\boldsymbol{Q}_1, \dots, \boldsymbol{Q}_j]\bar{\boldsymbol{J}}^T \\
&= [\hat{\boldsymbol{Q}}_1, \dots, \hat{\boldsymbol{Q}}_j]
\end{aligned}
$$

Matrix $\hat{\boldsymbol{u}}_j$ is real because $\hat{\boldsymbol{Q}}_1$ to $\hat{\boldsymbol{Q}}_j$ are real and hence $\boldsymbol{\phi}$ and $\boldsymbol{z}$ are mutually conjugate, so that $\boldsymbol{z} = \bar{\boldsymbol{\phi}}$.

In connection with the selective orthogonalization of $\boldsymbol{Q}_{j+1}$ in step 6, the preferred procedure is described next. Denoting $\boldsymbol{q}$ as a column $\boldsymbol{Q}_{i+1}$ and $(\boldsymbol{\psi}, \boldsymbol{\phi})$ a Ritz pair, so that $\|\boldsymbol{\phi}\| = 1$, the usual orthogonalization process is given,

$$
\boldsymbol{q} := \boldsymbol{q} - (\bar{\boldsymbol{\phi}}^T\boldsymbol{q})\bar{\boldsymbol{\phi}} \tag{9.56}
$$

in which all computations are carried out in complex numbers. Furthermore, although the initial $\boldsymbol{q}$ is of the form $\hat{\boldsymbol{q}}\boldsymbol{J}$, $\hat{\boldsymbol{q}}$ being real, the orthogonalized $\boldsymbol{q}$ will not be, in general, of the same form. To avoid such undesirable features the process may be altered as follows. Writing $\boldsymbol{\phi} = \boldsymbol{\alpha} + i^*\boldsymbol{\beta}$ in which $\boldsymbol{\alpha}$ and $\boldsymbol{\beta}$ are real n vectors, it may be noted that they span the same subspace of complex n space as $\boldsymbol{\phi}$ and $\bar{\boldsymbol{\phi}}$. Further, because $\boldsymbol{\phi}$ and $\bar{\boldsymbol{\phi}}$ are mutually orthogonal, it implies that $\boldsymbol{\alpha}$ and $\boldsymbol{\beta}$ are also mutually orthogonal, so that they form an orthogonal basis for the subspace spanned by $\boldsymbol{\phi}$ and $\bar{\boldsymbol{\phi}}$. Thus it follows that, instead of directly orthogonalizing the complex vector $\boldsymbol{q}$ with respect to the complex vector $\boldsymbol{\phi}$, one may orthogonalize the real vector $\hat{\boldsymbol{q}}$ relative to the two real vectors $\boldsymbol{\alpha}$ and $\boldsymbol{\beta}$, obtaining a new real vector $\hat{\boldsymbol{q}}$ from which the orthogonalized $\boldsymbol{q}$ can be expressed as $\boldsymbol{q} = \hat{\boldsymbol{q}}\boldsymbol{J}$. Thus the procedure in Eq. (9.56) may be replaced by the following computations:

$$
\hat{\boldsymbol{q}} := \hat{\boldsymbol{q}} - \frac{(\boldsymbol{\alpha}^T\hat{\boldsymbol{q}})}{(\boldsymbol{\alpha}^T\hat{\boldsymbol{\alpha}})}\boldsymbol{\alpha} \tag{9.57}
$$

and

$$
\hat{\boldsymbol{q}} := \hat{\boldsymbol{q}} - \frac{(\boldsymbol{\beta}^T\hat{\boldsymbol{q}})}{(\boldsymbol{\beta}^T\hat{\boldsymbol{\beta}})}\boldsymbol{\beta} \tag{9.58}
$$

from which the orthogonalized $\boldsymbol{q}$ could be obtained as $\boldsymbol{q} = \hat{\boldsymbol{q}}\boldsymbol{J}$. It may be noted

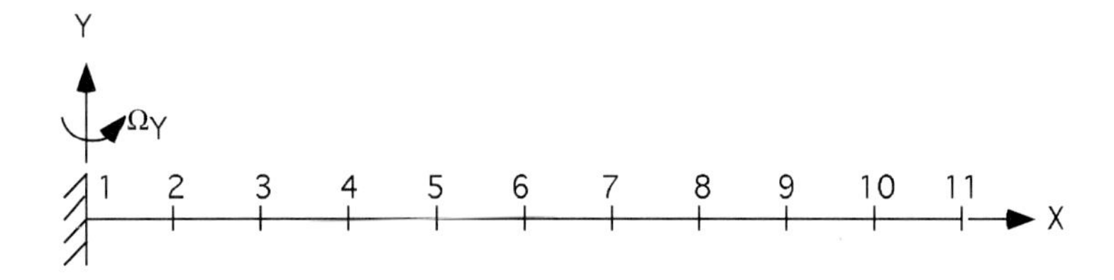

Fig. 9.6 Spinning cantilever beam.

that all computations in the entire procedure of the proposed Lanczos algorithm involve only real numbers.

An estimation of error of the present algorithm follows the same pattern as that of the usual Lanczos procedure.[12] Also, although the transformation matrix $\boldsymbol{J}$ is not explicitly used in any computation, it proves to be useful in demonstrating the transformation of complex into real matrices inherent in the current procedure. In this connection, it may be noted that the computer implementation of the algorithm involves only real matrices inherent in the respective final sets of equations at the end of each solution step and follows the same procedure as in Ref. 11. Furthermore, the procedure enables effective computation of multiple roots and only real numbers are involved in all computations.

9.3.4 Numerical Examples

Spinning cantilevers are considered in the two numerical examples that follow. Solution times pertain to a 1.7 GHz PC machine.

Example 9.3: Cantilever Beam

A spinning cantilever beam is chosen for this example problem. The beam (Fig. 9.6), spinning about the Y axis, is discretized with an increasing number of elements and vibration analyses are performed for each test case. Relevant data parameters are Young's modulus $E = 30 \times 10^6$, cross sectional area $A = 1.0$, moment of inertia about Y axis $= 1/12$, moment of inertia about Z axis $= 1/24$, mass density $\rho = 0.1666$, and the spinning rate $= 0.1$ Hz. Table 9.4 presents the analysis results in detail pertaining to the first six roots to establish the pattern of convergence as the model is refined.

Table 9.4 Natural frequencies of a spinning cantilever beam

Number of elements	Natural frequencies, rad/s					
	λ_1	λ_2	λ_3	λ_4	λ_5	λ_6
2	2.7809	3.7957	17.0150	23.9594	57.2405	80.8352
4	2.7659	3.7929	16.8622	23.7720	47.3586	66.8812
6	2.7633	3.7926	16.8427	23.7490	47.0775	66.4890
8	2.7624	3.7925	16.8384	23.7443	47.0190	66.4080
10	2.7522	3.7925	16.8369	23.7410	47.0017	66.3832
20	2.7614	3.7925	16.8355	23.7420	46.9897	66.3669

Table 9.5 Natural frequencies of a spinning cantilever plate

	Natural frequencies, rad/s			
			$\Omega_R = 100$ rad/s	
	$\Omega_Z = 100$ rad/s		$\Omega_X = \Omega_Y = \Omega_Z = 57.73$ rad/s	
Mode	PSI	Lanczos	PSI	Lanczos
1	245.1962	245.1962	162.0113	162.0113
2	549.7797	549.7797	515.0891	515.0891
3	1350.8408	1350.8408	1331.5199	1331.5199
4	1703.8664	1703.8664	1690.4811	1690.4811
5	1943.0787	1943.0787	1928.3656	1928.3656
6	3367.9677	3367.9677	3358.4562	3358.4562
7	3835.6040	3835.6049	3823.5349	3823.5356
8	3981.7147	3981.7197	3974.0801	3974.0804
9	4422.8750	4422.8751	4412.2784	4412.2786
10	5744.3196	5744.4048	5737.6322	5737.7588
CPU, s (NR = 40)	22.81	32.26	21.56	32.02

Example 9.4: Cantilever Plate

This example relates to the square, spinning cantilever plate (Fig. 9.5), which is discretized by a 20 × 20 mesh involving a 2646-degree-of-freedom problem. Natural frequency analyses were performed for various spin rates using PSI and Lanczos procedures, and such results are depicted in Table 9.5.

9.4 Quadratic Matrix Eigenvalue Problem for Free Vibration Analysis

When structural discretization is achieved by finite dynamic elements (FDEs)[16] the equation of free vibration takes the following form:

$$\left[\boldsymbol{K}_0 - \omega^2 \boldsymbol{M}_0 - \omega^4 (\boldsymbol{M}_2 - \boldsymbol{K}_4)\right]\boldsymbol{q} = 0 \tag{9.59}$$

which may be written in quadratic form[17] in $\lambda = \omega^2$:

$$(\boldsymbol{K} - \lambda \boldsymbol{M} - \lambda^2 \boldsymbol{C})\boldsymbol{q} = 0 \tag{9.60}$$

where $\boldsymbol{K}_0 = \boldsymbol{K}$ is the elastic stiffness matrix, $\boldsymbol{M}_0 = \boldsymbol{M}$ the usual inertia matrix, $\boldsymbol{M}_2$, $\boldsymbol{K}_4$ are the higher order dynamic correction terms, and $\boldsymbol{C} = \boldsymbol{M}_2 - \boldsymbol{K}_4$. A solution of this quadratic matrix eigenproblem may be achieved by first rearranging Eq. (9.60) as[18]

$$\left(\begin{bmatrix} \boldsymbol{C} & \boldsymbol{0} \\ \boldsymbol{0} & \boldsymbol{K} \end{bmatrix} - \lambda \begin{bmatrix} \boldsymbol{0} & \boldsymbol{C} \\ \boldsymbol{C} & \boldsymbol{M} \end{bmatrix}\right) \begin{Bmatrix} \dot{\boldsymbol{q}} \\ \boldsymbol{q} \end{Bmatrix} = \boldsymbol{0} \tag{9.61}$$

with $\dot{\boldsymbol{q}} = \lambda \boldsymbol{q}$ and which is of the form

$$(\boldsymbol{E} - \lambda \boldsymbol{F})\boldsymbol{y} = 0 \tag{9.62}$$

where $\boldsymbol{E}$ and $\boldsymbol{F}$ are both symmetric, $\boldsymbol{E}$ being also positive definite; the roots and vectors of the system are real in nature. These matrices are almost always highly banded and an efficient eigenproblem solver is expected to exploit this matrix sparsity fully. The analysis procedure involving higher-order correction terms will be referred to as the dynamic element method.

9.4.1 Single Vector Iteration

A combined Sturm sequence and inverse iteration procedure may also be employed for determining a number of specified roots only and associated vectors. Thus it is well known[18] that the Sturm sequence property is valid for the current formulation. For any specified value of $\lambda = \lambda^s$, the number of negative diagonal elements of the triangularized $(\boldsymbol{E} - \lambda^s \boldsymbol{F})$ matrix is equal to the number of roots of the system having an algebraic value less than λ^s. Exploitation of this property along with a bisection strategy and an inverse iteration scheme enables an effective solution of the current eigenvalue problem.[19]

Assuming that during a typical bisection step λ^u and λ^l are the current upper and lower root bounds, with NL and NU the corresponding number of roots, the following solution steps are performed.

Step 1: Calculate $\lambda^m = (\lambda^u + \lambda^l)/2$.

Step 2: Triangularize $\boldsymbol{K} - \lambda^m \boldsymbol{M} - (\lambda^m)^2 \boldsymbol{C}$ and count the number of roots NM of the system that has algebraic values less than λ^m.

Step 3: If NM < NL or NM > NU, then go to step 6.

Step 4A: Perform inverse iterations. At a typical $(i + 1)$th step iterate on $(\boldsymbol{E} - \lambda_m \boldsymbol{F})\boldsymbol{y}_{i+1} = N_{i+1}\boldsymbol{F}\boldsymbol{y}_i$, which is expressed in detail next,

$$\begin{bmatrix} \boldsymbol{C} & -\lambda\boldsymbol{C} \\ -\lambda\boldsymbol{C} & \boldsymbol{K} - \lambda\boldsymbol{M} \end{bmatrix} \begin{bmatrix} \dot{\boldsymbol{q}}_{i+1} \\ \boldsymbol{q}_{i+1} \end{bmatrix} = N_{i+1} \begin{bmatrix} \boldsymbol{C}\boldsymbol{q}_i \\ \boldsymbol{C}\dot{\boldsymbol{q}}_i + \boldsymbol{M}\boldsymbol{q}_i \end{bmatrix}$$

N_{i+1} being a normalizing factor, and $\dot{\boldsymbol{q}} = \lambda\boldsymbol{q}$. The process involves two distinct steps.

Lower half:

$$\left[\boldsymbol{K} - \lambda_m \boldsymbol{M} - \lambda^2 \boldsymbol{C}\right]\boldsymbol{q}_{i+1} = N_{i+1}[\boldsymbol{C}\dot{\boldsymbol{q}}_i + \boldsymbol{M}\boldsymbol{q}_i + \lambda\boldsymbol{C}\boldsymbol{q}_i]$$

Upper half:

$$\boldsymbol{C}\dot{\boldsymbol{q}}_{i+1} - \lambda\boldsymbol{C}\boldsymbol{q}_{i+1} = N_{i+1}\boldsymbol{C}\boldsymbol{q}_i$$

Step 4B: Check root convergence by comparing norms of $\boldsymbol{y}_{i+1}$ and $\boldsymbol{y}_i$.

Step 5A: If convergence has been achieved in the preceding step, then compute the converged root value comparing the Rayleigh quotient:

$$\lambda_{i+1}^r = \left(\boldsymbol{y}_{i+1}^r\right)^T \boldsymbol{E}\boldsymbol{y}_{i+1}^r \Big/ \left(\boldsymbol{y}_{i+1}^r\right)^T \boldsymbol{F}\boldsymbol{y}_{i+1}^r$$

where

$$\left(\boldsymbol{y}_{i+1}^r\right)^T \boldsymbol{E}\boldsymbol{y}_{i+1}^r = \left(\dot{\boldsymbol{q}}_{i+1}^r\right)^T \boldsymbol{C}\dot{\boldsymbol{q}}_{i+1}^r + \left(\boldsymbol{q}_{i+1}^r\right)^T \boldsymbol{K}\boldsymbol{q}_{i+1}^r$$

and

$$(\boldsymbol{y}_{i+1}^r)^T \boldsymbol{F}\boldsymbol{y}_{i+1} = \left(\boldsymbol{q}_{i+1}^r\right)^T \boldsymbol{C}\dot{\boldsymbol{q}}_{i+1}^r + \left(\dot{\boldsymbol{q}}_{i+1}^r\right)^T \boldsymbol{C}\boldsymbol{q}_{i+1}^r + \left(\boldsymbol{q}_{i+1}^r\right)^T \boldsymbol{M}\boldsymbol{q}_{i+1}^r$$

Step 5B: From information on λ_l, λ_m, λ_u and NM and other associated data, check if the converged root is the required one and store λ_{i+1} and $\boldsymbol{y}_{i+1}$.

Step 6: Adjust the bounds of all roots under consideration and reset values of λ_l, λ_u for convergence to the next root under consideration. Repeat the iteration procedure until all required roots within a specified bound have been computed.

9.4.2 PSI

A procedure is described next for the extraction of the first few roots and vectors. The solution process consists of the following basic steps pertaining to Eq. (9.62).

Step 1: Generate a NRT ($NRT > NR$) set of random vectors of order $2N$:

$$\hat{\boldsymbol{Y}}_1 = \begin{bmatrix} \hat{\boldsymbol{y}}_1^1 & \hat{\boldsymbol{y}}_1^2 & \cdots & \hat{\boldsymbol{y}}_1^{\mathrm{NRT}} \end{bmatrix}$$

Step 2: Perform Cholesky factorization on $\boldsymbol{E}$ matrix, initially only $\boldsymbol{E} = \boldsymbol{LDL}^T$.

Step 3: Start of iteration; at a ith iteration step, solve $\boldsymbol{EY}_{i+1} = \boldsymbol{F}\hat{\boldsymbol{Y}}_i$ using the progressive strategy involving only unconverged roots and simple individual constituent matrix and vector multiplication.

Step 4: Estimate the magnitude of the jth eigenvalue λ^j progressively on unconverged roots only as

$$\lambda_{i+1}^j = \|\boldsymbol{y}_i^j\| / \|\boldsymbol{y}_{i+1}^j\|$$

If this is not the first iteration, check for convergence using the test

$$\left(\lambda_{i+1}^j - \lambda_i^j\right) / \lambda_{i+1}^j \le \mathrm{EPSN}$$

Let NR1 be the number of leading consecutive roots that have converged. If NR1 = NR, go to step 7.

Step 5: Next, perform progressive vector orthonormalization with respect to the $\boldsymbol{F}$ matrix:

$$\tilde{\boldsymbol{y}}_{i+1}^k = \boldsymbol{y}_{i+1}^k - \sum_{l=1}^{\mathrm{NR1}} [(\boldsymbol{y}^l)^T \boldsymbol{E} \boldsymbol{y}_{i+1}^k] \boldsymbol{y}^l - \sum_{l=\mathrm{NR1}+1}^{k-1} \left[\left(\hat{\boldsymbol{y}}_{i+1}^l\right)^T \boldsymbol{E} \boldsymbol{y}_{i+1}^k\right] \hat{\boldsymbol{y}}_{i+1}^l$$

$$\hat{\boldsymbol{y}}_{i+1}^k = \tilde{\boldsymbol{y}}_{i+1}^k / \left[\left(\tilde{\boldsymbol{y}}_{i+1}^k\right)^T \boldsymbol{E} \tilde{\boldsymbol{y}}_{i+1}^k\right]^{\frac{1}{2}}, \qquad k = \mathrm{NR1} + 1, \mathrm{NRT}$$

Step 6: Go to step 3 to continue iteration with the set of vectors $\hat{\boldsymbol{Y}}_{i+1}$ defined as

$$\hat{\boldsymbol{Y}}_{i+1} = \begin{bmatrix} \hat{\boldsymbol{y}}_{i+1}^{\mathrm{NR1}+1} & \hat{\boldsymbol{y}}_{i+1}^{\mathrm{NR1}+2} & \cdots & \hat{\boldsymbol{y}}_{i+1}^{\mathrm{NRT}} \end{bmatrix}$$

Step 7: Employing the Rayleigh–Ritz method and using the progressive strategy, compute only the unconverged roots from the reduced set of equations:

$$\hat{\boldsymbol{E}} \boldsymbol{X}_{i+1} = \hat{\boldsymbol{F}} \boldsymbol{X}_{i+1} \hat{\boldsymbol{\Lambda}}$$

$\hat{\lambda}$ being an approximation for λ and in which

$$\hat{\boldsymbol{E}} = \hat{\boldsymbol{Y}}_{i+1}^T \boldsymbol{E} \hat{\boldsymbol{Y}}_{i+1}, \qquad \hat{\boldsymbol{F}} = \hat{\boldsymbol{Y}}_{i+1}^T \boldsymbol{F} \hat{\boldsymbol{Y}}_{i+1}$$

Step 8: Perform the test for convergence for each typical jth root under consideration:

$$\left|\left(\hat{\lambda}_{i+1}^{j} - \hat{\lambda}_{i}^{j}\right)\right| / \left|\hat{\lambda}_{i+1}^{j}\right| \leq \text{EPS}$$

Step 9: Recalculate the eigenvector:

$$\tilde{\boldsymbol{Y}}_{i+1} = \hat{\boldsymbol{Y}}_{i+1}\boldsymbol{X}_{i+1}$$

and set

$$\hat{\boldsymbol{Y}}_{i+1} = \tilde{\boldsymbol{Y}}_{i+1}$$

Step 10: Go to step 3 and perform steps 3, 5, and 7–9 only if convergence is achieved for NR1 < NR roots; otherwise, go to step 11.

Step 11: End of analysis.

The convergence characteristics and other relevant details of this algorithm are similar to that described in Sec. 9.2.2.

9.4.3 Lanczos Method

To achieve a solution it is first necessary to perform a Cholesky decomposition of the $\boldsymbol{E}$ matrix in Eq. (9.62): $\boldsymbol{E} = \boldsymbol{L}_E\boldsymbol{L}_E^T$, the lower triangular matrix being composed of the same, pertaining to the constituent matrices

$$\boldsymbol{L}_E = \begin{bmatrix} \boldsymbol{L}_C & 0 \\ 0 & \boldsymbol{L}_K \end{bmatrix} \tag{9.63}$$

Then performing some simple matrix manipulations, Eq. (9.62) may be written as

$$\left[\boldsymbol{L}_E^{-1}\left(\boldsymbol{L}_E\boldsymbol{L}_E^T\right)\boldsymbol{L}_E^{-T} - \lambda\boldsymbol{L}_E^{-1}\boldsymbol{F}\boldsymbol{L}_E^{-T}\right]\boldsymbol{y} = 0 \tag{9.64}$$

which also can be written in the form

$$(\tilde{\boldsymbol{A}} - \gamma\boldsymbol{I})\tilde{\boldsymbol{y}} = \boldsymbol{0} \tag{9.65}$$

where $\gamma = 1/\lambda = 1/\omega^2$, $\tilde{\boldsymbol{y}} = \boldsymbol{L}_E^T\boldsymbol{y}$, and

$$\begin{aligned} \tilde{\boldsymbol{A}} &= \boldsymbol{L}_E^{-1}\boldsymbol{F}\boldsymbol{L}_E^{-T} \\ &= \begin{bmatrix} \boldsymbol{L}_C^{-1} & 0 \\ 0 & \boldsymbol{L}_K^{-1} \end{bmatrix} \begin{bmatrix} 0 & \boldsymbol{C} \\ \boldsymbol{C} & \boldsymbol{M} \end{bmatrix} \begin{bmatrix} \boldsymbol{L}_C^{-T} & 0 \\ 0 & \boldsymbol{L}_K^{-T} \end{bmatrix} \\ &= \begin{bmatrix} 0 & \boldsymbol{L}_C^{-T}\boldsymbol{L}_K^{-T} \\ \boldsymbol{L}_K^{-1}\boldsymbol{L}_C^{-1} & \boldsymbol{L}_K^{-1}\boldsymbol{M}\boldsymbol{L}_k^{-T} \end{bmatrix} \end{aligned} \tag{9.66}$$

is utilized as a single matrix in the current Lanczos scheme.

9.4.3.1 Implementation of $\tilde{\boldsymbol{A}}\tilde{\boldsymbol{y}}$. A major amount of computational effort is expended in computing the product $\tilde{\boldsymbol{A}}\tilde{\boldsymbol{y}}$, and it is instructive to illustrate the various steps incurred in that process. Thus, for the upper half, the product $\boldsymbol{L}_C^T\boldsymbol{L}_K^{-T}\tilde{\boldsymbol{y}}_l$ can be achieved by adopting the following procedure. Let $\boldsymbol{L}_K^{-T}\tilde{\boldsymbol{y}}_l = \boldsymbol{v}$, then the solution for $\boldsymbol{v}$ is achieved by simple back substitution, starting from the top, using $\boldsymbol{L}_K\boldsymbol{v} = \tilde{\boldsymbol{y}}_l$.

Performing simple multiplication, $\boldsymbol{L}_C^T \boldsymbol{v}$ yields the required expression $\boldsymbol{L}_C^T \boldsymbol{L}_K^{-T} \tilde{\boldsymbol{y}}_l$. Similarly the expression in the lower half,

$$\boldsymbol{L}_K^{-1} \boldsymbol{L}_C^{-1} \tilde{\boldsymbol{y}}_u + \boldsymbol{L}_K^{-1} \boldsymbol{M} \boldsymbol{L}_K^{-T} \tilde{\boldsymbol{y}}_u$$

may be formed as follows.

Step 1: Solve for w_1, $\boldsymbol{L}_c \boldsymbol{w}_1 = \tilde{\boldsymbol{y}}_u$.

Step 2: Employing back substitution, starting from the top, solve for z in $\boldsymbol{L}_K^T \boldsymbol{z} = \tilde{\boldsymbol{y}}_l$ in which $\boldsymbol{z} = \boldsymbol{L}_K^{-T} \tilde{\boldsymbol{y}}_l$.

Step 3: Obtain by direct multiplication $\boldsymbol{w}_2 = \boldsymbol{M}\boldsymbol{z}$.

Step 4: With $\boldsymbol{w} = \boldsymbol{w}_1 + \boldsymbol{w}_2$, solve $\boldsymbol{L}_K \boldsymbol{r} = \boldsymbol{w}$, employing back substitution from the top, where $\boldsymbol{r}$ is the required expression. The procedure enables the computation of $\tilde{\boldsymbol{A}}\tilde{\boldsymbol{y}}$ without incurring any explicit inversion of a matrix.

9.4.3.2 Block Lanczos algorithm. The procedure starts by first forming an arbitrary $(n \times k)$ real matrix $\hat{\boldsymbol{Q}}_1$ with orthonormal columns; n is the order of $\tilde{\boldsymbol{A}}$, k being the block size. This can be achieved by forming columns of computer-generated random numbers followed by appropriate orthonormalization by a standard procedure. Then for a typical jth computational stage, for $j = 1, 2, \ldots,$ the following computations are performed.

Step 1: Perform matrix multiplication:

$$\hat{\boldsymbol{P}}_j = \tilde{\boldsymbol{A}}\hat{\boldsymbol{Q}}_j \qquad \text{for} \qquad j = 1$$

$$\hat{\boldsymbol{P}}_j = \tilde{\boldsymbol{A}}\hat{\boldsymbol{Q}}_j - \hat{\boldsymbol{Q}}_{j-1}\hat{\boldsymbol{B}}_{j-1}^T \qquad \text{for} \qquad j > 1$$

Step 2: Similarly obtain $\hat{\boldsymbol{A}}_j = \hat{\boldsymbol{Q}}_j^T \hat{\boldsymbol{P}}_j$.

Step 3: This step involves computing

$$\hat{\boldsymbol{R}}_j = \hat{\boldsymbol{P}}_j - \hat{\boldsymbol{Q}}_j \hat{\boldsymbol{A}}_j \tag{9.67}$$

Step 4: Compute the QR factors of $\hat{\boldsymbol{R}}_j$; i.e., find the orthogonal $\hat{\boldsymbol{Q}}_{j+1}$ and upper triangular $\hat{\boldsymbol{B}}_j$ such that $\hat{\boldsymbol{R}}_j = \hat{\boldsymbol{Q}}_{j+1}\hat{\boldsymbol{B}}_j$.

Step 5: Form block tridiagonal $\hat{\boldsymbol{T}}_j$ of order jk:

$$\hat{\boldsymbol{T}}_j = \begin{bmatrix} \hat{\boldsymbol{A}}_1 & \hat{\boldsymbol{B}}_1^T & & & \\ \hat{\boldsymbol{B}}_1 & \hat{\boldsymbol{A}}_2 & \hat{\boldsymbol{B}}_2 & & \\ & \ddots & \ddots & \ddots & \\ & & & & \hat{\boldsymbol{B}}_{j-1}^T \\ & & & \hat{\boldsymbol{B}}_{j-1} & \hat{\boldsymbol{A}}_j \end{bmatrix}$$

An eigenvalue solution of this reduced-order matrix is trivial in nature, and employment of the standard Sturm sequence procedure yields the eigenvalues and vectors of the system as the jth approximation.

Step 6: Conduct convergence tests on some of the roots and associated vectors obtained in the preceding step and perform selective orthogonalization,[8] if continuing.

An adequately converged solution like this yields the eigenvalues of the $\hat{\boldsymbol{T}}_j$ that are also the roots of the original matrix $\tilde{\boldsymbol{A}}$. Describing $\boldsymbol{U}_j$ as an $(n \times jk)$ matrix defined by

$$\boldsymbol{U}_j = [\hat{\boldsymbol{Q}}_1, \hat{\boldsymbol{Q}}_2, \ldots, \hat{\boldsymbol{Q}}_j] \tag{9.68}$$

then the approximate eigenvector $\tilde{\boldsymbol{y}}$ of $\tilde{\boldsymbol{A}}$ may be obtained:

$$\tilde{\boldsymbol{y}} = \boldsymbol{U}_j \boldsymbol{s} \tag{9.69}$$

where $\boldsymbol{s}$ is an eigenvector of $\hat{\boldsymbol{T}}_j$. This current algorithm has proved to be reliable and effective.

9.5 Structural Stability Problems

Examples of structural stability problems include elastic structural buckling and are characterized by the following matrix eigenvalue equation:

$$(\boldsymbol{K}_E + \lambda \boldsymbol{K}_G)\boldsymbol{\phi} = \boldsymbol{0} \tag{9.70}$$

in which $\boldsymbol{K}_E$ is the elastic stiffness matrix, and $\boldsymbol{K}_G$ incorporates the effects of in-plane forces on the out-of-plane deformations. Since both $\boldsymbol{K}_E$ and $\boldsymbol{K}_G$ matrices are symmetric and at least one of them is positive definite, standard analysis procedures such as those described in Sec. 9.2 are conveniently used for their solution. In Eq. (9.70) λ denotes the factor by which the existing external loading is to be multiplied to initiate the buckling, i.e., to reach the zero stiffness condition.

Example 9.5: Buckling Analysis

A buckling analysis was performed for the square plate in Fig. 9.5 with simply supported edges. Uniform unit stress was exerted, acting along the two edges parallel to the Y axis. Table 9.6 summarizes such solution results for increasing numbers of elements.

9.6 Vibration of Prestressed Structures

An example of this class of problem is a simple column under axial loading. The natural frequencies of the column are a function of the intensity of the axial

Table 9.6 Critical load of a simply supported square plate

Buckling load parameter for mode 1			
Numerical solution			Exact solution
Mesh			
4×4	8×8	14×14	
3530.695	3552.620	3570.561	3615.240

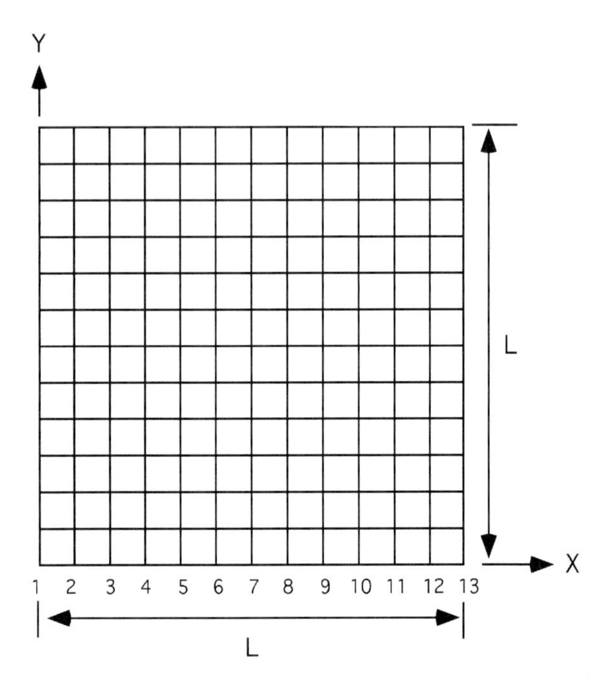

Fig. 9.7 Free-free composite square plate.

force. The matrix equation of motion of any such prestressed structure is given by

$$\left(\boldsymbol{K}_E + \boldsymbol{K}_G - \omega^2 \boldsymbol{M}\right)\boldsymbol{\phi} = \boldsymbol{0} \tag{9.71}$$

a solution that may be conveniently achieved by any of the analysis schemes given in Sec. 9.2.

Example 9.6: Thermally Prestressed Square Plate

A composite square plate (Fig. 9.7) subjected to temperatures varying along the X axis in the range of 29–82° was analyzed to yield the first 12 roots and vectors. The relevant data parameters are side length $L = 12$, plate thickness $t = 0.24$, mass density $\rho = 0.1475 \times 10^{-3}$, and composite stacking = $[30°/-30°/-30°/30°]$. The analysis results are shown in Table 9.7.

9.7 Vibration of Damped Structural Systems

In an ideal conservative system, although the total energy of the system remains constant, there is a continuous exchange of energy between the potential and the kinetic energies of the vibrating system. For a nonconservative system in the presence of damping, energy is dissipated from the system that ultimately stops the motion. Generation of damping forces is dependent on the nature of the oscillating body as well as the medium in which the motion takes place and the damping mechanism is characterized accordingly.[20]

Table 9.7 Natural frequencies of a free-free square composite plate

	Natural frequencies ω, rad/s			
	Quadrilateral element		Triangular element	
Mode number	Zero temperature	Varying temperature	Zero temperature	Varying temperature
1–6	0.00	0.00	0.00	0.00
7	1368.08	1566.65	1320.18	1503.57
8	1654.79	1762.14	1584.41	1683.26
9	3631.85	3749.15	3506.99	3610.58
10	3856.04	3857.21	3702.82	3705.50
11	3968.35	4192.31	3858.03	4074.09
12	5838.15	5888.37	5630.24	5673.17

The *structural* damping manifests itself in the presence of material internal friction as well as friction at joints connected to structural elements. For elastic bodies the jth damping force f_{Dj} is proportional to the internal displacement and hence the elastic force f_{Ej}, acting opposite to the direction of the velocity vector $\dot{u}_j$, and is expressed as

$$f_{Dj} = i^* g f_{Ej} \tag{9.72}$$

g being a constant. The viscous damping on the other hand, occurs because of the motion of a body in a fluid; typical examples are hydraulic dashpots and bodies sliding on a lubricated surface. Then the jth viscous damping force is expressed as

$$f_{Dj} = c_j \dot{u}_j \tag{9.73}$$

in which c_j is a constant representing the jth viscous damping mechanism. A special case of viscous damping is *Coulomb* damping, usually caused by a body sliding on a dry surface and can be expressed as

$$f_D = \mu N_p \tag{9.74}$$

where μ is the coefficient of kinetic friction, N_p being the normal pressure between the sliding body and the surface on which the motion takes place. Although for viscous damping the amplitudes of vibration decay exponentially, for Coulomb damping the decay is linear in nature, noting that for the latter case the damping force is almost constant. When energy is added to a vibrating system instead of the same dissipating from the system, it results in negative damping that is characterized by a growth of amplitudes resulting in eventual destruction of the system. A typical example of such a phenomenon is aeroelastic instability such as flutter, which will be discussed in a later chapter. Assuming linear damping, the equation of free vibration for structural damping may be obtained from consideration of Lagrange's equations of motion as

$$[\boldsymbol{K}(1 + i^* g)]\boldsymbol{q} + \boldsymbol{M}\ddot{\boldsymbol{q}} = \boldsymbol{0} \tag{9.75}$$

For viscous damping, a dashpot represents the phenomenon in which the damping force is linearly proportional to the velocity differential between the piston and the cylinder containing the fluid. The equation of motion may be written as

$$\boldsymbol{K}\boldsymbol{q} + \boldsymbol{C}\dot{\boldsymbol{q}} + \boldsymbol{M}\ddot{\boldsymbol{q}} = \boldsymbol{0} \tag{9.76}$$

in which $\boldsymbol{C} = \boldsymbol{C}_D$ is the symmetric viscous damping matrix.

For many practical problems the damping forces may not be linear functions of displacement or velocity, and also it is difficult to derive expressions for these values. It is customary to express the damping forces in an engineering system in terms of viscous damping calculated on the basis of equivalent energy dissipation. For dynamic response analysis it is customary and advantageous to express the matrix $\boldsymbol{C}$ as being proportional to $\boldsymbol{K}$ and $\boldsymbol{M}$ matrices as

$$\boldsymbol{C} = 2\alpha\boldsymbol{K} + 2\beta\boldsymbol{M} \tag{9.77}$$

in which α and β are real constants. The factor of 2 is used for the convenient solution of the associated differential equation of motion. Alternatively it may also be expressed in terms of a certain percentage of critical damping. By solving the equation of motion of a single degree-of-freedom vibrating system it can be shown that critical damping is given as $C_{\text{crit}} = 2\omega m$ and any damping value may be expressed as $2r\omega m$, r being the ratio of the actual damping to the critical damping value. If the damping has a value higher than the critical damping, the roots are negative with no vibration occurring in the system. For $C < C_{\text{crit}}$, the motion is oscillatory with gradual decay, the roots and vectors occurring in complex conjugate pairs; the real part of the roots relates to the exponential decay of the oscillation of amplitudes, whereas the imaginary part represents the oscillatory motion. Similar expressions can be derived for multiple degree-of-freedom systems in terms of the generalized mass matrix to be presented in a later section. Various forms of damping are shown in Fig. 9.8.

9.8 Solution of Damped Free Vibration Problem

To achieve an efficient solution for a structural system with general damping the related equation of free vibration

$$\boldsymbol{K}\boldsymbol{q} + \boldsymbol{C}\dot{\boldsymbol{q}} + \boldsymbol{M}\ddot{\boldsymbol{q}} = \boldsymbol{0} \tag{9.78}$$

may be combined with the identity $\boldsymbol{M}\dot{\boldsymbol{q}} - \boldsymbol{M}\dot{\boldsymbol{q}} = \boldsymbol{0}$

$$\left(\begin{bmatrix} -\boldsymbol{M} & \boldsymbol{0} \\ \boldsymbol{0} & \boldsymbol{K} \end{bmatrix} \begin{Bmatrix} \dot{\boldsymbol{q}} \\ \boldsymbol{q} \end{Bmatrix} + \begin{bmatrix} \boldsymbol{0} & \boldsymbol{M} \\ \boldsymbol{M} & \boldsymbol{C} \end{bmatrix}\right) \begin{Bmatrix} \ddot{\boldsymbol{q}} \\ \dot{\boldsymbol{q}} \end{Bmatrix} = \boldsymbol{0} \tag{9.79}$$

which can also be rewritten as

$$\tilde{\boldsymbol{E}}\boldsymbol{y} + \tilde{\boldsymbol{F}}\dot{\boldsymbol{y}} = \boldsymbol{0} \tag{9.80}$$

with $\boldsymbol{y} = [\dot{\boldsymbol{q}}\ \boldsymbol{q}]^T$.

A solution of Eq. (9.80) may be taken as

$$\boldsymbol{y} = \boldsymbol{\gamma}e^{pt} \tag{9.81}$$

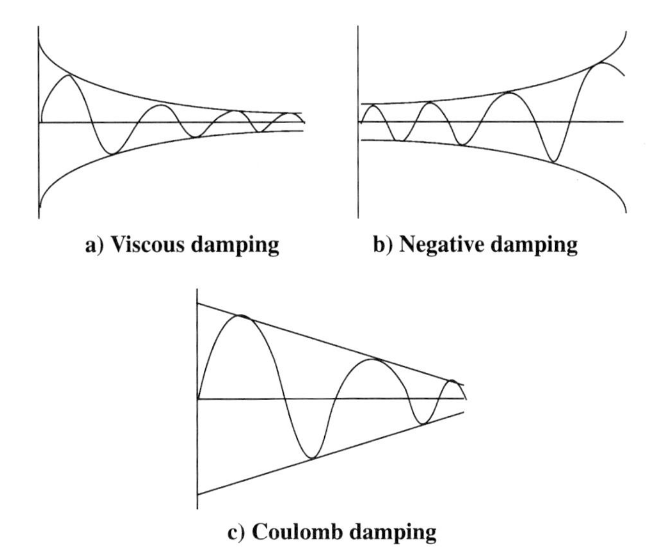

Fig. 9.8 Free vibration of damped system.

yielding the following eigenvalue problem:

$$(\tilde{\boldsymbol{E}} + p\tilde{\boldsymbol{F}})\boldsymbol{\gamma} = \boldsymbol{0} \tag{9.82}$$

The solution yields the natural frequencies p and the associated vectors $\boldsymbol{\gamma}$. As for the undamped case, the eigenvectors for the damped case are also orthogonal, and typically,

$$\boldsymbol{\gamma}_r^T \tilde{\boldsymbol{E}} \boldsymbol{\gamma}_s = \boldsymbol{0}, \qquad \boldsymbol{\gamma}_r^T \tilde{\boldsymbol{F}} \boldsymbol{\gamma}_s = \boldsymbol{0} \tag{9.83}$$

For an underdamped system the eigenvalues will have a negative real part representing damping and an imaginary part that represents the frequency of damped vibration. For a critically damped or overdamped system, roots will be real and negative. A relationship is also applicable to the two complex conjugate eigenvectors associated with a typical single mode, since the related eigenvalues are different in the sense that they are also complex conjugate in nature. Furthermore, assuming an underdamped system usually encountered in practice, the orthogonality relations also can be derived for these vectors in terms of its real and imaginary parts. Thus expressing two typical vectors as

$$\boldsymbol{\gamma}_r = \boldsymbol{\xi}_r + i^* \boldsymbol{\zeta}_r \tag{9.84}$$

$$\boldsymbol{\gamma}_s = \boldsymbol{\xi}_s + i^* \boldsymbol{\zeta}_s \tag{9.85}$$

it can be shown that

$$\boldsymbol{\xi}_r^T \tilde{\boldsymbol{F}} \boldsymbol{\xi}_r = 0 \tag{9.86}$$

$$\boldsymbol{\zeta}_r^T \tilde{\boldsymbol{F}} \boldsymbol{\zeta}_r = 0 \tag{9.87}$$

and also

$$\boldsymbol{\xi}_r^T \tilde{\boldsymbol{F}} \boldsymbol{\zeta}_s = 0 \tag{9.88}$$

$$\boldsymbol{\zeta}_r^T \tilde{\boldsymbol{F}} \boldsymbol{\xi}_s = 0 \tag{9.89}$$

and furthermore

$$\boldsymbol{\xi}_r^T \tilde{\boldsymbol{F}} \boldsymbol{\xi}_r = -\boldsymbol{\zeta}_r^T \tilde{\boldsymbol{F}} \boldsymbol{\zeta}_r \tag{9.90}$$

Similar relationships are obtained with $\tilde{\boldsymbol{E}}$ as the weighting matrix. Also, by virtue of the orthogonality conditions expressed by Eq. (9.83), the uncoupling of the dynamic equations of motion is assured. Thus expressing the first few typical eigenvectors of the damped system as

$$\boldsymbol{\Gamma} = [\boldsymbol{\gamma}_1 \bar{\boldsymbol{\gamma}}_1 \cdots \boldsymbol{\gamma}_r \bar{\boldsymbol{\gamma}}_r] \tag{9.91}$$

one may obtain

$$\hat{\boldsymbol{E}} = \boldsymbol{\Gamma}_r^T \tilde{\boldsymbol{E}} \boldsymbol{\Gamma} \tag{9.92}$$

$$\hat{\boldsymbol{F}} = \boldsymbol{\Gamma}_r^T \tilde{\boldsymbol{F}} \boldsymbol{\Gamma} \tag{9.93}$$

where $\hat{\boldsymbol{E}}$ and $\hat{\boldsymbol{F}}$ are now diagonal matrices. The procedure becomes even more complicated for damped spinning structures as the matrix $\tilde{\boldsymbol{F}}$ loses its symmetry. A general solution algorithm for damped spinning as well as nonspinning structures that employs the individual matrices in their original form and enables computation of roots and associated vectors lying within a specified range is described in Ref. 3. Thus rewriting the general equation of free vibration as in Eq. (9.26)

$$(\boldsymbol{E} - \lambda \boldsymbol{F})\boldsymbol{\phi} = \mathbf{0}$$

in which

$$\boldsymbol{E} = \begin{bmatrix} \boldsymbol{M} & \mathbf{0} \\ \mathbf{0} & \boldsymbol{K} \end{bmatrix} \tag{9.94}$$

$$\boldsymbol{F} = \begin{bmatrix} \mathbf{0} & -i^*\boldsymbol{M} \\ i^*\boldsymbol{M} & i^*\boldsymbol{C} \end{bmatrix} \tag{9.95}$$

$$\boldsymbol{\phi} = [\dot{\boldsymbol{\psi}} \; \boldsymbol{\psi}]^T \tag{9.96}$$

and $\boldsymbol{K} = \boldsymbol{K}_E(1 + i^*g)$, $\boldsymbol{C} = \boldsymbol{C}_C + \boldsymbol{C}_d$. The procedure employs a single-vector iteration scheme based on an inverse iteration procedure, and a modification of the same yields a multivector iteration process similar to that described in Sec. 9.2.

Previously, for undamped structures, an uncoupled system of free vibration was obtained from the modal orthogonality relationship. A similar uncoupling may also be obtained for the damped system by first writing the equation as

$$\boldsymbol{\Gamma}^T \boldsymbol{K} \boldsymbol{\Gamma} \boldsymbol{q} + \boldsymbol{\Gamma}^T \boldsymbol{C} \boldsymbol{\Gamma} \dot{\boldsymbol{q}} + \boldsymbol{\Gamma}^T \boldsymbol{M} \boldsymbol{\Gamma} \ddot{\boldsymbol{q}} = \mathbf{0} \tag{9.97}$$

$\boldsymbol{\Gamma}$ being the complex eigenvector of the damped system. However, for special cases the real vectors $\boldsymbol{\Phi}$ of the undamped system may be used for decoupling the equations of motion. Thus from Eq. (9.8), $\hat{\boldsymbol{M}} = \boldsymbol{\Phi}^T \boldsymbol{M} \boldsymbol{\Phi}$, $\hat{\boldsymbol{K}} = \boldsymbol{\Lambda} \hat{\boldsymbol{M}} = \boldsymbol{\Phi}^T \boldsymbol{K} \boldsymbol{\Phi}$ are diagonal matrices. Then

$$\hat{\boldsymbol{C}} = \boldsymbol{\Phi}^T \boldsymbol{C} \boldsymbol{\Phi} \tag{9.98}$$

can only be uncoupled if $\boldsymbol{C}$ is proportional to either $\boldsymbol{M}$ or $\boldsymbol{K}$ or a combination thereof as

$$\boldsymbol{C} = \alpha \boldsymbol{K} + 2\beta \boldsymbol{M} \tag{9.99}$$

α and β being constants of proportionality, and then $\boldsymbol{C}$ is referred to as proportional viscous damping. Alternatively, such a decoupling is also effected by the use of critical damping[21] as

$$\boldsymbol{C} = 2\boldsymbol{\mu} \hat{\boldsymbol{M}} \boldsymbol{\Omega} \tag{9.100}$$

where $\boldsymbol{\Omega}$ is the diagonal matrix containing frequencies ω_r, and $\boldsymbol{\mu}$ is the diagonal matrix in which a typical term represents a percentage of critical damping for the corresponding elastic mode; $\hat{\boldsymbol{M}}$ is the generalized mass matrix $\hat{\boldsymbol{M}} = \boldsymbol{\Phi}^T \boldsymbol{M} \boldsymbol{\Phi}$.

Conceptually a structure with proportional linear viscous damping vibrates freely in a set of uncoupled modes resembling in shape that of the undamped system with amplitudes decaying exponentially at the same rate over the entire body, until it comes to rest. In the case of nonproportional linear viscous damping the structure may also vibrate freely as uncoupled modes in which, however, every point in the body undergoes exponentially damped motion with differing phase angles, although at the same frequency.

Example 9.7: Damped Spinning Cantilever Beam

The spinning cantilever beam, shown in Fig. 9.6, is again chosen as an example. The basic elastic properties of the beam, divided in 10 discrete elements of length l and expressed in the inch-pound-second unit system, are as follows:

Moment of inertia (Y axis)	=	$\frac{1}{12}$
Moment of inertia (Z axis)	=	$\frac{1}{24}$
Area of cross section	=	1.0
Young's modulus	=	30×10^6
Nodal mass in translation	=	1
Nodal mass moment of inertia	=	$\frac{1}{35}$
Scalar viscous damping	=	0.628318
Structural damping parameter (g)	=	0.01
Element length (l)	=	6

The direction of the X axis is chosen along the length of the beam. Free vibration analyses were performed for a number of problem areas and results are summarized in Table 9.8. All numerical examples in this chapter were solved using the STARS[22] computer software.

Table 9.8 Natural frequencies of a spinning cantilever beam

Mode	Structure without damping	Structure with viscous damping	Structure with viscous and structural damping
1	2.526	$-0.3107 \pm 2.4885i^*$	$-0.3195 \pm 2.4820i^*$
2	3.449	$-0.3116 \pm 3.4207i^*$	$-0.3255 \pm 3.4132i^*$
3	15.397	$-0.3166 \pm 15.3900i^*$	$-0.3930 \pm 15.3831i^*$
4	21.706	$-0.3166 \pm 21.7015i^*$	$-0.4243 \pm 21.6914i^*$
5	43.161	$-0.3202 \pm 43.1551i^*$	$-0.4849 \pm 43.0627i^*$
6	60.951	$-0.3202 \pm 60.9495i^*$	$-0.6246 \pm 60.9391i^*$

9.9 Concluding Remarks

A number of eigenproblem solution techniques were presented in this chapter for nonspinning as well as spinning structures, with or without damping. Three solution techniques, namely PSI (progressive simultaneous iteration), Lanczos, and SS/II (combined Sturm sequence and inverse iteration) were described in detail, and comparisons were made on their related efficiencies by solving representative problems. These studies indicate much promise for the PSI technique.

References

[1]Wilkinson, J. H., *The Algebraic Eigenvalue Problem*, Clarendon Press, Oxford, England, U.K., 1965.

[2]Gupta, K. K., "Eigenproblem Solution by a Combined Sturm Sequence and Inverse Iteration Technique," *International Journal for Numerical Methods in Engineering*, Vol. 7, 1973, pp. 17–42.

[3]Gupta, K. K., "Development of a Unified Numerical Procedure for Free Vibration Analysis of Structures," *International Journal for Numerical Methods in Engineering*, Vol. 8, 1974, pp. 877–911.

[4]Jennings, A., and Orr, D. R. L., "Application of the Simultaneous Iteration Method to Undamped Vibration Problem," *International Journal for Numerical Methods in Engineering*, Vol. 3, 1971, pp. 13–24.

[5]Bathe, K. J., and Wilson, E. L., "Solution Methods for Eigenvalue Problems in Structural Mechanics," *International Journal for Numerical Methods in Engineering*, Vol. 6, 1973, pp. 213–226.

[6]Gupta, K. K., and Lawson, C. L., "Structural Vibration Analysis by a Progressive Simultaneous Iteration Method," *Proceedings of the Royal Society of London, Series A: Mathematical and Physical Sciences*, Vol. 455, 1999, pp. 3415–3424.

[7]Lanczos, C., "An Iteration Method for the Solution of the Eigenvalue Problem of Linear Differential and Integral Operation," *Journal of Research of the National Bureau of Standards*, Vol. 45, 1950, pp. 255–282.

[8]Parlett, B. N., and Scott, D. S., "The Lanczos Algorithm with Selective Orthonormalization," *Math. Comp.*, Vol. 33, 1979, pp. 217–238.

[9]Gupta, K. K., "Free Vibration Analysis of Spinning Structural Systems," *International Journal for Numerical Methods in Engineering*, Vol. 5, 1973, pp. 395–418.

[10]Gupta, K. K., "Formulation of Numerical Procedures for Dynamic Analysis of Spinning

Structures," *International Journal for Numerical Methods in Engineering*, Vol. 23, 1986, pp. 2347–2357.

[11]Parlett, B. N., *The Symmetric Eigenvalue Problem*, Prentice–Hall, Englewood Cliffs, NJ, 1980.

[12]Nour-Omid, B., "Lanczos Method for Heat Conduction Analysis," *International Journal for Numerical Methods in Engineering*, Vol. 24, 1987, pp. 251–261.

[13]Nour-Omid, B., Parlett, B. N., and Taylor, R. L., "Lanczos Versus Subspace Iteration for Solution of Eigenvalue Problems," *International Journal for Numerical Methods in Engineering*, Vol. 19, 1983, pp. 859–871.

[14]Bauchau, O. A., "A Solution of the Eigenproblem for Gyroscopic Systems with the Lanczos Algorithm," *International Journal for Numerical Methods in Engineering*, Vol. 23, 1986, pp. 1705–1713.

[15]Gupta, K. K., and Lawson, C. L., "Development of a Block Lanczos Algorithm for Free Vibration Analysis of Spinning Structures," *International Journal for Numerical Methods in Engineering*, Vol. 26, 1988, pp. 1029–1037.

[16]Gupta, K. K., "On a Finite Dynamic Element Method for Free Vibration Analysis of Structures," *Computer Methods in Applied Mechanics and Engineering*, Vol. 9, 1976, pp. 105–120.

[17]Przemieniecki, J. S., "Quadratic Matrix Equations for Determining Vibration Modes and Frequencies of Continuous Elastic System," *Proceedings of the Conference on Matrix Methods in Structural Mechanics*, AFFDL TR 66-8, Wright–Patterson AFB, OH, 1966.

[18]Gupta, K. K., "Solution of Quadratic Matrix Equations for Free Vibration Analysis of Structures," *International Journal for Numerical Methods in Engineering*, Vol. 6, 1973, pp. 129–135.

[19]Gupta, K. K., "Development of a Unified Numerical Procedure for Free Vibration Analysis of Structures," *International Journal for Numerical Methods in Engineering*, Vol. 17, 1981, pp. 187–198.

[20]Hurty, W. C., and Rubinstein, M. F., *Dynamics of Structures*, Prentice–Hall, Englewood Cliffs, NJ, 1968.

[21]Przemieniecki, J. S., *Theory of Matrix Structural Analysis*, McGraw–Hill, New York, 1968.

[22]Gupta, K. K., "STARS-An Integrated, Multidisciplinary, Finite Element, Structural, Fluids, Aeroelastic, and Aeroservoelastic Analysis Computer Program," NASA TM 4795, May 1997.

10
Dynamic Response of Elastic Structures

10.1 Introduction

This chapter deals with the computation of the response of a structural system subjected to time-dependent dynamic loading. Usually the required response includes displacements and stresses. The choice of a relevant analysis technique is dictated by the nature of the external excitation shown in Fig. 10.1. For deterministic loading, when the forcing functions are known, the response is also deterministic and can be directly calculated either by the modal superposition method or by a step-by-step direct integration process. Associated external excitation may be general in nature, including periodic and aperiodic forcing functions.

In the modal superposition method any arbitrarily varying load function may be considered as a summation of a series of small loading steps and the response for the total force is taken as the cumulative response of the individual steps, which is permissible for any linear system. Response due to a small individual loading step can be calculated using Duhamel's integrals, which have been determined for a number of typical forcing functions. In the direct integration method, a step-by-step recurrence formulation is adapted for each incremental time step yielding a procedure that is capable of solving a variety of related linear and nonlinear problems. In the frequency response method, the system is characterized by its response to a simple forcing function that is harmonic in nature. This technique can be used for both deterministic as well as random excitation. In practice, random loads occur from a variety of effects. For example, for aeronautical problems the structure may be subjected to dynamic loads caused by rapid maneuvering, gusts, and landing.

10.2 Method of Modal Superposition

10.2.1 Undamped System

The differential equation of motion for an undamped system is written as

$$\boldsymbol{K}\boldsymbol{q} + \boldsymbol{M}\ddot{\boldsymbol{q}} = \boldsymbol{p}(t) \tag{10.1}$$

in which $\boldsymbol{p}$ is the time-dependent external forcing function. These equations are coupled because of the existence of large numbers of nonzero off-diagonal terms in the $\boldsymbol{K}$ and $\boldsymbol{M}$ matrices that render its solution impractical. However, uncoupling of these matrices can be easily achieved by employing the modal method. A free vibration equation of the system is obtained by substituting

$$\boldsymbol{q} = \boldsymbol{\phi} e^{i\omega t} \tag{10.2}$$

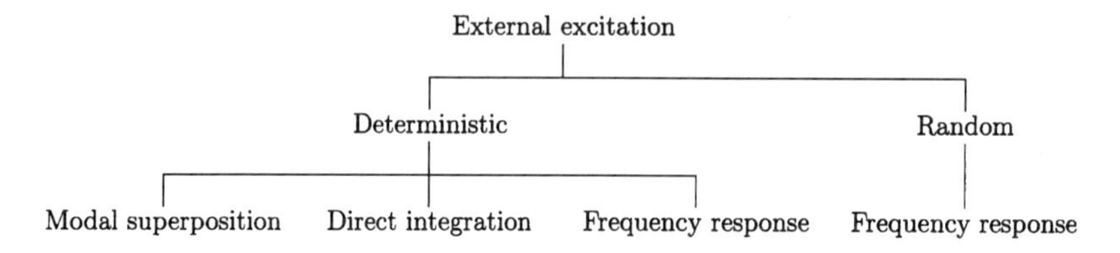

Fig. 10.1 Analysis procedures for different excitation types.

in Eq. (10.1), yielding the eigenvalue problem

$$(\boldsymbol{K} - \omega^2 \boldsymbol{M})\boldsymbol{\phi} = \boldsymbol{0} \tag{10.3}$$

an appropriate solution of which yields eigenvalues ω and the vectors $\boldsymbol{\phi}$. For unconstrained structures, e.g., aircraft, there are typically NZ number of modes $\boldsymbol{\Phi}_R$ of frequencies zero (up to six rigid-body modes) and a set of NE number of elastic frequencies and modes $\boldsymbol{\Phi}_E$.

A modal transformation is next effected as

$$\boldsymbol{q} = \boldsymbol{\Phi}\boldsymbol{\eta} = [\boldsymbol{\Phi}_R \quad \boldsymbol{\Phi}_E]\begin{bmatrix} \eta_R \\ \eta_E \end{bmatrix} \tag{10.4}$$

and the equations of motion become

$$\boldsymbol{K}(\boldsymbol{\Phi}_R\eta_R + \boldsymbol{\Phi}_E\eta_E) + \boldsymbol{M}(\boldsymbol{\Phi}_R\ddot{\eta}_R + \boldsymbol{\Phi}_E\ddot{\eta}_E) = \boldsymbol{p}(t) \tag{10.5}$$

To assess the effect of rigid-body motion, Eq. (10.5) is premultiplied by $\boldsymbol{\Phi}_R^T$, yielding

$$\boldsymbol{\Phi}_R^T\boldsymbol{K}\boldsymbol{\Phi}_R\eta_R + \boldsymbol{\Phi}_R^T\boldsymbol{K}\boldsymbol{\Phi}_E\eta_E + \boldsymbol{\Phi}_R^T\boldsymbol{M}\boldsymbol{\Phi}_R\ddot{\eta}_R + \boldsymbol{\Phi}_R^T\boldsymbol{M}\boldsymbol{\Phi}_E\ddot{\eta}_E = \boldsymbol{\Phi}_R^T\boldsymbol{p}(t) \tag{10.6}$$

which reduces to

$$\ddot{\eta}_R = \boldsymbol{M}_R^{-1}\boldsymbol{\Phi}_R^T\boldsymbol{p}(t) \tag{10.7}$$

because of the orthogonality conditions

$$\boldsymbol{\Phi}_R^T\boldsymbol{M}\boldsymbol{\Phi}_E = 0, \qquad \boldsymbol{\Phi}_R^T\boldsymbol{M}\boldsymbol{\Phi}_R = \boldsymbol{I} \tag{10.8}$$

$$\boldsymbol{\Phi}_R^T\boldsymbol{K}\boldsymbol{\Phi}_R = \boldsymbol{\Phi}_R^T\boldsymbol{K}\boldsymbol{\Phi}_E = 0 \tag{10.9}$$

since rigid-body modes do not produce any restraining force. Also

$$\boldsymbol{M}_R = \boldsymbol{\Phi}_R^T\boldsymbol{M}\boldsymbol{\Phi}_R \tag{10.10}$$

is defined as the generalized mass. Similarly for elastic modes, Eq. (10.5) is premultiplied by $\boldsymbol{\Phi}_E^T$ yielding

$$\boldsymbol{\Phi}_E^T\boldsymbol{K}\boldsymbol{\Phi}_R\eta_R + \boldsymbol{\Phi}_E^T\boldsymbol{K}\boldsymbol{\Phi}_E\eta_E + \boldsymbol{\Phi}_E^T\boldsymbol{M}\boldsymbol{\Phi}_R\ddot{\eta}_R + \boldsymbol{\Phi}_E^T\boldsymbol{M}\boldsymbol{\Phi}_E\ddot{\eta}_E = \boldsymbol{\Phi}_E^T\boldsymbol{p}(t) \tag{10.11}$$

which also reduces to

$$\ddot{\eta}_E + \boldsymbol{\Omega}^2\eta_E = \boldsymbol{M}_E^{-1}\boldsymbol{\Phi}_E^T\boldsymbol{p}(t) \tag{10.12}$$

by using orthogonality and other relationships as for the rigid-body case and also noting that $\boldsymbol{\Phi}_E^T\boldsymbol{K}\boldsymbol{\Phi}_E = \boldsymbol{M}_E\boldsymbol{\Omega}^2$; the generalized mass matrix $\boldsymbol{M}_E = \boldsymbol{\Omega}_E^T\boldsymbol{M}\boldsymbol{\Omega}_E$,

$\boldsymbol{\Omega}^2$ is a diagonal matrix storing all ω_i^2, and $i = 1$, NR, with NR being the number of required roots.

Solution of Eq. (10.7) can be obtained by direct integration, assuming an incremental pulse load as

$$\boldsymbol{\eta}_R = \boldsymbol{M}_R^{-1} \int_{\tau_2=0}^{\tau_2=t} \int_{\tau_1=0}^{\tau_1=\tau_2} \boldsymbol{p}_R(\tau_1)\,\mathrm{d}\tau_1\,\mathrm{d}\tau_2 + \boldsymbol{\eta}_R(0) + \dot{\boldsymbol{\eta}}_R(0)t \tag{10.13}$$

in which $\boldsymbol{\eta}_R(0)$ and $\dot{\boldsymbol{\eta}}_R(0)$ are integration constants to be determined from initial conditions; $\boldsymbol{p}_R(t) = \boldsymbol{\Phi}_R^T \boldsymbol{p}(t)$ are the generalized forces for the rigid-body modes. Effect of elastic modes, on the other hand, is obtained by solving Eq. (10.9), employing Duhamel's integral for a unit step increase

$$\boldsymbol{\eta}_E = \boldsymbol{M}_E^{-1}\boldsymbol{\Omega}^{-1} \int_0^t \mathbf{sin}[\omega(t-\tau)]\boldsymbol{p}_E(\tau)\,\mathrm{d}\tau + \boldsymbol{\Omega}^{-1}\mathbf{sin}(\omega t)\dot{\boldsymbol{\eta}}_E(0) + \mathbf{cos}(\omega t)\boldsymbol{\eta}_E(0) \tag{10.14}$$

in which the generalized force $\boldsymbol{p}_E(t) = \boldsymbol{\Phi}_E^T \boldsymbol{p}(t)$ and $\mathbf{sin}(\omega t)$, and so on, are typically diagonal matrices of order NE:

$$\mathbf{sin}(\omega t) = \lceil \sin \omega_{\mathrm{NZ}+1} t \, \sin \omega_{\mathrm{NZ}+2} t \cdots \sin \omega_{\mathrm{NZ}+\mathrm{NE}} t \rfloor \tag{10.15}$$

Substituting Eqs. (10.13) and (10.14) into Eq. (10.4) yields the expression for displacement as

$$\begin{aligned}
\boldsymbol{q} &= \boldsymbol{\Phi}_R \boldsymbol{\eta}_R + \boldsymbol{\Phi}_E \boldsymbol{\eta}_E \\
&= \boldsymbol{\Phi}_R \boldsymbol{\eta}_R(0) + \boldsymbol{\Phi}_R \dot{\boldsymbol{\eta}}_R(0)t + \boldsymbol{\Phi}_E \boldsymbol{\Omega}^{-1}\mathbf{sin}(\omega t)\dot{\boldsymbol{\eta}}_E(0) + \boldsymbol{\Phi}_E \mathbf{cos}(\omega t)\boldsymbol{\eta}_E(0) \\
&\quad + \boldsymbol{M}_R^{-1}\boldsymbol{\Phi}_R \int_{\tau_2=0}^{\tau_2=t} \int_{\tau_1=0}^{\tau_1=\tau_2} \boldsymbol{p}_R(\tau_1)\,\mathrm{d}\tau_1\,\mathrm{d}\tau_2 \\
&\quad + \boldsymbol{\Phi}_E \boldsymbol{M}_E^{-1}\boldsymbol{\Omega}^{-1} \int_0^t \mathbf{sin}[\omega(t-\tau)]\boldsymbol{p}_E(\tau)\,\mathrm{d}\tau
\end{aligned} \tag{10.16}$$

The initial conditions may next be derived in terms of displacements as

$$\begin{aligned}
\boldsymbol{\eta}_R(0) &= \boldsymbol{M}_R^{-1}\boldsymbol{\Phi}_R^T \boldsymbol{M}\boldsymbol{q}(0) \qquad & \dot{\boldsymbol{\eta}}_R(0) &= \boldsymbol{M}_R^{-1}\boldsymbol{\Phi}_R^T \boldsymbol{M}\dot{\boldsymbol{q}}(0) \\
\boldsymbol{\eta}_E(0) &= \boldsymbol{M}_E^{-1}\boldsymbol{\Phi}_E^T \boldsymbol{M}\boldsymbol{q}(0) \qquad & \dot{\boldsymbol{\eta}}_E(0) &= \boldsymbol{M}_E^{-1}\boldsymbol{\Phi}_E^T \boldsymbol{M}\dot{\boldsymbol{q}}(0)
\end{aligned} \tag{10.17}$$

which when substituted in Eq. (10.16) yields the final form of displacement $\boldsymbol{q}$ expressed as[1]

$$\begin{aligned}
\boldsymbol{q} &= \boldsymbol{\Phi}_R \boldsymbol{M}_R^{-1}\boldsymbol{\Phi}_R^T \boldsymbol{M}\boldsymbol{q}(0) + \boldsymbol{\Phi}_R \boldsymbol{M}_R^{-1}\boldsymbol{\Phi}_R^T \boldsymbol{M}\dot{\boldsymbol{q}}(0)t \\
&\quad + \boldsymbol{\Phi}_E \boldsymbol{\Omega}^{-1}\mathbf{sin}(\omega t)\boldsymbol{M}_E^{-1}\boldsymbol{\Phi}_E^T \boldsymbol{M}\dot{\boldsymbol{q}}(0) + \boldsymbol{\Phi}_E \mathbf{cos}(\omega t)\boldsymbol{M}_E^{-1}\boldsymbol{\Phi}_E^T \boldsymbol{M}\boldsymbol{q}(0) \\
&\quad + \boldsymbol{\Phi}_R \boldsymbol{M}_R^{-1} \int_{\tau_2=0}^{\tau_2=t} \int_{\tau_1=0}^{\tau_1=\tau_2} \boldsymbol{p}_R(\tau_1)\,\mathrm{d}\tau_1\,\mathrm{d}\tau_2 \\
&\quad + \boldsymbol{\Phi}_E \boldsymbol{\Omega}^{-1}\boldsymbol{M}_E^{-1} \int_0^t \mathbf{sin}[\omega(t-\tau)]\boldsymbol{p}_E(\tau)\,\mathrm{d}\tau
\end{aligned} \tag{10.18}$$

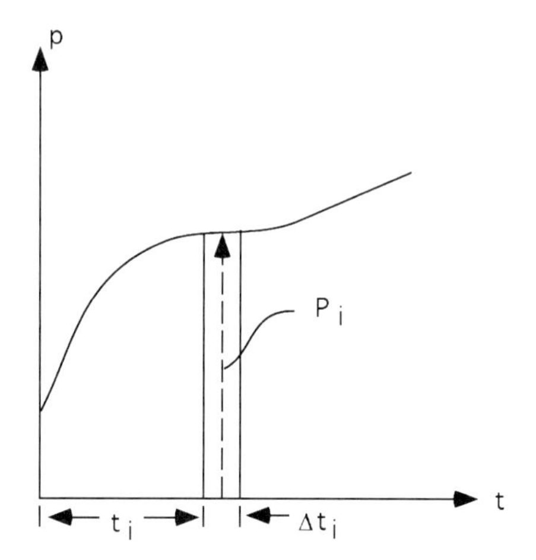

Fig. 10.2 Incremental applied load.

where the first two terms represent rigid-body initial conditions and the next two terms relate to the elastic-body initial conditions; the fifth and sixth terms are the rigid- and elastic-body contributions, respectively. For spinning structures the transpose of the eigenvectors are to be replaced by their tranjugate $\boldsymbol{\Phi}_E^T$ by $\bar{\boldsymbol{\Phi}}_E^T$, and so on. Any arbitrary loading function $\boldsymbol{p}(t)$ may be assumed to consist of a series of small load increments and the cumulative effect of each incremental load constitutes the total dynamic response. For a simple loading function, Duhamel's integrals can be determined directly. Figure 10.2 shows a typical loading function and for a typical incremental time step Duhamel's integral in Eq. (10.18) has the following form:

$$\frac{p_i}{\omega}(1 - \cos \omega t) \qquad \text{for} \qquad t < t_i + \Delta t_i \tag{10.19}$$

$$\frac{p_i}{\omega}[\cos \omega(t - \Delta t_i) - \cos \omega t] \qquad \text{for} \qquad t > t_i + \Delta t_i \tag{10.20}$$

and other expressions are also available in the literature[1] for a variety of step functions. Thus Fig. 10.3 depicts calculation of dynamic responses resulting from a series of incremental step loads approximating a time-dependent loading function. Since the response is inversely proportional to the frequencies, only the first few modes are usually retained in Eq. (10.4) for computation of the dynamic response.

10.2.2 Damped System

In the presence of damping forces, the equation of motion may be uncoupled for convenient solution by the conventional modal method, using undamped modes, only if the forces are proportional to either the mass or the stiffness matrices. For nonproportional damping, uncoupling may only be achieved by using the complex modes derived from the solution of the damped free vibration problem. Thus the equation of motion for a viscously damped structural system is written as

$$\boldsymbol{K}\boldsymbol{q} + \boldsymbol{C}\dot{\boldsymbol{q}} + \boldsymbol{M}\ddot{\boldsymbol{q}} = \boldsymbol{p}(t) \tag{10.21}$$

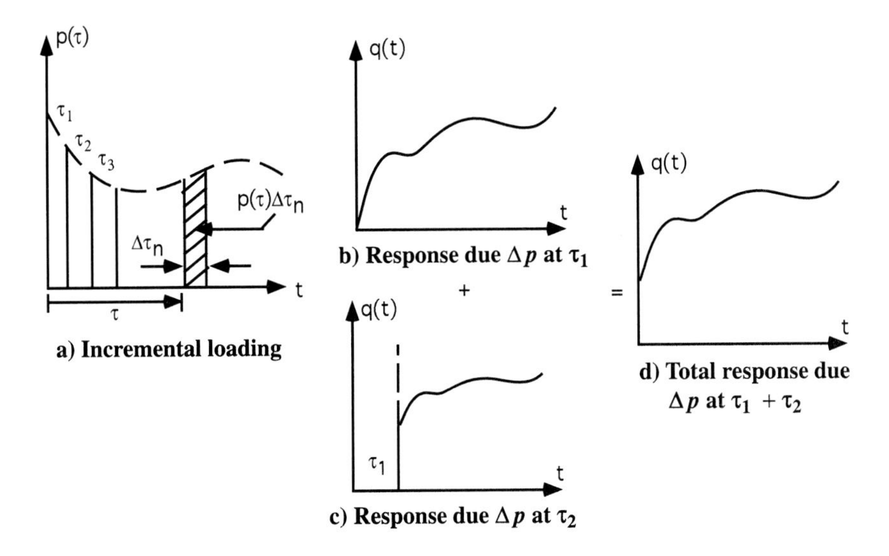

Fig. 10.3 Dynamic response as summation of responses for incremental loads.

in which $\boldsymbol{C}$ is the viscous damping matrix. Proceeding as before for the undamped case, Eq. (10.21) may be rewritten as

$$\boldsymbol{M}_R\ddot{\boldsymbol{\eta}}_R + \boldsymbol{\Phi}_R^T\boldsymbol{C}\boldsymbol{\Phi}_R\dot{\boldsymbol{\eta}}_R + \boldsymbol{\Phi}_R^T\boldsymbol{C}\boldsymbol{\Phi}_E\dot{\boldsymbol{\eta}}_E = \boldsymbol{p}_R(t) \tag{10.22}$$

$$\boldsymbol{M}_E\ddot{\boldsymbol{\eta}}_E + \boldsymbol{\Phi}_E^T\boldsymbol{C}\boldsymbol{\Phi}_R\dot{\boldsymbol{\eta}}_R + \boldsymbol{\Phi}_E^T\boldsymbol{C}\boldsymbol{\Phi}_E\dot{\boldsymbol{\eta}}_E + \boldsymbol{\Omega}^2\boldsymbol{M}_E\boldsymbol{\eta}_E = \boldsymbol{p}_E(t) \tag{10.23}$$

the two equations being uncoupled from each other provided $\boldsymbol{\Phi}_R^T\boldsymbol{C}\boldsymbol{\Phi}_E = \boldsymbol{0}$. Also equations in each set become uncoupled provided $\boldsymbol{\Phi}_R^T\boldsymbol{C}\boldsymbol{\Phi}_R$ and $\boldsymbol{\Phi}_E^T\boldsymbol{C}\boldsymbol{\Phi}_E$ are diagonal. This situation arises when $\boldsymbol{C}$ is proportional to either $\boldsymbol{K}$ or $\boldsymbol{M}$ and also when the generalized damping matrix $\boldsymbol{C}_E = \boldsymbol{\Phi}_E^T\boldsymbol{C}\boldsymbol{\Phi}_E$ is assumed to be a certain percentage of critical damping value.

10.2.2.1 Proportional damping. Assuming that the matrix $\boldsymbol{C}$ is proportional to the mass matrix

$$\boldsymbol{C} = 2\beta\boldsymbol{M} \tag{10.24}$$

and using the same modal transformation relationship of Eq. (10.4), proceeding as in the undamped case and using usual orthogonality relationship including $\boldsymbol{\Phi}_R^T\boldsymbol{C}\boldsymbol{\Phi}_E = \boldsymbol{0}$, the following are obtained from Eqs. (10.22) and (10.23):

$$\boldsymbol{M}_R\ddot{\boldsymbol{\eta}}_R + 2\beta\boldsymbol{M}_R\dot{\boldsymbol{\eta}}_R = \boldsymbol{p}_R(t) \tag{10.25}$$

$$\boldsymbol{M}_E\ddot{\boldsymbol{\eta}}_E + 2\beta\boldsymbol{M}_E\dot{\boldsymbol{\eta}}_E + \boldsymbol{\Omega}^2\boldsymbol{M}_E\boldsymbol{\eta}_E = \boldsymbol{p}_E(t) \tag{10.26}$$

in which the generalized mass and load matrices are defined again as $\boldsymbol{M}_R = \boldsymbol{\Phi}_R^T\boldsymbol{M}\boldsymbol{\Phi}_R$, $\boldsymbol{M}_E = \boldsymbol{\Phi}_E^T\boldsymbol{M}\boldsymbol{\Phi}_E$, $\boldsymbol{p}_R = \boldsymbol{\Phi}_R^T\boldsymbol{p}(t)$, and $\boldsymbol{p}_E = \boldsymbol{\Phi}_E^T\boldsymbol{p}(t)$.

Solution of the preceding two sets of equations may be obtained by using the Laplace transformation and convolution theorem[1] as shown next:

$$\eta_R = \frac{M_R^{-1}}{2\beta}\int_0^t \left(1 - e^{-2\beta(t-\tau)}\right)p_R(\tau)\,\mathrm{d}\tau + \eta_R(0) + \frac{1-e^{-2\beta t}}{2\beta}\dot{\eta}_R(0) \tag{10.27}$$

$$\begin{aligned}\eta_E = {} & \frac{M_E^{-1}}{(\boldsymbol{\Omega}^2 - \beta^2 \boldsymbol{I})^{\frac{1}{2}}}\int_0^t \left(e^{-\beta(t-\tau)}\mathbf{sin}\left[(\omega^2-\beta^2)^{\frac{1}{2}}(t-\tau)\right]p_E(\tau)\,\mathrm{d}\tau\right. \\ & + e^{-\beta t}\mathbf{cos}\left[(\omega^2-\beta^2)^{\frac{1}{2}}t\right]\eta_E(0) + e^{-\beta t}(\boldsymbol{\Omega}^2-\beta^2\boldsymbol{I})^{-\frac{1}{2}} \\ & \times \mathbf{sin}\left[(\omega^2-\beta^2)^{\frac{1}{2}}t\right] \times [\dot{\eta}_E(0) + \beta\eta_E(0)]\end{aligned} \tag{10.28}$$

The expression for the displacement is next obtained by substituting Eqs. (10.27) and (10.28) into Eq. (10.4):

$$q = \boldsymbol{\Phi}_R\eta_R + \boldsymbol{\Phi}_E\eta_E$$

as before. Similarly if the damping matrix is proportional to the stiffness matrix

$$C = 2\alpha K \tag{10.29}$$

in which α is a proportionality constant, the following are obtained using Eqs. (10.22) and (10.23):

$$M_R\ddot{\eta}_R = p_R \tag{10.30}$$

$$M_E\ddot{\eta}_E + 2\alpha M_E\boldsymbol{\Omega}^2\dot{\eta}_E + \boldsymbol{\Omega}^2 M_E\eta_E = p_E \tag{10.31}$$

The solutions of the preceding equations are obtained by application of the Laplace transformation and convolution theorem[1]:

$$\eta_R = M_R^{-1}\int_0^t (t-\tau)p_R(\tau)\,\mathrm{d}\tau + \eta_R(0) + \dot{\eta}_R(0)t \tag{10.32}$$

$$\begin{aligned}\eta_E = {} & \frac{\boldsymbol{\Omega}^{-1}M_E^{-1}}{(\boldsymbol{I}^2 - \alpha^2\boldsymbol{\Omega}^2)^{\frac{1}{2}}}\int_0^t \mathbf{exp}[-\alpha\omega^2(t-\tau)]\mathbf{sin}\left[\omega(1-\alpha^2\omega^2)^{\frac{1}{2}}(t-\tau)\right]p_E(\tau)\,\mathrm{d}\tau \\ & + \mathbf{exp}(-\alpha\omega^2 t)\mathbf{cos}\left[\omega(1-\alpha^2\omega^2)^{\frac{1}{2}}t\right]\eta_E(0) + \mathbf{exp}(-\alpha\omega^2 t) \\ & \times(\boldsymbol{I}-\alpha^2\boldsymbol{\Omega}^2)^{-\frac{1}{2}}\boldsymbol{\Omega}^{-1}\mathbf{sin}\left[\omega(1-\alpha^2\omega^2)^{\frac{1}{2}}t\right][\dot{\eta}_E(0) + \alpha\boldsymbol{\Omega}^2\eta_E(0)]\end{aligned} \tag{10.33}$$

in which the exponential terms are of the form

$$\mathbf{exp}(-\alpha\omega^2 t) = \left[\exp\left(-\alpha\omega^2_{\mathrm{NZ}+1}t\right)\exp\left(-\alpha\omega^2_{\mathrm{NZ}+2}t\right)\cdots\exp\left(-\alpha\omega^2_{\mathrm{NZ}+\mathrm{NE}}t\right)\right] \tag{10.34}$$

and so on, where NZ, NE are the number of rigid-body and elastic modes, respectively, under consideration. Again, the displacement may be calculated by substituting Eqs. (10.32) and (10.33) into Eq. (10.4). For many practical problems the assumption of a single constant α or β to define damping is inadequate for structures characterized by multiple degrees of freedom. It is then more appropriate to express damping in terms of the elastic or generalized degrees of freedom. Thus, the generalized damping may be expressed as

$$\hat{C} = \Phi^T C \Phi = 2\nu \hat{M} \lceil \mathbf{0} \quad \Omega \rfloor = 2 \lceil \mathbf{0} \quad \nu_E \rfloor \hat{M} \lceil \mathbf{0} \quad \Omega \rfloor \tag{10.35}$$

in which

$$\nu_e = \lceil \nu_{\mathrm{NZ}+1} \nu_{\mathrm{NZ}+2} \cdots \nu_{\mathrm{NZ}+\mathrm{NE}} \rfloor$$

The diagonal terms ν_i represent a percentage of critical damping for the ith elastic mode $\Phi_{\mathrm{NZ}+i}$ and $\hat{M}$ is the generalized mass. Applying the relationship of Eq. (10.35) into Eqs. (10.22) and (10.23), the two sets of uncoupled equations are obtained:

$$M_R \ddot{\eta}_R = p_R \tag{10.36}$$

$$M_E \ddot{\eta}_E + 2\nu_E M_E \Omega \dot{\eta}_E + \Omega^2 M_E \eta_E = p_E \tag{10.37}$$

The solution of these equations may be obtained using the Laplace transform procedure:

$$\eta_R = M_R^{-1} \int_0^t (t-\tau) p_R(\tau)\,\mathrm{d}\tau + \eta_0(0) + \dot{\eta}_0(0)t \tag{10.38}$$

$$\begin{aligned} \eta_E = {} & \frac{\Omega^{-1} M_E^{-1}}{(I - \nu_E^2)^{\frac{1}{2}}} \int_0^t \mathbf{exp}[-\nu\omega(t-\tau)]\mathbf{sin}\big[\omega(1-\nu^2)^{\frac{1}{2}}(t-\tau)\big] p_E(\tau)\,\mathrm{d}\tau \\ & + \mathbf{exp}(-\nu\omega t)\mathbf{cos}\big[\omega(1-\nu^2)^{\frac{1}{2}} t\big]\eta_E(0) + \mathbf{exp}(-\nu\omega t) \\ & \times \big(I - \nu_E^2\big)^{-\frac{1}{2}} \Omega^{-1} \mathbf{sin}\big[\omega(1-\nu^2)^{\frac{1}{2}} t\big][\dot{\eta}_E(0) + \nu\Omega\eta_E(0)] \end{aligned} \tag{10.39}$$

As before, displacements are determined by substituting Eqs. (10.38) and (10.39) into Eq. (10.4). The damping matrix C may easily be derived using Eq. (10.35), although it is not required for computation of displacements and stresses by the current procedure.

10.2.2.2 Nonproportional damping. In Chapter 9 it was shown that the modes of free damped motion may be conveniently used to uncouple the equation of motion

$$Kq + C\dot{q} + M\ddot{q} = p(t)$$

To achieve an efficient solution of the preceding equation, it is firstly rewritten as $\tilde{E}y + \tilde{F}\dot{y} = \tilde{p}$, in which

$$\tilde{E} = \begin{bmatrix} -M & \mathbf{0} \\ \mathbf{0} & K \end{bmatrix}, \qquad \tilde{F} = \begin{bmatrix} \mathbf{0} & M \\ M & \mathbf{C} \end{bmatrix}, \qquad y = \begin{Bmatrix} \dot{q} \\ q \end{Bmatrix}, \qquad \tilde{p} = \begin{Bmatrix} \mathbf{0} \\ p \end{Bmatrix} \tag{10.40}$$

and substitution of $\boldsymbol{y} = \boldsymbol{\gamma} e^{\lambda t}$ in the homogeneous equation yields the eigenvalue problem

$$(\tilde{\boldsymbol{E}} + \lambda \tilde{\boldsymbol{F}})\boldsymbol{\gamma} = \boldsymbol{0} \tag{10.41}$$

the solution of which yields the required eigenvalues λ and vectors $\boldsymbol{\gamma}$, occurring in complex conjugate pairs. A transformation of the form

$$\boldsymbol{y} = \boldsymbol{\Upsilon}\boldsymbol{\eta} \tag{10.42}$$

in which the eigenvectors

$$\boldsymbol{\Upsilon} = [\boldsymbol{\gamma}_1 \bar{\boldsymbol{\gamma}}_1 \quad \boldsymbol{\gamma}_2 \bar{\boldsymbol{\gamma}}_2 \quad \cdots \quad \boldsymbol{\gamma}_{\mathrm{NR}} \bar{\boldsymbol{\gamma}}_{\mathrm{NR}}] \tag{10.43}$$

is then applied to the reconstructed dynamic equations of motion:

$$\boldsymbol{\Upsilon}^T \tilde{\boldsymbol{E}} \boldsymbol{\Upsilon} \boldsymbol{\eta} + \boldsymbol{\Upsilon}^T \tilde{\boldsymbol{F}} \boldsymbol{\Upsilon} \dot{\boldsymbol{\eta}} = \boldsymbol{\Upsilon}^T \tilde{\boldsymbol{p}}$$

or

$$\hat{\boldsymbol{E}}\boldsymbol{\eta} + \hat{\boldsymbol{F}}\dot{\boldsymbol{\eta}} = \hat{\boldsymbol{p}} \tag{10.44}$$

resulting in uncoupled equations of motion because $\hat{\boldsymbol{E}}$ and $\hat{\boldsymbol{F}}$ are diagonal matrices of order $2n$. The solution of Eq. (10.44) may again be achieved by using Laplace transforms, and a coordinate transformation of Eq. (10.42) yields the response $\boldsymbol{y}(t)$. Using the relationship of Eq. (10.42) an expression for the required real part only of the jth component of $\boldsymbol{q}(t)$ is obtained as

$$q^{(j)}(t) = 2\sum_{r=1}^{\mathrm{NR}} \frac{|q_r^{(j)}|}{|\hat{F}_r|} \sum_{k=1}^{\mathrm{NR}} |q_r^{(k)}| \int_0^t e^{\alpha_r (t-\tau)} \cos\big[\beta_r (t-\tau) - \psi_r + \theta_r^{(j)} + \theta_r^{(k)}\big] p^{(k)}(\tau)\, \mathrm{d}\tau \tag{10.45}$$

in which $\psi^{(r)}$, $\theta_j^{(r)}$, and $\theta_k^{(r)}$ are phase angles pertaining to $\hat{F}_r$, $q_j^{(r)}$, and $q_k^{(r)}$, respectively, where

$$\begin{aligned} \hat{F}_r &= \boldsymbol{y}_r^T \hat{\boldsymbol{F}} \boldsymbol{y}_r \\ &= |\hat{F}_r| e^{i\psi_r} \\ \hat{E}_r &= \boldsymbol{y}_r^T \hat{\boldsymbol{E}} \boldsymbol{y}_r \end{aligned} \tag{10.46}$$

and $q_r^{(j)} = j$th component of rth vector $\boldsymbol{q}_r = |q_r^{(j)}| e^{i\theta_r^{(j)}}$, and similarly $q_k^{(j)}$, $p^{(k)} = k$th component of $\boldsymbol{p}$, $p_r = -\hat{E}_r/\hat{F}_r = \alpha_r + i\beta_r$.

10.3 Direct Integration Methods

While the modal superposition method is limited for calculation of responses to dynamic excitation in which the lower modes are dominant, the time integration methods (Fig. 10.4) can be effectively used for high frequency eigenspectrum and also for nonlinear systems. This solution technique can be either implicit or explicit and in general can prove to be rather expensive, requiring relatively large computing resources. Furthermore, much care is needed to ensure the stability of the solution convergence. Recurrence formulation, involving step-by-step forward integration, is used for the solution when the time differentials present in the

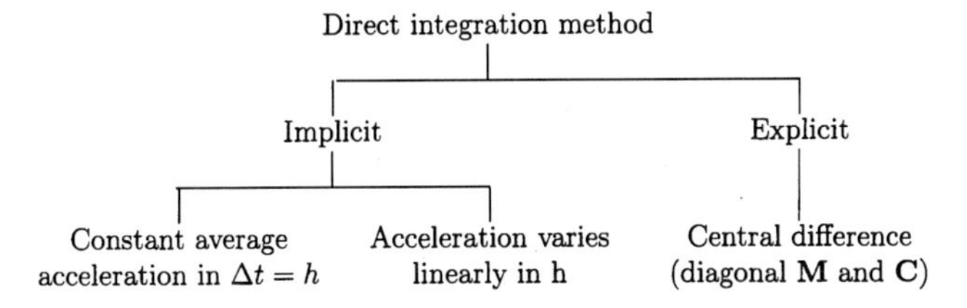

Fig. 10.4 Solution methods using direct integration.

equation of motion are replaced by suitable differences of displacement states at a typical time interval and at time intervals preceding it. Thus the numerical solution of the differential equation involves the replacement of derivatives by finite difference equivalents, in which the continuous lapse of time may be divided in small finite and generally equal time intervals. The choice of the finite difference equivalents directly governs the accuracy and performance of the procedure. In an implicit procedure a set of linear simultaneous algebraic equations needs to be solved for each incremental time step, whereas an explicit procedure involves the solution of a set of uncoupled algebraic equations.

10.3.1 Implicit Methods

In this procedure the solution at $(n+1)$th step for a small incremental time step h is sought using the equilibrium equation at the $(n+1)$th step

$$\boldsymbol{M}\ddot{\boldsymbol{q}}_{n+1} + \boldsymbol{C}\dot{\boldsymbol{q}}_{n+1} + \boldsymbol{K}\boldsymbol{q}_{n+1} = \boldsymbol{p}(t)_{n+1} \tag{10.47}$$

assuming a linear structural system in which $\boldsymbol{M}$, $\boldsymbol{C}$, and $\boldsymbol{K}$ matrices remain invariant, $\boldsymbol{p}$ being the total load. The Newmark method[2] is based on the assumption that the average acceleration $\ddot{\boldsymbol{q}}_{\text{av}} = \frac{1}{2}(\ddot{\boldsymbol{q}}_{n+1} + \ddot{\boldsymbol{q}}_n)$ remains constant within the small incremental time step h. Thus a Taylor's series expansion of the displacements yields

$$\boldsymbol{q}_{n+1} = \boldsymbol{q}_n + h\dot{\boldsymbol{q}}_n + \frac{h^2}{2}\ddot{\boldsymbol{q}}_{\text{av}} + \cdots = \boldsymbol{q}_n + h\dot{\boldsymbol{q}}_n + \frac{h^2}{4}\ddot{\boldsymbol{q}}_n + \frac{h^2}{4}\ddot{\boldsymbol{q}}_{n+1} \tag{10.48}$$

which yields

$$\ddot{\boldsymbol{q}}_{n+1} = \frac{4}{h^2}\left[\boldsymbol{q}_{n+1} - \boldsymbol{q}_n - h\dot{\boldsymbol{q}}_n - \frac{h^2}{4}\ddot{\boldsymbol{q}}_n\right] \tag{10.49}$$

Similarly the velocity vector may also be expressed in Taylor's series as

$$\begin{aligned}\dot{\boldsymbol{q}}_{n+1} &= \dot{\boldsymbol{q}}_n + h\ddot{\boldsymbol{q}}_{\text{av}} \\ &= \dot{\boldsymbol{q}}_n + \frac{h}{2}\ddot{\boldsymbol{q}}_{n+1} + \frac{h}{2}\ddot{\boldsymbol{q}}_n \\ &= \dot{\boldsymbol{q}}_n + \frac{2}{h}[\boldsymbol{q}_{n+1} - \boldsymbol{q}_n - h\dot{\boldsymbol{q}}_n]\end{aligned} \tag{10.50}$$

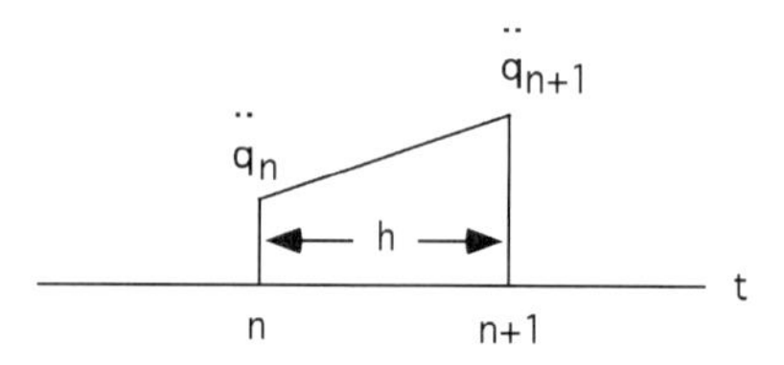

Fig. 10.5 Linear acceleration field.

Substituting Eqs. (10.49) and (10.50) into the equation of motion [Eq. (10.47)] yields the following:

$$\left[\boldsymbol{K} + \frac{2}{h}\boldsymbol{C} + \frac{4}{h^2}\boldsymbol{M}\right]\boldsymbol{q}_{n+1} = \boldsymbol{p}_{n+1} + \boldsymbol{C}\left[\frac{2}{h}\boldsymbol{q}_n + \dot{\boldsymbol{q}}_n\right] + \boldsymbol{M}\left[\frac{4}{h^2}\boldsymbol{q}_n + \frac{4}{h}\dot{\boldsymbol{q}}_n + \ddot{\boldsymbol{q}}_n\right] \tag{10.51}$$

or

$$\hat{\boldsymbol{K}}\boldsymbol{q}_{n+1} = \hat{\boldsymbol{p}}_{n+1} \tag{10.52}$$

which may be conveniently solved from known solution results in nth step to yield the required unknown displacement $\boldsymbol{q}_{n+1}$. It is noted in this connection that, for a varying time step, $\hat{\boldsymbol{K}}$ has to be inverted at each time step, whereas for uniform incremental time steps it is inverted only once in the beginning.

A similar procedure[3] is based on the assumption that acceleration varies linearly (see Fig. 10.5) within the incremental time interval under consideration. This assumption, being a special case of the general Newmark method, results in the following expression for velocity and deformation:

$$\dot{\boldsymbol{q}}_{n+1} = \dot{\boldsymbol{q}}_n + \frac{h}{2}[\ddot{\boldsymbol{q}}_{n+1} + \ddot{\boldsymbol{q}}_n] \tag{10.53}$$

$$\boldsymbol{q}_{n+1} = \boldsymbol{q}_n + h\dot{\boldsymbol{q}}_n + \frac{h^2}{3}\ddot{\boldsymbol{q}}_n + \frac{h^2}{6}\ddot{\boldsymbol{q}}_n \tag{10.54}$$

Substitution of Eqs. (10.53) and (10.54) into Eq. (10.47) yields

$$\left[\frac{h^2}{6}\boldsymbol{K} + \frac{h}{2}\boldsymbol{C} + \boldsymbol{M}\right]\ddot{\boldsymbol{q}}_{n+1} = \boldsymbol{p}_{n+1} - \boldsymbol{C}\left[\dot{\boldsymbol{q}}_n + \frac{h}{2}\ddot{\boldsymbol{q}}_n\right] + \boldsymbol{K}\left[\boldsymbol{q}_n + h\dot{\boldsymbol{q}}_n + \frac{h^2}{3}\ddot{\boldsymbol{q}}_n\right] \tag{10.55}$$

which can be solved for $\ddot{\boldsymbol{q}}_{n+1}$ from known solution results at the preceding time step.

The implicit solution procedures are stable for even large time steps. However, the size of the incremental time step affects the solution accuracy and, in general, should be chosen small enough so that the response in all critical modes is calculated accurately.

10.3.2 Explicit Methods

In any explicit scheme, solution for the $(n+1)$th step is sought from equilibrium equations written for the nth time step. The central difference method based on Taylor's series expansion at nth step is used to obtain the following expressions:

$$\ddot{\boldsymbol{q}}_n = \frac{1}{h^2}[\boldsymbol{q}_{n+1} - 2\boldsymbol{q}_n + \boldsymbol{q}_{n-1}] \tag{10.56}$$

$$\dot{\boldsymbol{q}}_n = \frac{1}{2h}[\boldsymbol{q}_{n+1} - \boldsymbol{q}_{n-1}] \tag{10.57}$$

which is substituted in the equation of motion at nth step

$$\boldsymbol{M}\ddot{\boldsymbol{q}}_n + \boldsymbol{C}\dot{\boldsymbol{q}}_n + \boldsymbol{K}\boldsymbol{q}_n = \boldsymbol{p}_n \tag{10.58}$$

yielding

$$\left[\frac{1}{h^2}\boldsymbol{M} + \frac{1}{2h}\boldsymbol{C}\right]\boldsymbol{q}_{n+1} = \boldsymbol{p}_n - \left[\boldsymbol{K} - \frac{2}{h^2}\boldsymbol{M}\right]\boldsymbol{q}_n - \left[\frac{1}{h^2}\boldsymbol{M} - \frac{1}{2h}\boldsymbol{C}\right]\boldsymbol{q}_{n-1} \tag{10.59}$$

which may be solved to yield the unknown $\boldsymbol{q}_{n+1}$. Assuming diagonal $\boldsymbol{M}$ and $\boldsymbol{C}$ matrices, factorization of the left-hand side requires trivial computational effort. However, the procedure is only conditionally stable for $h \leq h_{\text{cr}} = \omega_1/\pi$. Also, an explicit technique usually needs a special starting procedure as

$$\boldsymbol{q}_{(-1)} = \boldsymbol{q}_0 - h\dot{\boldsymbol{q}}_0 + \frac{h^2}{2}\ddot{\boldsymbol{q}}_0 \tag{10.60}$$

which is essential to obtain a reasonable solution. Backward, central, and forward difference approximations for derivatives are obtained by writing series expressions for $\boldsymbol{q}$ at $t-h$, t, $t+h$, respectively.

A variety of other techniques for dynamic response analysis are given in Refs. 4–6.

10.4 Frequency Response Method

The frequency response method is an alternative procedure for calculating the response of a structure subjected to external excitation. In this method, a system is characterized by its response to a simple harmonic forcing function, which can then be used to determine the structural response due to any nonharmonic periodic or nonperiodic forces, as well as random excitation.

Thus, the Laplace transform that is of the form

$$\mathcal{L}f(t) = \int_0^\infty e^{-st} f(t)\,\mathrm{d}t = f(s) \tag{10.61}$$

is applied to the equation of motion

$$\boldsymbol{K}\boldsymbol{q} + \boldsymbol{C}\dot{\boldsymbol{q}} + \boldsymbol{M}\ddot{\boldsymbol{q}} = \boldsymbol{p}(t)$$

resulting in

$$\boldsymbol{K}\boldsymbol{q}(s) + s\boldsymbol{C}\boldsymbol{q}(s) + s^2\boldsymbol{M}\boldsymbol{q}(s) = \boldsymbol{p}(s) + s\boldsymbol{M}\boldsymbol{q}(0) + \boldsymbol{M}\dot{\boldsymbol{q}}(0) + \boldsymbol{C}\boldsymbol{q}(0) \tag{10.62}$$

in which $\boldsymbol{q}(0)$ and $\dot{\boldsymbol{q}}(0)$ are the initial displacements and velocities at time $t = 0$. An impedance matrix is next defined as

$$\boldsymbol{Z}(s) = \boldsymbol{K} + s\boldsymbol{C} + s^2\boldsymbol{M} \tag{10.63}$$

and then Eq. (10.62) is written as

$$\boldsymbol{Z}(s)\boldsymbol{q}(s) = \boldsymbol{p}(s) + (s\boldsymbol{M} + \boldsymbol{C})\boldsymbol{q}(0) + \boldsymbol{M}\dot{\boldsymbol{q}}(0) \tag{10.64}$$

Solution of Eq. (10.64) is then obtained as

$$\boldsymbol{q}(s) = \boldsymbol{Z}(s)^{-1}\boldsymbol{p}(s) + \boldsymbol{Z}(s)^{-1}(s\boldsymbol{M} + \boldsymbol{C})\boldsymbol{q}(0) + \boldsymbol{Z}(s)^{-1}\boldsymbol{M}\dot{\boldsymbol{q}}(0) \tag{10.65}$$

and the required response $\boldsymbol{q}(t)$ can be obtained by evaluating the inverse transforms of the three terms on the right-hand side of this equation.

It is instructive to determine the steady-state response of an undamped structure to harmonic forces without any initial displacement or velocities, so that $\boldsymbol{q}(0) = 0$ and $\dot{\boldsymbol{q}}(0) = 0$. Thus, the jth component of the applied force $\boldsymbol{p}(t)$ may be expressed as

$$p_j(t) = p_{0j} f_j(t) \tag{10.66}$$

in which

$$f_j(t) = e^{i(\gamma_j t - \psi_j)} \tag{10.67}$$

where γ_j and ψ_j are, respectively, the frequency and phase angle of the jth harmonic force, p_{0j} being its amplitude. Laplace transform of the applied force is

$$p_j(s) = p_{0j} f_j(s) = p_{0j} \frac{e^{-i\psi_j}}{s - i\gamma_j} \tag{10.68}$$

and the transform of the response, in the absence of initial displacement and velocities, can be written as

$$\boldsymbol{q}(s) = \boldsymbol{Z}(s)^{-1}\boldsymbol{p}(s) = \frac{\boldsymbol{A}(s)}{|\boldsymbol{Z}(s)|}\boldsymbol{p}(s) \tag{10.69}$$

where $\boldsymbol{A}(s)$ is the adjoint $\boldsymbol{Z}(s)$. Equation (10.69) can also be written as a summation of each of its force components as

$$\boldsymbol{q}(s) = \sum_{j=1}^{n} \frac{1}{|\boldsymbol{Z}(s)|} \boldsymbol{A}(s)^{(j)} p_{0j} \frac{e^{-i\psi_j}}{s - i\gamma_j} \tag{10.70}$$

Although poles of the preceding equation exists as $s = i\gamma_j$, as well as $s = \pm i\omega_j$, ω_j being the jth mode, the steady-state vibration is the product of the externally impressed force and such are calculated by using inverse transforms that involve determining residues only at $s = i\gamma_j$ and summing the same, yielding

$$\boldsymbol{q}(t) = \sum_{j=1}^{n} \frac{p_{0j}}{|\boldsymbol{Z}(i\gamma_j)|} \boldsymbol{A}(i\gamma_j)^{(j)} e^{i(\gamma_j t - \psi_j)} \tag{10.71}$$

It is also useful to write the kth component of the response resulting from the application of the jth force only as

$$q_k(t) = \frac{p_{0j}}{|Z(i\gamma_j)|} A_{kj}(i\gamma_j) e^{i(\gamma_j t - \psi_j)} \tag{10.72}$$

and the dynamic magnification factor, defined as the ratio of maximum dynamic to the static deflection as

$$(MF)_{kj} = \frac{A_{kj}(i\gamma_j)}{|\boldsymbol{Z}(i\gamma_j)| a_{kj}} \tag{10.73}$$

where a_{kj} is the deflection influence coefficient.

These expressions may also be derived for responses to initial displacements and velocities as

$$\boldsymbol{q}(t) = \frac{1}{B} \sum_{r=1}^{n} \frac{1}{\prod_{k=1,k\neq r}^{n} \left(\omega_k^2 - \omega_r^2\right)} \boldsymbol{A}(\pm i\omega_r) \boldsymbol{M}\boldsymbol{q}(0) \mathbf{cos}\, \omega_r(t) \tag{10.74}$$

and

$$\boldsymbol{q}(t) = \frac{1}{B} \sum_{r=1}^{n} \frac{1}{\prod_{k=1,k\neq r}^{n} \left(\omega_k^2 - \omega_r^2\right)} \boldsymbol{A}(\pm i\omega_r) \boldsymbol{M}\dot{\boldsymbol{q}}(0) \frac{\mathbf{sin}\, \omega_r t}{\omega_r} \tag{10.75}$$

respectively; B is a constant, determined from the elements of the mass matrix[4].

The influence of structural damping may be evaluated, in which case the elements of $\boldsymbol{q}(t)$ in Eq. (10.71) are suitably modified to include this effect. Response of the structure to nonharmonic periodic forces may be obtained by resolving these forces into a set of harmonic components using Fourier series analyses and further extended to nonperiodic forces by using Fourier integrals.

10.5 Response to Random Excitation

A random process[7] consists of a collection of records that are nondeterministic and can only be characterized by statistical properties. If the expected value of this process is independent of time t, it is called stationary. Additionally, if its expected value can be replaced by a time average of a single representative record, then the process is known as ergodic. Response to these random excitations is also characterized through the appropriate use of statistical theory. The mean square value $\bar{y}^2(t)$ of a random response variable $y(t)$ in a random process is used effectively to make probability statements on the variable. Thus, as an example, a statement can be made toward the probability of the random response $y(t)$ exceeding a specified response, if $\bar{y}^2(t)$ is known, a priori, by using the central limit theorem. The mean square response $\bar{y}^2(t)$ can be obtained as a function of the mean square value $\bar{f}^2(t)$ of the random excitation $f(t)$ and the complex frequency response $H(\gamma)$ of a linear system. Distribution of $y(t)$ is assumed to be approximately normal being completely characterized by two parameters μ_y and σ_y^2, the mean and variance of y. The relevant definitions are given by

$$\mu_y = \int_{-\infty}^{\infty} y f(y)\, \mathrm{d}y \tag{10.76}$$

$$\sigma_y^2 = \int_{-\infty}^{\infty} y^2 f(y)\,\mathrm{d}y \tag{10.77}$$

$$\bar{y}^2(t) = \lim_{T\to\infty} \frac{1}{2T} \int_{-T}^{T} y^2(t)\,\mathrm{d}t \tag{10.78}$$

In practice $\bar{y}^2(t)$ is calculated approximately from a sample of records of finite duration yielding an estimate of mean and variance:

$$\hat{\mu}_y = \frac{1}{2T} \int_{-T}^{T} y(t)\,\mathrm{d}t \tag{10.79}$$

$$\hat{\sigma}_y^2 = \frac{1}{2T} \int_{-T}^{T} y^2(t)\,\mathrm{d}t \tag{10.80}$$

Next distribution of $y(t)$ can be obtained approximately by the normal distribution function

$$f(y) = \frac{1}{\sqrt{2\pi\sigma_y}} e^{-(y-\mu_y)^2/2\sigma_y^2} = \frac{1}{\sqrt{2\pi}} \frac{1}{\sigma_y} e^{-(y-\mu_y)^2/2\sigma_y^2} \tag{10.81}$$

where μ_y and σ_y^2 are the mean and variance of y. A probability statement

$$Pr(|y(t)| > y_k) = \int_{y_k}^{\infty} f(y)\,\mathrm{d}y + \int_{-\infty}^{-y_k} f(y)\,\mathrm{d}y = \epsilon \tag{10.82}$$

in which it is expected that $y(t)$ may exceed y_k by only a very small positive number ϵ, can then be made with reference to the response exceeding a specified value. It is also useful to express the mean square value in terms of the power spectral density function of $y(t)$ defined as

$$y(\gamma) = \frac{|Y(\gamma)|^2}{T} \tag{10.83}$$

in which $y(\gamma)$, $Y(\gamma)$ form a Fourier transform pair. The mean square value may then be expressed as

$$\bar{y}^2(t) = \lim_{T\to\infty} \frac{1}{2T} \int_{-T}^{T} y^2(t)\,\mathrm{d}t = \frac{1}{2\pi} \int_{0}^{\infty} y(\gamma)\,\mathrm{d}\gamma \tag{10.84}$$

For multidegree-of-freedom problems, a matrix formulation needs to be initiated with an appropriate assumption for damping values. For the simple case of proportional damping the equation of motion has the following form:

$$\boldsymbol{M}\ddot{\boldsymbol{y}} + \boldsymbol{C}\dot{\boldsymbol{y}} + \boldsymbol{K}\boldsymbol{y} = \boldsymbol{p}(t) \tag{10.85}$$

which can be easily decoupled by employing the modal method. Thus the rth equation is obtained as

$$\ddot{\eta}_r + 2\beta\dot{\eta}_r + \omega_r^2\eta_r = \boldsymbol{\Phi}^T\boldsymbol{p}(t) \tag{10.86}$$

using the modal transformation $\boldsymbol{y} = \Phi\boldsymbol{\eta}$, Φ being a set of mass-orthonormalized eigenvectors. Equation (10.86) can also be written in terms of critical damping ζ_r as

$$\ddot{\eta}_r + 2\zeta_r\omega_r\dot{\eta}_r + \omega_r^2\eta_r = \boldsymbol{\Phi}_r^T\boldsymbol{p}(t) \tag{10.87}$$

ζ_r being a fraction of critical damping. When the forcing function $\boldsymbol{p}(t)$ is expressed as a simple harmonic motion as $\boldsymbol{p}(t) = e^{i\gamma t}$, then solution of Eq. (10.87) is obtained as

$$\eta_r(t) = \frac{H_r(\gamma)}{\omega_r^2}\boldsymbol{\Phi}_r^T\boldsymbol{p}(t) \tag{10.88}$$

in which

$$H_r(\gamma) = 1\bigg/\left[1 - \left(\frac{\gamma}{\omega_r}\right)^2 + i2\zeta_r\frac{\gamma}{\omega_r}\right] \tag{10.89}$$

Then the response is simply obtained from $\boldsymbol{y} = \boldsymbol{\Phi}\boldsymbol{\eta}$ and the mean square of response from Eqs. (10.84) and (10.88), which may be simplified by disregarding phase relations and also assuming that the forcing function is a record of an ergodic random process. The mean square response is then obtained in the frequency domain as

$$\bar{y}^2(t) = \sum_{r=1}^{n}\boldsymbol{\Phi}_r^2\frac{\left(\boldsymbol{\Phi}_r^T\boldsymbol{p}\right)^2}{\omega_r^4}\frac{p(\omega_r)\omega_r}{8\zeta_r} \tag{10.90}$$

in which $p(\gamma)$ is replaced by its discrete value $p(\omega_r)$ at the natural frequencies ω_r. For a system with structural damping, an appropriate value for mean square response is obtained as

$$\bar{y}^2(t) = \sum_{r=1}^{n}\boldsymbol{\Phi}_r^2\frac{\left(\boldsymbol{\Phi}_r^T\boldsymbol{p}\right)^2 p(\omega_r)\omega_r}{4g\omega_r^4} \tag{10.91}$$

Using results obtained by either Eq. (10.90) or Eq. (10.91), as the case may be, the variance σ_y^2 at any point in the structure can be calculated. Then also using the approximation of a normal distribution, a probability statement for a response not exceeding a certain specified value y_k at some point in the structural system is made by using Eqs. (10.84), (10.81), and (10.82). More detailed descriptions of this phenomenon are available in Ref. 4.

10.6 Numerical Examples

A rocket subjected to a pulse loading function and a cantilever beam subjected to time-dependent forcing functions are discussed in the following two examples.

Example 10.1: Rocket Structure—Dynamic-Response Analysis

A rocket idealized by four line elements, as shown in Fig. 10.6,[1] is subjected to a pulse loading function at its base. Data parameters consist of arbitrary element and material properties used to correlate results with available ones expressed in

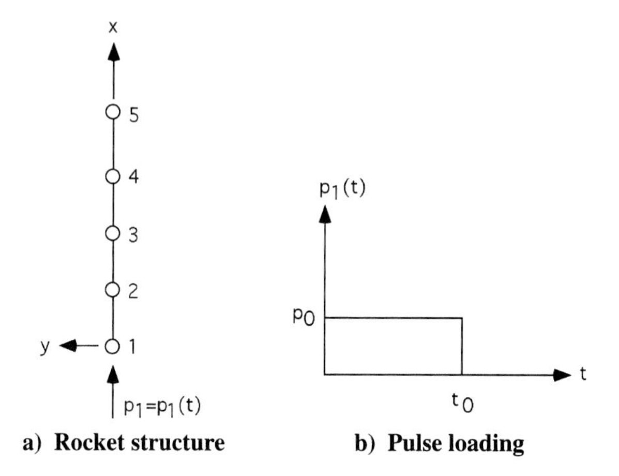

a) **Rocket structure** b) **Pulse loading**

Fig. 10.6 Rocket subjected to dynamic loading.

parametric form. The relevant data parameters are

Young's modulus E $= 100$
Poisson's ratio ν $= 0.3$
Cross-sectional area A $= 1.0$
Mass density ρ $= 1.0$
Member length ℓ $= 2.5$
Pulse load intensity P_0 $= 10.0$
Duration of load $= 1.0$ s
Total time period for response evaluation $= 2.0$

For the modal analysis the first three natural frequencies used are 0.0, 3.2228, and 6.9282 rad/s, respectively; the first frequency is the rigid-body mode and the beam is constrained to move in the x direction only. Figures 10.7 and 10.8 show results of the modal dynamic response analysis. The problem was also solved using the direct integration method. Table 10.1 provides a comparison of these results obtained from the two procedures.

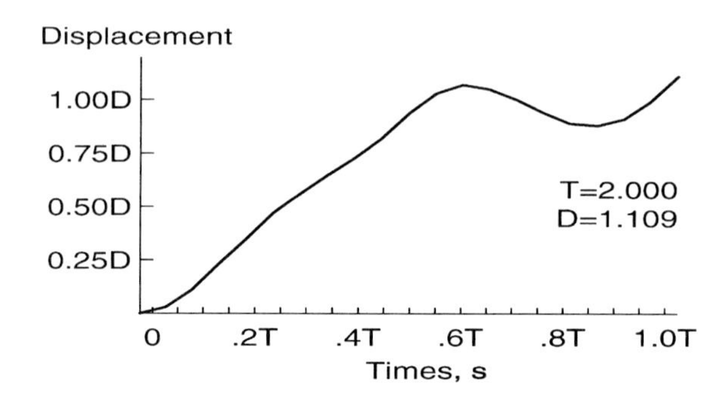

Fig. 10.7 Rocket nodal displacement as a function of time, node 1.

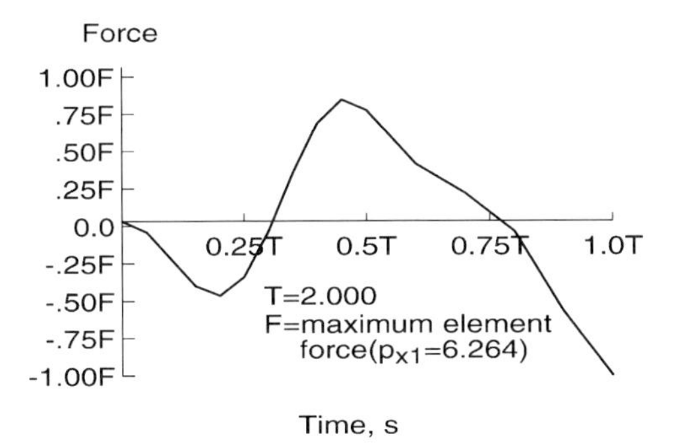

Fig. 10.8 Rocket element force as a function of time, element 4.

Example 10.2: Cantilever Beam—Frequency Response Analysis

The cantilever beam in Fig. 10.9 is subjected to time-dependent forcing functions at nodes 2 and 3, respectively, and is assumed to have motion in the Y direction only. The following properties are relevant to the present analysis: Young's modulus E, moment of inertia I_Z, the nodal lumped mass parameter m, and cross sectional area A. The stiffness and mass matrices may be written as

$$\boldsymbol{K} = \frac{12EI_Z}{L^3}\begin{bmatrix} 2 & -1 \\ -1 & 1 \end{bmatrix}, \qquad \boldsymbol{M} = m\begin{bmatrix} 2 & 0 \\ 0 & 1 \end{bmatrix}$$

Then the impedance matrix is given by

$$\begin{aligned} \boldsymbol{Z}(s) &= \boldsymbol{K} + s^2\boldsymbol{M} \\ &= m\begin{bmatrix} 2s^2 + 2\mu & -\mu \\ -\mu & s^2 + \mu \end{bmatrix} \end{aligned}$$

in which

$$\mu = 12EI_Z/L^3m$$

Table 10.1 Comparison of dynamic response analysis results of the rocket structure at $t = 1.2$ s

Node number	Nodal deflections	
	Modal method	Direct integration
1	1.0680	1.0000
2	0.9330	0.9310
3	0.6610	0.6370
4	0.4668	0.4500
5	0.4090	0.4200

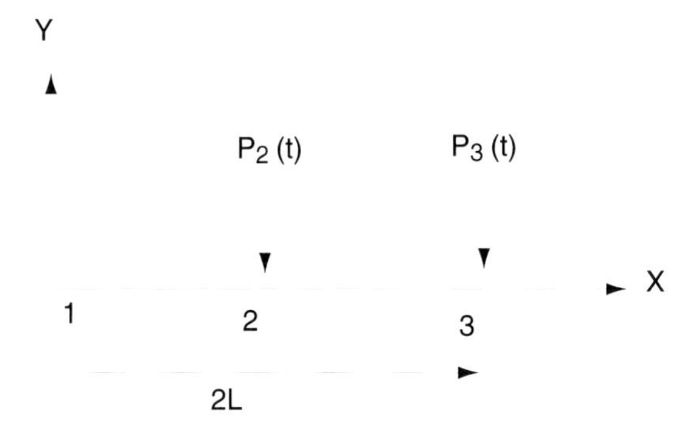

Fig. 10.9 Cantilever beam subjected to harmonic loading.

Assuming that the applied forces are harmonic

$$p_1(t) = p_{01}e^{i\gamma_1 t}$$

$$p_2(t) = p_{02}e^{i(\gamma_2 t - \psi_2)}$$

the steady-state amplitudes may be obtained from Eq. (10.71) as

$$\boldsymbol{q}(t) = \sum_{j=1}^{2} \frac{p_{0j}}{|\boldsymbol{Z}(i\gamma_j)|}\boldsymbol{A}(i\gamma_j)^{(j)}e^{i(\gamma_j t - \psi_j)}$$

Supposing that for the present problem the force frequencies are given by

$$\gamma_1^2 = \tfrac{1}{6}\mu \qquad \gamma_2^2 = \tfrac{1}{3}\mu$$

approximately bracketing the first frequency. Expressions for relevant terms for the force $p_1(t)$ are as follows:

$$s^2 = (i\gamma_1)^2 = -\gamma_1^2 = -\frac{\mu}{6}$$

$$\boldsymbol{Z}(i\gamma_1) = m\mu\begin{bmatrix} 5/3 & -1 \\ -1 & 5/6 \end{bmatrix}$$

$$\boldsymbol{A}(i\gamma_1)^{(1)} = m\mu\begin{bmatrix} 5/6 \\ 1 \end{bmatrix}$$

$$|\boldsymbol{Z}(i\gamma_1)| = m^2\mu^2\left[\frac{7}{18}\right]$$

and similarly for $p_2(t)$,

$$s^2 = -\gamma_2^2 = -\frac{\mu}{3}$$

$$\boldsymbol{Z}(i\gamma_2) = m\mu\begin{bmatrix} 4/3 & -1 \\ -1 & 2/3 \end{bmatrix}$$

$$A(i\gamma_2)^{(2)} = m\mu\begin{bmatrix} 1 \\ 4/3 \end{bmatrix}$$

$$|Z(i\gamma_2)| = -m^2\mu^2\begin{bmatrix} \frac{1}{9} \end{bmatrix}$$

Then using Eq. (10.71) the amplitude of motion becomes

$$\begin{aligned} q(t) &= q(t)^{(1)} + q(t)^{(2)} \\ &= \frac{18p_{01}}{7m\mu}\begin{bmatrix} 5/6 \\ 1 \end{bmatrix} e^{i\gamma_1 t} - \frac{8p_{02}}{m\mu}\begin{bmatrix} 1 \\ 4/3 \end{bmatrix} e^{i(\gamma_2 t - \psi_2)} \\ &= \begin{bmatrix} 0.1786 \\ 0.2143 \end{bmatrix} \frac{p_{01}L^3}{EI_Z} e^{i\gamma_1 t} - \begin{bmatrix} 0.6667 \\ 0.8889 \end{bmatrix} \frac{p_{02}L^3}{EI_Z} e^{i(\gamma_2 t - \psi_2)} \end{aligned}$$

which is the dynamic response of the structure due to externally applied harmonic forces.

References

[1]Przemieniecki, J. S., *Theory of Matrix Structural Analysis*, McGraw–Hill, New York, 1968.

[2]Newmark, N. M., "A Method of Computation for Structural Dynamics," *Proceedings of the ASCE*, Vol. 85, 1959.

[3]Wilson, E. L., Farhoomand, I., and Bathe, K. J., "Nonlinear Dynamic Analysis of Complex Structures," *International Journal of Earthquake Engineering and Structural Dynamics*, Vol. 1, 1973.

[4]Houbolt, J. C., "A Recurrence Matrix Solution for the Dynamic Response of Elastic Aircraft," *Journal of Aeronautical Sciences*, Vol. 17, 1950, pp. 540–550.

[5]Argyris, J. H., and Sharpf, D. W., "Finite Elements in Time and Space," *Journal of the Royal Aeronautical Society*, Vol. 73, 1969, pp. 1041–1044.

[6]Chan, S. P., Cox, H. L., and Benfield, W. A., "Transient Analysis of Forced Vibration of Complex Structural-Mechanical System," *Journal of the Royal Aeronautical Society*, Vol. 66, 1962.

[7]Hurty, W. C., and Rubinstein , M. F., *Dynamics of Structures*, Prentice–Hall, Englewood Cliffs, NJ, 1964.

11
Nonlinear Analysis

11.1 Introduction

Structural nonlinearities may be caused by a number of factors and are broadly classified into two categories: geometric and material. In nonlinear response the total load is no longer proportional to the total displacement. Individual factors causing nonlinearities may be grouped as follows:

1) Large rotations—Result in a nonlinear force–displacement relationship, and incremental effect is computed as the geometric stiffness matrix.

2) Large displacements—Require updating of the original equilibrium equations and involve not only updating geometry at every computational time step but also calculating the initial stress stiffness matrix, representing the effect of realignment of current internal stresses because of displacements.

3) Nonlinear stress–strain law—Occurs in rubberlike materials as well as in most metals, elastomers, and some composites when properties are unequal in tension and compression.

4) Large strain—Occurs in plastics, some metals, rubbers, and elastomers.

The first two items belong to the large deformation category, involving geometrical nonlinearity, whereas the last two items belong to material nonlinearity. Descriptions of numerical algorithms for solution of nonlinear problems are given next with small strain assumption.

11.2 Geometric Nonlinearity

For geometric nonlinearity, a Newton–Raphson iterative procedure is used in which the elastic stiffness matrix is supplemented with the geometric stiffness matrix $\boldsymbol{K}_G$, so that both large displacements and rotations, as well as the effect of in-plane stretching, are taken into consideration; strains are assumed to be small in the formulation given herein.

For static analysis, the solution algorithm is as follows.

Let n equal the number of load increments and i equal the number of iterations within a loop. Then for each load increment ("n" loop), form $n\boldsymbol{K}_G$, the geometric stiffness matrix based on accumulated element stresses σ'(=0 for $n = 1$, and the prime denotes the local coordinate system); $n\boldsymbol{K}_E$, the elastic stiffness matrix $(=1\boldsymbol{K}_E^0$ for $n = 1; n\boldsymbol{K}_E^i$ for $n > 1)$; and $n\boldsymbol{r}_s$, residual forces (=0, for $n = 1; n\boldsymbol{r}_s^{(i)}$ for $n > 1$).

Then for each iteration ("i" loop) solve

$$\left[n\boldsymbol{K}_E^{(i-1)} + n\boldsymbol{K}_G\right]\Delta\boldsymbol{u}^{(i)} = \sum_n \Delta\boldsymbol{p} - n\boldsymbol{r}_s^{(i-1)} \tag{11.1}$$

to yield $\Delta \boldsymbol{u}^{(i)}$. Update geometry

$$\boldsymbol{u}^{(i)} = \boldsymbol{u}^{(i-1)} + \Delta \boldsymbol{u}^{(i)} \tag{11.2}$$

and obtain incremental deformations in LCS

$$\Delta \boldsymbol{u}'^{(i)} = \boldsymbol{\lambda}^{(i)} \Delta \boldsymbol{u}^{(i)} \tag{11.3}$$

where $\boldsymbol{\lambda}$ is the direction cosine matrix. Calculate the element stresses in the LCS and accumulate, based on $\Delta \boldsymbol{u}'^{(i)}$.

Obtain residual forces in GCS

$$\Delta \boldsymbol{s}^{(i)} = \left[n\boldsymbol{K}_E^{(i-1)} + n\boldsymbol{K}_G \right] \Delta \boldsymbol{u}^{(i)} \tag{11.4}$$

$$n\boldsymbol{r}_s^{(i)} = n\boldsymbol{r}_s^{(i-1)} + \Delta \boldsymbol{s}^{(i)} \tag{11.5}$$

Check convergence; if $|\Delta \boldsymbol{u}| > \text{EPS}$, EPS is a specified accuracy parameter. Compute $\boldsymbol{K}'^{(i)}_E$ and $\boldsymbol{K}^{(i)}_E$, element and global stiffness matrices, respectively. Repeat the iteration ("i" loop), if not converged. Continue on the incremental load loop ("n" loop) if convergence is achieved in "i" loop. Stop the computation on the completion of the load cycle.

For dynamic response analysis of geometrically nonlinear structures, a time integration scheme along with a Newton–Raphson iterative technique is adopted for the solution; the associated algorithm follows.

For a jth time increment Δt_j ("j" loop), form $j\boldsymbol{K}_G$, the geometric stiffness matrix based on accumulated element stresses σ'(=0, for $j = 1$). Perform the iteration ("i" loop) and form

$$\boldsymbol{K}^{(i-1)}, \boldsymbol{M}, \boldsymbol{C}, \boldsymbol{r}_s^{(i-1)} \left[\boldsymbol{K}^{(i-1)} = \boldsymbol{K}_E^{(i-1)} + j\boldsymbol{K}_G \right] \quad \left(\boldsymbol{r}_s^0 = 0, \qquad \text{for } j = 1 \right) \tag{11.6}$$

Calculate

$$\hat{\boldsymbol{K}} = \frac{4}{\Delta t^2} \boldsymbol{M} + \frac{2}{\Delta t} \boldsymbol{C} + \boldsymbol{K}^{(i-1)} \tag{11.7}$$

Calculate

$$\begin{aligned} \boldsymbol{F}^{\text{eff}} &= \boldsymbol{F}_{t+\Delta t} + \boldsymbol{M} \left[\frac{4}{\Delta t^2} \boldsymbol{u}_t + \frac{4}{\Delta t} \dot{\boldsymbol{u}}_t + \ddot{\boldsymbol{u}}_t - \frac{4}{\Delta t^2} \boldsymbol{u}_{t+\Delta t}^{(i-1)} \right] \\ &\quad + \boldsymbol{C} \left[\frac{2}{\Delta t} \boldsymbol{u}_t + \dot{\boldsymbol{u}}_t - \frac{2}{\Delta t} \boldsymbol{u}_{t+\Delta t}^{(i-1)} \right] - \boldsymbol{r}_s^{(i-1)} \end{aligned} \tag{11.8}$$

Solve

$$\hat{\boldsymbol{K}} \Delta \boldsymbol{u}^{(i)} = \boldsymbol{F}^{\text{eff}} \tag{11.9}$$

Update

$$\boldsymbol{u}_{t+\Delta t}^{(i)} = \boldsymbol{u}_{t+\Delta t}^{(i-1)} + \Delta \boldsymbol{u}^{(i)} \tag{11.10}$$

Obtain deformation in LCS

$$\Delta \boldsymbol{u}'^{(i)} = \boldsymbol{\lambda}^{(i)} \Delta \boldsymbol{u}^{(i)} \tag{11.11}$$

and calculate element stresses and accumulate based on $\Delta \boldsymbol{u}'^{(i)}$. Obtain residual forces in GCS

$$\Delta \boldsymbol{s}^{(i)} = \left[\boldsymbol{K}_E^{(i-1)} + j\boldsymbol{K}_G\right]\Delta \boldsymbol{u}^{(i)} \tag{11.12}$$

$$\boldsymbol{r}_s^{(i)} = \boldsymbol{r}_s^{(i-1)} + \Delta \boldsymbol{s}^{(i)} \tag{11.13}$$

Check convergence $|\Delta \boldsymbol{u}^{(i)}| = 0$; if not converged, continue the iteration ("i" loop). If converged, update

$$\dot{\boldsymbol{u}}_{t+\Delta t} = \frac{2}{\Delta t}(\boldsymbol{u}_{t+\Delta t} - \boldsymbol{u}_t) - \dot{\boldsymbol{u}}_t \tag{11.14}$$

$$\ddot{\boldsymbol{u}}_{t+\Delta t} = \frac{4}{\Delta t^2}(\boldsymbol{u}_{t+\Delta t} - \boldsymbol{u}_t) - \frac{4}{\Delta t}\dot{\boldsymbol{u}}_t - \ddot{\boldsymbol{u}}_t \tag{11.15}$$

If $t \neq t_{\text{final}}$, update time ("j" loop). Stop at the end of the cycle time.

11.3 Material Nonlinearity

In problems that exhibit material nonlinearity, the Prandtl–Reuss equation for the plastic strain increments is combined with the von Mises yield criteria for material characterization. An iterative solution procedure is then employed for the solution of the associated static problem. Initialize:

1) Set loads external ($\boldsymbol{R}^E$) and internal ($\boldsymbol{R}^I$) to zero.
2) Set all stresses and deformations ($\boldsymbol{u}_0$) to zero.
3) Set the yield function $F = -\sigma_{\text{yp}}$.
4) Set the slope of the strain hardening curve to its initial value H'_{in}.
5) Set the plastic strain ϵ_p to zero.
6) Set $\Delta \boldsymbol{p}$, the incremental load.

For each incremental load ("j" loop), calculate the total load at jth step:

$$\boldsymbol{R}_j^E = \boldsymbol{R}_{j-1}^E + \Delta \boldsymbol{p} \tag{11.16}$$

Calculate the norm

$$\|\boldsymbol{R}_j^E\| \tag{11.17}$$

and find the residual force ("i" loop)

$$\boldsymbol{R}_i^{\text{res}} = \boldsymbol{R}_j^E - \boldsymbol{R}_{i-1}^I \tag{11.18}$$

Calculate $\|\boldsymbol{R}_i^{\text{res}}\|$; if $\|\boldsymbol{R}_i^{\text{res}}\| < \text{EPS} \times \|\boldsymbol{R}_j^E\|$, go to the end of the "$j$" loop; calculate

$\boldsymbol{K}_{\text{ep}}$ (initially $\boldsymbol{K}_e$). Solve $\Delta \boldsymbol{u}_i = \boldsymbol{K}_{\text{ep}}^{-1} \boldsymbol{R}_i^{\text{res}}$, the increment of displacement, and calculate $\boldsymbol{u}_i = \boldsymbol{u}_{i-1} + \Delta \boldsymbol{u}_i$, the total elastoplastic displacement.

For each element ("n" loop), $\Delta \epsilon_i = \boldsymbol{B} \Delta \boldsymbol{u}_i$, the total increment in element strain (using $\boldsymbol{e} = \boldsymbol{B}\boldsymbol{u}$ for small strain). Check the last value of yield function F_{i-1}. If $F_{i-1} = 0$, the element is already on the yield surface; go to the elastoplastic loop. If $F_{i-1} < 0$, the element is elastic at the start of the loop.

Calculate $\Delta \boldsymbol{\sigma}_i = \boldsymbol{D} \Delta \epsilon_i$:

$$\boldsymbol{\sigma}_i = \boldsymbol{\sigma}_{i-1} + \Delta \boldsymbol{\sigma}_i \tag{11.19}$$

Then calculate $F_i = \bar{\sigma}_i + \sigma_{\text{yp}_{i-1}}$, in which the effective stress $\bar{\sigma} = \sqrt{3/2}\sqrt{\boldsymbol{\sigma}'^T \boldsymbol{\sigma}'}$; $\boldsymbol{\sigma}' = [\sigma'_x \; \sigma'_y \; \sigma'_z \; 2\tau_{xy} \; 2\tau_{yz} \; 2\tau_{zx}]^T$ in which $\sigma'_x = (\sigma_x - \sigma_m)$, and so on, and $\sigma_m = (\sigma_x + \sigma_y + \sigma_z)/3$. If $F_i \le 0$, then it is elastic and store $\boldsymbol{\sigma}_i$, F_i; go to the end of the "n" loop. If $F_i > 0$, set $\boldsymbol{\sigma}_A = \boldsymbol{\sigma}_{i-1}$, $\boldsymbol{\sigma}_B = \boldsymbol{\sigma}_i$. Loop to bring $\boldsymbol{\sigma}_C$ onto the yield surface

$$\boldsymbol{\sigma}_C = (\boldsymbol{\sigma}_A + \boldsymbol{\sigma}_B)/2 \tag{11.20}$$

Calculate $F_C = \bar{\sigma}_C - \sigma_{\text{yp}_{i-1}}$. If $F_C < 0, \boldsymbol{\sigma}_A = \boldsymbol{\sigma}_C$. If $F_C > 0, \boldsymbol{\sigma}_B = \boldsymbol{\sigma}_C$. If $|F_C| > \text{TOL}$, return to the top of the loop; TOL is the specified accuracy parameter to end the loop. Then calculate

$$\Delta \epsilon_{e_i} = \boldsymbol{D}_{\text{ep}}^{-1}(\boldsymbol{\sigma}_C - \boldsymbol{\sigma}_{i-1}) \tag{11.21}$$

$$\Delta \epsilon_{p_i} = \Delta \epsilon_i - \Delta \epsilon_{e_i} \tag{11.22}$$

$\boldsymbol{\sigma}_{i-1} = \boldsymbol{\sigma}_C$, and F_i becomes 0.

Here is the start of the elastoplastic loop. Set $\boldsymbol{\sigma}_i = \boldsymbol{\sigma}_C$, stress on yield surface

$$\text{d}\Delta \epsilon_{\text{ep}} = \frac{\Delta \epsilon_{p_i}}{20} \tag{11.23}$$

Loop, $k = 1, n$ (e.g., $n = 20$), and calculate $\boldsymbol{D}_{\text{ep}}$, based on $\boldsymbol{\sigma}_C$:

$$\Delta \boldsymbol{\sigma}_k = \boldsymbol{D}_{\text{ep}} \, \text{d}\Delta \epsilon_{\text{ep}} \tag{11.24}$$

$$\boldsymbol{\sigma}_{C+k} = \boldsymbol{\sigma}_C + \Delta \boldsymbol{\sigma}_k \tag{11.25}$$

Use $\boldsymbol{\sigma}_{C+k}$ to calculate the normal $\partial F/\partial \boldsymbol{\sigma} = \text{d}\bar{\sigma}/\text{d}\boldsymbol{\sigma}$ and F_k. Calculate

$$\text{d}(\Delta \boldsymbol{\sigma}) = \left[\frac{-\text{d}\bar{\sigma}/\text{d}\boldsymbol{\sigma}}{\sqrt{\frac{\text{d}\bar{\sigma}}{\text{d}\boldsymbol{\sigma}} \frac{\text{d}\bar{\sigma}}{\text{d}\boldsymbol{\sigma}}}} \right] \times F_k$$

$$\boldsymbol{\sigma}_{C+k} = \boldsymbol{\sigma}_{C+k} + \text{d}(\Delta \boldsymbol{\sigma}) \tag{11.26}$$

Set $\boldsymbol{\sigma}_C = \boldsymbol{\sigma}_{C+k}$, and end "$k$" loop. Calculate F_i, which should be 0; if not, set to 0 (it may be very small). Store $\boldsymbol{\sigma}_{C+k}$ and F_i, and calculate the elastic strain increment:

$$\Delta \epsilon_e = \boldsymbol{D}_e^{-1}\{\boldsymbol{\sigma}_{C+k} - \boldsymbol{\sigma}_{i-1}\} \tag{11.27}$$

$$\Delta \epsilon_p = \Delta \epsilon_i - \Delta \epsilon_e \tag{11.28}$$

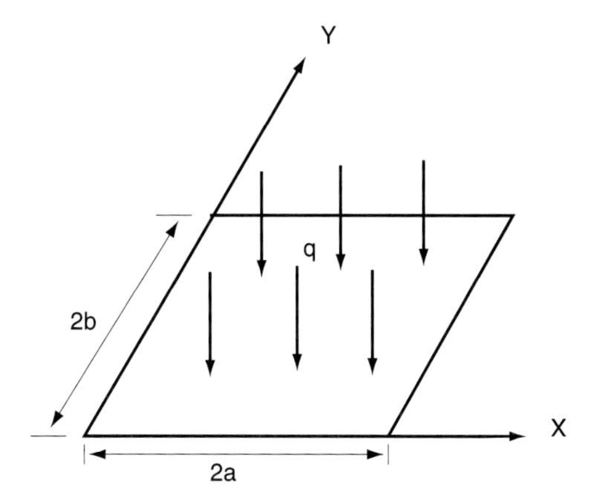

Fig. 11.1 Clamped plate with uniform load.

Store $\epsilon_{p_i} = \epsilon_{p_{i-1}} + \Delta\epsilon_p$. Calculate $H_i' = f(\epsilon_{p_i})$ from input data (which may be assumed to have a constant slope/value). End the element "n" loop; end the residual force "i" loop; and end the load "j" loop.

11.4 Numerical Examples

In this section, several typical nonlinear test cases are presented. The analyses were performed using the STARS computer program.[18]

Example 11.1: Clamped Square Plate—Uniform Load

Figure 11.1 shows a square plate with all edges fixed under uniform load. The results of the geometric nonlinear analysis involving large displacement and rotation are presented herein. The relevant data parameters are thickness $h = 1.0$, Young's modulus $E = 2.0 \times 10^{11}$, Poisson's ratio $\nu = 0.3$, length $a = b = 50$, and uniform pressure $q = 6.0 \times 10^4$, in 30 equal increments of $\Delta q = 2000.0$. Table 11.1 summarizes the results, and Fig. 11.2 shows the maximum plate displacement ($W_{\max}$) as a function of the applied load.

Table 11.1 Center deflection for a clamped square plate with a uniform load

	$W_{\max}/h$		
qa^4/Dh	Theory	STARS	Difference,%
109.3(P)	1.20	1.21	0.8
218.6(P)	1.66	1.70	2.4
327.9(P)	2.00	2.03	1.5

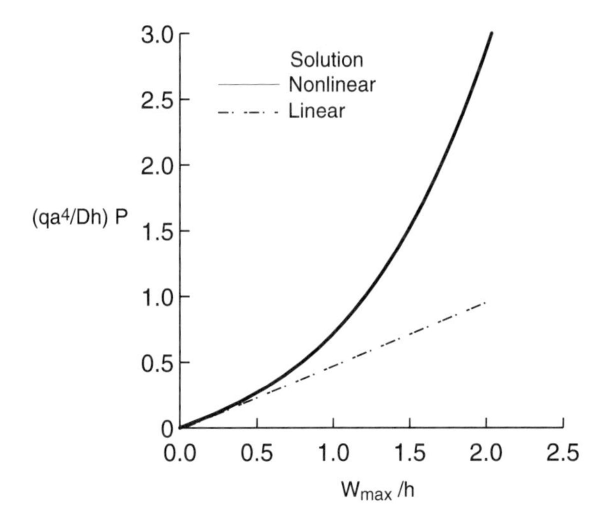

Fig. 11.2 Clamped plate center displacement as a function of load curve.

Example 11.2: Cantilever Beam—Moment at Tip

A static nonlinear analysis of a cantilever beam in Fig. 11.3 was performed using line elements. The relevant data parameters are Young's modulus $E = 2.0 \times 10^7$, cross-sectional area $A = 0.15$, moments of inertia about the y axis $= 2.813 \times 10^{-4}$, moments of inertia about the z axis $= 2.813 \times 10^{-4}$, member length $\ell = 1.0$, and tip moment $M = 3000.0$ in 600 equal increments of $\Delta M = 5.0$. Table 11.2 summarizes the output, and Figs. 11.4 and 11.5 show deflection curves.

Example 11.3: Shallow Spherical Cap—Dynamic-Response Analysis

Figure 11.6 shows a clamped spherical cap under apex load. Both triangular and quadrilateral elements, (Fig. 11.7) were used for modeling one-quarter of the cap for the nonlinear static analysis. The triangular element mesh was also used for the nonlinear response analysis. The relevant data parameters are

Young's modulus E $=$ 1.0×10^7
Poisson's ratio ν $=$ 0.3
Thickness t $=$ 0.01576
Radius R $=$ 4.76
Height H $=$ 0.08589
Length a $=$ 0.9
Mass density ρ $=$ 2.45×10^{-4}
Static load P $=$ 100
ΔP $=$ 1
Dynamic load P_o $=$ 100
Δt $=$ 0.5×10^{-6}

Table 11.2 Tip displacement v for a cantilever beam with end moment

	Tip Y displacement, v	
M	STARS	Theory
300	3.7048	3.7107
600	6.6675	6.6860
900	8.3599	8.3884
1200	8.5825	8.6043
1500	7.4971	7.4936
1800	5.5581	5.5183
2000	3.3671	3.2954
2400	1.4980	1.4146
2700	0.3452	0.2787
3000	0.0385	0.0128

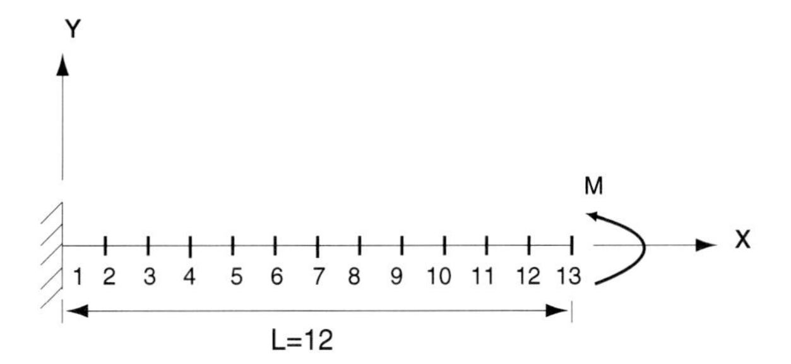

Fig. 11.3 Cantilever beam with moment at tip.

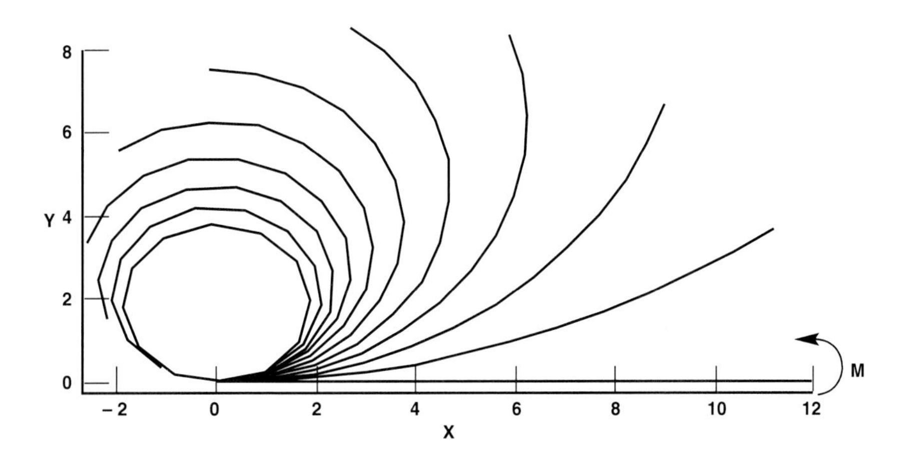

Fig. 11.4 Deflection curves for cantilever beam with end moment.

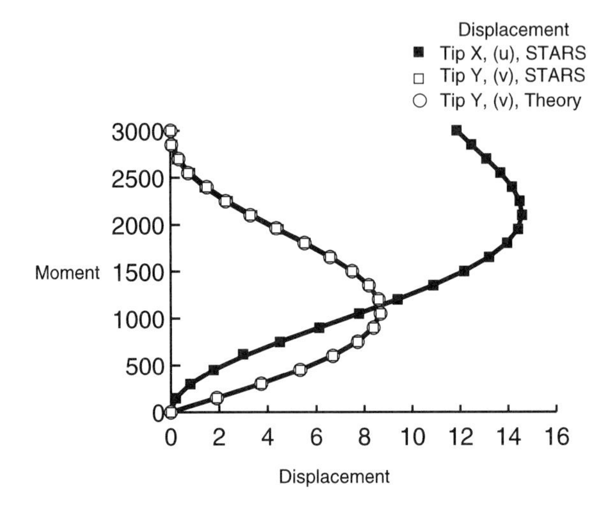

Fig. 11.5 Cantilever tip deflection as a function of moment curve.

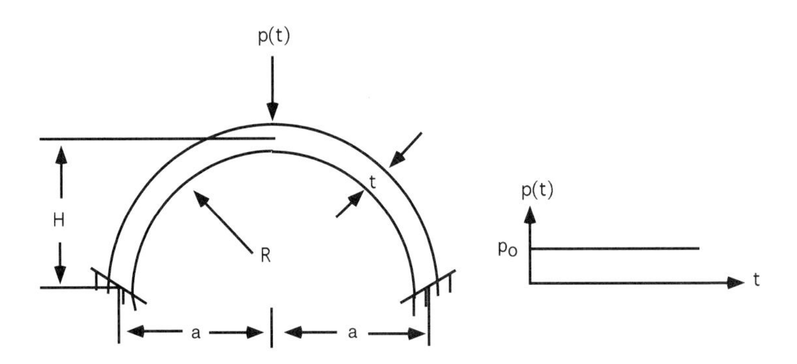

Fig. 11.6 Clamped spherical cap with apex load.

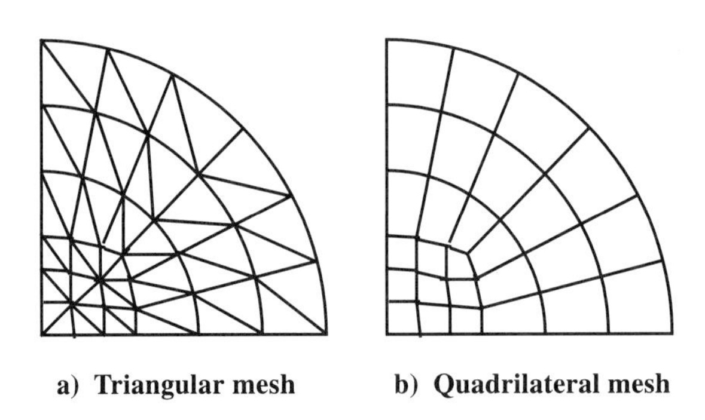

a) Triangular mesh **b) Quadrilateral mesh**

Fig. 11.7 One-quarter cap model.

Table 11.3 Center deflection for a clamped spherical cap under apex load

Load increment	Displacement W/H	
	Quadrilateral mesh	Triangular mesh
4	0.0384	0.0369
12	0.1601	0.1537
20	0.4542	0.4154
24	0.5859	0.5607
28	0.7007	0.6984
32	0.8103	0.8289
36	0.9186	0.9522
40	1.0308	1.0753
44	1.1575	1.2173
48	1.3470	1.4091
52	1.5054	1.5400
56	1.6313	1.6211
60	1.7058	1.6794
70	1.8159	1.7827
80	1.8882	1.8576
90	1.9444	1.9179
100	1.9891	1.9689

Table 11.3 shows the results for the nonlinear static solution. Figure 11.8 shows the displacement as a function of the applied load, and Fig. 11.9 plots the displacement as function of time for the nonlinear response solution.

Example 11.4: Clamped Beam—Elastoplastic Analysis

Figure 11.10 shows a clamped beam under a uniformly distributed load. A triangular shell-element mesh was used for modeling. The beam consists of several

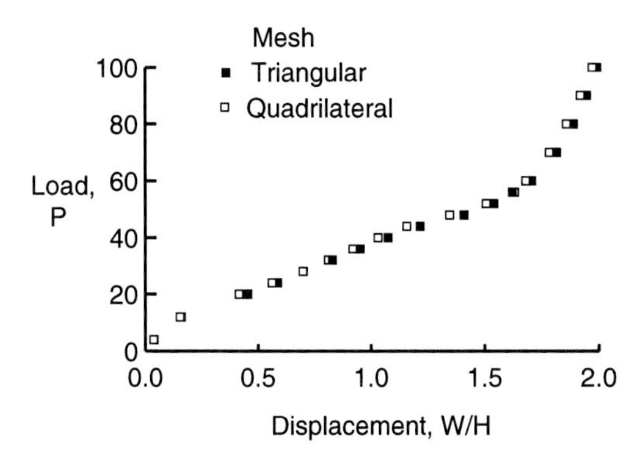

Fig. 11.8 Cap center displacement as a function of load.

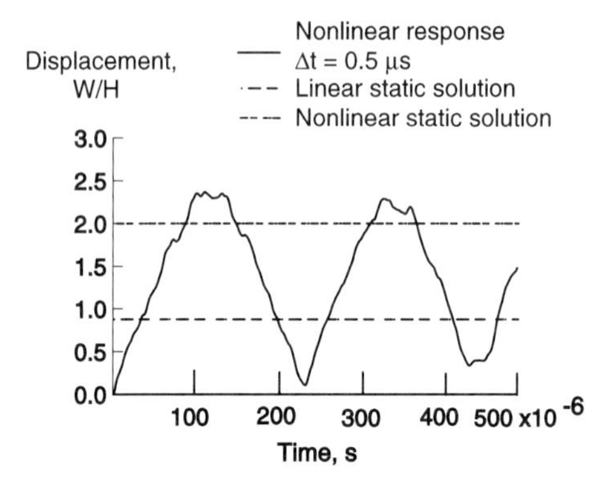

Fig. 11.9 Cap center displacement as a function of time.

layers of elements. The relevant data parameters are Young's modulus $E = 1.0 \times 10^7$, Poisson's ratio $\nu = 0.3$, thickness $t = 0.1$, length $L = 10.0$, width $B = 1.0$, yield stress $\sigma_y = 35{,}000$, and hardening $H' = 0$.

Table 11.4 shows the results for the elastoplastic analysis and Fig. 11.11 depicts the displacement as a function of load plot.

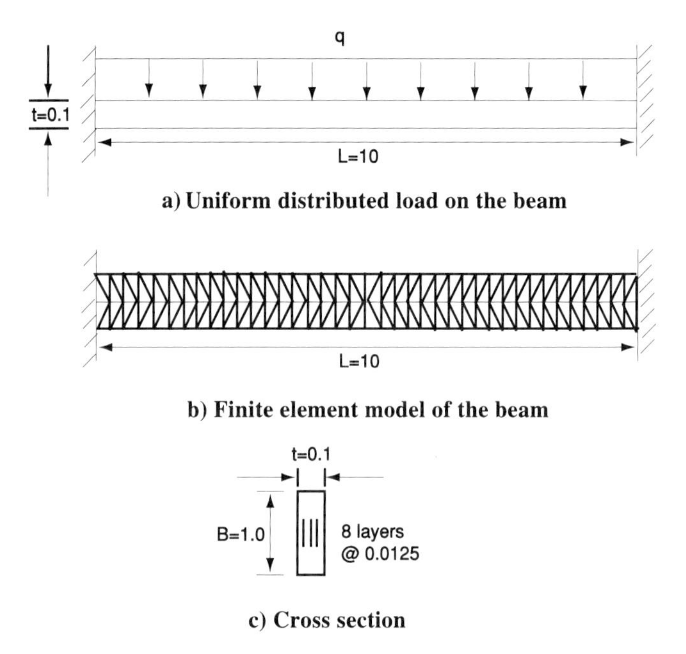

Fig. 11.10 Clamped beam under uniformly distributed load.

Table 11.4 Maximum uniformly distributed load for a clamped beam

q_{max}		
STARS	Theory	Difference, %
14.875	14.0	6.25

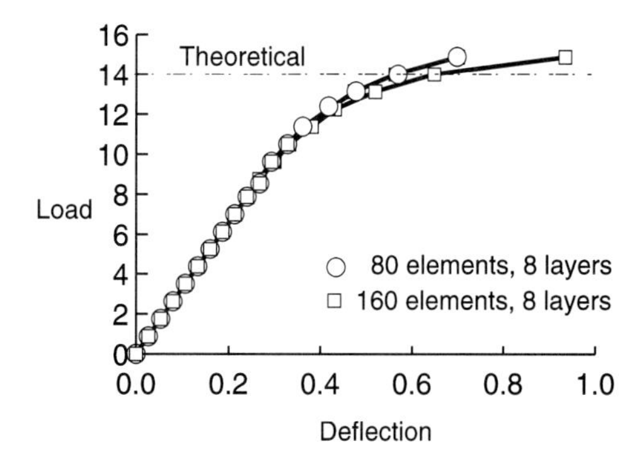

Fig. 11.11 Load-deflection curve for a clamped beam.

References

[1]Martin, H. C., "On the Derivation of Stiffness Matrices for the Analysis of Large Deflection and Stability Problems," *Proceedings of the Conference on Matrix Methods in Structural Mechanics*, Air Force Inst. of Technology, Wright–Patterson AFB, OH, Oct. 1965.

[2]Kapur, K. K., and Hartz, B. J., "Stability of Thin Plates Using the Finite Element Method," *Proceedings of the ASCE*, Vol. 92, EM2, 1966, pp. 177–195.

[3]Gallagher, R. H., and Padlog, J., "Discrete Element Approach to Structural Instability Analysis," *AIAA Journal*, Vol. 1, 1963, pp. 1537–1539.

[4]Martin, H. C., "Finite Elements and the Analysis of Geometrically Non-Linear Problems," *Recent Advances in Matrix Methods and Structural Analysis and Design*, Univ. of Alabama Press, 1971.

[5]Marcal, P. V., "Finite Element Analysis of Combined Problems of Material and Geometric Behavior," technical report 1, ONR, Brown Univ., 1969; also *Proceedings of the ASCE Conference on Computational Approaches in Applied Mechanics*, June 1969 (Paper 133).

[6]Zienkiewicz, O. C., and Nayak, G. C., "A General Approach to Problems of Plasticity and Large Deformation Using Isoparametric Elements," *Proceedings of the 3rd Conference on Matrix Methods in Structural Mechanics*, Wright–Patterson AFB, OH, 1971.

[7]Przemieniecki, J. S., "Stability Analysis of Complex Structures Using Discrete Element Techniques," *Symposium on Structure Stability and Optimization*, Loughborough Univ., U.K., March 1967.

[8]Hill, R., *The Mathematical Theory of Plasticity*, Clarendon Press, Oxford, England, U.K., 1950.

[9]Prager, W., *An Introduction to Plasticity*, Addison-Wesley, Reading, MA, 1959.

[10]Drucker, D. C., "Conventional and Unconventional Plastic Response and Representation," *Applied Mechanics Review*, Vol. 41, 1988, pp. 151–167.

[11]Yamada, Y., Yishimura, N., and Sakurai, T., "Plastic Stress-Strain Matrix and Its Application for the Solution of Elastic-Plastic Problems by the Finite Element Method," *International Journal of Mech. Sci.*, Vol. 10, 1968, pp. 343–354.

[12]Zienkiewicz, O. C., Valliappan, S., and King, I. P., "Elasto-Plastic Solutions of Engineering Problems. Initial-Stress, Finite Element Approach," *International Journal for Numerical Methods in Engineering*, Vol. 1, 1969, pp. 75–199.

[13]Argyris, J. H., "Elasto-Plastic Matrix Displacement Analysis of Three-Dimensional Continua," *Journal of the Royal Aeronautical Society*, Vol. 69, 1965, pp. 633–635.

[14]Marcal, P. V., and King, I. P., "Elastic-Plastic Analysis of Two Dimensional Stress Systems by the Finite Element Method," *International Journal of Mechanical Sciences*, Vol. 9, 1967, pp. 143–155.

[15]Stricklin, J. A., Heisler, W. E., and Von Rusman, W., "Evaluation of Solution Procedures for Material and/or Geometrically Non-Linear Structural Analysis," *AIAA Journal*, Vol. 11, 1973, pp. 292–299.

[16]Zienkiewicz, O. C., and Cormeau, I. C., "Visco-Plasticity, Plasticity and Creep in Elastic Solids—A Unified Numerical Solution Approach," *International Journal for Numerical Methods in Engineering*, Vol. 8, 1974, pp. 821–845.

[17]Belytschko, T., Fish, J., and Bayliss, A., "The Spectral Overlay on Finite Elements for Problems with High Gradients," *Computer Methods in Applied Mechanics and Engineering*, Vol. 6, 1990, pp. 71–76.

[18]Gupta, K. K., "STARS—An Integrated, Multidisciplinary, Finite-Element, Structural, Fluid, Aeroelastic, and Aeroservoelastic Analysis Computer Program," NASA TM 4795, May 1997.

12
Stress Computations and Optimization

12.1 Introduction

In the preceding chapters, computational procedures have been described that yield structural deformations for linear and nonlinear problems subjected to static and also dynamic time-dependent mechanical and thermal loading. Stresses in various elements in the structure due to nodal deflections may be calculated to check if the structure can safely withstand the externally applied load. Typically the process first includes computation of element nodal deflections in the LCS from values already computed in the GCS. These values are then used along with the element strain transformation and constitutive matrices derived in their LCS to compute element stresses. Thus element stress computation involves the following steps.[1,2,3]

Step 1: Obtain element node deflections from global values

$$\boldsymbol{u}^e = \boldsymbol{\lambda} \boldsymbol{q}^e \tag{12.1}$$

Step 2: Compute element strains from the strain displacement relationship

$$\boldsymbol{\epsilon} = \boldsymbol{B} \boldsymbol{u}^e \tag{12.2}$$

Step 3: Compute element stresses from strains with the appropriate constitutive law

$$\begin{aligned} \boldsymbol{\sigma} &= \boldsymbol{D} \boldsymbol{\epsilon} \\ &= \boldsymbol{D} \boldsymbol{B} \boldsymbol{u}^e \end{aligned} \tag{12.3}$$

The derivations for a number of elements are given next, followed by details on structural optimization.

12.2 Line Elements

The following steps are necessary for the computation of stresses in a line beam element.

Step 1: Obtain element nodal deformations in LCS

$$\boldsymbol{u}^e = \boldsymbol{\lambda} \boldsymbol{q}^e \tag{12.4}$$

in which

$$\boldsymbol{u}^e = \left\{ u_x^1 \, u_y^1 \cdots u_{\theta z}^1 \quad u_x^2 \, u_y^2 \cdots u_{\theta z}^2 \right\}^T \tag{12.5}$$

and $\boldsymbol{\lambda}$ is the element direction cosine matrix.

Step 2: Compute element nodal forces in LCS from

$$\boldsymbol{p}^e = \boldsymbol{K}^e \boldsymbol{u}^e \tag{12.6}$$

in which $\boldsymbol{K}^e$ is the element stiffness matrix in LCS.

Step 3: Calculate element stresses

$$f_x = \text{axial stress (typical)} = p_x/A$$

$$f_{by} = \text{bending stress (typical)} = (p_{\theta y}/I_y)(t_z/2)$$

The total stress is a suitable combination of the various components; I_y, t_z being the moment of inertia about local y axis and thickness in the local z direction, respectively.

12.3 Triangular Shell Elements

Computation of stresses in a triangular element, plane, plate bending, or shell may be obtained by following the standard procedure.

Step 1: Compute element node displacements

$$\boldsymbol{u}^e = \boldsymbol{\lambda q}^e \tag{12.7}$$

and form the node displacement vector

$$\boldsymbol{u} = \begin{bmatrix} u_{x_1} + zu_{\theta y_1} & u_{x_2} + zu_{\theta y_2} & u_{x_3} + zu_{\theta y_3} & u_{y_1} - zu_{\theta x_1} & u_{y_2} - zu_{\theta x_2} & u_{y_3} + zu_{\theta x_3} \end{bmatrix}^T \tag{12.8}$$

in which $z = t/2$ and $-t/2$ for top and bottom layers, respectively, where t is the shell thickness.

Step 2: Compute strain–displacement matrix $\boldsymbol{B}$ by forming three nodal matrices:

$$\boldsymbol{B}_2^i = \begin{bmatrix} X_1^i & q_X^i & q_{\theta X}^i \\ Y_1^i & q_Y^i & q_{\theta Y}^i \\ Z_1^i & q_Z^i & q_{\theta Z}^i \end{bmatrix} \qquad i = 1, 2, 3 \tag{12.9}$$

in which typically $X_1^i = X_i - X_1$, and so on, where the X and q are in GCS. Transform $\boldsymbol{B}_2^i$ into element LCS

$$\boldsymbol{B}_1^i = \boldsymbol{\lambda B}_2^i \tag{12.10}$$

Obtain $\boldsymbol{B}$ as

$$\boldsymbol{B} = \begin{bmatrix} \boldsymbol{B}_1^1 & \boldsymbol{B}_1^2 & \boldsymbol{B}_1^3 \end{bmatrix} \tag{12.11}$$

Step 3: Calculate stresses at top and bottom layer

$$[\sigma_x \quad \sigma_y \quad \tau_{xy}]^T = \frac{1}{2A}\boldsymbol{DBu} \tag{12.12}$$

in which A is the surface area of the element and $\boldsymbol{D}$ is the (3×3) material constitutive matrix for plane stress formulation.

This procedure may also be used to compute stresses in elements constructed of layered composite materials. Thus element stresses are first calculated in the LCS at a particular fiber level and then transfered in the coordinate axes of the laminate as

$$\boldsymbol{\sigma}^* = \boldsymbol{T}_\sigma \boldsymbol{\sigma} \tag{12.13}$$

in which $\boldsymbol{T}_\delta$ is the stress transformation matrix defined in an earlier section and here takes the following form:

$$\begin{bmatrix} \sigma_x^* \\ \sigma_y^* \\ \tau_{xy}^* \end{bmatrix} = \begin{bmatrix} \cos^2\theta & \sin^2\theta & \sin 2\theta \\ \sin^2\theta & \cos^2\theta & -\sin 2\theta \\ \dfrac{-\sin 2\theta}{2} & \dfrac{\sin 2\theta}{2} & \cos 2\theta \end{bmatrix} \begin{bmatrix} \sigma_x \\ \sigma_y \\ \tau_{xy} \end{bmatrix} \tag{12.14}$$

where θ is the angle between the element local x axis and the direction of the laminate fiber.

For quadrilateral elements, stresses are first computed in each of the four constituent triangles in the element coordinate system and multiplied with their respective areas, and next their sum is divided by the entire element area to yield the average stress.

12.4 Solid Elements

For a tetrahedron element, computation of stresses follow the usual procedure in which nodal deflections are first converted into the LCS and stresses are computed from

$$\boldsymbol{\sigma} = \boldsymbol{D}\boldsymbol{B}\boldsymbol{u}^e \tag{12.15}$$

using the material constitutive matrix $\boldsymbol{D}$ that pertains to three-dimensional stress distribution. A hexahedron, on the other hand, is divided into two sets of five suitable tetrahedrons, and stresses are calculated in each element and multiplied by their respective volumes. Their sum is next divided by the volume of the hexahedron to yield the average element stress. Similarly three tetrahedron elements are used for stress computation in a pentahedron and are then suitably combined to yield the average stress. For isoparametric element formulation, stresses are usually calculated at the Gauss points of integration of order at least one less than that used in element stiffness matrix generation.

Example 12.1: Stress Analysis of a Cantilever Shell

Figure 12.1 shows a cantilever curved shell subjected to a uniformly distributed load. The relevant data parameters are side length $A, B = 10$, radius $R = 20$, thickness $t = 0.1$, Young's modulus $E = 29.5 \times 10^6$, Poisson's ratio $\nu = 0.3$, mass density $\rho = 0.733 \times 10^{-3}$, and uniform distributed load $= 1.0$.

Figure 12.2 shows the maximum principal stress distribution in the shell due to external loading.

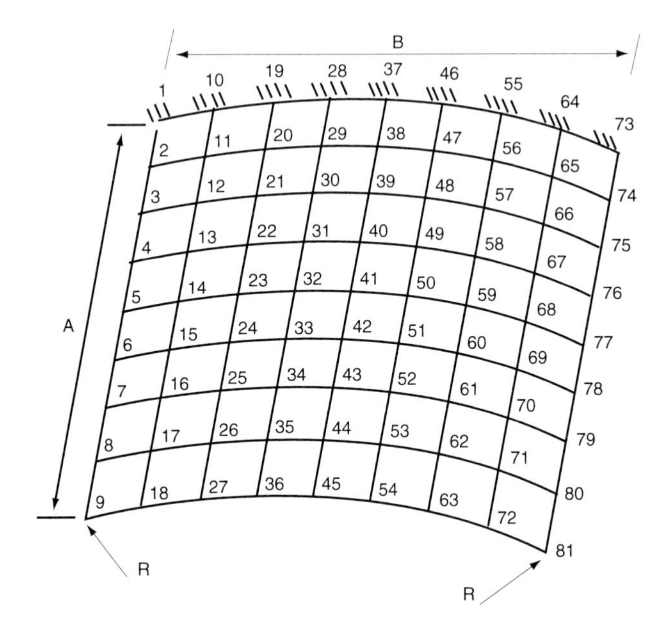

Fig. 12.1 Finite-element model of cylindrical shell.

Fig. 12.2 Maximum principal stress distribution in a cantilever shell.

12.5 Optimization

12.5.1 Introduction

The design and synthesis[3-7] of an engineering system involves repeated, iterative analysis to obtain an efficient solution based on the satisfaction of specified design criteria. In mathematical terms, it involves optimization of an objective function by varying a set of prescribed design variables subject to certain specified physical requirements known as regional constraints. The design variables may be related by a set of functional constraints defining their interrelationship and also, most important, may be subjected to side constraints that define the limits of variation of their values. The following definitions are pertinent:

Objective function [function to be optimized(minimized or maximized)]:

$$f(\boldsymbol{x}) = f(x_1, \ldots, x_n) \tag{12.16}$$

In this definition f is the objective function and the various x_i are the design variables. The optimization procedure involves finding $\boldsymbol{x}$, subject to a number of constraints, for which the value of the objective function f is a minimum (or a maximum). The chosen functional in structural mechanics may be the vehicle weight, whereas in fluid mechanics it may be the drag or prescribed pressure.

Regional constraints (upper and lower limits):

$$\phi_j^l \leq \phi_j(\boldsymbol{x}) \leq \phi_j^u \qquad j = 1, \ldots, q \tag{12.17}$$

A set (m) of lower (l) and/or upper (u) limits are specified, such as stresses and displacements in structural mechanics or drag and lift in fluid mechanics.

Side constraints (limits on some design variables):

$$x_i^l \leq x_i \leq x_i^u \qquad i = 1, \ldots, r \tag{12.18}$$

Some of the design variables, for example, shell thickness in structural mechanics or airfoil shape in fluid mechanics, are constrained to vary within certain limits.

Functional constraints (functional relationship between design variables):

$$\psi_p(\boldsymbol{x}) = 0 \qquad p = 1, \ldots, s \tag{12.19}$$

These functions define the relationship between the design variables. They may occur, for example, in structural mechanics in wing design.

Examples of structural optimization[8–12] mostly involve weight as the objective function to be minimized, subject to a combination of stress, displacement, or frequency restraints as well as imposed bounds (member cross sectional area, shell thickness, etc.), on design variables. Specified bounds may be imposed on structural material properties, and vehicle shape may be an added design variable. All of these requirements can complicate the optimization process. In aeronautics,[13,14] an optimum wing shape is sought to provide maximum efficiency for a representative number of flight conditions within the flight envelope. The related objective function may include maximum lift and minimum drag and pitching moments, each of which may have the others as constraints and in addition be further subjected to constraints such as aeroelastic flutter and aeroservoelastic instability. A multidisciplinary optimization[15,16] may involve structural design for minimum weight with aerodynamic data of drag and wing plan form as design variables, subjected to constraints of flutter and structural strength. Adoption of the finite element method for both fluid and structural discretization enables a seamless interface between the two disciplines for the purpose of design optimization.

12.5.2 Optimality Conditions

The present discussion of optimality conditions follows Ref. 17. The optimization problems previously described all can be stated in the following canonical form:

Minimize

$$f(\boldsymbol{x})$$

Subject to

$$\begin{aligned} c_i(\boldsymbol{x}) &= 0, \qquad & i &= 1, \ldots, k \\ c_i(\boldsymbol{x}) &\geq 0, \qquad & i &= k+1, \ldots, m \end{aligned} \tag{12.20}$$

where $\boldsymbol{x}$ is an n-dimensional vector of design parameters, f and c_i are real valued functions of $\boldsymbol{x}$, f is the objective function, and the various c_i are constraint functions. A point $\boldsymbol{x}$ that satisfies all the constraints is called a *feasible point*. The set of all such points is called the *feasible region*.

Only the identification of a local minimum for Eq. (12.20) will be considered. Methods for global minimization generally have two parts: 1) some method of partitioning the total area into smaller regions and 2) local minimization in the smaller regions.

Quite different solution algorithms have been developed for the local solution of Eq. (12.20) depending on the properties of the objective function and the constraint functions and also the presence or absence of the equality and/or inequality constraints. For example, the objective function may be linear, quadratic, or general nonlinear but continuous, or with continuous first partial derivatives, or with continuous second partial derivatives. These conditions must be considered in combination with the constraint functions being linear, or nonlinear but continuous, or with continuous first partial derivatives. In structural design one is most likely to be interested in a case of an objective function that is nonlinear with at least continuous second derivatives, and constraint functions that are either linear, or are nonlinear with at least continuous first derivatives.

Assume f has continuous second partial derivatives. Let $\boldsymbol{g}(\boldsymbol{x})$ denote the n-dimensional gradient vector of $f(\boldsymbol{x})$, and $\boldsymbol{G}(\boldsymbol{x})$ denote the $n \times n$ Hessian matrix of $f(\boldsymbol{x})$, i.e., the symmetric matrix of second partial derivatives of $f(\boldsymbol{x})$.

For the unconstrained case, i.e., with $m = 0$, necessary conditions for a point $\boldsymbol{x}^*$ to be a local minimum are that $\boldsymbol{g}(\boldsymbol{x}^*) = 0$ and $\boldsymbol{G}(\boldsymbol{x}^*)$ is positive semidefinite. With the stronger requirement that $\boldsymbol{G}(\boldsymbol{x}^*)$ is positive definite these become sufficient conditions.

Moving on to consider the constrained case, assume the constraint functions $c_i(\boldsymbol{x})$ have at least continuous first partial derivatives, and let $\boldsymbol{J}(x)$ denote the $m \times n$ matrix whose ith row is the n-dimensional gradient vector of $c_i(\boldsymbol{x})$. Let

$\mathcal{A}(\boldsymbol{x})$ denote the set of indices i such that $c_i(\boldsymbol{x}) = 0$. This set of constraints is called the *active set* at $\boldsymbol{x}$. Let $\boldsymbol{Z}(\boldsymbol{x})$ be a matrix whose columns form a basis for the subspace of n-dimensional vectors that are orthogonal to all rows of $\boldsymbol{J}(\boldsymbol{x})$ indexed in $\mathcal{A}(\boldsymbol{x})$.

To appreciate the significance of $\boldsymbol{Z}(\boldsymbol{x})$, note that if all of the constraint functions in the active set associated with $\boldsymbol{x}$ are linear, then all of these constraint functions will continue to have the value zero at any point $\boldsymbol{x} + \boldsymbol{h}$ such that the vector $\boldsymbol{h}$ is a linear combination of columns of $\boldsymbol{Z}(\boldsymbol{x})$. Thus, moves from $\boldsymbol{x}$ can be made in directions defined by such vectors $\boldsymbol{h}$ without violating any of the constraints indexed in $\mathcal{A}(\boldsymbol{x})$. Various algorithms make use of this property. However if the constraints indexed in $\mathcal{A}(\boldsymbol{x})$ are nonlinear, such $\boldsymbol{h}$ moves generally will not keep the active set constraints at a constant value of zero, but these directions are still of importance in algorithms, because they will be directions in which small moves can be expected to make relatively small changes in the values of the constraint functions.

Optimality conditions for Eq. (12.20) with linear constraints. Necessary conditions for $\boldsymbol{x}^*$ to be a local minimum are

$$c_i(\boldsymbol{x}^*) = 0, \qquad i = 1, \ldots, k$$

$$c_i(\boldsymbol{x}^*) \geq 0, \qquad i = k+1, \ldots, m$$

There exists an m-vector λ^* such that $\boldsymbol{g}(\boldsymbol{x}^*) = \boldsymbol{J}(\boldsymbol{x}^*)^T \lambda^*$ with

$$\lambda_i^* \geq 0, \qquad i = k+1, \ldots, m$$

For $i = k+1, \ldots, m$, if $c_i(\boldsymbol{x}^*) > 0$ then $\lambda_i^* = 0$; $\boldsymbol{Z}(\boldsymbol{x}^*)^T \boldsymbol{G}(\boldsymbol{x}^*) \boldsymbol{Z}(\boldsymbol{x}^*)$ is positive semidefinite.

The conditions involving $\boldsymbol{\lambda}^*$ imply

$$\boldsymbol{Z}(\boldsymbol{x}^*)^T \boldsymbol{g}(\boldsymbol{x}^*) = 0$$

Sufficient conditions for $\boldsymbol{x}^*$ to be a local minimum are as in the necessary conditions, but requiring

$$\lambda_i^* > 0, \qquad i = k+1, \ldots m$$

and $\boldsymbol{Z}(\boldsymbol{x}^*)^T \boldsymbol{G}(\boldsymbol{x}^*) \boldsymbol{Z}(\boldsymbol{x}^*)$ is positive definite.

Alternative sufficient conditions can be stated allowing for some of the λ_i^*'s associated with inequality constraints to be zero. Let $\tilde{\boldsymbol{Z}}(\boldsymbol{x}^*)$ be a matrix whose columns form a basis for the subspace of n-dimensional vectors that are orthogonal to all rows of $\boldsymbol{J}(\boldsymbol{x}^*)$ indexed in $\mathcal{A}(\boldsymbol{x}^*)$ and also to each row i of $\boldsymbol{J}(\boldsymbol{x}^*)$ for which $\lambda_i^* > 0$, with $k + 1 \leq i \leq m$. Then the sufficient conditions involving all λ_i^* and $\boldsymbol{G}(\boldsymbol{x}^*)$ are

$$\lambda_i^* \geq 0, \qquad i = k+1, \ldots, m$$

and $\tilde{\boldsymbol{Z}}(\boldsymbol{x}^*)^T \boldsymbol{G}(\boldsymbol{x}^*) \tilde{\boldsymbol{Z}}(\boldsymbol{x}^*)$ is positive definite.

Optimality Conditions for Eq. (12.20) with nonlinear constraints. Optimality conditions for the problem with nonlinear constraints are the same as already stated for linear constraints with the addition of constraint qualifications. Researchers have studied different constraint qualifications, but one that is straightforward to state is the linear independence constraint qualification, which is satisfied at a feasible point $\boldsymbol{x}$ if the set of gradient vectors of the equality constraints and the active inequality constraints are linearly independent at $\boldsymbol{x}$.

12.5.3 Numerical Solution Methods

For Eq. (12.20) without constraints, when f has at least a continuous first derivative, one could use a method of gradient descent. The negative gradient gives a direction of local steepest descent. Moves of various lengths in that direction can be tried. For sufficiently small moves the value of f will decrease, but for larger moves the value will generally increase. After sampling a few points along the negative gradient direction, various methods based on quadratic or cubic polynomial interpolation may be used to predict the lowest point along that path. From the lowest point found, the process can be started again by computing the gradient vector at this point. This can be a very slowly converging method.

If f has a continuous second derivative one can seek a zero of the gradient function by using some variation of Newton's method. In its simplest form, Newton's method for solving the (assumed nonlinear) problem $\boldsymbol{g}(\boldsymbol{x}) = 0$, would be as follows.

Give $\boldsymbol{x}$ an initial value.
Do
  Solve for $\mathrm{d}\boldsymbol{x}$ in $\boldsymbol{G}(\boldsymbol{x})\,\mathrm{d}\boldsymbol{x} = -\boldsymbol{g}(\boldsymbol{x})$
  $\boldsymbol{x} = \boldsymbol{x} + \mathrm{d}\boldsymbol{x}$
  Exit if $\|\mathrm{d}x\|$ is sufficiently small.
Enddo

The main attraction of Newton's method is its second-order convergence, i.e., when sufficiently close to the solution it approximately doubles the number of correct digits at each iteration. Many elaborations and variations on this method have been developed[18] to make this algorithm robust and to approximate the behavior of this method when it is inconvenient to compute values of second partial derivatives of f.

Now consider Eq. (12.20) with the equality and inequality constraints. If the constraint functions are all linear, there are very specialized and relatively efficient algorithms for the cases of f being linear (*linear programming*) and the case of f being quadratic (*quadratic programming*).

Approaches for more general problems in which f is at least assumed to have continuous second partial derivatives and the various $\boldsymbol{c}_i$ are at least assumed to have continuous first partial derivatives generally fall into one of three general types: 1) sequential quadratic programming (SQP); 2) methods that attempt to follow a descent path that bends to follow the boundary of the feasible region and generally tries to keep a changing set of already-encountered constraints (the active set) at zero values while continuing to descend; and 3) methods that construct a sequence of new *pseudo-objective* functions that combine the given objective function and the constraint functions in such a way that unconstrained minimization of a pseudo-objective function will move the process toward the solution of the original constrained problem.

In the SQP method, at a current trial point $\boldsymbol{x}^j$ one sets up a quadratic programming problem with the objective function being the second order Taylor series expansion of f at $\boldsymbol{x}^j$ and the constraint functions being the first order Taylor series expansion of c_i at $\boldsymbol{x}^j$. The solution of this QP problem defines the next trial point, $\boldsymbol{x}^{j+1}$.

Examples of type 2 are the feasible directions, projected gradient, and reduced gradient methods. Examples of type 3 are penalty, barrier, and interior point methods.

The theory and algorithmic details of the methods mentioned here, and others, are thoroughly presented in Ref. 19. An excellent survey of these and additional methods, with emphasis on considerations that arise in applying them to engineering applications, is given in Ref. 4. In 1984, Ref. 21 sparked a major renewal of interest in interior barrier methods by researchers in optimization algorithms. Exposition of this work is given in Refs. 17 and 22. Readers seeking optimization software will find Ref. 23 to be helpful.

Independent of research in optimization algorithms, but significantly motivated by its usefulness in optimization computations, extensive research has been carried out in computer methods for computing first and second derivatives of functions that are defined by source code, e.g., by FORTRAN or C/C++ programs. Much of this work is identified as AD, which can denote either *automatic* or *algorithmic differentiation*. Implementation approaches vary. For example, one approach is to have an AD software system take as input the source code for a function, or set of functions, and produce source code that will compute first and/or second partial derivatives of the function, or functions, with respect to a specified set of independent variables. Remarkable success has been achieved in extending these methods to very large and difficult cases. Further information is given in Ref. 24.

12.6 Examples of Applications of Optimization

Two structural optimization examples that have been run are included in this chapter. To perform the structural optimization, data in addition to that used for standard structural analysis is required, and this, together with the optimization results, is given for the two examples included in this chapter. The STARS software[25] includes optimization[26] and has been used for the cantilever shell structure Example 4.1.5 in Ref. 25, reinforced by beam elements along its three free edges. The first problem includes static analysis with stress and displacement constraints, while the second problem involves vibration analysis in which the shell is subject to constraints on the bounds of its natural frequencies.

Example 12.2: Cantilever Shell–Static Analysis Optimization

The shell is subdivided into four zones (Fig. 12.3), each with different design variables, which in this case are the thicknesses and the beam cross sections.

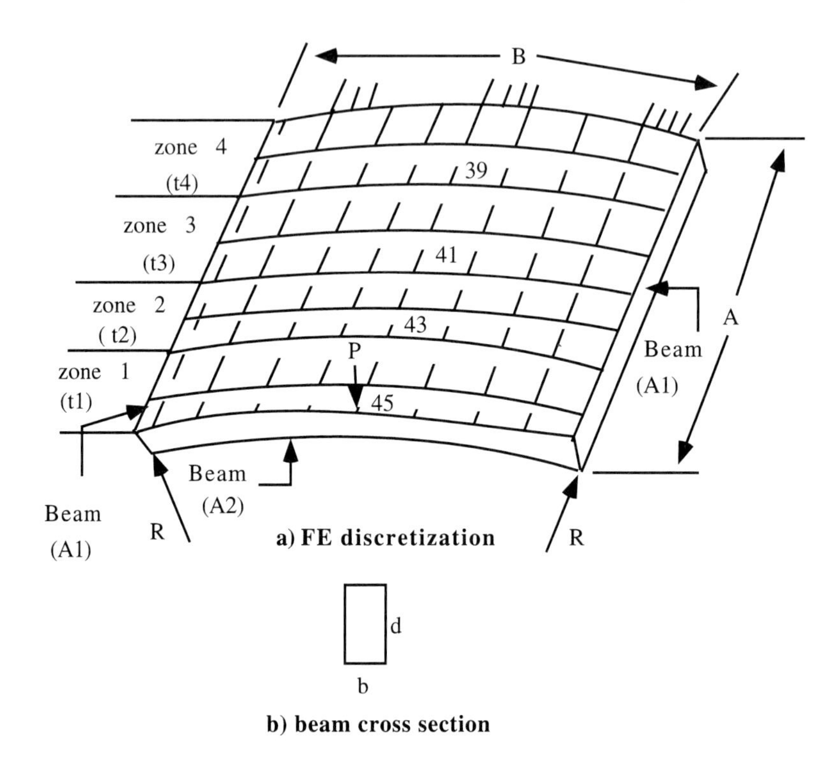

Fig. 12.3 Cantilever shell structure with edge beams.

Displacement constraints of 0.1 are imposed on the four central nodes in each of the four zones. Also, in-plane stresses in the structure are required to not exceed the yield stress value of 50,000. A static load is applied at the center of the tip of the cantilever edge. The optimization analysis was performed to produce a shell of minimum weight. Table 12.1 gives a summary of the results of this optimization. Note that the weight was reduced from the initial 15.7 to 3.7.

Table 12.1 Optimized static analysis—cantilever shell

Design variable	Initial values	Final values
A_1	1.5	0.078245
A_2	1.5	0.010011
t_1	0.1	0.128530
t_2	0.1	0.070448
t_3	0.1	0.044998
t_4	0.1	0.204520
Functional weight	15.6565	3.68200

The input parameters for optimization analysis are as follows.

Shell properties:

Side length AB	= 10
Radius R	= 20
Young's modulus E	$= 29.5 \times 10^6$
Poisson's ratio μ	= 0.3
Density ρ	$= 0.733 \times 10^{-3}$
shell thickness	
$t_1 = t_2 = t_3 = t_4$	= 0.1

Beam element properties:

Inertia I_1	= 0.08333
Inertia I_2	= 0.041667
Polar inertia J	= 0.125
Beam dimensions b, d	= 1.0, 1.5
Fixed aspect ratio b/d	= 2/3
Beam area A_1, A_2	= 1.5
concentrated load P	= 500

Example 12.3: Cantilever Shell—Optimization Analysis with Frequency Constraints

An optimization analysis of the structure (Fig. 12.3) was performed with specified frequency constraints. All basic data are the same as in Example 12.2. The first six frequencies were constrained to have specified lower limits with design variable side constraints as follows: 1) minimum beam cross section area = 0.25 and 2) minimum shell thickness = 0.025. Results of the optimization are given in Table 12.2

Table 12.2 Optimized free vibration analysis—cantilever shell

Design variable	Initial values	Final optimized values
A_1 (X1)	1.5	0.455
A_2 (X2)	1.5	0.250
t_1 (X3)	0.1	0.0887
t_2 (X4)	0.1	0.025
t_3 (X5)	0.1	0.211
t_4 (X6)	0.1	0.025
	constraints frequency (minimum)	
$\lambda_1 = 1000$	1144.20	1360.04
$\lambda_2 = 1600$	1807.54	1744.24
$\lambda_3 = 3000$	5070.27	3892.05
$\lambda_4 = 4000$	5251.39	5653.48
$\lambda_5 = 5000$	6246.96	6246.08
$\lambda_6 = 6000$	6929.61	7605.45
	functional weight (OBJ)	
	15.6565	5.79

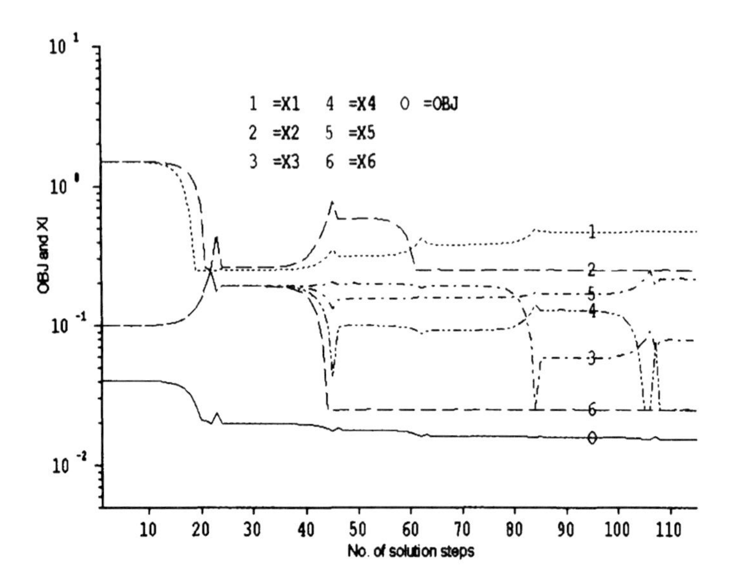

Fig. 12.4 Optimization convergence plots for the cantilever shell.

and the solution convergence features are shown in Fig. 12.4. Note in Table 12.2 that the weight was reduced from the initial 15.7 to 5.8.

References

[1]Timoshenko S., and Goodier, N., *Theory of Elasticity*, 3rd ed., McGraw-Hill, New York, 1951.

[2]Timoshenko S., and Woinsowsky-Krieger, S., *Theory of Plates and Shells*, 2nd ed., McGraw–Hill, New York, 1959.

[3]Przemieniecki, J. S., *Theory of Matrix Structural Analysis*, McGraw-Hill, New York, 1968.

[4]Vanderplaats, G. N., *Numerical Optimization Techniques for Engineering Design: With Applications*. McGraw–Hill, New York, 1984.

[5]Fox, R. L., *Optimization Methods for Engineering Design*, Addison-Wesley, Reading, MA, 1971.

[6]Haug, E. J., and Arora, J. S., *Applied Optimal Design* Wiley, New York, 1979.

[7]Kirch, V., *Optimum Structural Design*, McGraw–Hill, New York, 1981.

[8]Schmidt, L. A., Structural Synthesis: Its Genesis and Development, *AIAA Journal*, Vol. 10, No. 10, Oct. 1981, pp. 1249–1263.

[9]Gellatly, R. A., and Gallagher, R. H., "Development of Advanced Structural Optimization Programs and Their Application to Large Order Systems,' *Proceedings of Conference on Matrix Methods in Structural Mechanics*, Wright–Patterson AFB, OH, 1965.

[10]Razzni, R., "Behavior of Fully Stressed Design of Structures and Its Relationship to Minimum Weight Design," *AIAA Journal*, Vol. 3, No. 12, 1965, pp. 2262–2268.

[11]Venkayya, V. B., Structural Optimization: A Review and Some Recommendations. *International Journal of Numerical Methods in Engineering*, Vol. 13, No. 2, 1978, pp. 203–228.

[12]Miura, H., and Schmidt, L. A., "Second-Order Approximation of Natural Frequency Constraints in Structural Synthesis," *International Journal of Numerical Methods in Engineering*, Vol. 13, No. 2, 1978, pp. 337–351.

[13]Livne, E., "Integrated Aeroservoelastic Optimization: Status and Direction," *Journal of Aircraft*, Vol. 36, No. 1, 1999, pp. 122–143.

[14]Jameson, A., and Vassbreg, J., "Computational Fluid Dynamic Design: Its Current and Future Impact," *AIAA Paper 2001-0538*, 2001.

[15]Reuther, J. A., Martins, J., and Smith, S., "A Coupled Aerostructural Optimization Method for Complete Aircraft Configurations," *AIAA Paper 2001-0538*, 1999.

[16]Striz, A. G., and Venkayya, V. B.,"Influence of Structural and Aerodynamic Modeling on Flutter Analysis," *Journal of Aircraft*, Vol. 31, No. 9, 1994, pp. 1205–1211.

[17]Forsgren, A., Gill, P. E. and Wright, M. H., *Interior Method for Nonlinear Optimization, SIAM Review*, Vol. 44, No. 4, 2002, 525–597.

[18]Dennis, J. E., Jr., and Schnabel, R. B., *Numerical Methods for Unconstrained Optimization and Nonlinear Equations*, Prentice–Hall, Englewood Cliffs, NJ, 1983.

[19]Gill, P. E., Murray, D. B. and Wright, R. B., *Practical Optimization*, Academic Press, New York, 1981.

[20]Vanderplaats, G. N., *Numerical Optimization Techniques for Engineering Design*, 3rd ed., Vanderplaats Research and Development, Colorado Springs, CO, 2001.

[21]Karmarkar, N., *A New Polynomial–Time Algorithm for Linear Programming, Combinatorica*, Vol. 4, 1984, 373–395.

[22]Wright, S. J., *Primal-Dual Interior-Point Methods*, SIAM, Philadelphia, PA, 1997.

[23]Moré, J. J. and Wright, S. J., *Optimization Software Guide*, SIAM, Philadelphia, PA, 1993.

[24]Griewank, A., *Evaluating Derivatives: Principles and Techniques of Algorithmic Differentiation*, SIAM, Philadelphia, PA, 2000.

[25]Gupta, K. K., "STARS—An Integrated General-Purpose Finite Element Structural, Aeroelastic and Aeroservoelastic Analysis Computer Program," NASA TM 4795, May 1997, revised Jan. 2003.

[26]Vanderplaats, G. N., *ADS-A Fortran Program for Automated Design Synthesis-Version 1.10*, NASA Contractor Rept. 177985, Langley Research Center, Hampton, VA, 1985.

13
Heat Transfer Analysis of Solids

13.1 Introduction

The process of heat transfer in a general medium may be categorized as conduction, convection, or radiation. In conduction,[1–3] transfer of thermal energy through a solid or fluid medium is caused by a temperature gradient between two locations that does not involve any material mass motion. The transfer of this thermal energy through a fluid because of motion of the fluid is known as convection. In this process, transfer of energy from one fluid particle to another occurs by conduction, but the thermal energy is transported by the fluid motion. The related convective heat transfer analysis is achieved by combining conduction heat transfer with the equations of fluid mechanics. This process depends on the temperature difference between two locations. In radiation,[4–6] a transfer between two locations is effected by electromagnetic wave propagation, which depends on the temperature difference between the two locations. Figure 13.1 provides some basic features of each of the phenomena; v, p, and T are the velocity, pressure, and temperature, respectively.

The heat conduction problem is primarily driven by the set of boundary conditions of the region being considered, whereas in radiation, the pattern of transfer is either between two radiating surfaces or from a single surface. Convection heat transfer phenomenon for a viscous flow, on the other hand, combines the effect of fluid flow heat transport with that due conduction and is beyond the scope of this text.

13.2 Heat Conduction

13.2.1 Background

Heat conduction occurs at atomic and molecular levels without any material mass motion and is caused by temperature gradients in the solid or fluid. The transfer of thermal energy in an anisotropic solid domain D, bounded by a surface S, is governed by the following energy equation:

$$-\frac{\partial q_{X_i}}{\partial X_i} + Q = \rho c \frac{\partial T}{\partial t} \tag{13.1}$$

in which $X_1 = X$, $X_2 = Y$, $X_3 = Z$ are the coordinate axes, q_{X_i} is the heat flow rate vector per unit area having components q_X, q_Y, q_Z, $Q(X, Y, Z, t)$ is internal heat generation rate per unit volume, ρ is density, c is specific heat, T is temperature, and t is time. This equation is next coupled with Fourier's law of heat conduction that defines the rate equation describing heat transfer mode:

$$q_{X_i} = -\bar{K}_{ij} \frac{\partial T}{\partial X_i}, \qquad i = 1, 2, 3 \qquad j = 1, 2, 3 \tag{13.2}$$

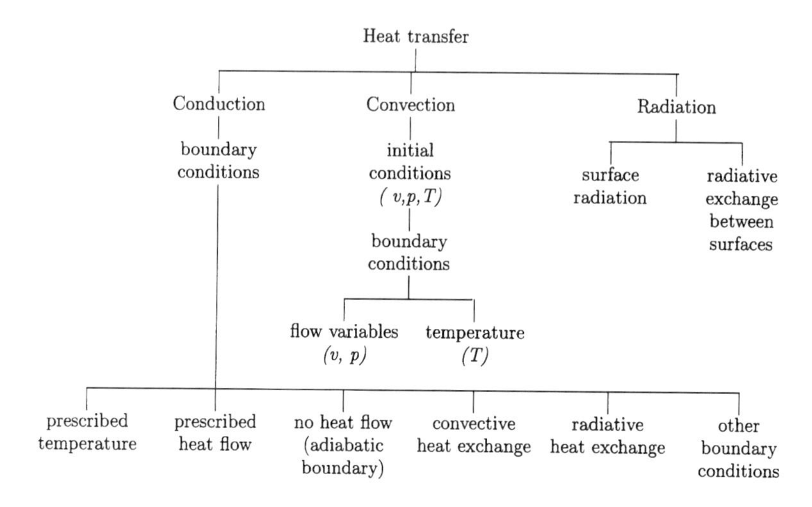

Fig. 13.1 Various heat transfer processes.

where q_X, q_Y, q_Z are the components of the heat flow rate vector per unit area, and K_{ij} is the symmetric thermal conductivity tensor. This yields the parabolic heat conduction equation to be solved subject to initial conditions as well as boundary conditions on all portions of the surface S. These conditions are defined as follows.

Initial condition:

$$T(X, Y, Z, t) = T_0(X, Y, Z) \qquad \text{specifying temperature distribution at time zero}$$

Boundary conditions:

$$T_{s_T} = T_{s_T}(X, Y, Z, t), \qquad \text{specified temperature on portion of surface } S_T$$

$$q_X n_X + q_Y n_Y + q_Z n_Z = -q_s, \qquad \text{specified surface heat flow on } S_S$$

$$q_X n_X + q_Y n_Y + q_Z n_Z = h(T_S - T_e), \qquad \text{convective heat exchange on } S_h$$

$$q_X n_X + q_Y n_Y + q_Z n_Z = \sigma \epsilon T_S^4 - \alpha q_r, \qquad \text{radiation heat exchange on } S_r$$

where

n_X, n_Y, n_Z = direction cosines of outward normal to the surface
q_s = specified heat flow rate per unit area, positive into the surface
T_e = convective exchange temperature of an adjacent fluid
h = convective heat transfer coefficient that may be a function of T_e or T
T_s = unknown surface temperature
σ = Stefan–Boltzmann constant
ϵ, α = surface emissivity and absorptivity, respectively, usually being a function of T_s
q_r = incident radiant heat flow rate per unit area

13.2.2 Finite Element Formulation

The solution domain[7] D is divided into r number of typical elements, one, two, or three dimensional as the case may be. Each element has p number of vertices. Then the temperature T and its gradient $\partial T/\partial X_i$ within each element may be expressed in terms of its nodal unknown temperatures as

$$T = \boldsymbol{N}\boldsymbol{T} \tag{13.3}$$

$$\frac{\partial T}{\partial X_i} = \frac{\partial \boldsymbol{N}}{\partial X_i}\boldsymbol{T} \tag{13.4}$$

in which $\boldsymbol{N}$ is a suitable shape function row matrix and $\boldsymbol{T}$ is the element nodal temperature vector. Then

$$\frac{\partial T}{\partial X_i} = \boldsymbol{B}_i\boldsymbol{T} \tag{13.5}$$

and Fourier's law can then be expressed using Eq. (13.2):

$$\boldsymbol{q} = -\boldsymbol{K}\frac{\partial \boldsymbol{T}}{\partial \boldsymbol{X}_i} = -\boldsymbol{K}\boldsymbol{B}\boldsymbol{T} \tag{13.6}$$

For a single finite element, the Galerkin weighted residual for Eq. (13.1) takes the following form:

$$\int_{D^{(e)}} \left(\frac{\partial q_{X_i}}{\partial X_i} - Q + \rho c \frac{\partial T}{\partial t} \right) \boldsymbol{N}^{(e)}\, \mathrm{d}D = 0 \tag{13.7}$$

Using Gauss's theorem, Eq. (13.7) can be split into surface and volume integrals, and also using Eqs. (13.3) and (13.6) and with proper consideration of boundary conditions, the element equilibrium equation is obtained as

$$\boldsymbol{C}^{(e)}\dot{\boldsymbol{T}} + \boldsymbol{K}^{(e)}\boldsymbol{T} = \boldsymbol{p}^{(e)} \tag{13.8}$$

in which the element conductance matrix is

$$\boldsymbol{K}^{(e)} = \left[\boldsymbol{K}_c^{(e)}(T) + \boldsymbol{K}_h^{(e)}(T,t) + \boldsymbol{K}_r^{(e)}(T)\right]\boldsymbol{T}(t) \tag{13.9}$$

the heat input vector is given as

$$\boldsymbol{p}^{(e)} = \boldsymbol{p}_Q(T,t) + \boldsymbol{p}_q(T,t) + \boldsymbol{p}_h(T,t) + \boldsymbol{p}_r(T,t) \tag{13.10}$$

the capacitance matrix is defined as

$$\boldsymbol{C}^{(e)} = \int_{D^{(e)}} \rho c \boldsymbol{N}^T\boldsymbol{N}\, \mathrm{d}D \tag{13.11}$$

the conductance matrix related to conduction is

$$\boldsymbol{K}_c^{(e)} = \int_{D^{(e)}} \boldsymbol{B}^T\boldsymbol{K}\boldsymbol{B}\, \mathrm{d}D \tag{13.12}$$

the conduction matrix related to convection, acting on the surface S_c, is

$$K_h^{(e)} = \int_{S_c}^{(e)} hN^T N \, \mathrm{d}S \tag{13.13}$$

and the conduction matrix related to radiation, acting on the surface S_r, is

$$K_r = \int_{S_r} \sigma \epsilon T^4 N^T \, \mathrm{d}S \tag{13.14}$$

The heat input vectors arising from various sources are given by specified nodal temperatures on the surface S_T:

$$p_T = \int_{S_T} (q \cdot \hat{n}) N^T \, \mathrm{d}S \tag{13.15}$$

due to internal heat generation

$$p_Q = \int_D Q N^T \, \mathrm{d}D \tag{13.16}$$

due to specified surface heating on surface S_s

$$p_q = \int_{S_s} q_s N^T \, \mathrm{d}S \tag{13.17}$$

due to surface convection

$$p_h = \int_{S_c} hT_e N^T \, \mathrm{d}S \tag{13.18}$$

and due to incident surface radiation

$$p_r = \int_{S_r} \alpha q_r N^T \, \mathrm{d}S \tag{13.19}$$

The various element matrices, defined in Eqs. (13.11–13.19), are easily derived for any finite element by choosing appropriate shape function N for each element type as described in Chapter 4. Thus for a line element of length ℓ

$$N = \left[1 - \frac{x}{\ell} \quad \frac{x}{\ell} \right] \tag{13.20}$$

and for a plane triangular element

$$N = \left[\left(\frac{x_3}{x_2} \frac{y}{y_3} - \frac{y}{y_3} - \frac{x}{x_2} + 1 \right) \left(\frac{x}{x_2} - \frac{x_3}{x_2} \frac{y}{y_3} \right) \frac{y}{y_3} \right] \tag{13.21}$$

Similarly for other elements, shape functions defined in Chapter 4 may be adopted. These shape functions enable computation of the various element heat transfer matrices. Equation (13.8) is the general finite element nonlinear formulation for transient heat conduction in an anisotropic medium. For temperature-dependent material,[8,9] elements of the K_c matrix as well as ρ, c, and h are nonlinear in nature. Also the problem of radiation heat exchange characterized by Eq. (13.19) is inherently nonlinear in form. The element matrices are combined as usual to yield the system matrix equation

$$\boldsymbol{C}\dot{\boldsymbol{T}} + \boldsymbol{K}\boldsymbol{T} = \boldsymbol{p} \tag{13.22}$$

which can be solved appropriately to yield desired results.

13.3 Solution of System Equations

A heat conduction analysis may pertain either to the steady-state or transient analysis. A steady-state analysis involves the solution of a set of simultaneous equations, whereas the transient analysis requires solution of first-order ordinary differential equations. These problems may be either linear or nonlinear in nature.

13.3.1 Linear Analysis

Figure 13.2 shows the analysis procedures for solution of various linear heat conduction problems. These techniques are now described in detail.

13.3.1.1 Linear steady-state analysis. The linear steady-state problem involves solution of the system equations

$$\boldsymbol{K}\boldsymbol{T} = \boldsymbol{p} \tag{13.23}$$

which may be achieved either by the standard Gaussian elimination or by the usual Cholesky decomposition, given in Chapter 7, noting that the element conductance matrices as well as the heat load vectors are constant. In Eq. (13.23),

$$\boldsymbol{K} = \boldsymbol{K}_c + \boldsymbol{K}_h \quad \text{and} \quad \boldsymbol{p} = \boldsymbol{p}_Q + \boldsymbol{p}_q + \boldsymbol{p}_h \tag{13.24}$$

13.3.1.2 Linear transient analysis. The equation of motion is written as[10–12]

$$\boldsymbol{C}\dot{\boldsymbol{T}} + \boldsymbol{K}\boldsymbol{T} = \boldsymbol{p} \tag{13.25}$$

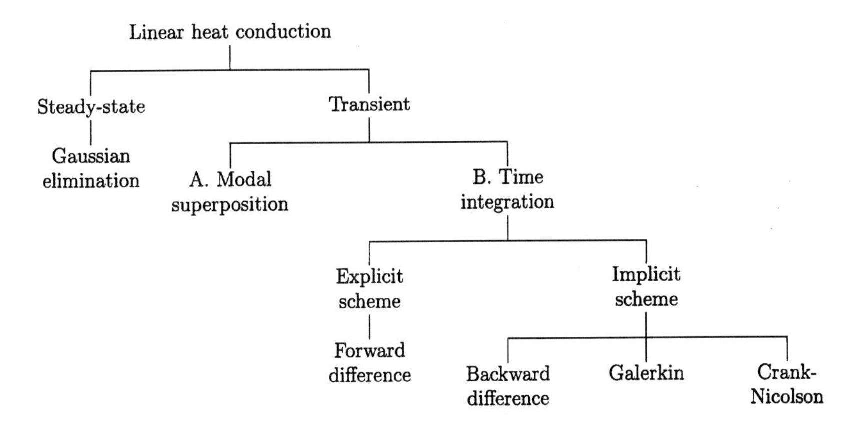

Fig. 13.2 Linear heat conduction analysis.

in which the element matrices as well as the heat load vectors are time dependent. A solution of the first-order ordinary differential equation (ODE) given by Eq. (13.25) may be achieved either by modal superposition or by a time-marching scheme, as shown in Fig. 13.2.

Modal Superposition Method. The equation of motion has the following form:

$$\boldsymbol{C}[\dot{\boldsymbol{T}}(t)] + [\boldsymbol{K}_c + [\boldsymbol{K}_h(t)]][\boldsymbol{T}(t)] = [\boldsymbol{p}_Q(t)] + [\boldsymbol{p}_q(t)] + [\boldsymbol{p}_h(t)] \quad (13.26)$$

or

$$\boldsymbol{C}\dot{\boldsymbol{T}} + \boldsymbol{K}\boldsymbol{T} = \boldsymbol{p} \quad (13.27)$$

The solution of Eq. (13.27) may be taken as $\boldsymbol{T} = \boldsymbol{\phi} e^{-\lambda t}$, where $\boldsymbol{\phi}$ is a number of modal vectors. Then the modal equation

$$(\boldsymbol{K} - \lambda \boldsymbol{C})\boldsymbol{\phi} = \boldsymbol{0} \quad (13.28)$$

The solution steps are as follows.

Step 1: Solve Eq. (13.28) for λ and $\boldsymbol{\phi}$; $\boldsymbol{C}$ and $\boldsymbol{K}$ matrices need to be modified appropriately for finite temperature boundary conditions.

Step 2: Using Duhamel's integration technique the responses are evaluated. An alternative procedure involves modification of the right-hand side of Eq. (13.27) for finite temperature boundary conditions. Then generalized equations are formed using $\boldsymbol{T} = \boldsymbol{\Phi}\eta$ as

$$[\boldsymbol{\Phi}^T \boldsymbol{K} \boldsymbol{\Phi}]\eta + [\boldsymbol{\Phi}^T \boldsymbol{C} \boldsymbol{\Phi}]\eta = \boldsymbol{\Phi}^T \boldsymbol{p} \quad (13.29)$$

or

$$\hat{\boldsymbol{K}}\eta + \hat{\boldsymbol{C}}\dot{\eta} = \hat{\boldsymbol{p}} \quad (13.30)$$

Step 3: Perform integration at each time step, using

$$\dot{\eta}_{n+1} = \frac{\eta_{n+1} - \eta_n}{\Delta t}; \qquad \eta_{n+1} = \frac{\eta_{n+1} + \eta_n}{2} \quad (13.31)$$

so that

$$\left(\frac{1}{\Delta t}\hat{\boldsymbol{C}} + \frac{1}{2}\hat{\boldsymbol{K}}\right)\eta_{n+1} = \left(\frac{1}{\Delta t}\hat{\boldsymbol{C}} - \frac{1}{2}\hat{\boldsymbol{K}}\right)\eta_n + \frac{\hat{\boldsymbol{p}}_{n+1} + \hat{\boldsymbol{p}}_n}{2} \quad (13.32)$$

yielding the solution η_{n+1}. Then recover $\boldsymbol{T}$ from $\boldsymbol{T} = \boldsymbol{\Phi}\eta$. In this process Q, q, and T_e are functions of time, and $\boldsymbol{p}_Q$, $\boldsymbol{p}_q$, and $\boldsymbol{p}_h$ may all vary.

Time Integration Method. The relevant matrix equation

$$\boldsymbol{C}\dot{\boldsymbol{T}}_{n+1} + \boldsymbol{K}\boldsymbol{T}_{n+1} = \boldsymbol{p}_{n+1} \quad (13.33)$$

may adopt a backward differencing scheme at $(n+1)$th step for its solution as

$$\dot{\boldsymbol{T}}_{n+1} = \frac{\boldsymbol{T}_{n+1} - \boldsymbol{T}_n}{\Delta t} \quad (13.34)$$

and substituting Eq. (13.34) into Eq. (13.33) yields the solution scheme

$$\left[\boldsymbol{K} + \frac{\boldsymbol{C}}{\Delta t}\right]\boldsymbol{T}_{n+1} = \frac{\boldsymbol{C}}{\Delta t}\boldsymbol{T}_n + \boldsymbol{p}_{n+1} \tag{13.35}$$

which may be solved for $\boldsymbol{T}_{n+1}$.

A general solution algorithm may now be developed by setting $t_\theta = t_n + \theta\Delta t, 0 < \theta < 1$, which yields the iterative scheme

$$\left[\theta\boldsymbol{K} + \frac{\boldsymbol{C}}{\Delta t}\right]\boldsymbol{T}_{n+1} = \left[-(1-\theta)\boldsymbol{K} + \frac{\boldsymbol{C}}{\Delta t}\right]\boldsymbol{T}_n + (1-\theta)\boldsymbol{p}_n + \theta\boldsymbol{p}_{n+1} \tag{13.36}$$

or

$$\tilde{\boldsymbol{K}}\boldsymbol{T}_{n+1} = \tilde{\boldsymbol{p}}_{n+1} \tag{13.37}$$

Depending on the prescribed value of Θ, the analysis may be described as

a) $\theta = 0$ forward (Euler) difference method
b) $\theta = \frac{1}{2}$ Crank–Nicholson method
c) $\theta = \frac{2}{3}$ Galerkin method
d) $\theta = 1$ backward difference method

The following comments are relevant in this connection:

1) For varying time steps $\tilde{\boldsymbol{K}}$ has to be inverted at each step.

2) For uniform time steps $\tilde{\boldsymbol{K}}$ needs to be inverted only once at the beginning of the analysis.

3) Conditions b–d are unconditionally stable; condition d always predicts smooth decay. Conditions b and c each predicts oscillatory response. These procedures are implicit methods.

4) Equation (13.35) becomes explicit for $\theta = 0$; then $\Delta t_{\text{cr}} = 2/\lambda_{\max}$ and is stable only for $\Delta t < \Delta t_{\text{cr}}$. The procedure is efficient if $\boldsymbol{C}$ is assumed lumped.

5) Step sizes Δt for conditions b–d may be greater than Δt_{cr} but must be small enough for accurate response calculations.

13.3.2 Nonlinear Analysis

For nonlinear systems,[13,14] steady-state problems are characterized by nonlinear simultaneous equations, whereas nonlinear ODEs represent the transient problem. Figure 13.3 shows problem classification and the associated analysis techniques. The nonlinearities may be caused by the following: 1) temperature-dependent anisotropic material properties: ρ, c values modify $\boldsymbol{C}$ for transient problems, and also elements of $\boldsymbol{K}_c$ become nonlinear; 2) temperature-dependent convection coefficients: h modifies $\boldsymbol{K}_h$, $\boldsymbol{p}_h$; 3) temperature-dependent heating: q_s modifies $\boldsymbol{p}_q$; and 4) temperature-dependent internal heat generation: Q modifies $\boldsymbol{p}_Q$.

These conditions result in nonlinear element matrices and heat load vectors, being temperature dependent for steady-state problems and temperature as well as time dependent for transient problems. Whereas the solution for steady-state

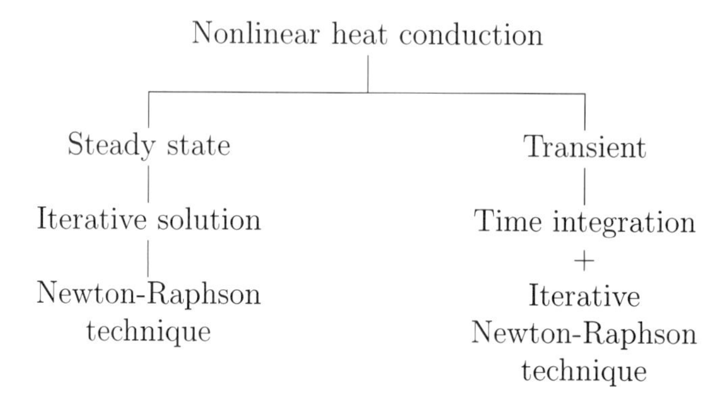

Fig. 13.3 Nonlinear heat conduction analysis.

problems requires an iterative scheme, a transient problem may be solved by an iterative, time-marching procedure.

13.3.2.1 Nonlinear steady-state analysis. The matrix equation of motion is written as

$$[\boldsymbol{K}_c(T) + \boldsymbol{K}_h(T) + \boldsymbol{K}_r(T)][\boldsymbol{T}] = \boldsymbol{p}_Q(T) + \boldsymbol{p}_q(T) + \boldsymbol{p}_h(T) + \boldsymbol{p}_r(T) \quad (13.38)$$

or

$$\boldsymbol{K}(T)[\boldsymbol{T}] = \boldsymbol{p}(T) \quad (13.39)$$

In the special case, in the presence of radiation when material properties are not a function of temperature, the equations of motion are simplified:

$$[\boldsymbol{K}_c + \boldsymbol{K}_h + \boldsymbol{K}_r(T)][\boldsymbol{T}] = [\boldsymbol{p}_Q] + [\boldsymbol{p}_q] + [\boldsymbol{p}_h] + [\boldsymbol{p}_r] \quad (13.40)$$

which can be solved by the Newton–Raphson method. Thus a typical iteration step is as follows.

Step 1: Solve

$$[\boldsymbol{K}_c + \boldsymbol{K}_h + \boldsymbol{K}_{rt}(T)]^{m+1}\Delta\boldsymbol{T}^{m+1} = \boldsymbol{p} - \Sigma[\boldsymbol{K}_c + \boldsymbol{K}_h + \boldsymbol{K}_r(t)]\boldsymbol{T}^m \quad (13.41)$$

Step 2: Update temperature

$$\boldsymbol{T}^{m+1} = \boldsymbol{T}^m + \Delta\boldsymbol{T}^{m+1} \quad (13.42)$$

Step 3: Continue iteration until the solution converges to a steady-state value.

In Eq. (13.40), $\boldsymbol{K}_r$ is a function of the input temperature, $\boldsymbol{K}_{rt} = 4\boldsymbol{K}_r$. This process can be extended for the more general case of Eq. (13.36) when material properties are also functions of temperature.

13.3.2.2 Nonlinear transient analysis. The equation of motion has the general form[10,15] of

$$\boldsymbol{C}(T)\dot{\boldsymbol{T}} + [\boldsymbol{K}_c(T) + \boldsymbol{K}_h(T) + \boldsymbol{K}_r(T)]\boldsymbol{T}(t) = \boldsymbol{p}_Q(T,t) + \boldsymbol{p}_q(T,t) + \boldsymbol{p}_h(T,t) + \boldsymbol{p}_r(T,t) \quad (13.43)$$

For the special case of a nonlinear radiation boundary condition only, a simplified Eq. (13.41) may be solved with Newton–Raphson iteration at each time step in which the response equation has the following form:

$$\left(\frac{1}{\Delta t}\boldsymbol{C}+\theta[\boldsymbol{K}_c+\boldsymbol{K}_h+\boldsymbol{K}_r(T)]\right)\boldsymbol{T}_{t+\Delta t}$$
$$=\left(\frac{1}{\Delta t}\boldsymbol{C}-[1-\theta][\boldsymbol{K}_c+\boldsymbol{K}_h+\boldsymbol{K}_r(T)]\boldsymbol{T}_t\right)+\theta\boldsymbol{p}_{t+\Delta t}+(1-\theta)\boldsymbol{p}_t \quad (13.44)$$

This nonlinear equation is cast next as a Newton–Raphson iterative scheme for a small incremental time step Δt.

Step 1: Solve for $\Delta\boldsymbol{T}$

$$\left(\frac{1}{\Delta t}\boldsymbol{C}+\theta[\boldsymbol{K}_c+\boldsymbol{K}_h+\boldsymbol{K}_{rt}]\right)\Delta\boldsymbol{T}=\theta\boldsymbol{p}_{t+\Delta t}+(1-\theta)\boldsymbol{p}_t$$
$$+\left(\frac{1}{\Delta t}\boldsymbol{C}-(1-\theta)[\boldsymbol{K}_c+\boldsymbol{K}_h+\boldsymbol{K}_r]\right)\boldsymbol{T}_t-\left(\frac{1}{\Delta t}\boldsymbol{C}+\theta[\boldsymbol{K}_c+\boldsymbol{K}_h+\boldsymbol{K}_r]\right)$$
$$\times\boldsymbol{T}^m_{t+\Delta t} \quad (13.45)$$

Step 2: Update temperature

$$\boldsymbol{T}^{m+1}_{t+\Delta t}=\boldsymbol{T}^m_{t+\Delta t}+\Delta\boldsymbol{T} \quad (13.46)$$

and iterate until the solution converges. This procedure is repeated for the entire time period. An extension of this technique can be used for the solution of the more general Eq. (13.41).

Much of the emphasis in this chapter has been placed on simple heat conduction analysis. For convection analysis, the heat conduction and diffusion processes need to be combined with the viscous flow concept. Such a procedure is beyond the scope of the present text and can be studied in the relevant literature.[15]

13.4 Numerical Examples

Heat-transfer analysis is elaborated in the following examples.

Example 13.1: Cooling Fin—Convection Boundary Condition

A linear steady-state heat-transfer analysis of a cooling fin in Fig. 13.4 was performed using a heat-transfer line element. Arbitrary available element and material properties data are used for the analysis to correlate results with existing

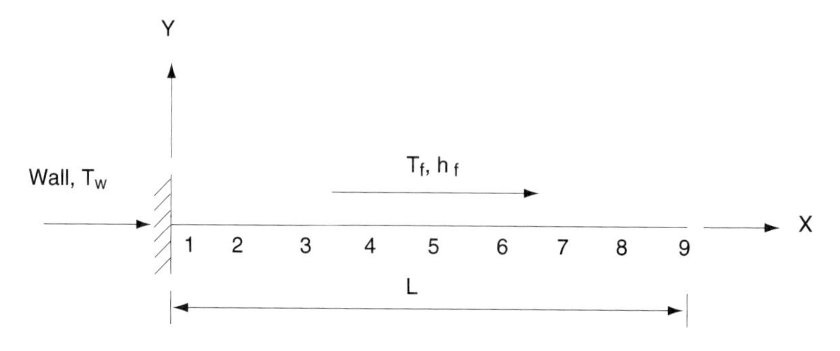

Fig. 13.4 Cooling fin with convection.

ones expressed in parametric form. Relevant data parameters are

Coefficient of conductivity k	$= 132.0$
Convective heat transfer coefficient h_f	$= 1.6$
Fluid temperature T_f	$= 70$
Wall temperature T_w	$= 250$
Length L	$= 1$
Area A	$= 0.001365$
Perimeter P	$= 0.13091$
Specific heat c_p	$= 0.2$

The results are given in Table 13.1.

Example 13.2: Square Plate—Transient Heating

A heat-transfer analysis of a square plate in Fig. 13.5 with transient internal heating, heat flow, and convective heating was performed. Solution results are presented in Table 13.2. The relevant data parameters are

Coefficient of conductivity k	$= 1.0$
Internal heat generation rate Q	$= 1.0$
Surface heat flow rate q_s	$= 10.0$
Convective heat-transfer coefficient h_a	$= 3.0$
Air temperature T_a	$= 20$
Edge temperature T_{edge}	$= 10$
Length L	$= 1$
Thickness t	$= 0.1$
Time step Δt	$= 0.05$
Total time period for response	$= 4.0$

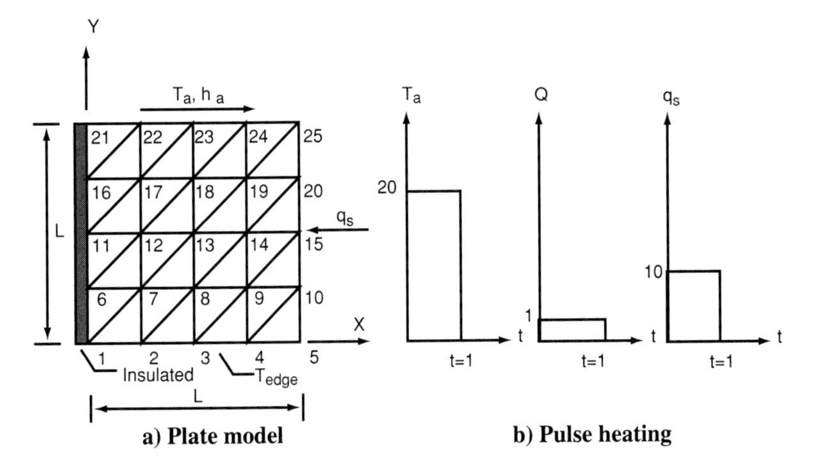

Fig. 13.5 Square plate with transient heating.

Example 13.3: Two-Dimensional Strip—Transient Heating with a Radiation Boundary Condition

A radiation heat-transfer analysis of a square isotropic two-dimensional strip in Fig. 13.6 with a transient heat flow was performed. Table 13.3 shows the results. The relevant data parameters are coefficient of conductivity $k = 10.5$, length $L = 1.0$, thickness $t = 0.1$, emissivity $\epsilon = 0.6$, Stefan–Boltzmann constant $\sigma = 0.1713 \times 10^{-3}$, density $\rho = 658$, and capacity $c = 0.038$.

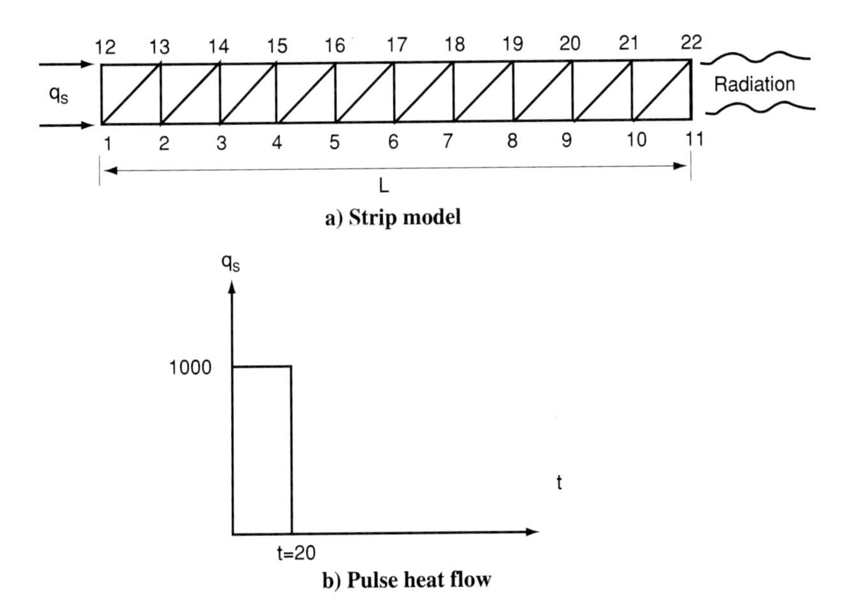

Fig. 13.6 Two-dimensional strip with transient heat flow and radiation boundary condition.

Table 13.1 Cooling fin heat transfer analysis results

Node	Temperature
1	250.00
2	232.33
3	217.62
4	205.61
5	196.06
6	188.80
7	183.72
8	180.70
9	179.70

Table 13.2 Heat-transfer analysis results of a square plate with transient heating

	Temperature	
Node	Time = 1.0	Time = 4.0
1	10.0000	10.0000
2	10.0000	10.0000
3	10.0000	10.0000
4	10.0000	10.0000
5	10.0000	10.0000
6	12.4740	8.1250
7	12.5685	8.1250
8	12.9048	8.1250
9	13.6278	8.1250
10	15.1516	8.1250
11	14.6934	6.2500
12	14.8553	6.2500
13	15.3559	6.2500
14	16.4015	6.2500
15	18.3025	6.2500
16	16.5866	4.3750
17	16.7153	4.3750
18	17.2313	4.3750
19	18.2784	4.3750
20	20.1826	4.3750
21	18.0622	2.5000
22	18.1970	2.5000
23	18.5183	2.5000
24	19.2186	2.5000
25	20.8599	2.5000

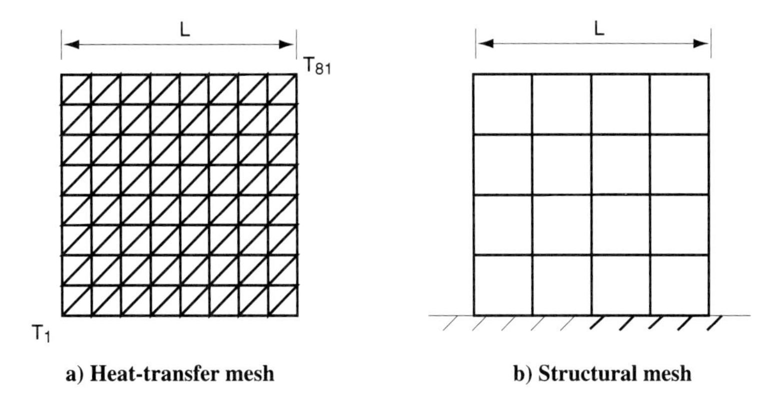

a) Heat-transfer mesh **b) Structural mesh**

Fig. 13.8 Finite-element model of a plate.

data parameters for structural analysis are Young's modulus $E = 30 \times 10^6$, cross-sectional area $A = 1.0$, the moment of inertia about the y axis $= 1/12$, the moment of inertia about the z axis $= 1/24$, length $L = 60$, mass density $\rho = 0.1666$, and the coefficient of thermal expansion $\alpha = 6.6 \times 10^{-6}$. Tables 13.4a–13.4c provide a summary of analysis results; accuracy of a heat-transfer input procedure is verified by an alternative procedure involving direct input of thermal data.

Table 13.5a Nodal deformation for static analysis

Node	X displacement	Y displacement
6	−0.809997E-04	0.631047E-04
7	−0.242790E-04	0.734264E-04
8	0.719179E-05	0.650178E-04
9	0.346446E-04	0.580581E-04
10	0.735603E-04	0.354295E-04
11	−0.771033E-04	0.125418E-03
12	−0.299133E-04	0.120472E-03
13	0.154035E-04	0.113178E-03
14	0.555337E-04	0.993192E-04
15	0.919275E-04	0.842963E-04
16	−0.615285E-04	0.177819E-03
17	−0.163020E-04	0.164476E-03
18	0.276178E-04	0.151509E-03
19	0.669176E-04	0.136094E-03
20	0.959403E-04	0.124534E-03
21	−0.435429E-04	0.222953E-03
22	0.108579E-05	0.206311E-03
23	0.452413E-04	0.188285E-03
24	0.834809E-04	0.164283E-03
25	0.104933E-03	0.146110E-03

Table 13.5b Thermally prestressed vibration analysis results for a cantilever plate with varying temperature distribution

	Natural frequencies ω, rad/s		
Mode	Without temperature	With temperature from heat-transfer mesh	With temperature input from data file
1	214.02	213.92	213.92
2	506.62	507.58	507.58
3	1248.40	1249.69	1249.69
4	1538.29	1537.72	1537.72
5	1765.53	1766.13	1766.13
6	2889.11	2888.35	2888.35

Table 13.5c Buckling load parameter for thermal load buckling analysis

Mode	Buckling load parameter
1	19.048

Example 13.5: Cantilever Plate

A cantilever plate model, described in Section 9.2.4, has been used for vibration and static analysis, and the same plate with two opposite edges fixed has been chosen for buckling analysis. Figure 13.8 shows finite element meshes for heat-transfer and structural analysis; The relevant data parameters for heat-transfer analysis are coefficient of conductivity $k = 1.0$, length $L = 10$, and thickness $t = 0.1$. The nodal temperatures T_1 and T_{81} are 10.0 and 0.0, respectively. The relevant data parameters for structural analysis are Young's modulus $E = 10 \times 10^6$, thickness $t = 0.1$, Poisson's ratio $\nu = 0.3$, length $L = 10$, mass density $\rho = 0.259 \times 10^{-3}$, and the coefficient of thermal expansion $\alpha = 3.5 \times 10^{-6}$. Tables 13.5a–13.5c present a summary of analysis results.

References

[1]Zienkiewicz, O. C., and Cheung, Y. K., "Finite Elements in the Solution of Field Problems," *The Engineer*, Sept. 1965, pp. 507–510.

[2]Wilson, E. L., and Nickell, R. E., "Application of the Finite Element Method to Heat Conduction Analysis," *Nucl. Eng. Des.*, Vol. 4, 1966.

[3]Gallagher, R. H., and Mallett, R. H., "Efficient Solution Processes for Finite Element Analysis of Transient Heat Conduction," *Journal of Heat Transfer*, Vol. 93, 1971, pp. 257–263.

[4]Lee, H. P., "Application of Finite Element Method in the Computation of Temperature with Emphasis on Radiative Exchange," *Thermal Control and Radiation*, edited by C. L. Tien, Progress in Astronautics and Aeronautics, Vol. 31, MIT Press, Cambridge, MA, 1973, pp. 491–520.

[5]Beckett, R. E., and Chu, S. C., "Finite Element Method Applied to Heat Conduction with Nonlinear Boundary Conditions," *Journal of Heat Transfer*, Vol. 95, 1973, pp. 126–129.

[6]Lee, H. P., and Jackson, C. E., Jr., "Finite Element Solution for Radiative Conductive Analyses with Mixed Diffuse-Specular Surfaces," *Radiative Transfer and Thermal Control*, edited by A. M. Smith, Progress in Astronautics and Aeronautics, Vol. 49, AIAA, New York, 1976, pp. 25–46.

[7]Zienkiewicz, O. C., *The Finite Element Method*, 3rd ed., McGraw–Hill, New York, 1977.

[8]Thornton, E. A., and Wieting, A. R., "A Finite Element Thermal Analysis Procedure for Several Temperature-Dependent Parameters," *Journal of Heat Transfer*, Vol. 100, 1978, pp. 551–553.

[9]Thornton, E. A., and Wieting, A. R., "Finite Element Methodology for Thermal Analysis of Convectively Cooled Structures," *Heat Transfer and Thermal Control Systems*, edited by L. S. Fletcher, Progress in Astronautics and Aeronautics, Vol. 60, AIAA, New York, 1978, pp. 171–189.

[10]Huebner, K. H., and Thornton, E. A., *The Finite Element Method for Engineers*, Wiley, New York, 1982.

[11]Thornton, E. A., and Wieting, A. R., "Finite Element Methodology for Transient Conduction/Forced-Convection Thermal Analysis," *Heat Transfer, Thermal Control, and Heat Pipes*, edited by W. B. Olstead, Progress in Astronautics and Aeronautics, Vol. 70, AIAA, New York, 1980, pp. 77–103.

[12]Adelman, H. M., and Haftka, R., "On the Performance of Explicit Transient Algorithms for Transient Thermal Analysis of Structures," NASA TM-81880, NASA Langley Research Center, Hampton, VA, Sept. 1980.

[13]Fernandes, R., Francis, J., and Reddy, J. N., "A Finite Element Approach to Combined Conductive and Radiative Heat Transfer in a Planar Medium," AIAA 15th Thermophysics Conference, Snowmass, CO, AIAA Paper 80-1487, 14–16 July, 1980.

[14]Orivuori, S.,"Efficient Method for Solution of Nonlinear Heat Conduction Problems," *International Journal for Numerical Methods in Engineering*, Vol. 14, 1979, pp. 1461–1476.

[15]Hogge, M. A., "A Survey of Direct Integration Procedures for Nonlinear Transient Heat Transfer," International Conference on Numerical Methods in Thermal Problems, Univ. College, Swansea, Wales, U.K., 2–6 July, 1979.

14
Computational Linear Aeroelasticity and Aeroservoelasticity

14.1 Introduction

Modern aircraft and other aerospace vehicles have become increasingly complex because of the integration of distinct technologies used to attain numerous objectives in the areas of performance, control, flying qualities, maneuver techniques, fuel efficiency, and various mission requirements. The design process must respond to specifications from all disciplines to achieve these diverse goals and integrate accordingly. Design procedures must account for conflicting objectives and the interaction of aerodynamics, structures, and dynamics of control systems. Coupling between these dynamic elements of the model can be treated passively with structural modification and passive filtering or actively with control mechanisms driven by appropriate control laws. Analysis of the consequences of the design is essential to perform safe and effective mission tasks.

The term *aeroelasticity*[1,2] refers to the study of the phenomenon of mutual interaction of aerodynamic and structural dynamic forces and their effect on the design of an aerospace vehicle. In the absence of inertial forces this interaction is termed as static aeroelasticity. Instability of a vehicle occurs when elastic deformation induces additional aerodynamic forces that in turn produce further structural deformations. When continued, this process may cause eventual destruction of the vehicle. Flutter and divergence are examples of these phenomena, the latter occurring in the absence of inertial forces. Calculation of response for moving shock waves relates to dynamic aeroelasticity. The static aeroelastic phenomenon includes control effectiveness that relates to the influence of structural deformations on controllability of the vehicle. Also the phenomenon control surface reversal refers to the speed at which effect of a control surface motion is entirely negated by elastic structural deformation. The structural dynamics phenomenon includes buffeting and refers to transient vibration of an aircraft component because of aerodynamic impulses produced by the wake behind another major component such as a wing.

Other related dynamic response analysis involves computing the transient response of the vehicle subjected to rapidly applied dynamic loads such as gusts or sudden motion of control surfaces. Aeroservoelastic analysis, involving interaction of aerodynamics, controls, and elastic forces, is used to investigate the potential problems arising from high bandwidth control of relatively flexible aircraft by combining linear models of structure, unsteady aerodynamics, and control system into one dynamic system. The models are augmented to address stability and performance issues. Frequency responses may then be calculated from this model yielding phase and gain margins. Associated damping and frequency values are also calculated from the same model. These results are indicative of the state of

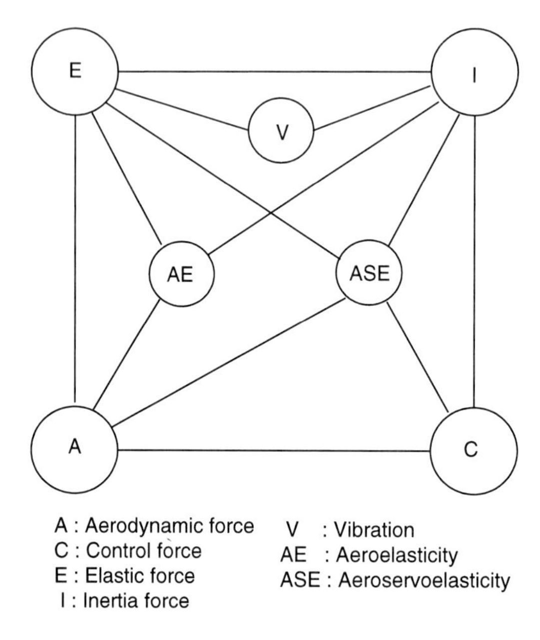

Fig. 14.1 Aeroelastic and aeroservoelastic forces.

stability of the flight vehicle. The various forces and their interaction involved in aeroelastic and aeroservoelastic analysis are shown in Fig. 14.1.

Numerical formulations for the linear aeroelastic and aeroservoelastic phenomena are presented next in matrix form that enables effective solution of a broad class of related stability problems. A general Lagrangian formulation of the equations of motion for an aircraft maneuver analysis is given in Ref. 3.

14.2 Formulation of Numerical Procedure

In the ensuing formulation[3] structural discretization is achieved by the usual FEM, whereas the panel methods, such as the doublet lattice method[4] and the constant pressure method,[5] are used for computation of unsteady aerodynamic forces for the subsonic and supersonic flows, respectively. The aerostructural problem is recast in the Laplace domain and the unsteady generalized aerodynamic forces are curve fitted with Padé and least-square approximations to generate a state-space aerostructural dynamic model. This model is transformed from the inertial to body-axis coordinate system for control system augmentation. Aeroservoelastic analysis is achieved for either analog or digital controllers, in addition to various analog elements such as actuators and filters with hybrid frequency responses and eigenvalue solutions for closed-loop modal behavior. The aeroservoelastic analysis using linear aerodynamic methods is performed in terms of the generalized coordinates using the normal modes of an elastic structure. This simplification is rendered possible because the unsteady aerodynamic forces are computed for each normal mode and a number of reduced frequencies.

14.2.1 Aeroservoelastic Equation of Motion

The matrix equation of motion of a structure, such as an aircraft, relevant to the current analysis has the following form:

$$\boldsymbol{M}\ddot{\boldsymbol{q}} + \boldsymbol{C}_d\dot{\boldsymbol{q}} + \boldsymbol{K}\boldsymbol{q} + \bar{q}\boldsymbol{A}_e(k)\boldsymbol{q} = \boldsymbol{p}(t) \tag{14.1}$$

where $\bar{q} = 1/2\rho V^2$, $k = \omega\bar{c}/2V$, and $\boldsymbol{p}(t)$ is the external forcing function. The free vibration solution is first affected in matrix formulation

$$\boldsymbol{M}\ddot{\boldsymbol{q}} + \boldsymbol{K}\boldsymbol{q} = \boldsymbol{0} \tag{14.2}$$

yielding the frequencies ω and mode shapes $\boldsymbol{\Phi}$. Applying a transformation

$$\boldsymbol{q} = \boldsymbol{\Phi}\eta \tag{14.3}$$

to Eq. (14.1) and premultiplying both sides by $\boldsymbol{\Phi}^T$ yields the generalized equation of motion

$$\hat{\boldsymbol{M}}\ddot{\eta} + \hat{\boldsymbol{C}}\dot{\eta} + \hat{\boldsymbol{K}}\eta + \bar{q}\boldsymbol{Q}(k)\eta = \hat{\boldsymbol{p}}(t) \tag{14.4}$$

where

$$\hat{\boldsymbol{M}} = \boldsymbol{\Phi}^T\boldsymbol{M}\boldsymbol{\Phi}, \quad \hat{\boldsymbol{C}} = \boldsymbol{\Phi}^T\boldsymbol{C}\boldsymbol{\Phi}, \quad \hat{\boldsymbol{K}} = \boldsymbol{\Phi}^T\boldsymbol{K}\boldsymbol{\Phi}, \quad \boldsymbol{Q} = \boldsymbol{\Phi}^T\boldsymbol{A}_e\boldsymbol{\Phi}, \quad \hat{\boldsymbol{p}} = \boldsymbol{\Phi}^T\boldsymbol{p} \tag{14.5}$$

and the modal matrix is

$$\boldsymbol{\Phi} = [\boldsymbol{\Phi}_r \quad \boldsymbol{\Phi}_e \quad \boldsymbol{\Phi}_\delta] \tag{14.6}$$

and the generalized coordinate is

$$\eta = [\eta_r \quad \eta_e \quad \eta_\delta] \tag{14.7}$$

thereby incorporating rigid-body, elastic, and control surface motions, respectively. An aerodynamic force matrix is computed by the panel method[4,5] and is of order (NAEL × NAEL), NAEL being the number of aerodynamic elements. An appropriate transformation then yields the generalized force matrix $\boldsymbol{Q}(k)$ that is computed for a number of k values for a particular Mach number. The $\boldsymbol{Q}(k)$ matrix may be approximated[6] with Padé polynomials in $i^*k(=i^*\omega\bar{c}/2V = s\bar{c}/2V$, where the Laplace variable $s = i^*\omega$, i^* being $\sqrt{-1}$) as

$$\boldsymbol{Q}(k) = \boldsymbol{A}_0 + i^*k\boldsymbol{A}_1 + (i^*k)^2\boldsymbol{A}_2 + \frac{i^*k}{i^*k + \beta_1}\boldsymbol{A}_3 + \frac{i^*k}{i^*k + \beta_2}\boldsymbol{A}_4 + \cdots \tag{14.8}$$

with aerodynamic lag terms β_j (assume $j = 1, 2$), and

$$\frac{i^*k}{i^*k + \beta_j} = \frac{k^2}{k^2 + \beta_j^2} + \frac{i^*k\beta_j}{k^2 + \beta_j^2} \tag{14.9}$$

The rigid airload coefficients assume the following form:

$$\boldsymbol{A}_0 = \boldsymbol{Q}_R(k_1) \tag{14.10}$$

$$\boldsymbol{A}_1 = \frac{\boldsymbol{Q}_I(k_1)}{k_1} - \frac{\boldsymbol{A}_3}{\beta_1} - \frac{\boldsymbol{A}_4}{\beta_2} \tag{14.11}$$

where k_1 is the smallest reduced frequency, with a value near zero, used to compute $\boldsymbol{A}_j$ for $j = 0, 1, 2, \ldots$ Separating real and imaginary parts in Eq. (14.8) yields

$$
\begin{aligned}
\tilde{\boldsymbol{Q}}_R(k) &= \boldsymbol{Q}_R(k) - \boldsymbol{A}_0 \\
&= \begin{bmatrix} -k^2\boldsymbol{I} & \dfrac{k^2}{k^2+\beta_1^2}\boldsymbol{I} & \dfrac{k^2}{k^2+\beta_2^2}\boldsymbol{I} \end{bmatrix} \begin{bmatrix} \boldsymbol{A}_2 \\ \boldsymbol{A}_3 \\ \boldsymbol{A}_4 \end{bmatrix} \\
&= \boldsymbol{S}_R(k)\tilde{\boldsymbol{A}}
\end{aligned} \tag{14.12}
$$

$$
\begin{aligned}
\tilde{\boldsymbol{Q}}_I(k) &= \frac{\boldsymbol{Q}_I(k)}{k} - \boldsymbol{A}_1 \\
&= \begin{bmatrix} \boldsymbol{0} & \dfrac{\beta_1}{k^2+\beta_1^2}\boldsymbol{I} & \dfrac{\beta_2}{k^2+\beta_2^2}\boldsymbol{I} \end{bmatrix} \begin{bmatrix} \boldsymbol{A}_2 \\ \boldsymbol{A}_3 \\ \boldsymbol{A}_4 \end{bmatrix} \\
&= \boldsymbol{S}_I(k)\tilde{\boldsymbol{A}}
\end{aligned} \tag{14.13}
$$

The unknown coefficients $\boldsymbol{A}_3$ and $\boldsymbol{A}_4$ can be determined by substituting the expression for $\boldsymbol{A}_1$ in Eq. (14.11) into Eq. (14.13). However, the resulting solution is sensitive to the choice of β_j for approximating rigid airloads. If the elements of the aerodynamic damping matrix $\boldsymbol{A}_1$ are replaced with known damping coefficients (steady aerodynamic derivatives), then the solution for rigid airloads becomes insensitive to the β_j values. For a chosen number of values of reduced frequencies k_i, Eqs. (14.12) and (14.13) may be combined as

$$
\begin{bmatrix} \tilde{\boldsymbol{Q}}_R(k_2) \\ \tilde{\boldsymbol{Q}}_I(k_2) \\ \vdots \\ \tilde{\boldsymbol{Q}}_R(k_{NF}) \\ \tilde{\boldsymbol{Q}}_I(k_{NF}) \end{bmatrix} = \begin{bmatrix} \boldsymbol{S}_R(k_2) \\ \boldsymbol{S}_I(k_2) \\ \vdots \\ \boldsymbol{S}_R(k_{NF-1}) \\ \boldsymbol{S}_I(k_{NF-1}) \end{bmatrix} \begin{bmatrix} \boldsymbol{A}_2 \\ \boldsymbol{A}_3 \\ \boldsymbol{A}_4 \end{bmatrix} \tag{14.14}
$$

or

$$
\tilde{\tilde{\boldsymbol{Q}}} = \boldsymbol{S}\tilde{\boldsymbol{A}} \tag{14.15}
$$

and a least-square solution

$$
\tilde{\boldsymbol{A}} = [\boldsymbol{S}^T \quad \boldsymbol{S}]^{-1}\boldsymbol{S}^T\tilde{\tilde{\boldsymbol{Q}}} \tag{14.16}
$$

yields the required coefficients $\boldsymbol{A}_2$, $\boldsymbol{A}_3$, $\boldsymbol{A}_4$. The procedure is easily extended for a larger number of lag terms if desired. Equation (14.4) may then be rewritten, assuming simple harmonic motion as

$$
\hat{\boldsymbol{M}}\ddot{\eta} + \hat{\boldsymbol{C}}\dot{\eta} + \hat{\boldsymbol{K}}\eta + \bar{q}\left[\boldsymbol{A}_0\eta + \boldsymbol{A}_1\frac{s\bar{c}}{2V}\eta + \boldsymbol{A}_2\left(\frac{s\bar{c}}{2V}\right)^2\eta + \boldsymbol{A}_3\boldsymbol{x}_1 + \boldsymbol{A}_4\boldsymbol{x}_2 + \cdots\right] = \boldsymbol{0} \tag{14.17}
$$

such that

$$\boldsymbol{x}_j = \frac{s\boldsymbol{\eta}}{s + (2V/\bar{c})\beta_j} \tag{14.18}$$

from which

$$\dot{\boldsymbol{x}}_j + (2V/\bar{c})\beta_j \boldsymbol{x}_j = \dot{\boldsymbol{\eta}} \tag{14.19}$$

Collecting like terms gives

$$(\hat{\boldsymbol{K}} + \bar{q}\boldsymbol{A}_0)\boldsymbol{\eta} + \left(\hat{\boldsymbol{C}} + \bar{q}\frac{\bar{c}}{2V}\boldsymbol{A}_1\right)\dot{\boldsymbol{\eta}} + \left(\hat{\boldsymbol{M}} + \bar{q}\left(\frac{\bar{c}}{2V}\right)^2 \boldsymbol{A}_2\right)\ddot{\boldsymbol{\eta}} + \bar{q}\boldsymbol{A}_3\boldsymbol{x}_1 + \bar{q}\boldsymbol{A}_4\boldsymbol{x}_2 + \cdots = \boldsymbol{0} \tag{14.20}$$

or

$$\hat{\hat{\boldsymbol{K}}}\boldsymbol{\eta} + \hat{\hat{\boldsymbol{C}}}\dot{\boldsymbol{\eta}} + \hat{\hat{\boldsymbol{M}}}\ddot{\boldsymbol{\eta}} + \bar{q}\boldsymbol{A}_3\boldsymbol{x}_1 + \bar{q}\boldsymbol{A}_4\boldsymbol{x}_2 + \cdots = \boldsymbol{0} \tag{14.21}$$

Rewrite Eqs. (14.18), (14.19), and (14.21) as one matrix equation

$$\begin{bmatrix} \boldsymbol{I} & & & \\ & \hat{\hat{\boldsymbol{M}}} & & \\ & & \boldsymbol{I} & \\ & & & \boldsymbol{I} \end{bmatrix} \begin{bmatrix} \dot{\boldsymbol{\eta}} \\ \ddot{\boldsymbol{\eta}} \\ \dot{\boldsymbol{x}}_1 \\ \dot{\boldsymbol{x}}_2 \end{bmatrix} = \begin{bmatrix} \boldsymbol{0} & \boldsymbol{I} & \boldsymbol{0} & \boldsymbol{0} \\ -\hat{\hat{\boldsymbol{K}}} & -\hat{\hat{\boldsymbol{C}}} & -\bar{q}\boldsymbol{A}_3 & -\bar{q}\boldsymbol{A}_4 \\ \boldsymbol{0} & \boldsymbol{I} & -\frac{V}{b}\beta_1\boldsymbol{I} & \boldsymbol{0} \\ \boldsymbol{0} & \boldsymbol{I} & \boldsymbol{0} & -\frac{V}{b}\beta_2\boldsymbol{I} \end{bmatrix} \begin{bmatrix} \boldsymbol{\eta} \\ \dot{\boldsymbol{\eta}} \\ \boldsymbol{x}_1 \\ \boldsymbol{x}_2 \end{bmatrix} \tag{14.22}$$

or

$$\boldsymbol{M}'\dot{\boldsymbol{x}}' = \boldsymbol{K}'\boldsymbol{x}' \tag{14.23}$$

and

$$\begin{aligned} \dot{\boldsymbol{x}}' &= (\boldsymbol{M}')^{-1}\boldsymbol{K}'\boldsymbol{x}' \\ &= \boldsymbol{R}\boldsymbol{x}' \end{aligned} \tag{14.24}$$

where b is the aerodynamic semichord. Now, rearranging the state-space vector $\boldsymbol{x}'$ as

$$\begin{aligned} \boldsymbol{x}'' &= [(\eta_r \quad \eta_e \quad \dot{\eta}_r \quad \dot{\eta}_e \quad \boldsymbol{x}_1 \quad \boldsymbol{x}_2) \quad (\eta_\delta \quad \dot{\eta}_\delta)] \\ &= [\hat{\boldsymbol{x}} \quad \boldsymbol{u}] \end{aligned} \tag{14.25}$$

Eq. (14.24) may be partitioned as

$$\begin{bmatrix} \dot{\hat{\boldsymbol{x}}} \\ \dot{\boldsymbol{u}} \end{bmatrix} = \begin{bmatrix} \boldsymbol{R}_{1,1} & \boldsymbol{R}_{1,2} \\ \boldsymbol{R}_{2,1} & \boldsymbol{R}_{2,2} \end{bmatrix} \begin{bmatrix} \hat{\boldsymbol{x}} \\ \boldsymbol{u} \end{bmatrix} \tag{14.26}$$

where the first matrix equation denotes the plant dynamics and the second represents the dynamics of control modes. In the case of plant dynamics, the state-space equations become

$$\dot{\hat{x}} = \hat{A}\hat{x} + \hat{B}u \tag{14.27}$$

the associated matrices and vectors being defined as

$\hat{A}$ = plant dynamics matrix
$\hat{B}$ = control surface influence matrix
$\hat{x}$ = generalized coordinates in inertial frame
$\hat{u}$ = control surface motion input into plant

The terms $\hat{A}\hat{x}$ and $\hat{B}u$ represent the airplane dynamics and forcing function due to control surface motion, respectively.

14.2.2 Coordinate Transformation

To incorporate control laws designed to control body-axis motions, it is necessary to transform Eq. (14.27) from the Earth-fixed (inertial) to the body-fixed coordinate system. Because no transformations are applied to elastic and aerodynamic lag state vectors, a transformation of the form

$$\begin{aligned} \dot{x} &= \tilde{T}_2^{-1}(\hat{A}\tilde{T}_1 - \tilde{T}_3)x + \tilde{T}_2^{-1}\hat{B}u \\ &= Ax + Bu \end{aligned} \tag{14.28}$$

in which

$$\tilde{T}_1 = \begin{bmatrix} T_1 & 0 \\ 0 & I \end{bmatrix} \tag{14.29}$$

and so on, where T_1 is the (12×12) coordinate transformation matrix, and yields the required state-space equation in the body coordinate system. A detailed description of the transformation procedure is given in Ref. 3.

14.2.3 Determination of Sensor Outputs

The structural nodal displacements are related to the generalized coordinates by Eq. (14.3), and the related sensor motion can be expressed as

$$\begin{aligned} q_s &= T_s \Phi \eta \\ &= C_0 x \end{aligned} \tag{14.30}$$

where $C_0 = [T_s \ \ \Phi \ \ 0 \ \ 0 \ \ 0]$, and in which T_s is an interpolation matrix.[3] Similar relations may be expressed for sensor velocities and acceleration as

$$\begin{aligned} \begin{bmatrix} \dot{q}_s \\ \ddot{q}_s \end{bmatrix} &= \begin{bmatrix} T_s \Phi \dot{\eta} \\ T_s \Phi \ddot{\eta} \end{bmatrix} \\ &= C_1 \dot{x} \end{aligned} \tag{14.31}$$

where

$$C_1 = \begin{bmatrix} T_s \Phi & 0 & 0 & 0 \\ 0 & T_s \Phi & 0 & 0 \end{bmatrix} \tag{14.32}$$

Premultiplying Eq. (14.28) by $\boldsymbol{C}_1$,

$$\begin{aligned} \boldsymbol{C}_1\dot{\boldsymbol{x}} &= \boldsymbol{C}_1\boldsymbol{A}\boldsymbol{x} + \boldsymbol{C}_1\boldsymbol{B}\boldsymbol{u} \\ &= \boldsymbol{C}_2\boldsymbol{x} + \boldsymbol{D}_2\boldsymbol{u} \end{aligned} \tag{14.33}$$

Adjoining Eqs. (14.30) and (14.31),

$$\boldsymbol{y} = \begin{bmatrix} \boldsymbol{q}_s \\ \dot{\boldsymbol{q}}_s \\ \ddot{\boldsymbol{q}}_s \end{bmatrix} = \begin{bmatrix} \boldsymbol{C}_0 \\ \boldsymbol{C}_2 \end{bmatrix} \boldsymbol{x} + \begin{bmatrix} \boldsymbol{0} \\ \boldsymbol{D}_2 \end{bmatrix} \boldsymbol{u} \tag{14.34}$$

or

$$\boldsymbol{y} = \boldsymbol{C}\boldsymbol{x} + \boldsymbol{D}\boldsymbol{u} \tag{14.35}$$

which is the required sensor output relationship signifying motion at the sensors because of body motion $\boldsymbol{C}$ and control surface motions $\boldsymbol{D}$.

14.2.4 Augmentation of Analog Elements and Controller

Equations (14.28) and (14.35) represent the complete state-space formulation for the aircraft incorporating structural and aeroelastic effects. To conduct an aeroservoelastic analysis, it is essential to augment formulation with associated analog elements such as actuators, sensors, notches, and prefilters along with the controller. Denote the state-space equation of a typical element in a series as follows:

$$\begin{aligned} \dot{\boldsymbol{x}}^{(i)} &= \boldsymbol{A}^{(i)}\boldsymbol{x}^{(i)} + \boldsymbol{B}^{(i)}\boldsymbol{u}^{(i)} \\ \boldsymbol{y}^{(i)} &= \boldsymbol{C}^{(i)}\boldsymbol{x}^{(i)} + \boldsymbol{D}^{(i)}\boldsymbol{u}^{(i)} \end{aligned} \tag{14.36}$$

This can be augmented to the plant Eqs. (14.28) and (14.35) as

$$\begin{bmatrix} \dot{\boldsymbol{x}} \\ \dot{\boldsymbol{x}}^{(i)} \end{bmatrix} = \begin{bmatrix} \boldsymbol{A} & \boldsymbol{0} \\ \boldsymbol{B}^{(i)}\boldsymbol{C} & \boldsymbol{A}^{(i)} \end{bmatrix} \begin{bmatrix} \boldsymbol{x} \\ \boldsymbol{x}^{(i)} \end{bmatrix} + \begin{bmatrix} \boldsymbol{B} \\ \boldsymbol{B}^{(i)}\boldsymbol{D} \end{bmatrix} [\boldsymbol{u}] \tag{14.37}$$

or

$$\dot{\boldsymbol{x}}_{(i)} = \boldsymbol{A}_{(i)}\boldsymbol{x}_{(i)} + \boldsymbol{B}_{(i)}\boldsymbol{u} \tag{14.38}$$

noting that $\boldsymbol{u}^{(1)} = \boldsymbol{y}$. Also

$$\begin{bmatrix} \boldsymbol{y} \\ \boldsymbol{y}^{(i)} \end{bmatrix} = \begin{bmatrix} \boldsymbol{C} & \boldsymbol{0} \\ \boldsymbol{D}^{(i)}\boldsymbol{C} & \end{bmatrix} \begin{bmatrix} \boldsymbol{x} \\ \boldsymbol{x}^{(i)} \end{bmatrix} + \begin{bmatrix} \boldsymbol{D} \\ \boldsymbol{D}^{(i)}\boldsymbol{D} \end{bmatrix} [\boldsymbol{u}] \tag{14.39}$$

or

$$\boldsymbol{y}_{(i)} = \boldsymbol{C}_{(i)}\boldsymbol{x}_{(i)} + \boldsymbol{D}_{(i)}\boldsymbol{u} \tag{14.40}$$

becomes the new sensor output expression.

Any analog element, including a controller, can be augmented similarly at the input and the output of the plant. Figure 14.2 shows a typical feedback control system with controller $\boldsymbol{G}$. For this system the three sets of relevant matrix equations are

$$\dot{\boldsymbol{x}} = \boldsymbol{A}\boldsymbol{x} + \boldsymbol{B}\boldsymbol{u} \tag{14.41}$$

$$\boldsymbol{y} = \boldsymbol{C}\boldsymbol{x} + \boldsymbol{D}\boldsymbol{u} \tag{14.42}$$

$$\boldsymbol{u} = \boldsymbol{r} - \boldsymbol{G}\boldsymbol{y} \tag{14.43}$$

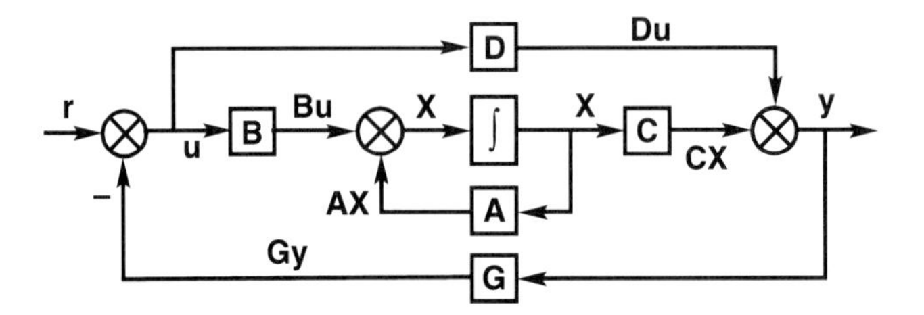

Fig. 14.2 Feedback control system.

where Eq. (14.43) is the feedback equation. The required transfer functions may be obtained by Laplace transformation

$$s\boldsymbol{x}(s) = \boldsymbol{A}\boldsymbol{x}(s) + \boldsymbol{B}\boldsymbol{u}(s) \tag{14.44}$$

$$\boldsymbol{y}(s) = \boldsymbol{C}\boldsymbol{x}(s) + \boldsymbol{D}\boldsymbol{u}(s) \tag{14.45}$$

$$\boldsymbol{u}(s) = \boldsymbol{r}(s) - \boldsymbol{G}(s)\boldsymbol{y}(s) \tag{14.46}$$

From Eq. (14.44)

$$\boldsymbol{x}(s) = [s\boldsymbol{I} - \boldsymbol{A}]^{-1}\boldsymbol{B}\boldsymbol{u}(s) \tag{14.47}$$

and substituting Eq. (14.47) into Eq. (14.45) yields the required open-loop frequency response relationship

$$\begin{aligned} \boldsymbol{y}(s) &= [\boldsymbol{C}(s\boldsymbol{I} - \boldsymbol{A})^{-1}\boldsymbol{B} + \boldsymbol{D}]\boldsymbol{u}(s) \\ &= \boldsymbol{H}(s)\boldsymbol{u}(s) \end{aligned} \tag{14.48}$$

where $\boldsymbol{H}(s)$ is the equivalent open-loop (loop-gain) transfer function with the analog controller or the open-loop transfer function without the controller. To obtain the closed-loop frequency response relationship, Eq. (14.48) is first substituted into Eq. (14.46), resulting in

$$\begin{aligned} \boldsymbol{u}(s) &= \boldsymbol{r}(s) - \boldsymbol{G}(s)\boldsymbol{H}(s)\boldsymbol{u}(s) \\ &= [\boldsymbol{I} + \boldsymbol{G}(s)\boldsymbol{H}(s)]^{-1}\boldsymbol{r}(s) \end{aligned} \tag{14.49}$$

and again substitution of Eq. (14.48) yields

$$\begin{aligned} \boldsymbol{y}(s) &= \boldsymbol{H}(s)[\boldsymbol{I} + \boldsymbol{G}(s)\boldsymbol{H}(s)]^{-1}\boldsymbol{r}(s) \\ &= \hat{\boldsymbol{H}}(s)\boldsymbol{r}(s) \end{aligned} \tag{14.50}$$

in which $\hat{\boldsymbol{H}}(s)$ is the desired closed-loop transfer function. The frequency response plots can be obtained from the transfer matrices $\boldsymbol{H}(s)$ or $\hat{\boldsymbol{H}}(s)$ as the case may be. Associated damping and frequency values of the system for the loop-gain or open-loop case, may also be calculated by solving the eigenvalue problem of the relevant $\boldsymbol{A}$ matrix for various k_i values or dynamic pressures and observing the changes in sign of the real part of an eigenvalue to detect instabilities.

In the presence of a digital controller, a hybrid approach is adopted for the frequency response solution. Thus if $\boldsymbol{A}'$, $\boldsymbol{B}'$, $\boldsymbol{C}'$, and $\boldsymbol{D}'$ are the state-space matrices associated with the controller, the related transfer function is simply given by

$$G(z) = C'[zI - A']^{-1}B' + D' \tag{14.51}$$

and the frequency response formulation[7] for the hybrid analog–digital system with time delay τ and sample time T can be written as

$$\begin{aligned} \tilde{y}(s) &= G(z)_{[z=e^{sT}]}\left(\frac{H(s)[\text{ZOH}]}{T}\right)u(s) \\ &= G(s, T)H^*(s, \tau, T)u(s) \\ &= \tilde{H}(s, \tau, T)u(s) \end{aligned} \tag{14.52}$$

in which $H(s)$ is the transfer function of the plant and all analog elements and [ZOH] is the zero-order hold complex expression $[=e^{-s\tau}(1 - e^{-sT}/s)]$, and where $\tilde{H}(s, \tau, T)$ is now the equivalent open-loop (loop-gain) transfer function of the hybrid system. Noting that $\tilde{y}(s) = G(s, T)y(s)$, the closed-loop frequency response relationship may be obtained as before by using Eqs. (14.52) and (14.46):

$$\begin{aligned} y(s) &= \{H^*(s, \tau, T)[I + G(s, T)H^*(s, \tau, T)]^{-1}\}r(s) \\ &= \hat{H}^*(s, \tau, T)r(s) \end{aligned} \tag{14.53}$$

To calculate the damping and frequencies, modes with natural frequencies much beyond the Nyquist frequency are truncated. The analog plant dynamics matrix A is then transformed into the z plane by standard discretization procedures and augmented to controller dynamics A'. Appropriate eigenproblem solution of the final matrix yields the required results, as previously discussed.

14.3 Numerical Example

A representative practical example problem and its solution are described next.

Example 14.1: Forward-Swept Wing Aircraft

The integrated aerostructural–control analysis capability of the STARS program was used extensively to solve related aeroelastic and aeroservoelastic problems of the X-29A forward-swept wing research aircraft.[8,9] The aircraft in Fig. 14.4 has thin composite wings that are aeroelastically tailored to eliminate structural divergence within the flight envelope. Full-span, double-hinged, variable-camber flaperons and strake flaps operate with full-authority, variable-incidence canards to yield minimum trim drag. The supercritical airfoil provides efficient transonic cruise performance and high transonic maneuvering capability. The canard-configured aircraft is up to 35% statically unstable, requiring appropriate feedback flight controls for augmented static stability.

Although these combined technologies result in significant improvement in overall aerodynamic and structural performance, they may also cause adverse dynamic interaction of the flight control with the flexible structure if not integrated properly. Therefore an ASE analysis assumes a very important role in the design process. Figure 14.5a shows the finite element symmetric dynamic half model of the aircraft consisting of about 3500 degrees of freedom, whereas Fig. 14.5b depicts the corresponding aerodynamic panel model.

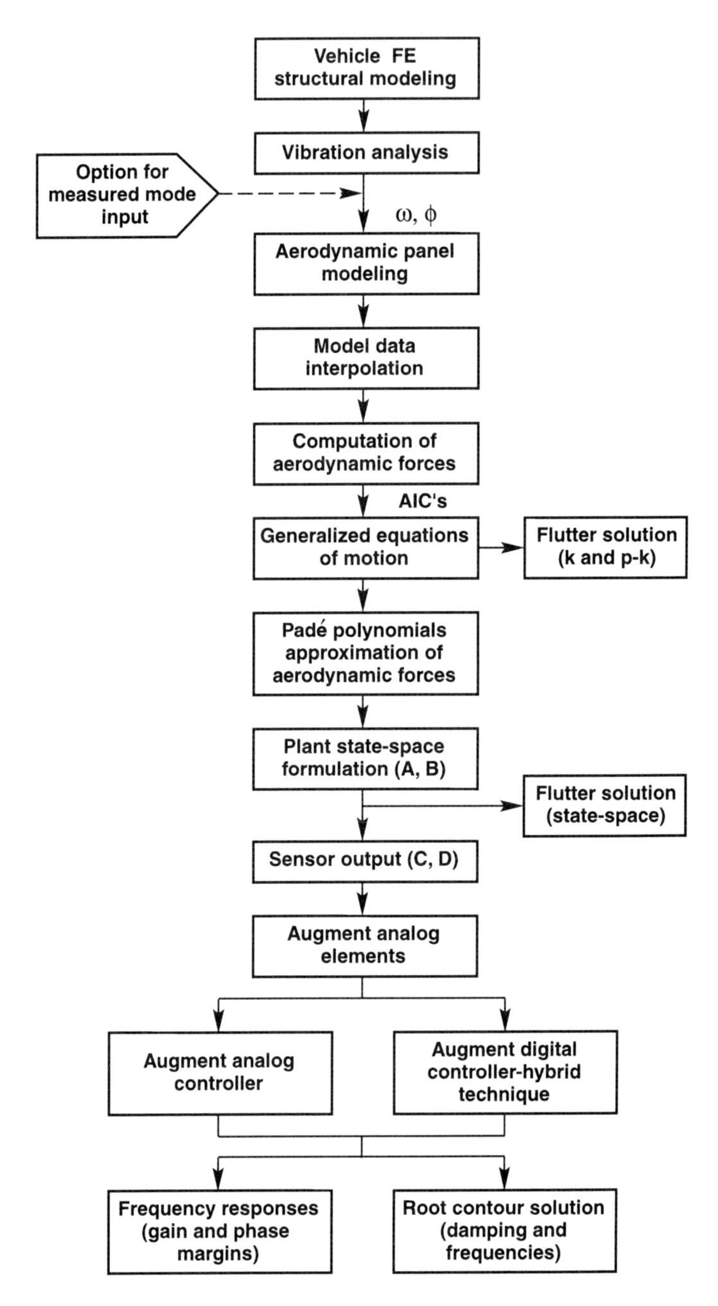

Fig. 14.3 STARS linear aeroservoelastic analysis flowchart.

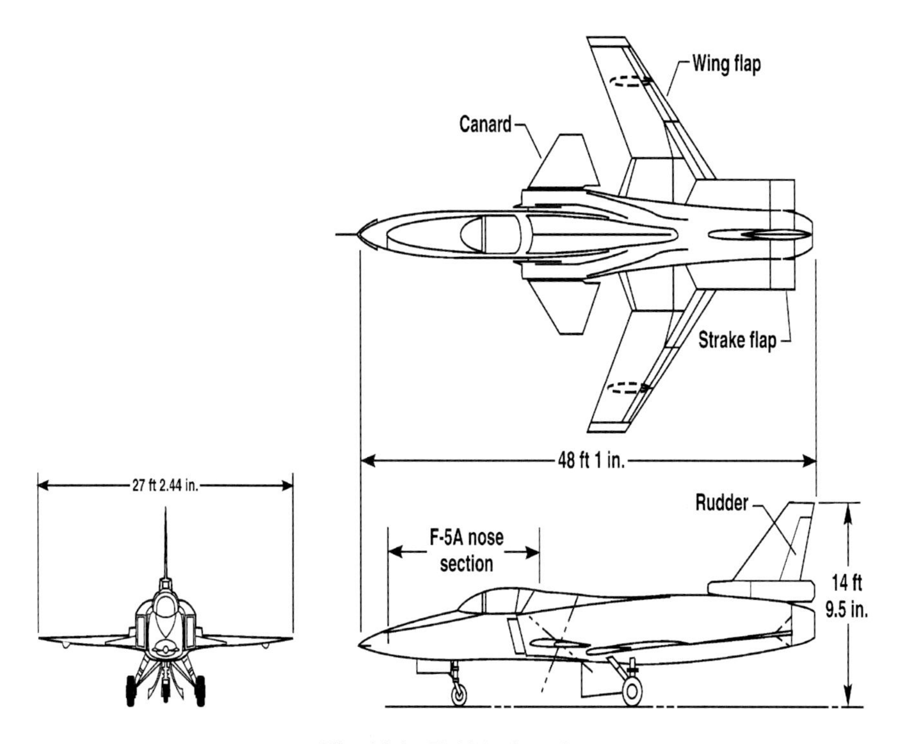

Fig. 14.4 X-29A aircraft.

As a first step toward effecting a linear ASE analysis, a free vibration analysis of the aircraft was performed to yield the first few natural frequencies and mode shapes that were then compared with test results from a ground vibration survey (GVS). These results are compared in Table 14.1, and results of subsequent aeroelastic flutter analysis are shown in Table 14.2. The control law basic stability loop is shown in Fig. 14.6. Figure 14.7 depicts the closed-loop frequency response, whereas Fig. 14.8 shows the damping and frequency values that are compared with flight-measured data.

14.4 Concluding Remarks

A brief outline of the procedures used for the determination of static and dynamic stability derivatives is given in Ref. 3. The numerical procedure presented herein has also been implemented in the NASA STARS[10] finite element based multidisciplinary analysis program, some details of which are also presented in Ref. 11. Figure 14.3 presents a computer flowchart of the analysis algorithm adapted for such an analysis. The aerodynamic influence coefficients are computed by standard panel methods described in Refs. 4 and 5. A detailed description of the coordinate transformation procedure is given in Ref. 3, which is consistent with that presented in Ref. 12. The analysis results were verified by comparing the same with flight test data obtained for a number of test vehicles.[3,8] A survey of aeroservoelastic analysis procedures is also given in Refs. 13 and 14.

Table 14.1 X-29A aircraft vibration frequency comparison (symmetric case)

Mode	GVS (measured) Frequency, Hz	(STARS) (calculated) Frequency, Hz	difference, %
Wing first bending (W1B)	8.61	8.96	+4.1
Fuselage first bending (F1B)	11.65	12.87	+10.5
Fuselage second bending (F2B)	24.30	19.03	−21.7
Canard pitch (CP)	21.07	21.02	−0.2
Wing second bending (W2B)	26.30	26.28	−0.1
Wing first torsion (W1T)	36.70	30.30	−17.4
Canard bending pitch (CBP)	42.20	47.70	+13.0
Wing third bending (W3B)	51.50	49.52	−3.8

The aeroelastic instability phenomenon known as flutter is analyzed by solving the eigenvalue problem for the plant state-space matrix $\boldsymbol{A}$. Such eigenvalues are complex, occurring usually as complex conjugate pairs $-\alpha \pm \beta$. At a certain flight speed, i.e., Mach number, the $\boldsymbol{A}$ matrix is calculated for a number of k values pertaining to a variation of flight altitude. The advent of flutter is indicated by a change of sign of α, and if the corresponding frequency, being the imaginary part of the root, also approaches the zero value, it is indicative of divergence (static aeroelastic instability), and this procedure is called the state-space method. Alternative flutter analysis procedures include the k method as well as the $p - k$ method,[15,16] which has become the industry standard and is used rather routinely for the solution of practical problems. A survey of aeroelastic analysis procedures is given in Ref. 17. Aeroservoelastic instability is assessed by computation of phase and gain margins as well as damping and frequency of a natural mode.[18] In this connection it may be noted that only elastic modes are used for aeroelastic analysis, whereas additionally, the rigid body and control modes are included for aeroservoelastic analysis.

Table 14.2 Comparison of calculated flutter speeds for the X-29A aircraft (symmetric case)

		Critical speed, KEAS		
		STARS modes		Contractor
Instability	Crossing mode	k method	Root-contour method	GVS k method
Divergence	(W1B)	838.10	833.50	808.00
Divergence	(CP)	913.01	917.91	980.00
Flutter	(F1B)	848.06	797.31	924.00
Flutter	(W2B)	1142.95	1156.98	1315.00

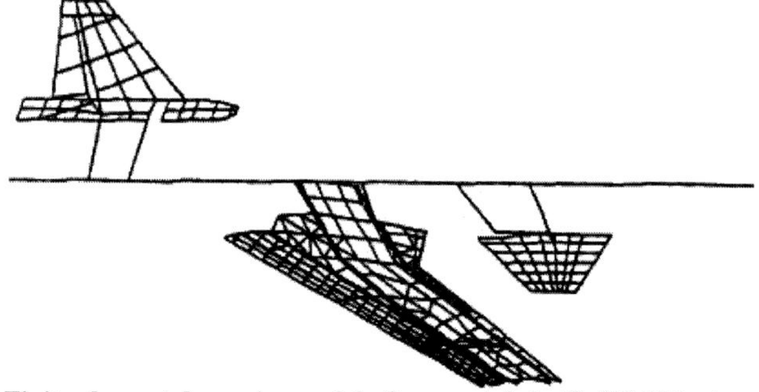

a) Finite element dynamics model of symmetric half of X-29A aircraft

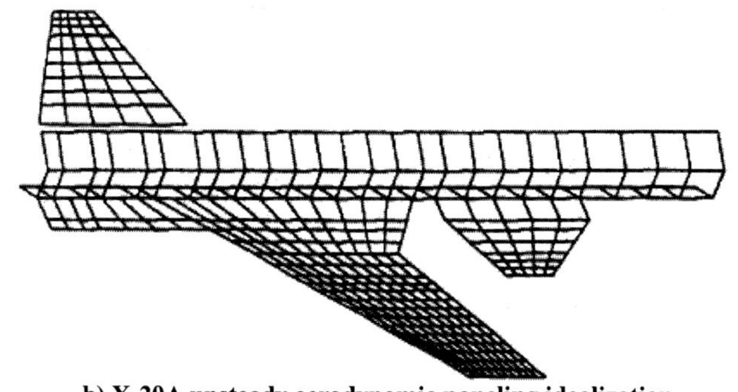

b) X-29A unsteady aerodynamic paneling idealization

Fig. 14.5 X-29A structural and linear aerodynamic models.

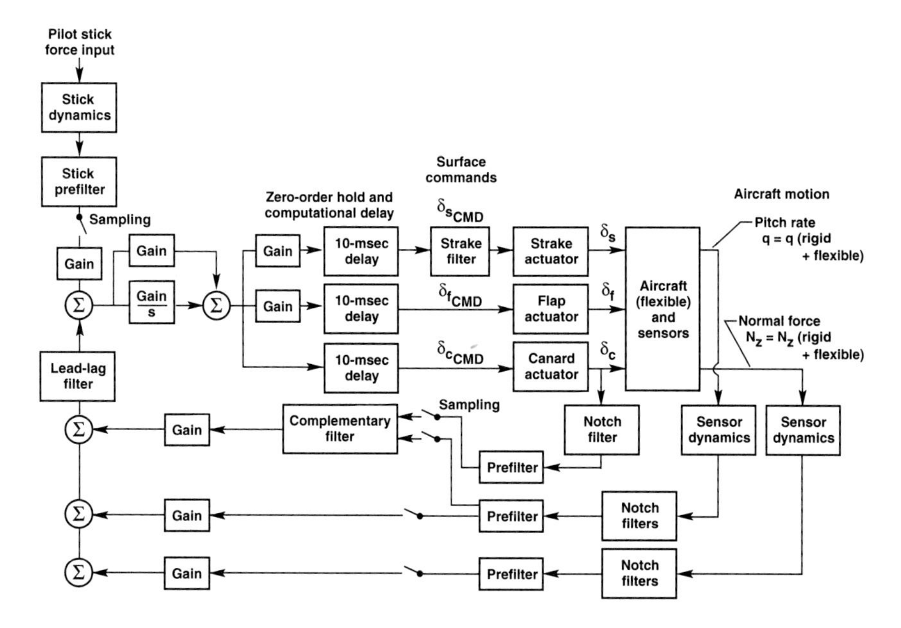

Fig. 14.6 X-29A normal mode longitudinal axis control law basic stability loop.

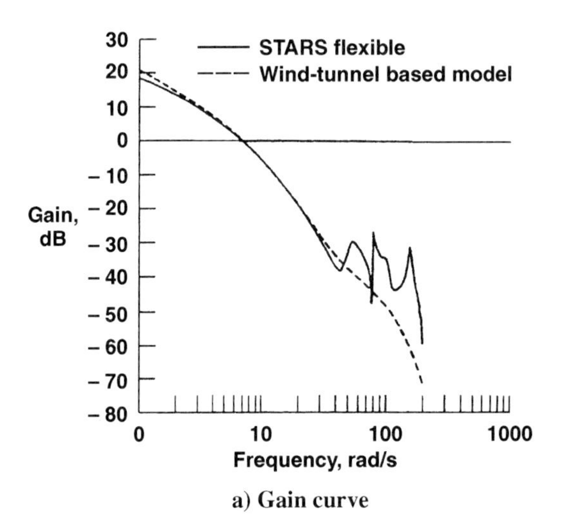

a) Gain curve

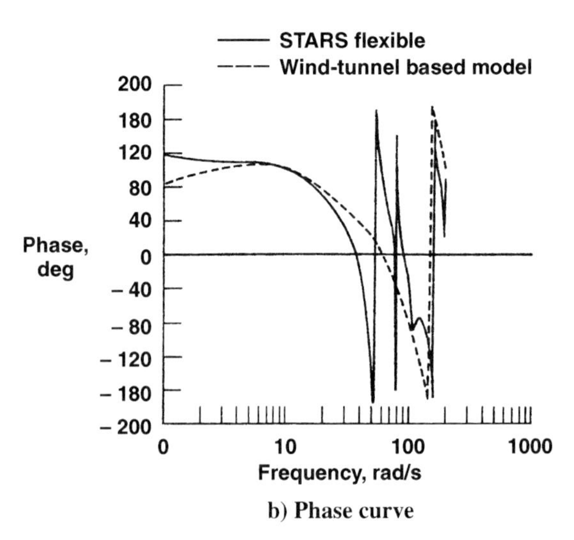

b) Phase curve

Fig. 14.7 Closed-loop frequency response for X-29A normal digital mode; STARS flexible and wind-tunnel dynamics (Mach 0.9, density ratio 1.0).

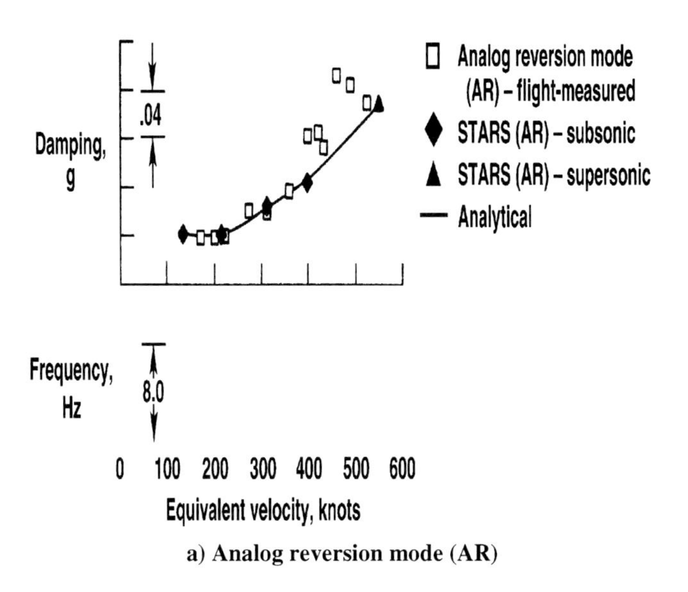

a) Analog reversion mode (AR)

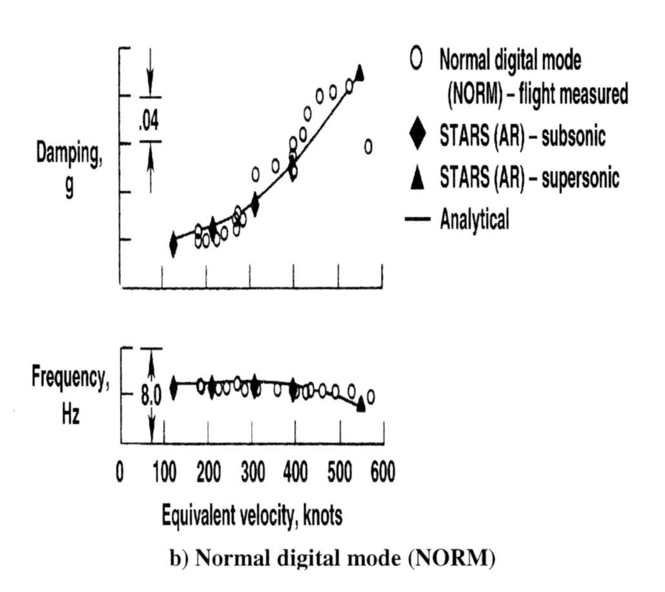

b) Normal digital mode (NORM)

Fig. 14.8 Comparison of X-29A closed-loop modal damping and frequencies between flight-measured data and STARS for symmetric first bending mode at 20,000 ft.

References

[1]Collar, A. R.,"The Expanding Domain of Aeroelasticity," *Journal of the Royal Aeronautical Society*, Vol. L, Aug. 1946, pp. 613–636.

[2]Bishinghoff, R. L., Ashley, H., and Halfman, R. L., *Aeroelasticity*, Dover, New York, 1983.

[3]Gupta, K. K., Brenner, M. J., and Voelker, L. S., "Development of an Integrated Aeroservoelastic Analysis Program and Correlation with Test Data," NASA TP 3120, May 1991.

[4]Geising, J. P., Kalman, T. P., and Rodden, W. P., "Subsonic Unsteady Aerodynamics for General Configurations," Vols. 1 and 2, Air Force Flight Dynamics Directorate, AFFDL-TR-71-5, 1971.

[5]Appa, K., "Constant Pressure Panel Method for Unsteady Airload Analysis," *Journal of Aircraft*, Vol. 24, No. 10, 1987, pp. 696–702.

[6]Abel, I., "An Analytical Technique for Predicting the Characteristics of a Flexible Wing Equipped with an Active Flutter-Suppression System and Comparison with Wind-Tunnel Data," NASA TP-1367, 1979.

[7]Whitbeck, R. F., Didaleusky, D. G. L., and Hofman, L. G., "Frequency Response of Digitally Controlled Systems," *Journal of Guidance and Control*, Vol. 4, No. 4, 1981, pp. 423–427.

[8]Gupta, K. K., Brenner, M. J., and Voelker, L. S., "Integrated Aeroservoelastic Analysis Capability with X-29A Comparison," *Journal of Aircraft*, Vol. 26, No. 1, 1989, pp. 84–90.

[9]Sefic, W. J., and Maxwell, C. M., "X-29A Technology Demonstration Flight Test Program Overview," NASA TM-86809, 1986.

[10]Gupta, K. K., "STARS—An Integrated Multidisciplinary Finite Element Structural, Fluids, Aeroelastic and Aeroservoelastic Analysis Computer Program," NASA TM 4795, May 1997.

[11]Gupta, K. K., "Development and Application of an Integrated Multidisciplinary Analysis Capability," *International Journal for Numerical Methods in Engineering*, Vol. 40, 1997, pp. 533–550.

[12]Miller, R. D., Fraser, R. J., Hirayama, M. Y., and Clemmons, R. E., "Equation Modifying Program L219(EQMOD)," Vol. 1, Engineering and Usage, NASA CR-2855, 1979.

[13]Zimmerman, H., "Aeroservoelasticity," *Computer Methods in Applied Mechanics and Engineering*, Vol. 90, 1991, pp. 719–735.

[14]Livne, E., "Integrated Aeroservoelastic Optimization: Status and Direction," *Journal of Aircraft*, Vol. 36, No. 1, 1999, pp. 122–143.

[15]Taylor, R. F., Miller, R. F., and Brockman, R. A., "A Procedure for Flutter Analysis of FASTOP-3 Compatible Mathematical Models, Vol. 1—Theory and Applications," Air Force Wright Aeronautics Lab., AFWAL-TR81-3063, Wright–Patterson AFB, OH, June 1981.

[16]Hassig, W., and Herrman, J., "An Approximate True Damping Solution of the Flutter Equation by Determinant Iteration," *Journal of Aircraft*, Vol. 8, No. 11, 1971, pp. 885–889.

[17]Friedmann, P. P., "Renaissance of Aeroelasticity and Its Future," *Journal of Aircraft*, Vol. 36, No. 1, 1999, pp. 105–121.

[18]Etkin, B., *Dynamics of Flight-Stability and Control*, Wiley, New York, 1982.

15
CFD-Based Aeroelasticity and Aeroservoelasticity

15.1 Introduction

In the previous chapter a fully integrated aeroelasticity and also aeroservoelasticity analysis procedure was described for a complete aircraft configuration. This approach is limited because the linearized aerodynamic theories used in this connection are not valid in the transonic range and further simplifying assumptions regarding the vehicle geometry need to be made to conform to the linearizing assumptions.

Modern, high performance aerospace vehicles are characterized by unprecedented levels of multidisciplinary interactions of a number of major technical disciplines including structures, aerodynamics, heat transfer, and controls engineering, which may impose considerable constraint on dynamic stability and controls performance margins required for flight safety. Accurate prediction of the flight characteristics prior to flight testing is thus of vital importance and requires employment of more accurate computational fluid dynamics (CFD) procedures for the aerodynamic computations. There are a number of recently published results of research work in the application of CFD to aeroelastic analysis involving unsteady flows. Reference 1 summarizes activities that utilize finite difference methods for the solution of lower-order CFD methods as the transonic small disturbance and the full potential equations in the aeroelastic analysis and also for effective control of flutter[2]; finite difference solution for practical problems have also been reported.[3,4] Extensive efforts have been made recently using structured aerodynamic grids and a finite volume solution of Euler equations[5,6] has also been reported. Further extensions of this work employing unstructured meshes are described in Refs. 7 and 8. In particular, Ref. 8 provides a detailed description of flutter analysis for some representative problems and compares results with available experimental data. Also another paper[9] presents an overview of the application of these techniques to National Aerospace Plane–like hypersonic vehicles, whereas Ref. 10 provides an alternative approach to the solution of the aeroelasticity problem.

The analyses, as discussed previously, effect the simultaneous time-integration of the structural equations of motion with the governing flow equations for the computation of unsteady aerodynamic forces. In this time-marching unsteady flow and aeroelastic calculations, the aerodynamic mesh needs to be updated at every time level, particularly for large structural deformations, so that it follows the deformed structural configuration. However, considering the time required to generate a single domain mesh for a flow solution, rediscretization of the computational domain at each time step appears to be rather impractical.

The number of calculations can be substantially reduced if the surface deformation can be simulated without having to update the aerodynamic mesh. This may be achieved by applying a transpiration boundary condition[11,12] at the surface nodal points. In this process, the body normals are rotated in the same directions as they would be in the actual deflected shape, thus leaving the original aerodynamic mesh unaffected throughout the aeroelastic stability investigation. This procedure has been implemented in a multidisciplinary analysis code[13] for the solution of practical aeroelasticity problems.[14] The same technique may also be used effectively for aeroservoelastic analysis involving motion of a control surface.

An aeroservoelastic analysis can be effected by computing the dynamic transient response due to an impulse load. The resulting response is then analyzed to determine the damping in the vibration modes and consequently the critical flutter speed of an open- or closed-loop aerospace vehicle system. For most aeroelastic solutions static nonlinear effects due to shock are of prime importance. Thus a nonlinear static flowfield is first captured using the Euler or Navier–Stokes method and a subsequent linear dynamic analysis about the mean flow condition proves adequate for predicting aeroelastic effects. Once a steady nonlinear flow solution is achieved, a linearized dynamic perturbation solution about this mean flow is computed, due to an impulsive force of known spectral content. A recursive procedure is next used to identify the unknown parameters in the dynamic perturbation model; the linear model for the generalized forces is then used for all subsequent aeroservoelastic solution runs. This procedure enables rapid and accurate predictions of aerodynamic instability boundaries and the modeling and computation of aeroservoelastic effects over the entire range of Mach numbers.

15.2 Computational Fluid Dynamics

15.2.1 Introduction

An accurate simulation of aerodynamic phenomena around a vehicle in flight involves computation of drag, lift, and pressure distribution by employing a suitable CFD solution strategy. This can be achieved by solving the Euler or Navier–Stokes equations, numerically in space and time. The vehicle surface as well as its surrounding aerodynamic domain are discretized by an unstructured finite-element mesh requiring very fine subdivision on the surface to enable accurate prediction of skin friction and pressure distribution. Unstructured mesh is the natural choice because of its versatility in mapping irregular shapes encountered in the aerospace vehicle geometry and also for seamless coupling with the finite element structural grid that is generally unstructured in nature. In addition the boundary conditions need to be accurately defined; the far field, symmetry, wall, engine inlet, and engine outlet are some of the common examples.

Aerodynamic flows are classified in terms of Mach number defined as the ratio of the vehicle velocity to the velocity of sound. The flow is subsonic, supersonic, or hypersonic for Mach number, < 1, > 1, and > 3, respectively. Around Mach 1.0, the flow is termed transonic and is difficult to predict. Shocks and waves are of common occurrence in high speed flows. Turbulence and shock boundary layer interaction as well as vortex shedding are quite common in viscous compressible flows. The pattern of fluid flow is primarily governed by the Reynolds number defined as $Re = \rho u \ell / \mu$; ρ, u, ℓ, and μ being the density, velocity, characteristic flow length,

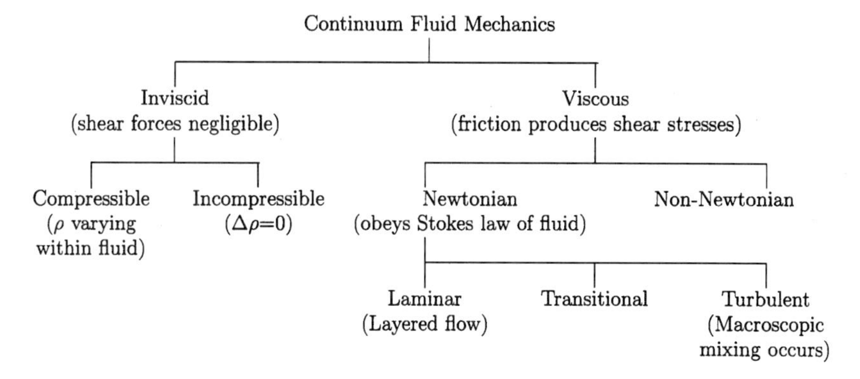

Fig. 15.1 Fluid flow characterization.

and viscosity, respectively. Flow is turbulent for high Reynolds number, and laminar for low Reynolds number. For high Reynolds number flows, the boundary layer is thin and inertial terms dominate except in regions of separation; viscosity has little or no effect(Euler flow). In the case of low Reynolds number flows, viscous terms predominate in comparison to inertia terms. Most practical aerodynamic problems involve compressible flow at high Reynolds number around complex shapes and are governed by nonlinear partial differential equations(Navier–Stokes or Euler). Figure 15.1 depicts the various flow characterizations.

For compressible flow involving turbulence, any one of the following increasingly complex methods may be adopted for the solution: 1) Reynolds averaged Navier–Stokes solution with turbulence modeling, 2) large eddy simulations with sub-grid turbulence modeling or 3) direct numerical simulation that does not require any turbulence modeling. Full Navier–Stokes flow simulation[15] involves very intensive computational effort and thus various approximations are usually adopted in the development of practical finite element CFD codes. Figure 15.2 depicts the full range of these analysis methods.

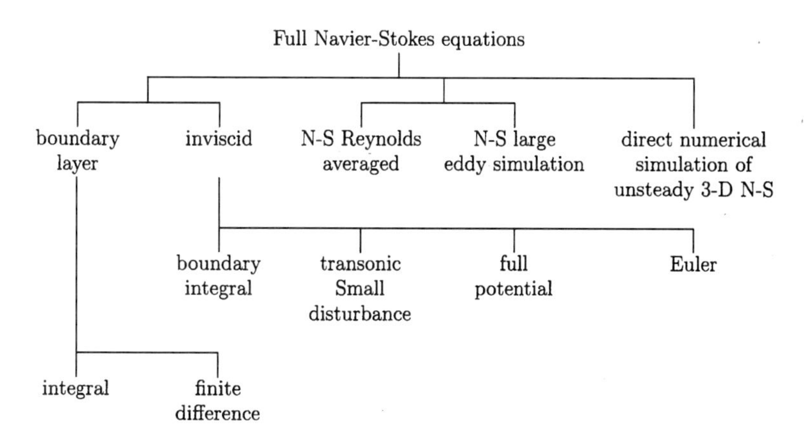

Fig. 15.2 Fluid flow solution approximations.

15.2.2 Equations of Motion for Fluids

Every fluid particle must obey Newton's second law of motion at all times. Additionally, the continuity equation relating the mass flow must be satisfied throughout the fluid domain, and, furthermore, appropriate kinematic boundary conditions must also be satisfied at every point on solid–fluid or differing fluid boundary.

The dynamic behavior of a viscous, heat-conducting, compressible fluid obeying conservation of mass, momentum, and energy may be expressed by a set of nonlinear partial differential equations, known as the Navier–Stokes (N–S) equation[16] in conservation form,

$$\frac{\partial \boldsymbol{v}}{\partial t} + \frac{\partial \boldsymbol{F}_j}{\partial x_j} = \boldsymbol{f}_b, \qquad j = 1, 2, 3 \tag{15.1}$$

which also may be written as

$$\frac{\partial \boldsymbol{v}}{\partial t} + \frac{\partial \boldsymbol{f}_j}{\partial x_j} + \frac{\partial \boldsymbol{g}_j}{\partial x_j} = \boldsymbol{f}_b \tag{15.2}$$

in which the flux vector is split into convection $\boldsymbol{f}_j$ and diffusion $\boldsymbol{g}_j$ terms, and the diffusion may be expressed in terms of velocity gradients and viscosity. The conservation variables, flux, and body force column vectors, as well as the viscous stress are defined as

$$\boldsymbol{v} = [\rho \;\; \rho u_i \;\; \rho E]^T \qquad i = 1, 2, 3 \tag{15.3}$$

$$\boldsymbol{f}_j = [\rho u_j \;\; (\rho u_i u_j + p\delta_{ij}) \;\; u_j(p + \rho E)]^T, \qquad j = 1, 2, 3 \tag{15.4}$$

$$\boldsymbol{g}_j = \left[0 \;\; \sigma_{ij} \;\; \left(u_i \sigma_{ij} + k\frac{\partial T}{\partial x_j}\right)\right]^T \tag{15.5}$$

$$\boldsymbol{f}_b = [0 \;\; f_{b_i} \;\; u_i f_{b_i}]^T \tag{15.6}$$

$$\sigma_{ij} = \mu\left[\frac{\partial u_i}{\partial x_j} + \frac{\partial u_j}{\partial x_i} - \frac{2}{3}\frac{\partial u_l}{\partial x_l}\delta_{ij}\right] \qquad l = 1, 2, 3 \tag{15.7}$$

in which u_i are velocity components in the x_i coordinate system; ρ, p, and E are the density, pressure, and total energy, respectively; μ is the dynamic viscosity; k is the thermal conductivity, the heat flux q_j being $-k\partial T/\partial x_j$; T is the temperature; $\boldsymbol{f}_b$ represents the body forces, and δ_{ij} is the Kronecker delta ($= 1, i = j, = 0, i \neq j$). In compressible flow, the energy equation is coupled with other governing flow equations. The coupling is accomplished through the perfect gas law

$$p = \rho RT \tag{15.8}$$

where R is the gas constant. For the isentropic assumption the speed of sound and the density and pressure are related as

$$c^2 = \frac{\gamma p}{\rho} = \gamma RT \tag{15.9}$$

and density and pressure are related by the equation

$$\frac{\partial \rho}{\partial t} = \frac{\partial \rho}{\partial p}\frac{\partial p}{\partial t} = \frac{1}{c^2}\frac{\partial p}{\partial t} \tag{15.10}$$

in which $\gamma = c_p/c_v$ and c_p is the specific heat at constant pressure, c_v being the specific heat at constant volume.

The preceding equations are in dimensional form and may be nondimensionalized for numerical calculations. Thus each related variable is replaced by ρ/ρ_∞, u_i/u_∞, $p/(\rho_\infty u_\infty^2)$, μ/μ_∞, c/u_∞,x_i/L,E/u_∞^2, $T/(u_\infty^2/c_p)$, and tL/u_∞; L is a characteristic dimension and the suffix ∞ indicates a reference value(free stream value). These scales are substituted in the governing equations, which remain in the same form excepting $\boldsymbol{g}_j$, which becomes

$$\boldsymbol{g}_j = [0 \;\; \sigma_{ij} \;\; (u_i\sigma_{ij} - q_j)]^T \tag{15.11}$$

and also the viscous stress tensor and heat flux take the following forms:

$$\sigma_{ij} = \frac{\mu}{Re}\left[\frac{\partial u_i}{\partial x_j} + \frac{\partial u_j}{\partial x_i} - \frac{2}{3}\frac{\partial u_l}{\partial x_l}\delta_{ij}\right]$$

$$q_j = \frac{1}{RePr}\frac{\partial T}{\partial x_j} \tag{15.12}$$

in which the Reynolds number is defined as $Re = u_\infty L/\nu_\infty$; $\nu_\infty = \mu_\infty/\rho_\infty$ is termed the kinematic viscosity; Pr is the Prandtl number, $Pr = \nu_\infty/\alpha_\infty$, with $\alpha_\infty = k/(\rho_\infty c_p)$ is the thermal diffusivity. Thus the normalized form of Eqs.(15.3), (15.4), and (15.10) and Eqs. (15.11) and (15.12) then may be used for numerical solution of the Navier–Stokes equations. Additionally the temperature-dependent dynamic viscosity variation may be incorporated through Sutherland's semiempirical formula and its nondimensional form is given as

$$\mu = \left[\frac{c_1 + S_0 c_1/T_\infty}{\theta + S_0 c_1/T_\infty}\right] \times \left[\frac{\theta}{c_1}\right]^{1.5} \tag{15.13}$$

in which $c_1 = 1/(M_\infty^2(\gamma - 1))$, Sutherland's constant $S_0 = 198.6R$ for air, and θ is the nondimensional local temperature. The first of the five Navier–Stokes equations represents conservation of mass, the next three represent conservation of momentum, whereas the fifth represents conservation of energy.

15.2.3 Discretization of the Euler Equations

The inviscid form of Eq. (15.1), known as the Euler equations, is often used for the solution of flow problems where viscous effects are confined to thin boundary layers. Solving the Euler equations involves much less computational effort, although it is incapable of simulating separated flows and other viscous effects, e.g., skin friction drag. Table 15.1 provides a comparison of the two flow characteristics. In the absence of body forces the Euler equations may be written as

$$\frac{\partial \boldsymbol{v}}{\partial t} + \frac{\partial \boldsymbol{f}_i}{\partial x_i} = 0 \tag{15.14}$$

Table 15.1 Flow simulation characteristic

Navier–Stokes solution	Euler solution
Compressible	Compressible
Viscous	Nonviscous
Vortex flows	Vortex flows
Separated flows	Sharp edges only
Heat-conducting	Limited

in which the flux vector for $\boldsymbol{v}$ has the following form:

$$\boldsymbol{f}_i = [\rho u_i \quad \rho u_i u_j + p\delta_{ij} \quad u_i(p + \rho E)] \tag{15.15}$$

The presence of derivatives of the $\boldsymbol{f}_i$ functions in Eq. (15.14) is an obstacle to computational methods. They are also undesirable from a theoretical viewpoint because they do not exist where shocks occur. A standard way to avoid these derivatives is to reformulate Eq. (15.14) as integrals involving test functions followed by integration by parts to move the spatial differentiation from the $\boldsymbol{f}_i$ functions to the test functions.

Let Ω denote the spacial region of interest and Γ denote the boundary surface of this region. Let Φ be a continuous, piecewise differentiable function defined over the region Ω. Then, for each fixed point in time, it follows that

$$\int_{\Omega} \left(\frac{\partial \boldsymbol{v}}{\partial t} + \sum_{i=1}^{3} \frac{\partial \boldsymbol{f}_i}{\partial x_i} \right) \Phi \, \mathrm{d}\Omega = 0 \tag{15.16}$$

Applying the Gauss theorem (a generalization of integration by parts), it follows that

$$\int_{\Omega} \frac{\partial \boldsymbol{v}}{\partial t} \Phi \, \mathrm{d}\Omega = \int_{\Omega} \sum_{i=1}^{3} \boldsymbol{f}_i \frac{\partial \Phi}{\partial x_i} \, \mathrm{d}\Omega - \int_{\Gamma} \left(\sum_{i=1}^{3} \boldsymbol{f}_i n_i \right) \Phi \, \mathrm{d}\Gamma \tag{15.17}$$

where $\boldsymbol{n}_i = (n_1, n_2, n_3)$ is the outward pointing normal vector on the boundary surface Γ. Given Eq. (15.14), Eq. (15.17) will hold for any Φ that is continuous and piecewise differentiable over Ω. Conversely, if Eq. (15.17) holds for every Φ that is continuous and piecewise differentiable over Ω, then Eq. (15.18) holds, except possibly over a set of measure zero in Ω where the $\boldsymbol{f}_i$ functions are not differentiable.

To move from theory to a Galerkin-weighted residual finite element computational method, it is possible to approximate Ω by a collection of tetrahedra, introduce shape functions $\boldsymbol{N}_k$, and use the shape function both as basis functions for approximating the $\boldsymbol{v}$ and $\boldsymbol{f}_i$ functions and as the test functions in Eq. (15.17). Denote the nodes of the tetrahedral grid by x_k and the tetrahedra by e_α. Let f_β denote the triangular faces of tetrahedra whose union is the boundary surface Γ. With each nodal point x_k a piecewise linear shape function $\boldsymbol{N}_k$ is associated, which is nonzero only over the tetrahedra incident on x_k and takes the value of 1 at x_k and 0 at all other nodal points. Use the notation $\alpha \in x_k$ to denote the indices α

of tetrahedra incident in the node x_k, and $k \in e_\alpha$ to denote the indices k of nodes incident on the tetrahedron e_α.

Let $\boldsymbol{v}^*$ and $\boldsymbol{f}^*$ denote, respectively, the computed approximations for $\boldsymbol{v}$ and $\boldsymbol{f}$. The values of $\boldsymbol{v}^*$ and $\boldsymbol{f}^*$ at the nodal point x_k will be denoted by $\boldsymbol{v}_k^*$ and $\boldsymbol{f}_k^*$, respectively. Given $\boldsymbol{v}_k^*(t)$, $\boldsymbol{f}_k^*(t)$ will be computed as $\boldsymbol{f}_k^*(t) = \boldsymbol{f}(\boldsymbol{v}_k^*(t))$ using the formulas of Eq. (15.15). In terms of $\boldsymbol{v}_k^*(t)$ and $\boldsymbol{f}_k^*(t)$, express $\boldsymbol{v}^*$ and $\boldsymbol{f}^*$ at any point x within a tetrahedron e_α as

$$\boldsymbol{v}^*(x,t) = \sum_{k_2 \in e_\alpha} \boldsymbol{v}_{k_2}^*(t) \boldsymbol{N}_{k_2}(x) \tag{15.18}$$

and

$$\boldsymbol{f}^*(x,t) = \sum_{k_2 \in e_\alpha} \boldsymbol{f}_{k_2}^*(t) \boldsymbol{N}_{k_2}(x) \tag{15.19}$$

Replacing $\boldsymbol{v}$ and $\boldsymbol{f}$ in Eq. (15.17) by Eqs. (15.18) and (15.19) and Φ by $\boldsymbol{N}_{k_1}$, and letting NN denote the total number of nodes, the following relationship is obtained:

$$\begin{aligned}
&\sum_\alpha \int_{e_\alpha} \left[\sum_{k_2 \in e_\alpha} \frac{\mathrm{d}\boldsymbol{v}_{k_2}^*(t)}{\mathrm{d}t} \boldsymbol{N}_{k_2}(x) \right] \boldsymbol{N}_{k_1}(x)\,\mathrm{d}x \\
&\quad = \sum_\alpha \int_{e_\alpha} \sum_{i=1}^{3} \left[\sum_{k_2 \in e_\alpha} \boldsymbol{f}_{ik_2}^*(t) \boldsymbol{N}_{k_2}(x) \right] \frac{\partial \boldsymbol{N}_{k_1}(x)}{\partial x_i}\,\mathrm{d}x \\
&\quad - \sum_\beta \int_{f_\beta} \sum_{i=1}^{3} \left[\sum_{k_2 \in f_\beta} \boldsymbol{f}_{ik_2}^*(t) \boldsymbol{N}_{k_2}(x) \right] n_i \boldsymbol{N}_{k_1}(x)\,\mathrm{d}x, \quad \text{for} \quad k_1 = 1, \ldots, \mathrm{NN}
\end{aligned} \tag{15.20}$$

The boundary normal vector $\boldsymbol{n}$ will be defined computationally as being normal to the boundary surface of the tetrahedral grid and thus constant over each triangular face f_β of the boundary surface. Let $n_{i\beta}^*$ denote the constant value of the ith component of the normalized outward pointing vector normal to face f_β. Because $\boldsymbol{N}_k(x)$ is linear over each tetrahedron, $\partial \boldsymbol{N}_k(x)/\partial x_i$ is constant over each tetrahedron. Let $\lambda_{ik\alpha}$ denote the constant value of $\partial \boldsymbol{N}_k(x)/\partial x_i$ over the tetrahedron e_α. All terms except $\boldsymbol{N}_{k_1}(x)$ and $\boldsymbol{N}_{k_2}(x)$ can be moved outside the integral signs in Eq. (15.20), giving

$$\begin{aligned}
&\sum_\alpha \sum_{k_2 \in e_\alpha} \left[\int_{e_\alpha} \boldsymbol{N}_{k_1}(x) \boldsymbol{N}_{k_2}(x)\,\mathrm{d}x \right] \frac{\mathrm{d}\boldsymbol{v}_{k_2}^*(t)}{\mathrm{d}t} \\
&\quad = \sum_\alpha \sum_{k_2 \in e_\alpha} \sum_{i=1}^{3} \lambda_{ik_1\alpha} \left[\int_{e_\alpha} \boldsymbol{N}_{k_2}(x)\,\mathrm{d}x \right] \boldsymbol{f}_{ik_2}^*(t) \\
&\quad - \sum_\beta \sum_{k_2 \in e_\alpha} \sum_{i=1}^{3} \left[\int_{f_\beta} \boldsymbol{N}_{k_1}(x) \boldsymbol{N}_{k_2}(x)\,\mathrm{d}x \right] n_{i\beta}^* \boldsymbol{f}_{ik_2}(t), \quad \text{for} \quad k_1, k_2 = 1, \ldots, \mathrm{NN}
\end{aligned} \tag{15.21}$$

The quantities in square brackets in Eq. (15.21) depend only on the subscripts involved, thus defining

$$\mu_{\alpha k_1 k_2} = \int_{e_\alpha} \boldsymbol{N}_{k_1}(x)\boldsymbol{N}_{k_2}(x)\,\mathrm{d}x \tag{15.22}$$

$$= \begin{cases} \dfrac{2}{20}V_\alpha & \text{if} \quad k_1 \in e_\alpha \quad \text{and} \quad k_2 = k_1 \\ \dfrac{1}{20}V_\alpha & \text{if} \quad k_1 \in e_\alpha, \quad k_2 \in e_\alpha, \quad \text{and} \quad k_2 \neq k_1 \\ 0 & \text{if} \quad k_1 \notin e_\alpha \quad \text{or} \quad k_2 \notin e_\alpha \end{cases}$$

in which V_α denotes the volume of e_α,

$$v_{\alpha k} = \int_{e_\alpha} \boldsymbol{N}_k(x)\,\mathrm{d}x \tag{15.23}$$

$$= \begin{cases} \dfrac{1}{4}V_\alpha & \text{if} \quad k \in e_\alpha \\ 0 & \text{if} \quad k \notin e_\alpha \end{cases}$$

$$\Psi_{\beta k_1 k_2} = \int_{f_\beta} \boldsymbol{N}_{k_1}(x)\boldsymbol{N}_{k_2}(x)\,\mathrm{d}x \tag{15.24}$$

$$= \begin{cases} \dfrac{2}{12}A_\beta & \text{if} \quad k_1 \in f_\beta \quad \text{and} \quad k_2 = k_1 \\ \dfrac{1}{12}A_\beta & \text{if} \quad k_1 \in f_\beta, \quad k_2 \in f_\beta, \quad \text{and} \quad k_2 \neq k_1 \\ 0 & \text{if} \quad k_1 \notin f_\beta \quad \text{or} \quad k_2 \notin f_\beta \end{cases}$$

where A_β denotes the area of f_β. Now Eq. (15.21) can be rewritten as

$$\sum_\alpha \sum_{k_2 \in e_\alpha} \mu_{\alpha k_1 k_2} \frac{\mathrm{d}\boldsymbol{v}^*_{k_2}(t)}{\mathrm{d}t} = \sum_\alpha \sum_{k_2 \in e_\alpha} \sum_{i=1}^{3} \lambda_{ik_1\alpha} v_{\alpha k_2} \boldsymbol{f}^*_{ik_2}(t)$$

$$- \sum_\beta \sum_{k_2 \in f_\beta} \sum_{i=1}^{3} \Psi_{\beta k_1 k_2} n^*_{i\beta} \boldsymbol{f}^*_{ik_2}(t) \qquad \text{for} \qquad k_1 = 1, \ldots, \mathrm{NN} \tag{15.25}$$

Rewrite Eq. (15.25) as

$$\sum_{k_2=1}^{\mathrm{NN}} a_{k_1 k_2} \frac{\mathrm{d}\boldsymbol{v}^*_{k_2}(t)}{\mathrm{d}t} = \sum_{k_2=1}^{\mathrm{NN}} \sum_{i=1}^{3} \left(b_{ik_1k_2} - c_{ik_1k_2}\right) \boldsymbol{f}^*_{ik_2}(t) \qquad \text{for} \qquad k_1, k_2 = 1, \ldots, \mathrm{NN} \tag{15.26}$$

where

$$a_{k_1k_2} = \sum_{\alpha\in(x_{k_1} \text{ and } x_{k_2})} \mu_{\alpha k_1 k_2} \tag{15.27}$$

$$b_{ik_1k_2} = \sum_{\alpha\in(x_{k_1} \text{ and } x_{k_2})} \lambda_{ik_1\alpha} v_{\alpha k_2} \tag{15.28}$$

$$c_{ik_1k_2} = \sum_{\beta\in(x_{k_1} \text{ and } x_{k_2})} \Psi_{\beta k_1 k_2} n^*_{i\beta} \tag{15.29}$$

Define matrices of order NN

$$\boldsymbol{A} = [a_{k_1k_2}] \tag{15.30}$$

$$\boldsymbol{B} = [b_{ik_1k_2}], \qquad i = 1, 2, 3 \tag{15.31}$$

$$\boldsymbol{C} = [c_{ik_1k_2}], \qquad i = 1, 2, 3 \tag{15.32}$$

Let $\boldsymbol{v}^*(t)$ and $\boldsymbol{f}^*_i(t)(i = 1, 2, 3)$, denote vectors of length NN having components $\boldsymbol{v}^*_{k_2}(t)$ and $\boldsymbol{f}^*_{ik_2}(t)$, respectively, for $k_2 = 1, \ldots,$ NN. Write $\dot{\boldsymbol{v}}(t)$ for $\mathrm{d}\boldsymbol{v}^*(t)/\mathrm{d}t$. Then Eq. (15.20) can be written as

$$\boldsymbol{A}\dot{\boldsymbol{v}}^*(t) = \sum_{i=1}^{3}(\boldsymbol{B}_i - \boldsymbol{C}_i)\boldsymbol{f}^*_i(t) \tag{15.33}$$

The matrices $\boldsymbol{A}$ and $\boldsymbol{C}_i$ are symmetric, whereas the matrix $\boldsymbol{B}_i$ is not symmetric. All of the matrices are sparse. The coefficients $a_{k_1k_2}$ and $b_{ik_1k_2}$ can be nonzero only if x_{k_1} and x_{k_2} are directly connected by some edge of the tetrahedral grid. Then $a_{k_1k_2}$ will be positive, whereas $b_{ik_1k_2}$ may be positive, negative, or zero. A coefficient $c_{ik_1k_2}$ will be nonzero—in fact, positive—if and only if x_{k_1} and x_{k_2} are boundary nodes and are directly connected by an edge that lies in the boundary.

Equation (15.33) is a system of (5 × NN) nonlinear ordinary differential equations (ODEs), which along with appropriate initial conditions, can be used to determine the (5 × NN) functions of time $v^*_{k_2}(t)$; $k_2 = 1, \ldots,$ NN. Note that Eq. (15.33) is coupled across different values of $j = 1, 5$ because, in general, each $\boldsymbol{f}^*_{ik_2}(t)$ is a function of $\boldsymbol{v}^*_{1k_2}(t), \ldots, \boldsymbol{v}^*_{5k_2}(t)$ as specified in Eqs. (15.3) and (15.4). If one is only interested in the steady-state solution, one can replace the left side of Eq. (15.33) by zeros and have a system of (5 × NN) nonlinear algebraic equations to solve for the steady-state values of the $\boldsymbol{v}^*_{k_2}$. There are various iterative methods that could be tried to solve this large, nonlinear system, but it appears that the usual approach is to use the ODE system of Eq. (15.33) as a means for approaching the steady-state solution.

The elements of the matrices $\boldsymbol{A}$, $\boldsymbol{B}_i$, and $\boldsymbol{C}_i$ depend only on the grid. Thus they are usually computed and stored in a sparse storage scheme after the grid is computed and before starting the iterative solution process. One way to order this computation is by tetrahedra. For each tetrahedron, say e_α, loop through the 12 ordered pairings (k_1, k_2) of the four vertex indices. For each pairing, compute $\mu_{\alpha k_1 k_2}$ and add it into $a_{k_1k_2}$. Also compute $\lambda_{ik_1\alpha} v_{\alpha k_2}$ and add it into $b_{ik_1k_2}$ for $i = 1, 2, 3$. For each face of e_α that is a boundary face (there may be none), say f_β, loop through the six ordered pairing (k_1, k_2) of the three vertex indices and compute $\Psi_{\beta K_1 k_2} n^*_{i\beta}$ and add it into $c_{ik_1k_2}$ for $i = 1, 2, 3$. Considering efficiency of computation, because $\boldsymbol{A}$ and $\boldsymbol{C}_i$ are symmetric, it suffices to compute only the diagonal elements and half of the off-diagonal elements. For example, one could compute only the elements for which $k_1 \geq k_2$. Thus, of the 12 ordered pairings (k_1, k_2) of vertices in a particular tetrahedron e_α, one could compute $\mu_{\alpha k_1 k_2}$ only for the 6 pairs for which $k_1 \geq k_2$. This would require half as much work as computing for all 12 pairs.

To solve Eq. (15.33) by time stepping, one must solve for $\boldsymbol{v}^*(t)$ at many time points and apply a suitable method of numerical solution of a system of ODEs. Define

$$\boldsymbol{R}(t) = \sum_{i=1}^{3} (\boldsymbol{B}_i - \boldsymbol{C}_i) \boldsymbol{f}^*_i(t) \tag{15.34}$$

It is necessary to solve for $\boldsymbol{v}^*(t)$ from the linear equations

$$\boldsymbol{A}\dot{\boldsymbol{v}}^*(t) = \boldsymbol{R}(t) \tag{15.35}$$

The matrix $\boldsymbol{A}$ is positive definite and can be written as

$$\boldsymbol{A} = \sum_{\alpha} V_\alpha \boldsymbol{P}^T_\alpha \tilde{\boldsymbol{M}} \boldsymbol{P}_\alpha \tag{15.36}$$

where V_α is the volume of tetrahedron e_α as in Eq. (15.22), $\tilde{\boldsymbol{M}}$ is a constant fourth-order matrix whose elements are the constants given in Eq. (15.22):

$$\tilde{\boldsymbol{M}} = \frac{1}{20} \begin{bmatrix} 2 & 1 & 1 & 1 \\ 1 & 2 & 1 & 1 \\ 1 & 1 & 2 & 1 \\ 1 & 1 & 1 & 2 \end{bmatrix} \tag{15.37}$$

and $\boldsymbol{P}_\alpha$ is a $(4 \times \mathrm{NN})$ matrix containing all zeros except for the four columns whose indices are the indices of the nodal points that are the vertices of e_α. The four columns are columns of the fourth-order identity matrix (in any order, but the natural order is easiest to understand).

15.2.4 Numerical Solution of the Flow Equations

A widely used method for iterative solution of a system of linear equations having a sparse positive-definite symmetric matrix is the (preconditioned) conjugate gradient method. However, a variety of other techniques adopted in this connection by various authors are available in the literature for the solution of the present problem. In one such procedure[17,18] an explicit time-stepping iterative scheme employing a Taylor expansion of the solution $\boldsymbol{v}(x_i, t)$ is adopted until steady conditions are achieved. Because the computed solution $\boldsymbol{v}^{n+1}$ is expected to show oscillations in the vicinity of high gradients in flow phenomena such as shocks, it is smoothed at such locations by the introduction of appropriate artificial viscosity before proceeding to the next time step.[19] This scheme may be further augmented by a suitable flux-corrected transport algorithm to achieve enhanced definition of discontinuities. Further improvement[20] of the Euler solution has been effected by implementing an Aitken acceleration technique. In this procedure, a solution variable result is sampled at three discrete time steps, and the Aitken technique then is used to obtain a revised estimate of the solution.

For the solution of practical problems involving large-scale computation, both spatial and temporal approximations are applied to the flow equations. A number of finite element (FE) schemes have been developed that are based on the Taylor–Galerkin formulation. These schemes are characteristized by somewhat higher CPU requirements primarily because of the use of predictor-corrector–type solution algorithms as well as evaluation of Jacobian-type matrices inherent in the formulation. The solution presented herein also employs the Taylor–Galerkin principles but without the higher order stabilization terms. In this procedure, a forward Euler time discretization is carried out and the scheme is complimented by the addition of artificial dissipation based on a pressure switch method that aids in stabilizing the solution process and in capturing shocks. For viscous flow the inviscid solution is augmented with the viscous terms.

The Navier–Stokes Eq. (15.1), in the absence of body forces takes the form

$$\frac{\partial \boldsymbol{v}}{\partial t} + \frac{\partial \boldsymbol{F}_i}{\partial x_i} = 0, \qquad i = 1, 2, 3 \tag{15.38}$$

which may also be written as

$$\frac{\partial \boldsymbol{v}}{\partial t} + \frac{\partial \boldsymbol{f}_i}{\partial x_i} + \frac{\partial \boldsymbol{g}_i}{\partial x_i} = 0 \tag{15.39}$$

Applying Taylor's expansion of the solution $\boldsymbol{v}(x, t)$ in the time domain,

$$\boldsymbol{v}(t + \Delta t) = \boldsymbol{v}(t) + \frac{\partial \boldsymbol{v}(t)}{\partial t}\Delta t + \frac{1}{2}\frac{\partial^2 \boldsymbol{v}(t)}{\partial t^2}\Delta t^2 + \cdots \tag{15.40}$$

Neglecting the second order term, Eq.(15.40) is written as

$$\begin{aligned} \Delta \boldsymbol{v} &= -\Delta t \left[\frac{\partial \boldsymbol{F}_i}{\partial x_i}\right]_{(t)} \\ &= -\Delta t \left[\frac{\partial \boldsymbol{f}_i}{\partial x_i} + \frac{\partial \boldsymbol{g}_i}{\partial x_i}\right]_{(t)} \end{aligned} \tag{15.41}$$

where $\Delta \boldsymbol{v} = \boldsymbol{v}(t + \Delta t) - \boldsymbol{v}(t)$. Applying Galerkin's spatial idealization $\boldsymbol{v} = \boldsymbol{N}\tilde{\boldsymbol{v}}$, $\tilde{\boldsymbol{v}}$ being the nodal values, and assuming absence of body forces, Eq. (15.41) takes the following form:

$$\int_V \boldsymbol{N}^T \left[\boldsymbol{N}\frac{\partial \tilde{\boldsymbol{v}}}{\partial t} + \frac{\partial \boldsymbol{f}_i}{\partial x_i} + \frac{\partial \boldsymbol{g}_i}{\partial x_i}\right] \mathrm{d}V = 0 \tag{15.42}$$

which results in

$$\int_V \boldsymbol{N}^T \boldsymbol{N} \Delta \tilde{\boldsymbol{v}}\, \mathrm{d}V = -\Delta t \int_V \boldsymbol{N}^T \frac{\partial \boldsymbol{N}}{\partial x_i} \tilde{\boldsymbol{f}}_i\, \mathrm{d}V - \Delta t \int_V \boldsymbol{N}^T \frac{\partial \boldsymbol{N}}{\partial x_i} \tilde{\boldsymbol{g}}_i\, \mathrm{d}V \tag{15.43}$$

The flow equation then may be expressed as

$$\boldsymbol{M}\Delta\tilde{\boldsymbol{v}} = -\Delta t\left[\frac{\partial u_i}{\partial x_i}\boldsymbol{M} + \boldsymbol{K}\right]\tilde{\boldsymbol{v}} - \Delta t(\hat{\boldsymbol{f}}_1 + \hat{\boldsymbol{f}}_2) + \Delta t\hat{\boldsymbol{R}} + \Delta t[\boldsymbol{K}_\sigma + \boldsymbol{f}_\sigma] \quad (15.44)$$

in which $\boldsymbol{M}$ is the consistent mass matrix, $\boldsymbol{K}$ the convection matrix, $\hat{\boldsymbol{f}}_1$, $\hat{\boldsymbol{f}}_2$ the pressure matrices, $\boldsymbol{K}_\sigma$ the second-order matrix that includes viscous and heat flux effects, and $\boldsymbol{f}_\sigma$ the boundary integral matrix from second-order terms. Then,

$$\boldsymbol{M} = \int_V \boldsymbol{N}^T\boldsymbol{N}\,\mathrm{d}V; \quad \boldsymbol{K} = \int_V \boldsymbol{N}^T\bar{u}_i\frac{\partial \boldsymbol{N}}{\partial x_i}\,\mathrm{d}V; \quad \hat{\boldsymbol{f}}_1 = \int_V \boldsymbol{N}^T\bar{p}\frac{\partial \boldsymbol{e}_i}{\partial x_i}\,\mathrm{d}V;$$

$$\hat{\boldsymbol{f}}_2 = \int_V \boldsymbol{N}^T\bar{\boldsymbol{e}}_i\frac{\partial p}{\partial x_i}\,\mathrm{d}V$$

$$\boldsymbol{K}_\sigma = -\int_V \frac{\partial \boldsymbol{N}^T}{\partial x_j}\boldsymbol{e}_j\sigma_{ij}\,\mathrm{d}V - \int_V \frac{\partial \boldsymbol{N}^T}{\partial x_j}\boldsymbol{m}_j q_j\,\mathrm{d}V$$

$$\boldsymbol{f}_\sigma = \int_\Gamma \boldsymbol{N}^T\boldsymbol{e}_j\sigma_{ij}\hat{\boldsymbol{n}}\,\mathrm{d}\Gamma + \int_\Gamma \boldsymbol{N}^T\boldsymbol{m}_j q_j\hat{\boldsymbol{n}}\,\Gamma \quad (15.45)$$

In these equations, $\bar{p}_i, \bar{u}_i, \bar{\boldsymbol{e}}_i$ are the average values; $\boldsymbol{e}_1 = (0\ 1\ 0\ 0\ u_1)^T$, $\boldsymbol{e}_2 = (0\ 0\ 1\ 0\ u_2)^T$, $\boldsymbol{e}_3 = (0\ 0\ 0\ 1\ u_3)^T$, $\hat{\boldsymbol{R}}$ is the artifical dissipation, and $m_1 = m_2 = m_3 = [0\,0\,0\,0\,1]^T$. Turbulence terms also may be included by modifying the viscous effects.

A two-step solution procedure is adopted for the flow equation, the inviscid solution being augmented with the viscous term and stabilized with artificial dissipation terms. Assuming

$$\Delta\tilde{\boldsymbol{v}} = \tilde{\boldsymbol{v}}_{n+1} - \tilde{\boldsymbol{v}}_n \quad (15.46)$$

then,

$$\boldsymbol{M}(\tilde{\boldsymbol{v}}_{n+1} - \tilde{\boldsymbol{v}}_n) = \frac{-\Delta t}{2}[c\boldsymbol{M} + \boldsymbol{K}](\tilde{\boldsymbol{v}}_{n+1} + \tilde{\boldsymbol{v}}_n) - \Delta t(\hat{\boldsymbol{f}}_1 + \hat{\boldsymbol{f}}_2) \quad (15.47)$$

which becomes

$$\left[\left(1 + \frac{\Delta t}{2}c\right)\boldsymbol{M} + \frac{\Delta t}{2}\boldsymbol{K}\right]\tilde{\boldsymbol{v}}_{n+1} = \left[\left(1 - \frac{\Delta t}{2}\right)\boldsymbol{M} - \frac{\Delta t}{2}\boldsymbol{K}\right]\tilde{\boldsymbol{v}}_n + \Delta t\boldsymbol{R} \quad (15.48)$$

or

$$[\boldsymbol{M}_+]\tilde{\boldsymbol{v}}_{n+1} = [\boldsymbol{M}_-]\tilde{\boldsymbol{v}}_n + \Delta t\boldsymbol{R} \quad (15.49)$$

in which

$$\boldsymbol{R} = -(\hat{\boldsymbol{f}}_1 + \hat{\boldsymbol{f}}_2) \quad (15.50)$$

Let

$$\boldsymbol{M}_+ = \boldsymbol{D}_+ + \boldsymbol{M}'_+ \quad (15.51)$$

the matrix $\boldsymbol{D}_+$ having diagonal elements. Then Eq. (15.49) may be solved as follows.

Step 1: Form

$$[\boldsymbol{D}_+]\tilde{\boldsymbol{v}}_{n+1} = [\boldsymbol{M}_-]\tilde{\boldsymbol{v}}_n - [\boldsymbol{M}'_+]\tilde{\boldsymbol{v}}_{n+1} + \Delta t\boldsymbol{R} \tag{15.52}$$

Step 2: Solve $\tilde{\boldsymbol{v}}_{n+1}$ iteratively

$$\tilde{\boldsymbol{v}}_{n+1}^{(i+1)} = [\boldsymbol{D}_+]^{-1}\left\{[\boldsymbol{M}_-]\tilde{\boldsymbol{v}}_n - [\boldsymbol{M}'_+]\tilde{\boldsymbol{v}}_{n+1}^{(i)} + \Delta t(\boldsymbol{R} + \hat{\boldsymbol{R}} + \boldsymbol{K}_\sigma + \boldsymbol{f}_\sigma)\right\} \tag{15.53}$$

Step 3: If $\|\tilde{\boldsymbol{v}}_{n+1}^{(i+1)}\| \neq \text{EPS1}\|\tilde{\boldsymbol{v}}_{n+1}^{(i)}\|$ go to Step 2.
Step 4: If$\|\tilde{\boldsymbol{v}}_{n+1}\| \neq \text{EPS2}\|\tilde{\boldsymbol{v}}_n\|$ go to Step 1.
Step 5: Repeat Steps 1 to 4 NITER times until convergence is reached, that is until $\tilde{\boldsymbol{v}}_{n+1} \approx \tilde{\boldsymbol{v}}_n$; EPS1 and EPS2 are suitable convergence factors.

The iterative solution in Step 2 takes a small number of steps, usually 1 or 2, to achieve convergence.

The artificial dissipation term is applied in regions of high gradients to prevent oscillations in the vicinity of discontinuities by incorporating pressure-switched diffusion coefficients appropriately. Thus,

$$\hat{\boldsymbol{R}} = \frac{C_S S_e}{\Delta t}\boldsymbol{M}_L^{-1}[\boldsymbol{M}_c - \boldsymbol{M}_L]\tilde{\boldsymbol{v}}_n \tag{15.54}$$

where C_S is a shock capturing constant, S_e is the averaged element value of the nodal pressure switch given as

$$S_i = \frac{|\sum(p_i - p_j)|}{\sum(|p_i - p_j|)} \tag{15.55}$$

and $\boldsymbol{M}_c$ and $\boldsymbol{M}_L$ are the consistent and lumped mass matrices respectively; i is the node under consideration and j are the nodes connected to i.

To obtain the viscous components, σ_{ij} in Eq. (15.45) is written as

$$\sigma_{ij} = -\frac{2}{3}\frac{\mu}{Re}\frac{\partial u_l}{\partial x_l}\delta_{ij} + \frac{\mu}{Re}\left(\frac{\partial u_i}{\partial x_j} + \frac{\partial u_j}{\partial x_i}\right) \tag{15.56}$$

and the diffusion flux of the Navier–Stokes equation is

$$\boldsymbol{g}_i = \left(0\ \sigma_{i1}\ \sigma_{i2}\ \sigma_{i3}\ u_j\sigma_{ij} + \frac{1}{Re\,Pr}\frac{\partial T}{\partial x_i}\right)^T \qquad i = 1, 2, 3; \qquad j = 1, 2, 3 \tag{15.57}$$

in which μ is the nondimensional viscosity term. The terms Re and Pr are the Reynolds and Prandtl numbers, respectively. Then components of $\partial \boldsymbol{g}_i/\partial x_i$ are evaluated term by term and are then discretized by Galerkin approximation. As evidenced from Eq. (15.5) the second order terms occur only in momentum and energy equations. Each of the three momentum equations will have seven terms, whereas the energy equation will have 24 terms. Thus for example the x_1 momentum component equation can be written as

$$\frac{\partial \sigma_{i1}}{\partial x_i} = \frac{\partial \sigma_{11}}{\partial x_1} + \frac{\partial \sigma_{21}}{\partial x_2} + \frac{\partial \sigma_{31}}{\partial x_3}$$
$$= \frac{\mu}{Re}\frac{\partial}{\partial x_1}\left[\frac{4}{3}\frac{\partial u_1}{\partial x_1} - \frac{2}{3}\frac{\partial u_2}{\partial x_2} - \frac{2}{3}\frac{\partial u_3}{\partial x_3}\right] + \frac{\mu}{Re}\frac{\partial}{\partial x_2}\left[\frac{\partial u_2}{\partial x_1} + \frac{\partial u_1}{\partial x_2}\right]$$
$$+ \frac{\mu}{Re}\frac{\partial}{\partial x_3}\left[\frac{\partial u_1}{\partial x_3} + \frac{\partial u_3}{\partial x_1}\right] \tag{15.58}$$

which is then discretized by the Galerkin method to yield the viscous contribution. Thus, for example, the first term of Eq. (15.58) yields

$$\int_V N^T \frac{4}{3}\frac{\mu}{Re}\frac{\partial^2 u_1}{\partial x_1^2}\,\mathrm{d}V = \frac{4}{3}\frac{\mu}{Re}\frac{A}{3}[1\ 1\ 1\ 0]^T\left[\frac{\partial N_1}{\partial x_1}u_1 + \frac{\partial N_2}{\partial x_1}u_2 + \frac{\partial N_3}{\partial x_1}u_3 + \frac{\partial N_4}{\partial x_1}u_4\right]$$
$$-\frac{4}{3}\frac{\mu}{Re}V\left[\frac{\partial N_1}{\partial x_1}\ \frac{\partial N_2}{\partial x_1}\ \frac{\partial N_3}{\partial x_1}\ \frac{\partial N_4}{\partial x_1}\right]^T\left[\frac{\partial N_1}{\partial x_1}u_1 + \frac{\partial N_2}{\partial x_1}u_2\right.$$
$$\left. + \frac{\partial N_3}{\partial x_1}u_3 + \frac{\partial N_4}{\partial x_1}u_4\right] \tag{15.59}$$

yielding the components of boundary integral and primary viscous terms. In Eq. (15.59) A is the area of the boundary elements. Three components of the viscous terms for each of the four nodes can be written as

$$\begin{bmatrix} \text{VIN(I1)} \\ \text{VIN(I1}+1) \\ \text{VIN(I1}+2) \end{bmatrix} = \frac{\Delta t \mu}{Re} V \begin{bmatrix} (J_{1i}\ J_{2i}\ J_{3i})(\sigma_{11}\ \sigma_{12}\ \sigma_{13})^T \\ (J_{2i}\ J_{1i}\ J_{3i})(\sigma_{22}\ \sigma_{12}\ \sigma_{23})^T \\ (J_{3i}\ J_{2i}\ J_{1i})(\sigma_{33}\ \sigma_{32}\ \sigma_{31})^T \end{bmatrix} \qquad i = 1, \ldots, 4 \tag{15.60}$$

in which I1 $= 2, 7, 12,$ and 17, respectively, for the momentum equation. The remaining terms of the energy equation are as follows:

$$\text{VIN}(5) = \sum_{i=1}^{3} J_{i1}[X_i c_1 + (\sum_{j=1}^{4} T_{ij})c_2] \tag{15.61}$$

$$\text{VIN}(10) = \sum_{i=1}^{3} J_{i2}[X_i c_1 + (\sum_{j=1}^{4} T_{ij})c_2] \tag{15.62}$$

$$\text{VIN}(15) = \sum_{i=1}^{3} J_{i3}[X_i c_1 + (\sum_{j=1}^{4} T_{ij})c_2] \tag{15.63}$$

$$\text{VIN}(20) = \sum_{i=1}^{3} J_{i4}[X_i c_1 + (\sum_{j=1}^{4} T_{ij})c_2] \tag{15.64}$$

in which

$$X_1 = \bar{u}\sigma_{11} + \bar{v}\sigma_{12} + \bar{w}\sigma_{13}$$

$$X_2 = \bar{u}\sigma_{21} + \bar{v}\sigma_{22} + \bar{w}\sigma_{23}$$

$$X_3 = \bar{u}\sigma_{31} + \bar{v}\sigma_{32} + \bar{w}\sigma_{33} \tag{15.65}$$

and

$$c_1 = \frac{\Delta t \mu}{Re} V, \quad c_2 = \frac{\Delta t}{Re\, Pr} V, \quad T_{ij} = \left[\frac{\partial \boldsymbol{N}_j}{\partial x_i}\right]^T \left[\frac{\partial \boldsymbol{N}_j}{\partial x_i}\right] \tag{15.66}$$

with $\boldsymbol{J}$ being the Jacobian matrix. For turbulent flow, μ is replaced by $\mu + \mu_t$, k by $k + (\mu_t c_p)/(Pr_t)$, and c_1 and c_2 take the forms

$$c_1 = \frac{\Delta t}{Re} V(\mu + \mu_t); \qquad c_2 = \frac{\Delta t}{Re} V \left[\frac{1}{Pr_t} + \frac{\mu_t}{Pr_t}\right] \tag{15.67}$$

in which $\bar{u}$, $\bar{v}$, $\bar{w}$ are the average velocity values in an element; μ_t is the turbulent viscosity and Pr_t is the turbulent Prandtl number.

15.2.4.1 Turbulence modeling. For turbulent flow calculations, a vast amount of literature[21–32] is available that covers a wide variety of turbulence modeling procedures. Ref. 33 provides a current literature survey in the subject area. A study of turbulence is necessary in the understanding of flow physics involving phenomena such as boundary layer, vortex–body, and vortex–vortex interaction as well as separation, transition, and relamination of the flow. An accurate simulation of these phenomena is essential to control the pattern of flow around an aircraft, thereby enhancing its performance to a desirable level. The approach of direct numerical simulation[25–27] produces all details of turbulence flow. In practice, though, a simplified turbulence model essentially reduces the turbulence problem to the determination of a scalar eddy viscosity $\hat{v}$ from a simple differential equation. For the sake of simplicity, if the turbulent kinetic energy term is neglected then the Reynolds averaged stress term simply becomes a function of the turbulent viscosity μ_t, which can be calculated using the computed eddy viscosity scalar. The one equation model of Ref. 24 has been adopted for turbulence modeling in the STARS program. The viscous stresses σ_{ij} pertaining to laminar viscous flow is thus replaced by $\sigma_{ij}^t = \sigma_{ij} + \sigma_{ij}^R$. Likewise, the heat flux q_j in the energy equation is replaced by $q_j^t = q_j + q_j^R$. These Reynolds averaged stress and heat flux are given as

$$\sigma_{ij} = -\frac{2}{3}\mu_t \frac{\partial u_k}{\partial x_k}\delta_{ij} + \mu_t \left(\frac{\partial u_i}{\partial x_j} + \frac{\partial u_j}{\partial x_i}\right) \tag{15.68}$$

$$q_j^R = -\mu_t \frac{c_p}{Pr_t}\frac{\partial T}{\partial x_j} \tag{15.69}$$

The turbulent viscosity μ_t is next related[24] to eddy viscosity scalar as

$$\nu_t = \frac{\mu_t}{\rho} = \hat{v} f_{v_1} \tag{15.70}$$

in which

$$f_{v_1} = \frac{\chi^3}{\chi^3 + c_{v_1}^3}, \qquad \chi = \frac{\hat{v}}{\nu} \tag{15.71}$$

ν is the laminar viscosity, and the eddy viscosity $\hat{\nu}$ is obtained by solving the nondimensional form of the following differential equation:

$$\frac{1}{St_\infty}\frac{\partial \hat{\nu}}{\partial t} + u_j \frac{\partial \hat{\nu}}{\partial x_j} = c_{b_1}[1 - f_{t_2}]\hat{S}\hat{\nu} + \frac{1}{Re\sigma}\left[\frac{\partial}{\partial x_j}\left\{(\nu + \hat{\nu})\frac{\partial \hat{\nu}}{\partial x_j}\right\} + c_{b_2}\frac{\partial \hat{\nu}}{\partial x_j}\frac{\partial \hat{\nu}}{\partial x_j}\right]$$
$$-\frac{1}{Re}\left[c_{w_1} f_w - \frac{c_{b_1}}{k^2} f_{t_2}\right]\left[\frac{\hat{\nu}}{d}\right]^2 + Ref_{t_1}(\Delta\hat{u})^2 \tag{15.72}$$

in which

$$\hat{S} = \hat{\omega} + \frac{1}{Re}\frac{\hat{\nu}}{k^2 d^2} f_{v_2}, \qquad f_{v_2} = 1 - \frac{\chi}{1 + \chi f_{v_1}} \tag{15.73}$$

$$f_w = g\left[\frac{1 + c_{w_3}^6}{g^6 + c_{w_3}^6}\right]^{1/6}, \qquad g = r + c_{w_2}\left(r^6 - r\right) \tag{15.74}$$

$$r = \left[\min\left(\frac{1}{Re}\frac{\hat{\nu}}{\hat{S}k^2 d^2}, 10\right)\right], \qquad f_{t_2} = c_{t_3}\exp\left(-c_{t_4}\chi^2\right) \tag{15.75}$$

$$f_{t_1} = c_{t_1} g_t \exp\left(-c_{t_2}\frac{\hat{w}_t^2}{\Delta\hat{u}^2}\left[d^2 + g_t^2 d_t^2\right]\right), \qquad g_t = \min\left(0.1, \frac{\Delta\hat{u}}{\hat{w}_t \Delta x}\right) \tag{15.76}$$

In these definitions, each of these related variables are replaced by $\hat{\nu}/\nu_\infty$, $\hat{u}_i/u_\infty$, $\Delta\hat{u}/u_\infty$, d/L, d_t/L, $\Delta x/L$, x_j/L and t/t_c; t_c is the characteristic time, L is a reference length, $St_\infty = u_\infty t_c/L$ and $t_c = L/u_\infty$. The following definitions are also relevant to the current problem:

d = distance from given point to the nearest wall
$\Delta\hat{u}$ = difference in velocity between the point and trip
d_t = closest distance to such a trip, point or curve
$\hat{w}_t$ = the vorticity magnitude at trip point or curve
Δx = surface grid spacing at trip
$\hat{\omega}$ = vorticity

$$\hat{\omega} = \left[\left(\frac{\partial u_3}{\partial x_2} - \frac{\partial u_2}{\partial x_3}\right)^2 + \left(\frac{\partial u_1}{\partial x_3} - \frac{\partial u_3}{\partial x_1}\right)^2 + \left(\frac{\partial u_2}{\partial x_1} - \frac{\partial u_1}{\partial x_2}\right)^2\right]^{1/2}$$

The associated constants have the following values: $c_{b_1} = 0.1355$, $c_{b_2} = 0.622$, $\sigma = 2/3$, $k = 0.41$, $c_{w_1} = c_{b_1}/k^2 + (1 + c_{b_2})/\sigma$, $c_{w_2} = 0.3$, $c_{w_3} = 2$, $c_{v_1} = 7.1$, $c_{t_1} = 1$, $c_{t_2} = 2$, $c_{t_3} = 1.1$, $c_{t_4} = 2$, and $Pr_t = 0.9$.

Equation (15.72) is conveniently solved using a finite element spatial and a finite difference time discretization. Using a nondimensional form for the variables, the terms in Eq. (15.72) can be written as shown next, and in which some of these terms are reduced to form surface integrals (Γ).

Transient term:

$$\frac{\partial \hat{\nu}}{\partial t} = \frac{\hat{\nu}^{n+1} - \hat{\nu}^n}{\Delta t} \tag{15.77}$$

Convection terms:

$$\int_V N^T u_j \frac{\partial \hat{\nu}}{\partial x_j} \, \mathrm{d}V \tag{15.78}$$

Diffusion terms:

$$\frac{1}{Re\,\sigma} \int_V N^T \left[\frac{\partial}{\partial x_j} \left\{ (\nu + \hat{\nu}) \frac{\partial \hat{\nu}}{\partial x_j} \right\} + c_{b_2} \frac{\partial \hat{\nu}}{\partial x_j} \frac{\partial \hat{\nu}}{\partial x_j} \right] \mathrm{d}V \tag{15.79}$$

Source terms:

$$\int_V N^T c_{b_1} [1 - f_{t_2}] \hat{S} \hat{\nu} \, \mathrm{d}V - \int_V \frac{N^T}{Re} \left[c_{w_1} f_w - \frac{c_{b_1}}{k^2} f_{t_2} \right] \left[\frac{\hat{\nu}}{d} \right]^2 \mathrm{d}V$$

$$+ Re \int_V N^T f_{t_1} (\Delta \hat{u})^2 \, \mathrm{d}V \tag{15.80}$$

in which, typically, $\partial \hat{\nu} / \partial x_j = (\partial N / \partial x_j) \{ \tilde{\hat{\nu}} \}$.

15.2.4.2 Numerical example. A large number of relevant CFD example problems have been solved to assess the accuracy of the currently developed numerical formulation and software. Examples of representative test problems for both viscous and inviscid problems are given in Example 15.1.

Example 15.1: Flow Over NACA 0012 Airfoil

The first example considered is the inviscid flow over a NACA0012 airfoil. The inlet Mach number of this flow is 1.2 and angle of attack is zero. Figure 15.3 shows the unstructured mesh used in the calculation. It contains about one million unstructured tetrahedron elements. The coefficient of pressure distribution along the chord surface is shown in Fig. 15.4. The benchmark AGARD solution[34] also is included in this figure. As seen, the agreement between the two results is excellent. The u_1 velocity contours are shown in Fig. 15.5. The shock in front of the stagnation zone is predicted accurately as shown.

In the second example, we present the viscous supersonic flow past a NACA0012 airfoil. The details of the mesh used close to this solid surface are shown in Fig. 15.6. A total number of about one million elements are used in this calculation. The inlet Mach number of this problem is assumed to be equal to 2. The airfoil surface is assumed to be solid and no slip flow conditions are assumed on the surface. Figure 15.7 shows the comparison of coefficient of pressure distribution along the chord surface between the present and the results obtained on an adapted mesh[35]. It is noticed that good agreement between the results is obtained. Figure 15.8 shows the u_1 velocity contours. As expected the supersonic shock appears in front of the stagnation zone.

15.2.5 Mesh Generation

In CFD applications it is essential to create unstructured mesh on the surface and also in the space surrounding an aerospace vehicle. The surface mesh is composed of triangles, whereas the three-dimensional volume mesh consists of tetrahedrons. The partial differential equations of fluid dynamics is solved in this domain subject to a set of boundary conditions at the various boundaries. A related meshing

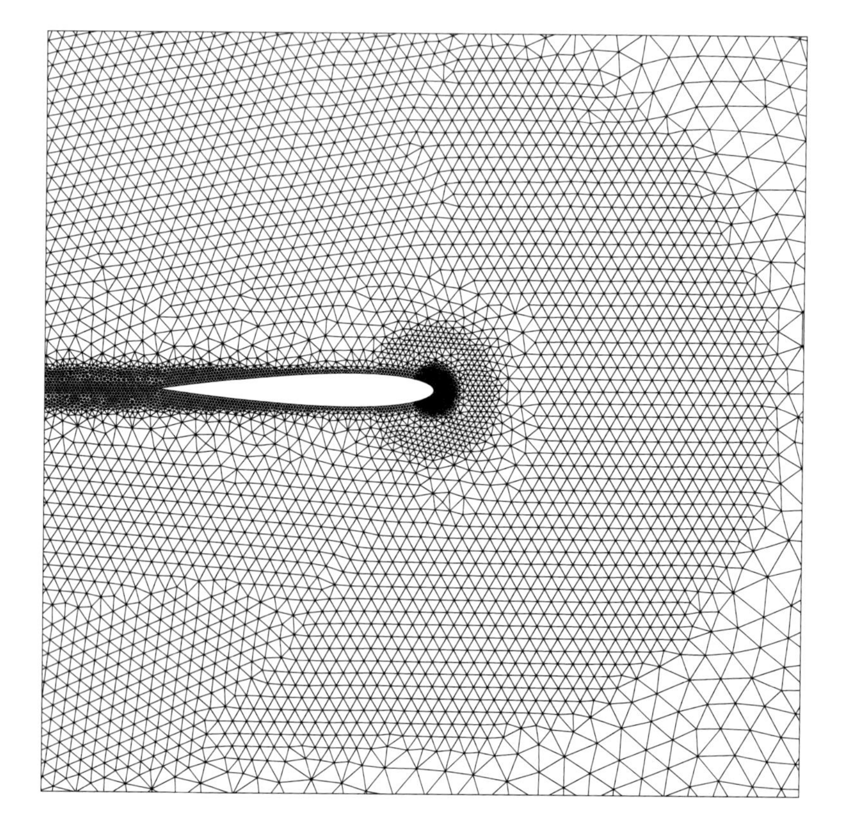

Fig. 15.3 CFD mesh for flow over NACA 0012 airfoil.

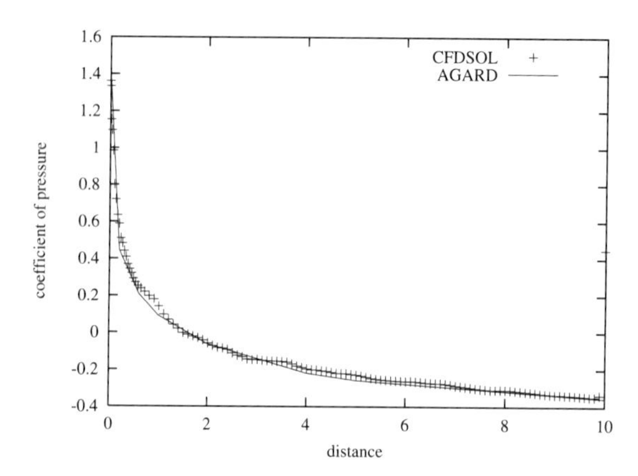

Fig. 15.4 C_p distribution along the airfoil surface.

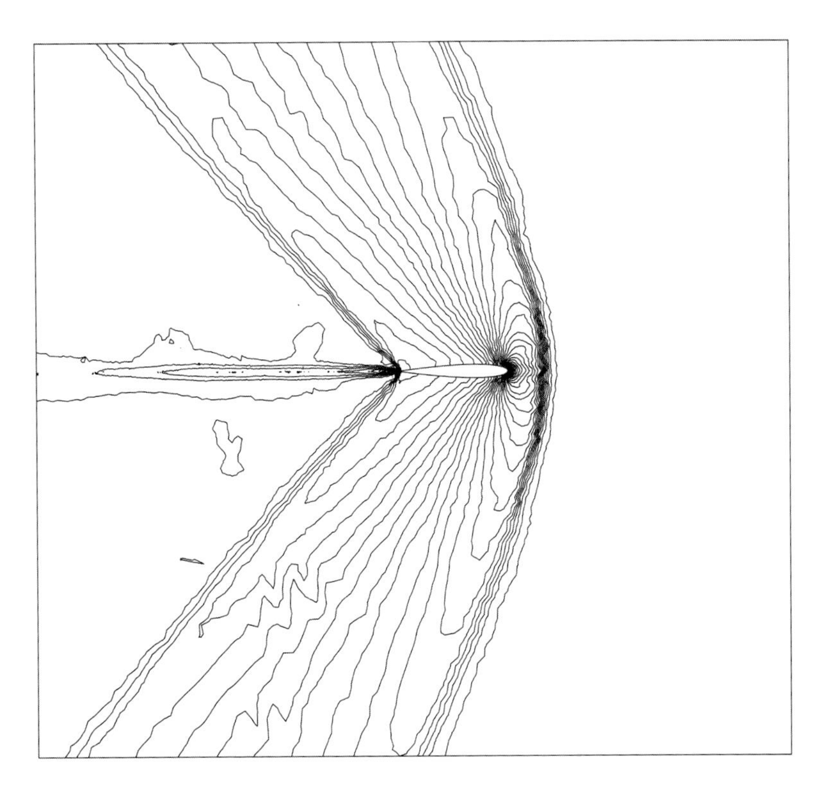

Fig. 15.5 Velocity u_1 distribution.

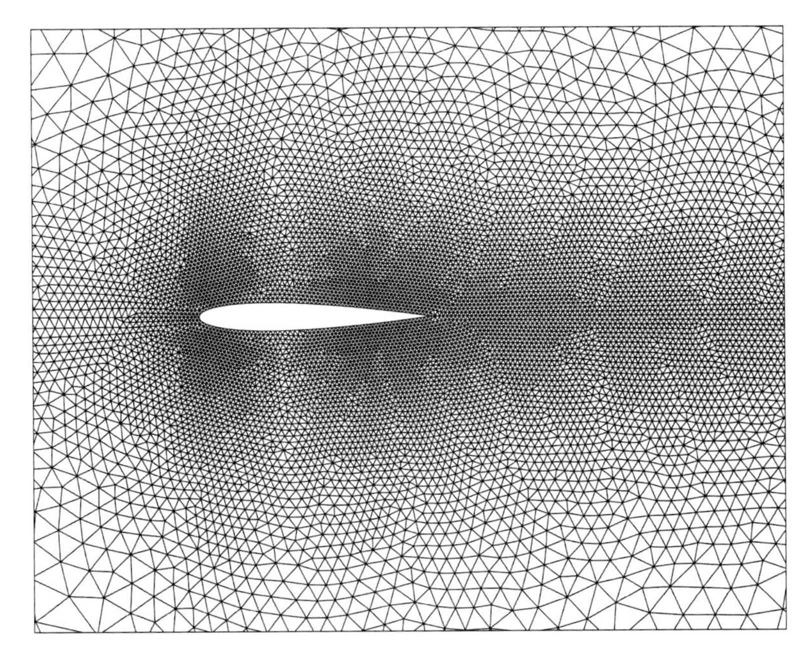

Fig. 15.6 Finite element mesh for supersonic viscous flow past a NACA 0012 airfoil.

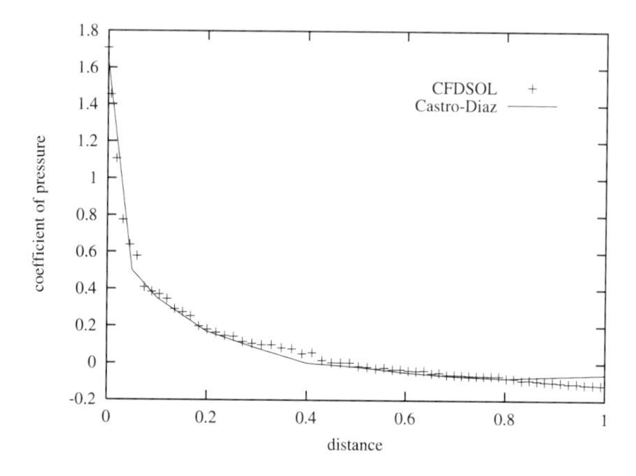

Fig. 15.7 Distribution of C_p for viscous supersonic flow: $Re = 10000$, $M = 2$.

software should be capable of automatic meshing of any practical domain in an accurate and efficient fashion. Furthermore, such an algorithm may have an option for adaptive refinement of the mesh, so that computational resources are optimized in concentrating calculations as necessary within the domain. A summary of various mesh generation techniques is given in Ref. 36.

The Delaunay triangulation[37–39] has the basic property that no other nodes pertaining to another element exist within the circumsphere that contains the four nodes of any element under consideration. Furthermore, the Delaunay triangulation

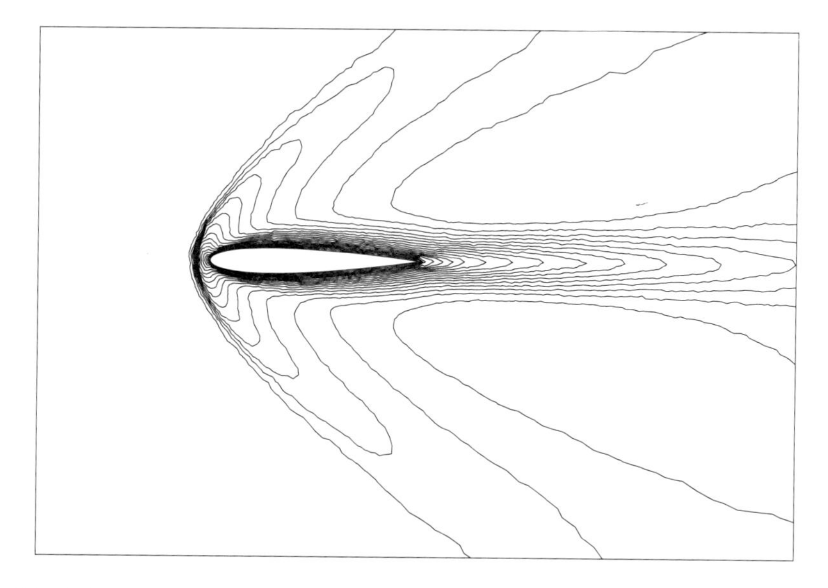

Fig. 15.8 Velocity u_1 contours for viscous supersonic flow: $Re = 10000$, $M = 2$.

minimizes a function that is derived from the linear finite element interpolation function pertaining to an element.

Advancing front algorithms were initially developed for arbitrary, multiconnected, planar domains in which the interior nodes are generated first, which are then suitably linked to yield the best possible triangulation. During this process, the generation front is continually updated each time a new element is constructed. Further improvement and extension of this technique in three dimensions is described in Refs. 18 and 40, in which the nodes and triangles are formed simultaneously for all boundary surfaces. This is followed by generation of tetrahedra by the advancing front approach to fill the entire solution domain. Suitable background grids are used to specify important mesh parameters defining node spacing, stretching parameters, and directions. This procedure has the following advantages: 1) It is flexible with regard to specification of arbitrary shapes and varying grid density throughout the domain, and 2) it facilitates adaptive mesh generation[40] in accordance with solution trend. This three-dimensional automated unstructured mesh generation scheme has been found to be rather versatile for modeling a practical CFD solution domain around complex structural form such as an aircraft.

Because the advancing front technique involves a rather extensive search for nodes and faces on the front, the grid generation time tends to be somewhat large for complex configurations. A simple modification of the procedure proves to be efficient and economical. In this procedure, the usual technique is first utilized to generate a grid whose cells have linear dimension about twice the desired size, and then each cell is reduced locally to reach desired cell sizes. Thus, Fig. 15.9a depicts an initial coarse mesh in which the location of the base starting nodes, lying on cubic curves generated from user input points, are determined on the basis of specified spatial weights. Each triangular cell in each surface segment is next replaced by four triangular cells, each having a surface area of one-quarter of the original cell. Such a subdivision is effected by introducing the midpoint of each edge of an original triangle as a new grid point and connecting the three midpoints with line segments as shown in Fig. 15.9b. In the particular case of a triangle side lying on a boundary edge, the midpoint of that edge is actually obtained by interpolation from the respective cubic curve defining that edge. Likewise, each

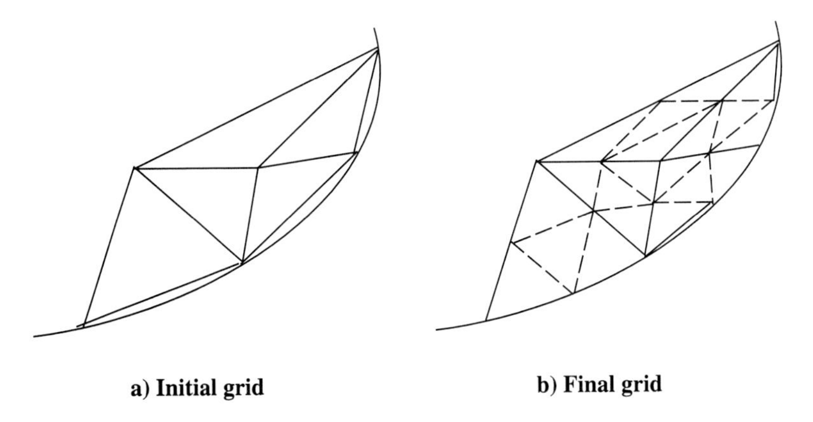

a) Initial grid **b) Final grid**

Fig. 15.9 Unstructured CFD grid generation.

tetrahedral cell of the three-dimensional grid is replaced by eight new cells by introducing six new nodes at the midpoint of each edge and connecting them appropriately with line segments; thus each tetrahedron is one-eighth the volume of the original tetrahedron. For viscous flow, a different grid generation strategy also may be employed in the boundary layer. Thus the surface normals of the unstructured grids,[41,42] suitably smoothed, are used to generate semistructured grids in the boundary layer that transit into an unstructured grid in the remainder of the solution domain.

15.3 Time-Marched Aeroelastic and Aeroservoelastic Analysis

A common finite element–based idealization for both the fluid and solid domains is adopted for the nonlinear aeroelastic and aeroservoelastic analyzes.[13,14] Figure 15.10 depicts the flow chart of the numerical algorithm adopted in the STARS program for the CFD-based, integrated, aerostructural-controls analysis of structures such as aerospace vehicles. This process starts with the finite-element structural modeling and subsequently computes the natural frequencies ω and modes ϕ that consist of rigid-body, elastic, and control-surface motions, by solving

$$\boldsymbol{M}\ddot{\boldsymbol{u}} + \boldsymbol{K}\boldsymbol{u} = \boldsymbol{0} \tag{15.81}$$

in which $\boldsymbol{M}$ and $\boldsymbol{K}$ are the inertial and stiffness matrices, repectively, and $\boldsymbol{u}$ is the displacement vector. This solution may be achieved by an efficient (progressive simultaneous iteration) technique or a block Lanczos procedure that fully exploits matrix sparsity. Next, a steady-state Euler/N–S solution is effected in which optimum solution convergence is achieved through an explicit or alternative quasi-implicit, local time-stepping solution procedure that also employs a residual smoothing strategy. The associated vehicle equation of motion then may be cast into the frequency domain as

$$\hat{\boldsymbol{M}}\ddot{\boldsymbol{q}} + \hat{\boldsymbol{C}}\dot{\boldsymbol{q}} + \hat{\boldsymbol{K}}\boldsymbol{q} + \boldsymbol{f}_a(t) + \boldsymbol{f}_I(t) = \boldsymbol{0} \tag{15.82}$$

in which the generalized matrices and vectors are as follows: $\boldsymbol{q}$ is the displacement vector $(= \Phi^T\boldsymbol{u})$, $\hat{\boldsymbol{M}}$ the inertia matrix $(= \Phi^T\boldsymbol{M}\Phi)$, and, similarly, $\hat{\boldsymbol{K}}, \hat{\boldsymbol{C}}$ the stiffness and damping matrices; $\boldsymbol{f}_a(t)$ the aerodynamic (CFD) load vector $(= \Phi^T pA)$, p is the fluid pressure at a structural node, and A is the appropriate surface area around the node; $\boldsymbol{f}_I(t)$ the applied generalized structural impulse force vector, to initiate response where $\boldsymbol{f}_I(t)$ is the appropriate user input containing a number of modes of interest (see Figure 15.11). The aerodynamic force vector $\boldsymbol{f}_a$ also may be generated by first deriving the modal vector Φ_a pertaining to aerodynamic nodes on the structural surface by suitable interpolation from the structural vector Φ; the Euler pressures at such nodes are next used directly to compute $\boldsymbol{f}_a$.

Equation (15.82) then is cast in the state-space matrix equation form as

$$\begin{bmatrix} \boldsymbol{I} & \boldsymbol{0} \\ \boldsymbol{0} & \boldsymbol{I} \end{bmatrix} \begin{bmatrix} \dot{\boldsymbol{q}} \\ \ddot{\boldsymbol{q}} \end{bmatrix} - \begin{bmatrix} \boldsymbol{0} & \boldsymbol{I} \\ -\hat{\boldsymbol{M}}^{-1}\hat{\boldsymbol{K}} & -\hat{\boldsymbol{M}}^{-1}\hat{\boldsymbol{C}} \end{bmatrix} \begin{bmatrix} \boldsymbol{q} \\ \dot{\boldsymbol{q}} \end{bmatrix} - \begin{bmatrix} \boldsymbol{0} \\ -\hat{\boldsymbol{M}}^{-1}\boldsymbol{f}_a(t) \end{bmatrix} - \begin{bmatrix} \boldsymbol{0} \\ -\hat{\boldsymbol{M}}^{-1}\boldsymbol{f}_I(t) \end{bmatrix} = \boldsymbol{0} \tag{15.83}$$

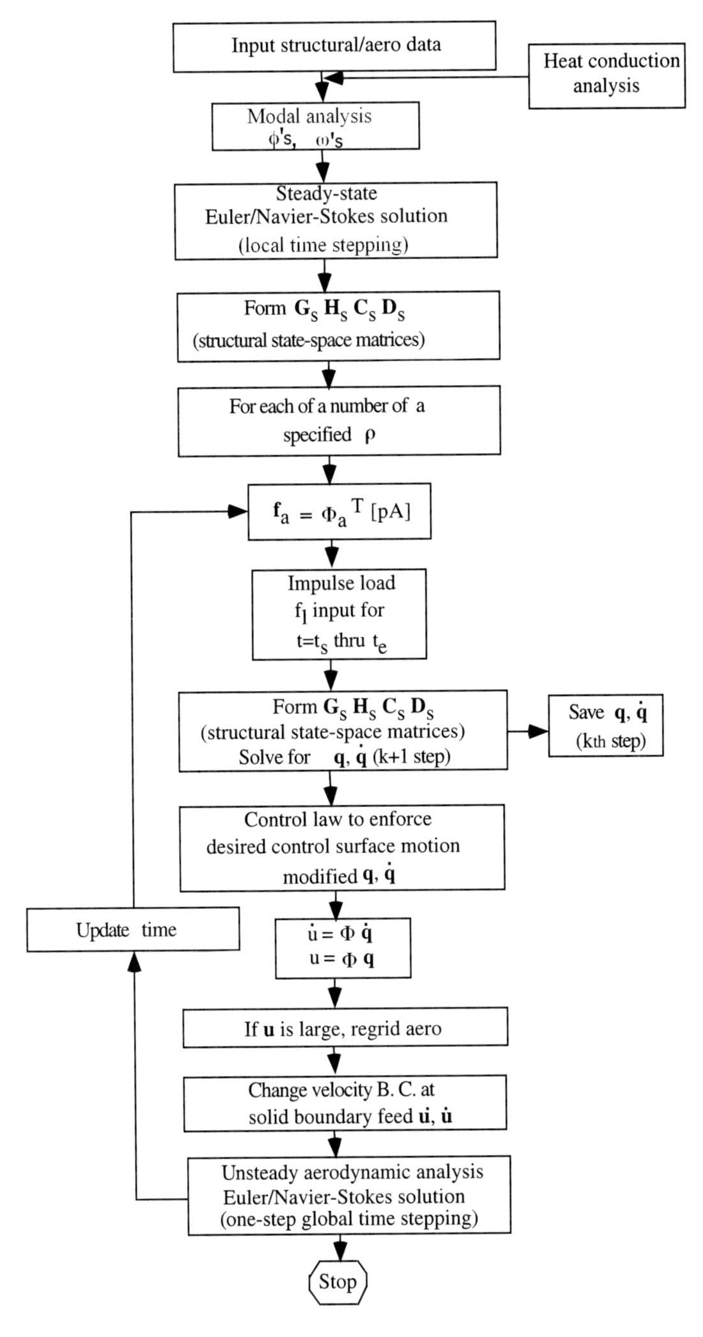

Fig. 15.10 Nonlinear aeroelastic/aeroservoelastic analysis methodology.

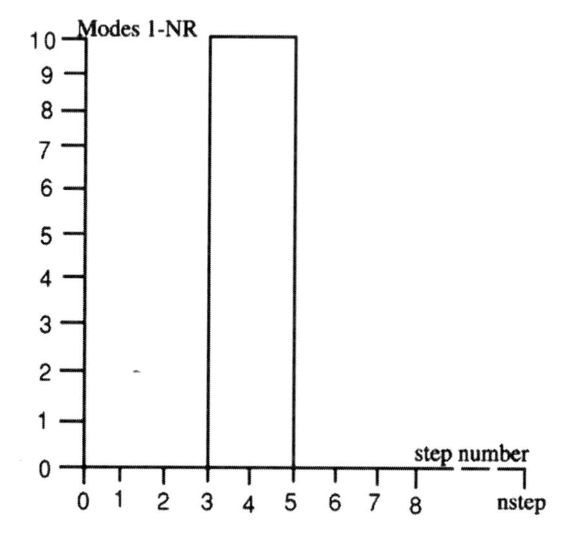

Fig. 15.11 Generalized impulse force vector $f_I(k)$ for each mode.

or

$$\dot{x}_s(t) = A_{st}x_s(t) + B_{st}f(t) \tag{15.84}$$

where

$$B_{st} = \begin{bmatrix} 0 \\ -\hat{M}^{-1} \end{bmatrix}, \quad f(t) = f_a(t) + f_I(t), \quad x_s(t) = \begin{bmatrix} q \\ \dot{q} \end{bmatrix} \tag{15.84a}$$

and

$$y_s(t) = C_{st}x_s(t) + D_{st}f(t) \tag{15.85}$$

in which $C_{st} = I$ and $D_{st} = 0$. A coupled aeroelastic model is shown in Fig. 15.12.

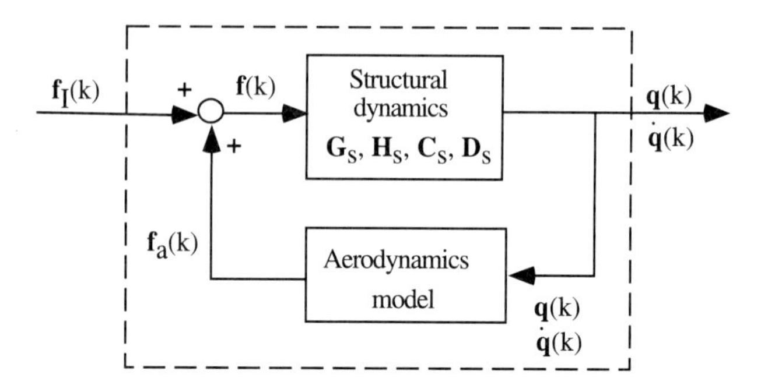

Fig. 15.12 Coupled aeroelastic model.

In the presence of sensors, Eq. (15.85), following the procedures of Sec. 14.2.3, takes the following form:

$$\begin{Bmatrix} \boldsymbol{q}(t) \\ \dot{\boldsymbol{q}}(t) \\ \boldsymbol{u}_{sn}(t) \\ \dot{\boldsymbol{u}}_{sn}(t) \\ \ddot{\boldsymbol{u}}_{sn}(t) \end{Bmatrix} = \begin{bmatrix} \boldsymbol{I} & \boldsymbol{0} \\ \boldsymbol{0} & \boldsymbol{I} \\ \boldsymbol{T}_{sn}\Phi & \boldsymbol{0} \\ \boldsymbol{0} & \boldsymbol{T}_{sn}\Phi \\ -\boldsymbol{T}_{sn}\Phi\hat{\boldsymbol{M}}^{-1}\hat{\boldsymbol{K}} & -\boldsymbol{T}_{sn}\Phi\hat{\boldsymbol{M}}^{-1}\hat{\boldsymbol{C}} \end{bmatrix} \begin{Bmatrix} \boldsymbol{q}(t) \\ \dot{\boldsymbol{q}}(t) \end{Bmatrix} + \begin{Bmatrix} 0 \\ 0 \\ 0 \\ 0 \\ -\boldsymbol{T}_{sn}\Phi\hat{\boldsymbol{M}}^{-1} \end{Bmatrix} \boldsymbol{f}(t) \tag{15.86}$$

or,

$$\boldsymbol{y}_{ss}(t) = \boldsymbol{C}_{ss}\boldsymbol{x}_s(t) + \boldsymbol{D}_{ss}\boldsymbol{f}(t) \tag{15.87}$$

in which the vectors $\boldsymbol{u}_{sn}$ denote the real, physical displacements of the sensors in the continuous time domain. Also the interpolation matrix $\boldsymbol{T}_{sn}$, having sensor location information, is of dimensions $(2 \times NS, N)$, NS being the number of sensors. The Eq. (15.87) is the sensor output relationship signifying motion at the sensors resulting from the body motion $\boldsymbol{C}_{ss}$ and control motion $\boldsymbol{D}_{ss}$. Converting the equations from inertial to body axis frame results in

$$\dot{\boldsymbol{x}}_s(t) = \tilde{\boldsymbol{T}}_2^{-1}(\boldsymbol{A}_{st}\tilde{\boldsymbol{T}}_1 - \tilde{\boldsymbol{T}}_3)\boldsymbol{x}_s(t) + \tilde{\boldsymbol{T}}_2^{-1}\boldsymbol{B}_{st}\boldsymbol{f}(t) \tag{15.88}$$

$$\boldsymbol{y}_s(t) = \boldsymbol{C}_{ss}\tilde{\boldsymbol{T}}_1\boldsymbol{x}_s(t) + \boldsymbol{D}_{ss}\boldsymbol{f}(t) \tag{15.89}$$

which now can be written as

$$\dot{\boldsymbol{x}}_s(t) = \boldsymbol{A}_s\boldsymbol{x}_s(t) + \boldsymbol{B}_s\boldsymbol{f}(t) \tag{15.90}$$

$$\boldsymbol{y}_s(t) = \boldsymbol{C}_s\boldsymbol{x}_s(t) + \boldsymbol{D}_s\boldsymbol{f}(t) \tag{15.91}$$

in which $\tilde{\boldsymbol{T}}_1$ and $\tilde{\boldsymbol{T}}_2$ are coordinate transformation matrices of dimensions $(NR2 \times NR2)$. These equations then are converted to the zero-order hold (ZOH) discrete time equivalent at the kth step:

$$\boldsymbol{x}_s(k+1) = \boldsymbol{G}_s\boldsymbol{x}_s(k) + \boldsymbol{H}_s\boldsymbol{f}(k) \tag{15.92}$$

$$\boldsymbol{y}_s(k) = \boldsymbol{C}_s\boldsymbol{x}_s(k) + \boldsymbol{D}_s\boldsymbol{f}(k) \tag{15.93}$$

where

$$\boldsymbol{G}_s = e^{\boldsymbol{A}_s\Delta t} \qquad \boldsymbol{H}_s = \left[e^{\boldsymbol{A}_s\Delta t} - \boldsymbol{I}\right]\left[\boldsymbol{A}_s^{-1}\boldsymbol{B}_s\right] \tag{15.94}$$

$$\boldsymbol{f}(k) = \boldsymbol{f}_a(k) + \boldsymbol{f}_I(k)$$

and $\Delta t = t_{k+1} - t_k$; $\boldsymbol{C}_s$ and $\boldsymbol{D}_s$ remain unaltered.

In related aeroservoelastic analysis, the procedure is depicted in Fig. 15.13, in which the generalized displacement $\boldsymbol{q}$ is defined as

$$\boldsymbol{q} = (\boldsymbol{q}_R\ \boldsymbol{q}_E\ \boldsymbol{q}_\delta)^T$$

where $\boldsymbol{q}_R$, $\boldsymbol{q}_E$, and $\boldsymbol{q}_\delta$ are generalized rigid body, elastic, and control surface displacements, respectively. Thus the continuous aerostructural outputs are converted to a discrete time equivalent form using the time interval Δt_{CFD}. Other continuous systems such as the sensor and the actuator transfer functions also are discretized

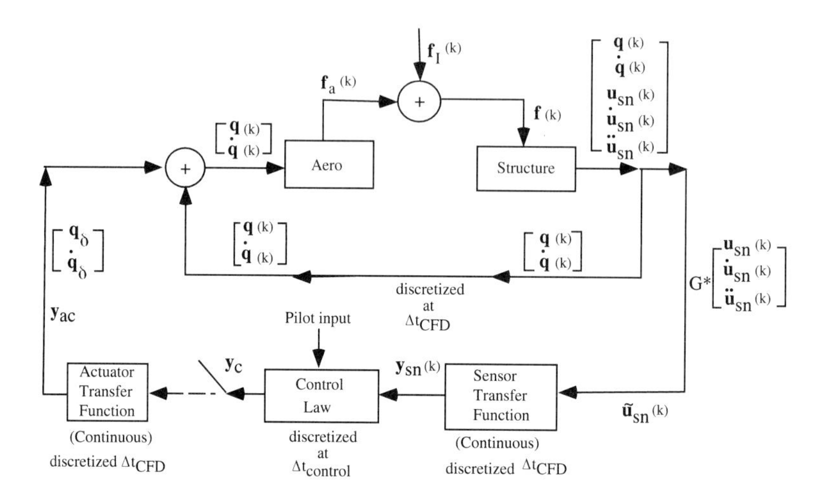

Fig. 15.13 Coupled aeroservoelastic model.

using the time interval Δt_{CFD}. The digital control law has its own sampling rate Δt_C and it only participates in the solution when the two sampling rates coincide. The sensor motions can be obtained from Eq. (15.93), $\boldsymbol{G}$ being the gain matrix, as

$$\tilde{\boldsymbol{u}}_{sn} = \boldsymbol{G}\ \boldsymbol{y}_s(k) \tag{15.95}$$

which yields the sensor output as follows:

$$\boldsymbol{x}_{sn}(k+1) = \boldsymbol{G}_{sn}\boldsymbol{x}_{sn}(k) + \boldsymbol{H}_{sn}\tilde{\boldsymbol{u}}_{sn}(k) \tag{15.96}$$

$$\boldsymbol{y}_{sn}(k) = \boldsymbol{C}_{sn}\boldsymbol{x}_{sn}(k) + \boldsymbol{D}_{sn}\tilde{\boldsymbol{u}}_{sn}(k) \tag{15.97}$$

Using $\boldsymbol{y}_{sn}(k)$ from Eq. (15.97), the control law output, corresponding to each sampling time interval Δt_C, is obtained as

$$\boldsymbol{x}_c(k+1) = \boldsymbol{G}_c\ \boldsymbol{x}_c(k) + \boldsymbol{H}_c\ \boldsymbol{y}_{sn}(k) \tag{15.98}$$

$$\boldsymbol{y}_c(k) = \boldsymbol{C}_c\ \boldsymbol{x}_c(k) + \boldsymbol{D}_c\ \boldsymbol{y}_{sn}(k) \tag{15.99}$$

The desired control surface motion is obtained from Eq. (15.99) and the actual motion of the control surface is obtained from the actuator transfer function

$$\boldsymbol{x}_{ac}(k+1) = \boldsymbol{G}_{ac}\boldsymbol{x}_{ac}(k) + \boldsymbol{H}_{ac}\boldsymbol{y}_c(k) \tag{15.100}$$

$$\boldsymbol{y}_{ac}(k) = \boldsymbol{C}_{ac}\boldsymbol{x}_{ac}(k) + \boldsymbol{D}_{ac}\boldsymbol{y}_c(k) \tag{15.101}$$

the control surface motion being

$$\boldsymbol{y}_{ac}(k) = (\boldsymbol{q}_{\delta_1}\ \boldsymbol{q}_{\delta_2}, \ldots, \dot{\boldsymbol{q}}_{\delta_1},\ \dot{\boldsymbol{q}}_{\delta_2}, \ldots,)^T \tag{15.102}$$

the position and velocity of the control surface are put back into the $\boldsymbol{x}_n$ vector, which is the $\boldsymbol{x}_s$ vector with new control surface definition.

The $\boldsymbol{u}$ and $\dot{\boldsymbol{u}}$ values are calculated next for $\boldsymbol{x}_n$ and modified to change the normal velocity boundary conditions at the solid boundary described earlier as the transpiration boundary condition. This is then followed by a one-step CFD solution using a global time-stepping scheme, and the entire solution process is then repeated for the required number of time steps. A time response solution of Eq. (15.92) in an interval ($\Delta t = t_{k+1} - t_k$), is obtained as,

$$\boldsymbol{x}_s(k+1) = e^{\boldsymbol{A}_s \Delta t}\boldsymbol{x}_s(k) + \boldsymbol{A}_s^{-1}[e^{\boldsymbol{A}_s \Delta t} - \boldsymbol{I}]\boldsymbol{B}_s \boldsymbol{f}(k) \tag{15.103}$$

The set of response data as already shown may be resolved into modal components using a fast Fourier transform (FFT):

$$\boldsymbol{q} = \sum_{m=1}^{r} e^{\zeta mt}(a_m \cos \omega_m t + b_m \sin \omega_m t) \tag{15.104}$$

to yield the damping ζ and frequency ω values. In practice, however, the generalized response is plotted against the Mach number. This plot depicting stability characteristics of the vehicle enables prediction of the onset of flutter or divergence occurring within the entire flight regime. A finite element multigrid Euler solver[43] has been successfully used in the present CFD-based aeroelastic analysis.

An earlier effort involving finite element CFD analysis for unsteady flow has been described in Ref. 44. The effect of aerodynamic heating[45] needs to be taken into consideration for aeroelastic response analysis of high speed vehicles such as the proposed National Aerospace Plane. In this procedure, once the steady-state CFD analysis is completed, the temperature distribution at various heat transfer mesh nodes is obtained by interpolation from such data at aerodynamic grid points on the structural surface. Subsequent heat transfer analysis of the structure then yields the temperature distribution in all structural elements by suitable interpolation at structural nodes, and the stiffness matrix is generated next, taking into account temperature-dependent material properties. Such a stiffness matrix then is used to perform a free vibration analysis, and the rest of the analysis continues as shown in Fig. 15.10. It may be noted that only elastic modes are used for aeroelastic analysis; for aeroservoelastic analysis the elastic modes are augmented by rigid body as well as control rigid body modes.

This is a very general formulation that allows for any digital control sampling rate that is different than the aeroelastic solution time steps. Such an analysis, pertaining to a specific Mach number, then may be repeated for a number of altitudes involving various dynamic pressure values, and the instability altitude signified by a zero damping value then may be extracted using simple interpolation of each desired state variable. An alternative, faster, procedure based on a system identification technique is described next, which also provides aerodynamic state-space matrices vital in the design of control laws.

15.4 ARMA Model in Aeroelastic and Aeroservoelastic Analysis

The model structure used in this system identification technique is the autoregressive moving average model (ARMA), which describes the modal response force at time k of a system as a summation of scaled previous outputs and scaled values of modal displacement inputs to the system as shown in Eq. (15.105).

Fig. 15.14 Aerodynamic model.

The ARMA model makes the assumptions that most aeroelastic systems can be treated as dynamically linear. That is, the aerodynamics respond linearly to small perturbations about a potential nonlinear steady-state mean flow.

The basic ARMA model in Fig. 15.14 at time k may be written as

$$f_a(k) = \sum_{i=1}^{na} \boldsymbol{A}_i \boldsymbol{f}_a(k-i) + \sum_{i=0}^{nb-1} \boldsymbol{B}_i \boldsymbol{q}(k-i) \tag{15.105}$$

in which $\boldsymbol{A}_i$ and $\boldsymbol{B}_i$ are unknowns to be determined from excitation of the structure through a prescribed motion containing the spectrum of calculated structural eigenmodes, and na and nb are the orders of the coefficients of $\boldsymbol{A}$ and $\boldsymbol{B}$.

It may be noted that the generalized forces may be scaled by the training density $\hat{\rho}$, and that further scaling the generalized forces on the right-hand side results in

$$\hat{\boldsymbol{f}}_a(k) = \sum_{i=1}^{na} \boldsymbol{A}_i \hat{\boldsymbol{f}}_a(k-i) + \frac{1}{\hat{\rho}} \sum_{i=0}^{nb-1} \boldsymbol{B}_i \boldsymbol{q}(k-i) \tag{15.106}$$

Next, a state vector $\mathbf{x}_a$ is defined for the scaled aerodynamic system that contains a total of $(na+nb\text{-}1)NR$ states:

$$\boldsymbol{x}_a(k) = \begin{bmatrix} \hat{\boldsymbol{f}}_a(k-1) \\ \vdots \\ \hat{\boldsymbol{f}}_a(k-na) \\ \boldsymbol{q}(k) \\ \vdots \\ \boldsymbol{q}(k-nb+1) \end{bmatrix} \tag{15.107}$$

The mathematical model of the unsteady CFD aerodynamic system can compute the aerodynamic force outputs based on any given structural displacement inputs. Thus the mathematical state-space representation of the aerodynamic system can replace the traditional Euler/N–S CFD step in the multidisciplinary aeroelastic or aeroservoelastic analysis.

Then the state-space form for the scaled aerodynamic model can be written as

$$\boldsymbol{x}_a(k+1) = \boldsymbol{G}_a \boldsymbol{x}_a(k) + \boldsymbol{H}_a \boldsymbol{q}(k) \tag{15.108}$$

$$\hat{\boldsymbol{f}}_a(k) = \boldsymbol{C}_a \boldsymbol{x}_a(k) + \boldsymbol{D}_a \boldsymbol{q}(k) + \hat{\boldsymbol{f}}_0 \tag{15.109}$$

in which

$$
G_a = \begin{bmatrix}
A_1 & A_2 & \cdots & A_{na-1} & A_{na} & \frac{1}{\hat{\rho}}B_1 & \frac{1}{\hat{\rho}}B_2 & \cdots & \frac{1}{\hat{\rho}}B_{nb-2} & \frac{1}{\hat{\rho}}B_{nb-1} \\
I & 0 & \cdots & 0 & 0 & 0 & 0 & \cdots & 0 & 0 \\
0 & I & \cdots & 0 & 0 & 0 & 0 & \cdots & 0 & 0 \\
\vdots & \vdots & \ddots & \vdots & \vdots & \vdots & \vdots & \ddots & \vdots & \vdots \\
0 & 0 & \cdots & I & 0 & 0 & 0 & \cdots & 0 & 0 \\
0 & 0 & \cdots & 0 & 0 & 0 & 0 & \cdots & 0 & 0 \\
0 & 0 & \cdots & 0 & 0 & I & 0 & \cdots & 0 & 0 \\
0 & 0 & \vdots & 0 & 0 & 0 & I & \cdots & 0 & 0 \\
\vdots & \vdots & \ddots & \vdots & \vdots & \vdots & \vdots & \ddots & \vdots & \vdots \\
0 & 0 & \cdots & 0 & 0 & 0 & 0 & \cdots & I & 0
\end{bmatrix}
\qquad
H_a = \begin{bmatrix} \frac{1}{\hat{\rho}}B_0 \\ 0 \\ 0 \\ \vdots \\ 0 \\ I \\ 0 \\ 0 \\ \vdots \\ 0 \end{bmatrix}
$$

$$
C_a = \begin{bmatrix} A_1 & A_2 & \cdots & A_{na-1} & A_{na} & \frac{1}{\hat{\rho}}B_1 & \frac{1}{\hat{\rho}}B_2 & \cdots & \frac{1}{\hat{\rho}}B_{nb-2} & \frac{1}{\hat{\rho}}B_{nb-1} \end{bmatrix}
$$

$$
D_a = \frac{1}{\hat{\rho}}B_0, \qquad q(k) = T_a y_s(k) = [I \quad 0] y_s(k) \tag{15.110}
$$

It may be noted that the output equation for the scaled aerodynamic model may include a known vector of static offsets, $\hat{f}_o = f_o/\hat{\rho}$. The static offsets are subtracted off of the time history data in the derivation of the aerodynamic model because the ARMA model structure only models the dynamics of the system. The generalized force vector $f_a (= \rho \hat{f}_a)$ is then fed back into structural state-space matrix Eqs. (15.92) and (15.93) in the solution iteration as shown in Fig. 15.15. The matrices $\boldsymbol{A}$ and $\boldsymbol{B}$ are of dimension $NR \times NR$.

15.4.1 Model Identification

As discussed in the previous section, the aerodynamic model is based on the ARMA structure of Eq. (15.106). This model structure describes the dynamic response of any multi-input multi-output system as a linear combination of scaled outputs and scaled inputs for the system. The discrete time model procedure uses this model structure to develop a simple algebraic model that is equivalent to the

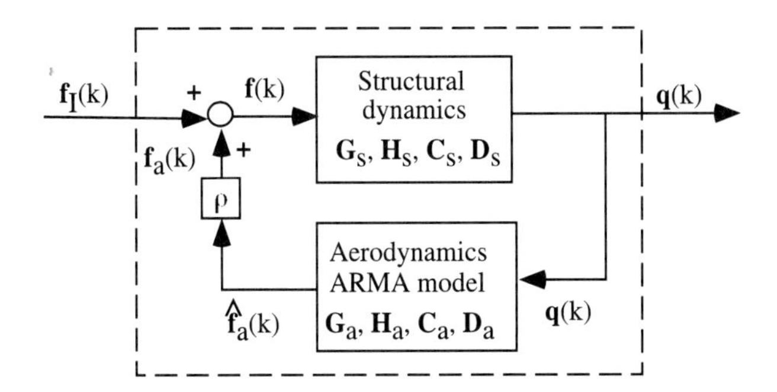

Fig. 15.15 Coupled ARMA aeroelastic model.

unsteady CFD solution for a given Mach number and structural geometry. The process for determining the unknown coefficient matrices in the ARMA model, $\boldsymbol{A}_i$ and $\boldsymbol{B}_i$, involves three steps:

1) Prescribe a known displacement time history, or input signal, through the unsteady CFD solution and record the aerodynamic response, or output signal.
2) Select a model size, na and nb in Eq. (15.105), and identify the ARMA coefficients that match the input–output data recorded in step 1.
3) Implement the computed aerodynamic model from step 2 for the input signal used in step 1 and compare the model output to the actual system output.

The system identification procedure just described is an iterative procedure. If the comparison in step 3 shows that the model does not match the actual CFD solution then step 2 is repeated for a different choice of na and nb. The coefficient identification mentioned in step 2 is accomplished by minimizing the error between the model output and the unsteady CFD output in a least squares sense. Because this will result in an overdetermined system of equations, STARS uses an implementation of singular value decomposition (SVD) to extract the ARMA coefficient matrices from a matrix of system equations assembled from the training data.

The success of this identification procedure will be highly dependent on the amount and quality of training data available in step 2. Hence, an optimum input signal should be used in step 1 that will excite a broad response spectrum in the unsteady CFD solution. Once an accurate model is identified, it then can be used in place of the CFD solution in the coupled aeroelastic simulation.

15.4.2 Input Optimization

The input of the ARMA model for the unsteady CFD solver is the generalized displacement of the structure in the flowfield. However, the Euler CFD solver in the STARS programs also requires the velocity of the structure to satisfy its boundary condition requirements. Therefore, any prescribed input signal for structural displacement must be uniquely differentiable to obtain the physical velocity of the structure for input to the CFD solver. Alternatively, a velocity may be specified that then can be integrated to obtain the consistent structural displacement. In either case, an input must be selected that will excite a wide range of dynamic frequencies in the flow field in order identify a practically useful model.

In STARS, a 3211 multistep input is implemented on the velocity boundary condition for the training process of the system. The 3211 multistep input is used widely by the flight test community because of its ease of implementation and excellent frequency content. As already mentioned, the prescribed velocity function is then integrated numerically to compute the displacement boundary condition that also becomes the training input for the ARMA model. Also implemented is a 753211 variable amplitude multistep velocity input training signal that proves to be effective for some practical problems.

Most aeroelastic problems are formulated with multiple structural modes and hence will require a separate multistep input signal for each mode. Intuitively, it is obvious that all modes may not be run simultaneously or the ARMA model will

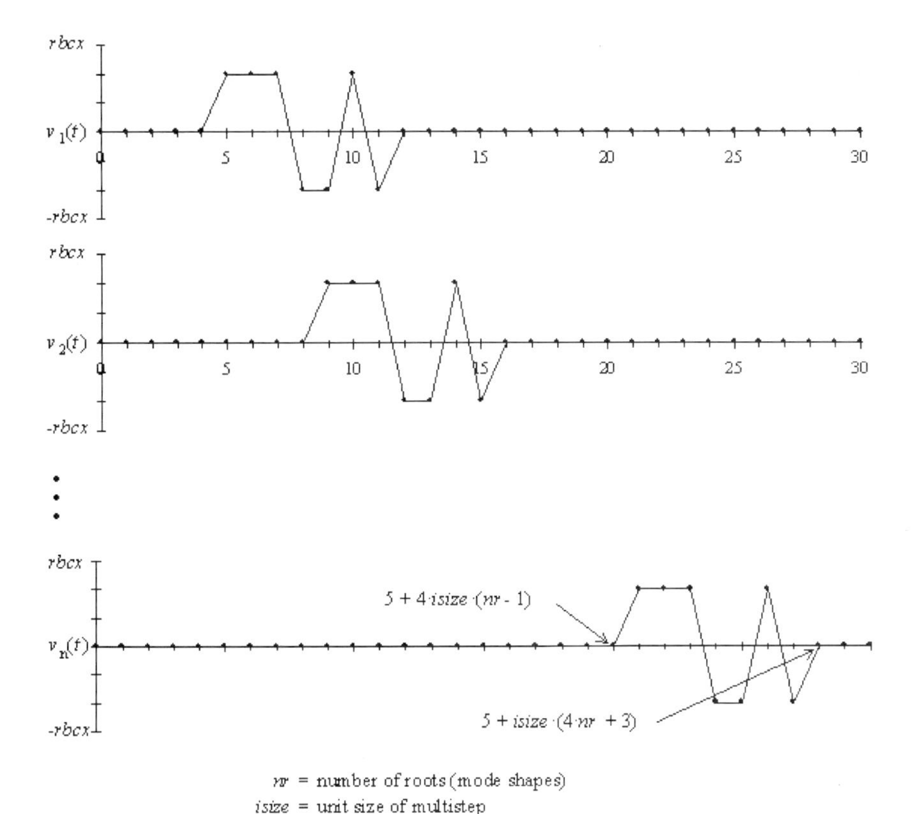

Fig. 15.16 A 3211 multistep velocity input signal for several modes.

fail to discriminate between the effects of the different mode shapes. Furthermore, it will take a large amount of computational effort if the input signal for each mode is run independently. To conserve time and still guarantee a unique solution, the multistep input signal for each mode is applied in a staggered fashion such that the signals slightly overlap but are still out of phase as shown in Fig. 15.16.

15.4.3 ARMA Aeroelastic and Aeroservoelastic Analysis

For the coupled aeroelastic model a combined structural and aerodynamic state–space matrix formulation is derived next that enables depiction of aeroelastic root-locus plots, which assists in the control law design. Thus the aerodynamic state–space equation's input can be written in terms of the structural output as follows:

$$\boldsymbol{x}_a(k+1) = \boldsymbol{G}_a\boldsymbol{x}_a(k) + \boldsymbol{H}_a\boldsymbol{T}_a\boldsymbol{C}_s\boldsymbol{x}_s(k) \tag{15.111}$$

$$\boldsymbol{f}_a(k) = \rho\boldsymbol{C}_a\boldsymbol{x}_a(k) + \rho\boldsymbol{D}_a\boldsymbol{T}_a\boldsymbol{C}_s\boldsymbol{x}_s(k) + \rho\hat{\boldsymbol{f}}_a \tag{15.112}$$

From Fig. 15.15, $\boldsymbol{f}_a$ can be obtained as follows:

$$\boldsymbol{f}(k) = \boldsymbol{f}_I(k) + \boldsymbol{f}_a(k) \tag{15.113}$$

Substitute Eqs. (15.112) and (15.113) into Eqs. (15.92) and (15.93) to have

$$\boldsymbol{x}_s(k+1) = (\boldsymbol{G}_s + \boldsymbol{H}_s \rho \boldsymbol{D}_a \boldsymbol{T}_a \boldsymbol{C}_s)\boldsymbol{x}_s(k) + \boldsymbol{H}_s \rho \boldsymbol{C}_a \boldsymbol{x}_a(k) + \boldsymbol{H}_s \boldsymbol{f}_I + \boldsymbol{H}_s \rho \hat{\boldsymbol{f}}_a \quad (15.114)$$

$$\boldsymbol{y}_s(k) = \begin{Bmatrix} \boldsymbol{q}(k) \\ \dot{\boldsymbol{q}}(k) \end{Bmatrix} = \boldsymbol{C}_s \boldsymbol{x}_s(k) \quad (15.115)$$

and the combined aeroelastic state–space matrix may be written as

$$\begin{Bmatrix} \boldsymbol{x}_s(k+1) \\ \boldsymbol{x}_a(k+1) \end{Bmatrix} = \begin{bmatrix} \boldsymbol{G}_s + \rho \boldsymbol{H}_s \boldsymbol{D}_a \boldsymbol{T}_a \boldsymbol{C}_s & \rho \boldsymbol{H}_s \boldsymbol{C}_a \\ \boldsymbol{H}_a \boldsymbol{T}_a \boldsymbol{C}_s & \boldsymbol{G}_a \end{bmatrix} \begin{Bmatrix} \boldsymbol{x}_s(k) \\ \boldsymbol{x}_a(k) \end{Bmatrix} + \begin{bmatrix} \boldsymbol{H}_s \\ \boldsymbol{0} \end{bmatrix} \boldsymbol{f}_I(k) + \begin{bmatrix} \rho \boldsymbol{H}_s \hat{\boldsymbol{f}}_o \\ \boldsymbol{0} \end{bmatrix} \quad (15.116)$$

$$\begin{bmatrix} \boldsymbol{q}(k) \\ \dot{\boldsymbol{q}}(k) \end{bmatrix} = [\boldsymbol{C}_s \quad \boldsymbol{0}] \begin{Bmatrix} \boldsymbol{x}_s(k) \\ \boldsymbol{x}_a(k) \end{Bmatrix} \quad (15.117)$$

or

$$\boldsymbol{x}_{sa}(k+1) = \boldsymbol{G}_{sa} \boldsymbol{x}_{sa}(k) + \boldsymbol{H}_{sa1} \boldsymbol{f}_I(k) + \boldsymbol{H}_{sa2} \quad (15.118)$$

$$\boldsymbol{y}_{sa}(k) = \boldsymbol{C}_{sa} \boldsymbol{x}_{sa}(k) \quad (15.119)$$

The relevant root locus plot, derived from the $\boldsymbol{G}_{sa}$ matrix, shows the location of roots as a function of density, which is analog to the plotting of $\boldsymbol{q}$ and $\dot{\boldsymbol{q}}$ contained in $\boldsymbol{y}_s$ (Eq. 15.85). This matrix $\boldsymbol{G}_{sa}$ can be used for control law design to yield the gain matrix $\tilde{\boldsymbol{K}}$, which can be used next to plot response of the controlled vehicle.

The preceding aeroelastic state–space matrices are valid for a certain Mach number and can be evaluated easily for any number of density values defining a new dynamic process. The requirement for stability is that all eigenvalues of $\mathbf{G}_{sa}$ [Eq. (15.116)] lie within the unit circle. Thus the following solution steps are relevant for an aeroelastic instability analysis for a given Mach number.

Step 1: Form $\boldsymbol{G}_a, \boldsymbol{H}_a, \boldsymbol{C}_a, \boldsymbol{D}_a$ for a specified training density within a typical flight envelope in Fig. 15.17.

Step 2: Solve eigenvalue problem for Eq. (15.116) for any given value of density matrix.

Step 3: Plot location of roots and watch for any root lying outside the unit circle signaling instability.

Step 4: Repeat preceding steps for a number of specified density values representative of critical flight points within the flight envelope given in Fig. 15.17, to detect possible instability.

The ARMA aeroservoelastic analysis procedure is described in Fig. 15.18 following the technique described earlier for the direct solution described in Sec. 15.3. The computer flowchart for the ARMA aeroservoelastic analysis is shown in Fig. 15.19.

For some pratical problems with analog controllers, the coupled aeroservoelastic model with a combined structural, aerodynamic, and control state–space matrix

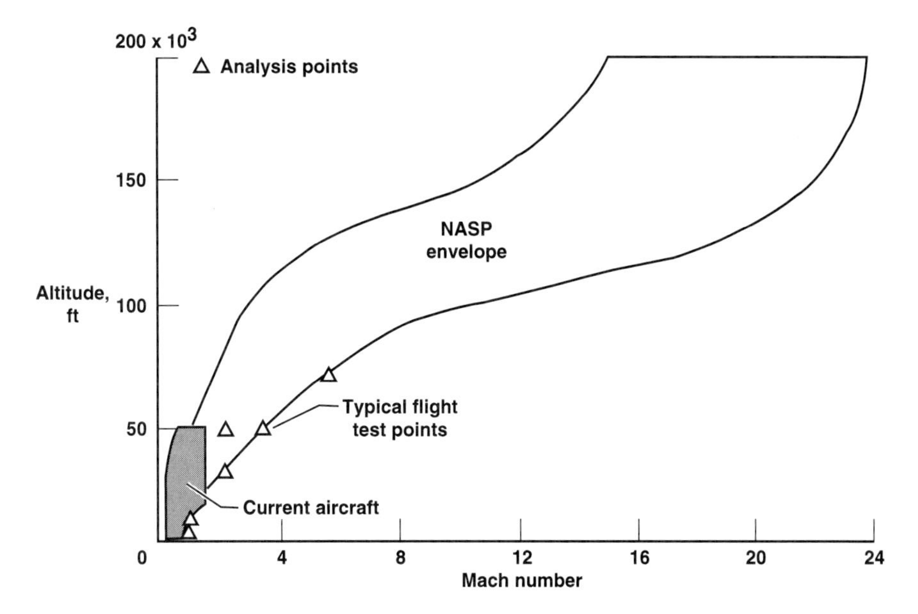

Fig. 15.17 Typical flight envelope for conventional and hypersonic vehicles.

formulation is derived next. This model enables depiction of aeroservoelastic root-locus plots, which assists in the control law design. In this simplified formulation we assume absence of sensors, which nevertheless can be augmented as described before. The control module will have the following state–space equation:

$$\boldsymbol{x}_c(k+1) = \boldsymbol{G}_c\boldsymbol{x}_c(k) + \boldsymbol{H}_c\boldsymbol{u}_c \tag{15.120}$$

$$\boldsymbol{y}_c(k) = \boldsymbol{C}_c\boldsymbol{x}_c(k) + \boldsymbol{D}_c\boldsymbol{u}_c \tag{15.121}$$

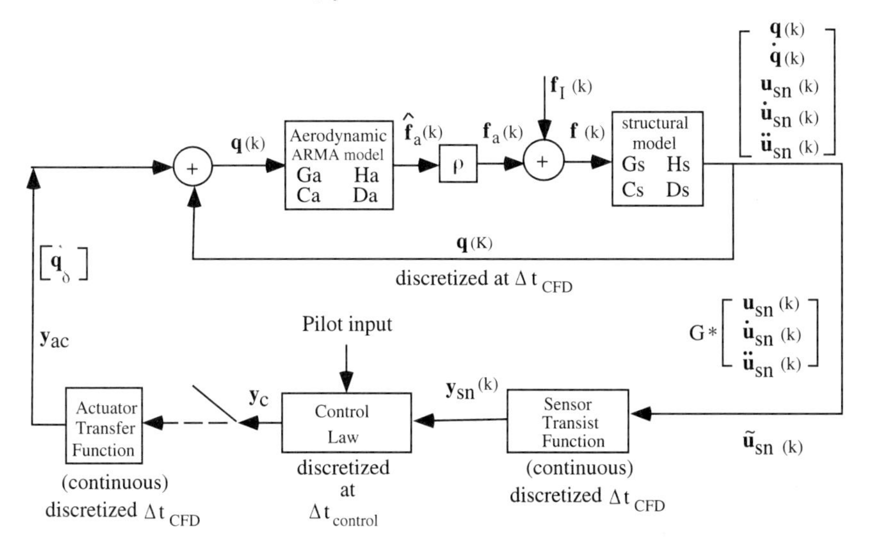

Fig. 15.18 Coupled ARMA aeroservoelastic simulation flowchart.

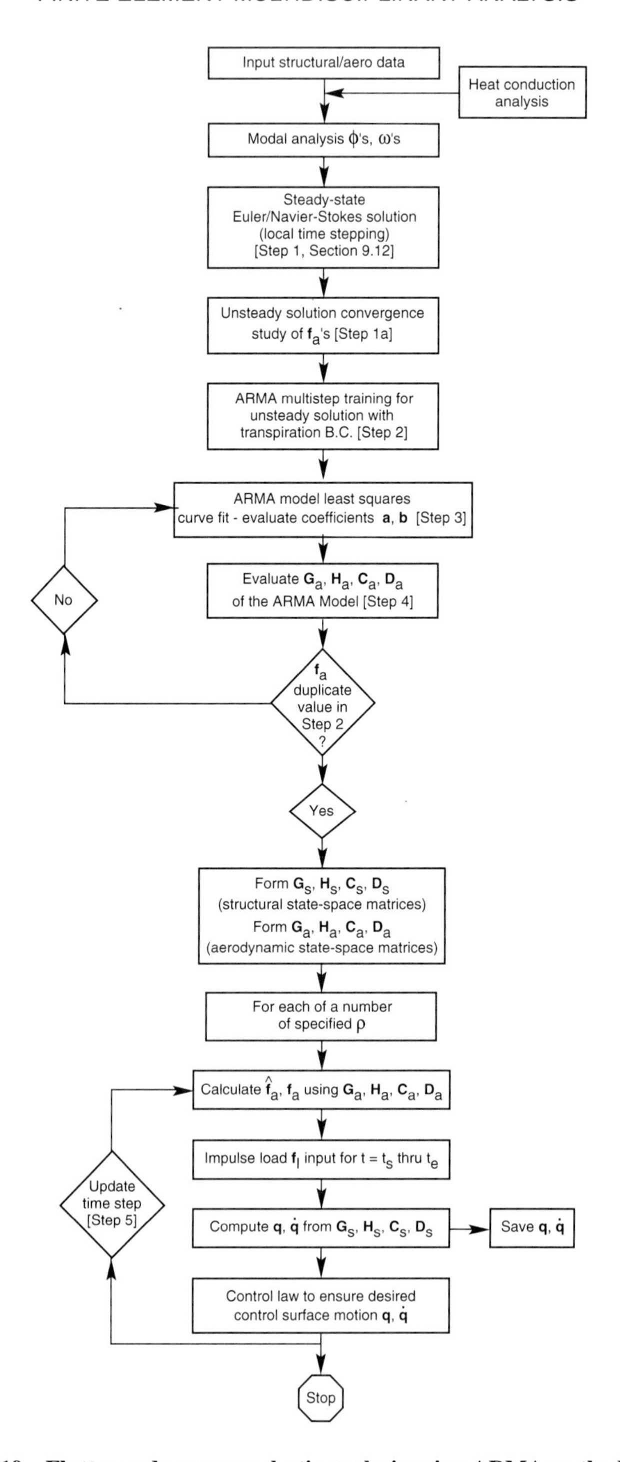

Fig. 15.19 Flutter and aeroservoelastic analysis using ARMA methodology.

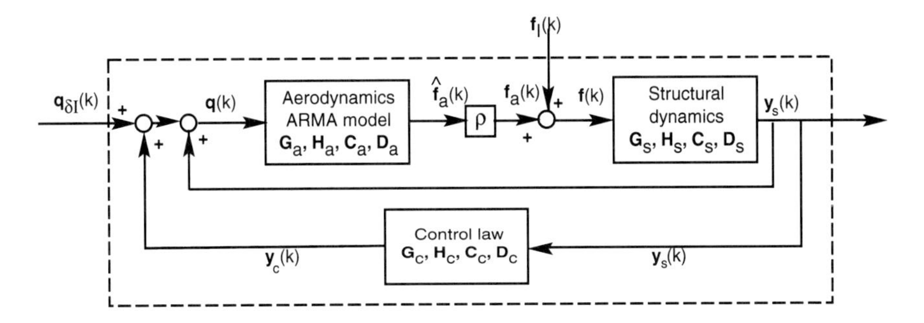

Fig. 15.20 Coupled ARMA aeroservoelastic model.

with

$$\boldsymbol{u}_c = \boldsymbol{T}_c \boldsymbol{y}_s(k) = \boldsymbol{T}_c \boldsymbol{C}_s \boldsymbol{x}_s(k) \tag{15.122}$$

$$\boldsymbol{x}_c(k+1) = \boldsymbol{G}_c \boldsymbol{x}_c(k) + \boldsymbol{H}_c \boldsymbol{T}_c \boldsymbol{C}_s \boldsymbol{x}_s(k) \tag{15.123}$$

$$\boldsymbol{y}(k) = \boldsymbol{C}_c \boldsymbol{x}_c(k) + \boldsymbol{D}_c \boldsymbol{T}_c \boldsymbol{C}_s \boldsymbol{x}_s(k) \tag{15.124}$$

$\boldsymbol{T}_c$ is a matrix of order NCOL, 2*NR; NCOL is the order of $\boldsymbol{u}_c$. Figure 15.20 shows that the feedback $\hat{\boldsymbol{q}}(k)$ signal is the summation of the following components:

$$\hat{\boldsymbol{q}}(k) = \boldsymbol{q}_{\delta I}(k) + \boldsymbol{y}_c(k) + \boldsymbol{y}_s(k) \tag{15.125}$$

where $\boldsymbol{q}_{\delta I}$ is the external generalized input. Substitute Eqs. (15.93), (15.124), and (15.125) into Eqs. (15.108) and (15.109). The aerodynamic equation can be written as

$$\boldsymbol{x}_a(k+1) = \boldsymbol{G}_a \boldsymbol{x}_a(k) + \boldsymbol{H}_a \boldsymbol{T}_a \boldsymbol{q}_{\delta I}(k) + \boldsymbol{H}_a \boldsymbol{T}_a \boldsymbol{C}_c \boldsymbol{x}_c(k) + \boldsymbol{H}_a \boldsymbol{T}_a (\boldsymbol{D}_c \boldsymbol{T}_c \boldsymbol{C}_s + \boldsymbol{C}_s) \boldsymbol{x}_s(k) \tag{15.126}$$

$$\boldsymbol{f}_a(k) = \rho \boldsymbol{C}_a \boldsymbol{x}_a(k) + \rho \boldsymbol{D}_a \boldsymbol{T}_a (\boldsymbol{C}_s + \boldsymbol{D}_c \boldsymbol{T}_c \boldsymbol{C}_s) \boldsymbol{x}_s(k) + \rho \boldsymbol{D}_a \boldsymbol{T}_a \boldsymbol{C}_c \boldsymbol{x}_c(k) + \rho \boldsymbol{D}_a \boldsymbol{T}_a \boldsymbol{q}_{\delta I}(k) + \rho \hat{\boldsymbol{f}}_o \tag{15.127}$$

Figure 15.15 shows that the signal $\boldsymbol{f}(k)$ can be obtained from

$$\boldsymbol{f}(k) = \boldsymbol{f}_I(k) + \boldsymbol{f}_a(k) \tag{15.128}$$

Substitute Eqs. (15.127) and (15.128) into Eqs. (15.92) and (15.93) to yield the structural state–space matrices with the following form:

$$\boldsymbol{x}_s(k+1) = \boldsymbol{H}_s \rho \boldsymbol{C}_a \boldsymbol{x}_a(k) + (\boldsymbol{G}_s + \boldsymbol{H}_s \rho \boldsymbol{D}_a \boldsymbol{T}_a (\boldsymbol{C}_s + \boldsymbol{D}_c \boldsymbol{T}_c \boldsymbol{C}_s)) \boldsymbol{x}_s(k) + \boldsymbol{H}_s \rho \boldsymbol{D}_a \boldsymbol{T}_a \boldsymbol{C}_c \boldsymbol{x}_c(k) + \boldsymbol{H}_s \rho \boldsymbol{D}_a \boldsymbol{T}_a \boldsymbol{q}_{\delta I}(k) + \boldsymbol{H}_s \rho \hat{\boldsymbol{f}}_o + \boldsymbol{H}_s \boldsymbol{f}_I \tag{15.129}$$

$$\boldsymbol{y}_s = \boldsymbol{C}_s \boldsymbol{x}_s(k) \tag{15.130}$$

The state–space matrix equation of the aeroservoelasticity may be written as follows:

$$\begin{bmatrix} \boldsymbol{x}_s(k+1) \\ \boldsymbol{x}_a(k+1) \\ \boldsymbol{x}_c(k+1) \end{bmatrix} = \begin{bmatrix} \boldsymbol{G}_s + \boldsymbol{H}_s\rho\boldsymbol{D}_a\boldsymbol{T}_a(\boldsymbol{C}_s + \boldsymbol{D}_c\boldsymbol{T}_c\boldsymbol{C}_s) & \boldsymbol{H}_s\rho\boldsymbol{C}_a & \boldsymbol{H}_s\rho\boldsymbol{D}_a\boldsymbol{T}_a\boldsymbol{C}_c \\ \boldsymbol{H}_a\boldsymbol{T}_a(\boldsymbol{C}_s + \boldsymbol{D}_c\boldsymbol{T}_c\boldsymbol{C}_s) & \boldsymbol{G}_a & \boldsymbol{H}_a\boldsymbol{T}_a\boldsymbol{C}_c \\ \boldsymbol{H}_c\boldsymbol{T}_c\boldsymbol{C}_s & \boldsymbol{0} & \boldsymbol{G}_c \end{bmatrix}$$

$$\begin{bmatrix} \boldsymbol{x}_s(k) \\ \boldsymbol{x}_a(k) \\ \boldsymbol{x}_c(k) \end{bmatrix} + \begin{bmatrix} \boldsymbol{H}_s & \boldsymbol{H}_s\rho\boldsymbol{D}_a\boldsymbol{T}_a & \boldsymbol{H}_s \\ \boldsymbol{0} & \boldsymbol{H}_a\boldsymbol{T}_a & \boldsymbol{0} \\ \boldsymbol{0} & \boldsymbol{0} & \boldsymbol{0} \end{bmatrix} \begin{bmatrix} \boldsymbol{f}_I(k) \\ \boldsymbol{q}_{\delta I} \\ \rho\hat{\boldsymbol{f}}_o \end{bmatrix} \tag{15.131}$$

and

$$\begin{bmatrix} \boldsymbol{q}(k) \\ \dot{\boldsymbol{q}}(k) \\ \boldsymbol{f}_a(k) \end{bmatrix} = \begin{bmatrix} \boldsymbol{I} & \boldsymbol{0} & \boldsymbol{0} \\ \boldsymbol{0} & \boldsymbol{I} & \boldsymbol{0} \\ \rho\boldsymbol{D}_a\boldsymbol{T}_a(\boldsymbol{C}_s + \boldsymbol{D}_c\boldsymbol{T}_c\boldsymbol{C}_s) & \rho\boldsymbol{C}_a & \rho\boldsymbol{D}_a\boldsymbol{T}_a\boldsymbol{C}_c \end{bmatrix} \begin{bmatrix} \boldsymbol{x}_s(k) \\ \boldsymbol{x}_a(k) \\ \boldsymbol{x}_c(k) \end{bmatrix}$$

$$+ \begin{bmatrix} \boldsymbol{0} & \boldsymbol{0} & \boldsymbol{0} \\ \boldsymbol{0} & \boldsymbol{0} & \boldsymbol{0} \\ \boldsymbol{0} & \rho\boldsymbol{D}_a\boldsymbol{T}_a & \boldsymbol{I} \end{bmatrix} \begin{bmatrix} \boldsymbol{f}_I(k) \\ \boldsymbol{q}_{\delta I} \\ \rho\hat{\boldsymbol{f}}_o \end{bmatrix} \tag{15.132}$$

or

$$\boldsymbol{x}_{sac}(k+1) = \boldsymbol{G}_{sac}\boldsymbol{x}_{sac}(k) + \boldsymbol{H}_{sac}\boldsymbol{u}_{sac}(k) \tag{15.133}$$

$$\boldsymbol{y}_{sac}(k) = \boldsymbol{C}_{sac}\boldsymbol{x}_{sac}(k) + \boldsymbol{D}_{sac}\boldsymbol{u}_{sac}(k) \tag{15.134}$$

The relevant root locus plot derived from the $\boldsymbol{G}_{sac}$ matrix shows the location of roots as a function of density, which is analogous to plotting of $\boldsymbol{q}$ and $\dot{\boldsymbol{q}}$ contained in $\boldsymbol{x}_s$ (Eq. 15.84). The response data for the aeroservoelastic analysis is obtained from Eq. (15.132), which pertains to a closed-loop control feedback. The phase and gain margins (Bode plot) also may be obtained from the state–space matrices. For open loop analyses the $\boldsymbol{G}_c$, $\boldsymbol{H}_c$, $\boldsymbol{C}_c$, and $\boldsymbol{D}_c$ matrices are set to zero in Eqs. (15.131) and (15.132). Also, for the root locus plot, the modified $\boldsymbol{G}_{sac}$ matrix is copied to a lower-order matrix avoiding zeroes in the diagonal position.

15.4.4 Summary

A discrete-time ARMA model described herein is a modification of time-marched CFD solution and considerably reduces a coupled aeroelastic and aeroservoelastic analysis computation time. This modeling procedure starts with the computation of the nonlinear mean flow around the geometry involving only one steady and one subsequent unsteady analysis. A system model is then developed representing small, linear perturbations about the nonlinear mean flow. A set of training data is extracted from the unsteady CFD solution by forcing a multistep input on the generalized velocity for each structural mode under consideration. This model next can be used for coupled analysis in place of the various unsteady flow solutions required in the time-marched method for each data point. Because the system model executes at a fraction of CPU time required for an unsteady solution, a significant saving is effected in detecting a flutter point. This model is dependent only on the structural dimensions and Mach numbers and as such can be efficiently utilized to determine the effect of varying density. This procedure also may be generalized by augmenting the various control elements into the state–space matrices as in the linear case.

15.5 Numerical Examples

A number of relevant example problems have been solved to assess the efficacy of the currently developed numerical algorithms and tools. Examples of aeroelastic flutter and aeroservoelastic analysis of some representative test problems are presented next.

Example 15.2: AGARD 45-Deg Swept-Back Wing 445.6

The wing is an AGARD standard aeroelastic configuration with a 45-deg, quarter-chord sweep angle, a panel aspect ratio of 1.65, a taper ratio of 0.06, and a NACA 65A004 airfoil section. Measured modal frequencies and wind-tunnel flutter test results of the wing are detailed elsewhere[49,50]; the model selected for analysis and correlation is referred to as the 2.5-ft weakened model 3.

A finite element model of the wing yielded natural frequencies and mode shapes similar to those derived experimentally.[49,50] The first four modes represent first bending, first torsion, second bending, and second torsion, respectively; the corresponding frequencies being 9.60, 38.20, 48.35, and 91.54 Hz, respectively. These data were used for subsequent aeroelastic analysis to obtain flutter characteristics of the wing at freestream Mach numbers 0.499, 0.6787, 0.901, 0.960, 1.072, and 1.140, respectively, to effect comparison with test results as well as that derived in Refs. 6 and 8 based on the Euler solution by the finite volume method and also Ref. 51, which extended this analysis to viscous flow. To bracket the flutter points for each Mach number a set of aeroelastic responses was computed for several values of dynamic pressure.[14] Figure 15.21 shows the partial view of the CFD grid on the surface of the wing and the symmetry plane; an associated three-dimensional CFD grid consists of over 240,000 tetrahedral fluid elements. For transonic and supersonic flow, an enhanced grid consisting of 373,798 elements was used for the respective solutions.

Figure 15.22 shows computed damping results against flutter speed index values for two typical Mach numbers. Table 15.2 presents a comparison of flutter solution results obtained by the STARS finite element aeroelastic analysis, wind-tunnel tests, and the finite volume procedures employing structured and unstructured grids, respectively; such results for viscous flow show improvement in solution accuracy for the Mach 1.14 case. These results are further depicted in Fig. 15.23. Although results quoted from Ref. 6 were available in tabular form, such results were extracted from graphical representation of the same results presented in Ref. 8. The associated v_f is defined as $V/(b_s\omega_a\sqrt{\bar{\mu}})$, in which V is the stream velocity, b_s is the root semichord, ω_a is the first torsional frequency, and $\bar{\mu}$ is the mass ratio. Figure 15.23 also depicts solution results obtained by the discrete time-marched (ARMA) procedure. Although such results are found to be very similar to the ones predicted by the time-marched analysis, the discrete time-marched solution involving analysis of only one unsteady flow calculation proves to be considerably more economical.

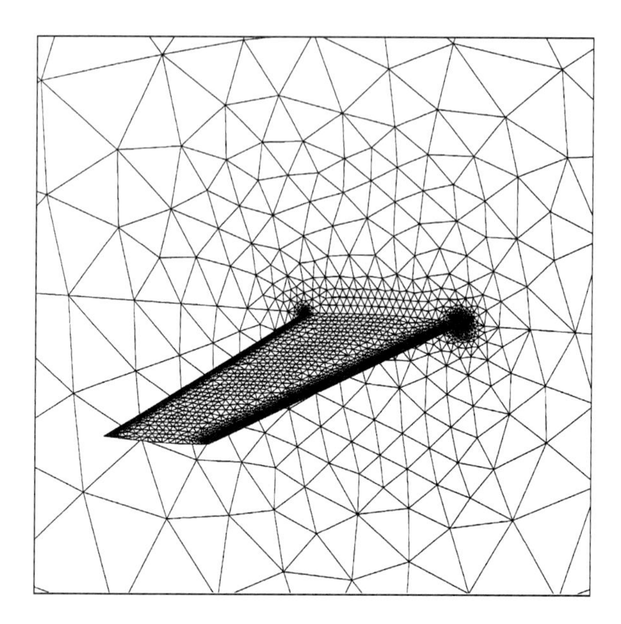

Fig. 15.21 AGARD wing: CFD surface grid on wing and plane of symmetry.

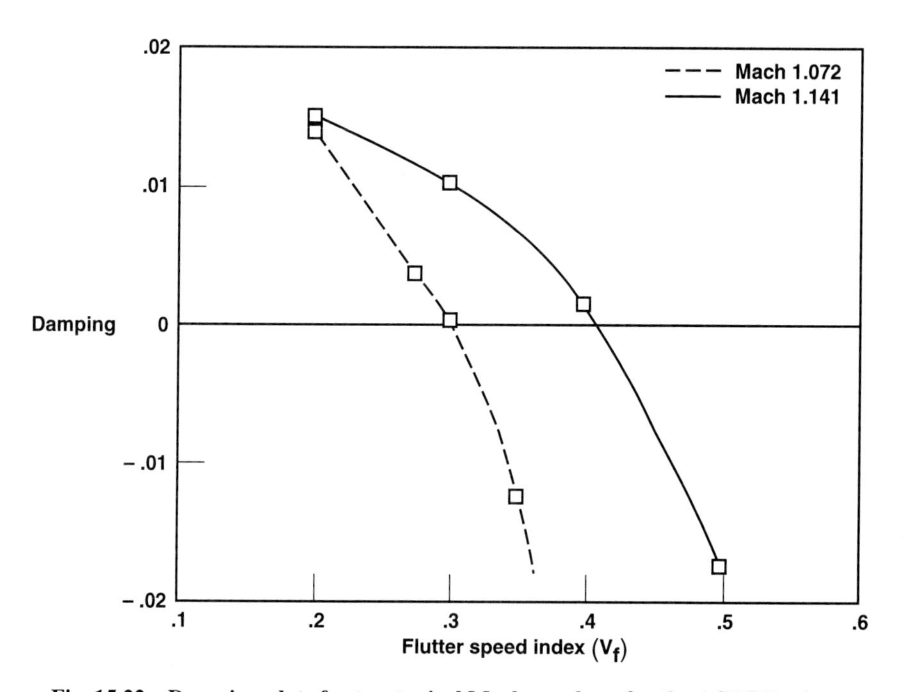

Fig. 15.22 Damping plots for two typical Mach numbers for the AGARD wing.

Table 15.2 Comparison of calculated and experimental flutter solutions for the AGARD wing

Freestream M	STARS	Ref. 6	Ref. 8	Test, Ref. 49
	Flutter speed index			
0.499	0.436	0.439	0.43	0.466
0.678	0.380	0.417	0.40	0.417
0.900	0.341	0.352	0.31	0.370
0.960	0.280	0.275	0.23	0.308
1.072	0.302	0.466	—	0.320
1.141	0.410	0.660	—	0.403
	(0.466, Ref. 51)			
	Flutter frequency ratio			
0.499	0.477	0.597	0.52	0.535
0.678	0.428	0.539	0.47	0.472
0.900	0.397	0.425	0.37	0.422
0.960	0.364	0.343	0.31	0.365
1.072	0.363	0.541	—	0.362
1.141	0.435	0.764	—	0.459
	(0.47, Ref. 51)			

Example 15.3: BACT Wing

The BACT (benchmark active controls technology) wing is analyzed to demonstrate the effectiveness of the transpiration boundary condition approach for steady and unsteady flow as well as control surface deflection and flutter prediction.[48,52] Control surface deflections pose a particularly difficult problem for mesh deformation and remeshing techniques because of the discontinuous nature of the surfaces and the new surfaces that are exposed. Figure 15.24 shows the BACT wing that has a NACA 0012 airfoil section and a rectangular platform with a 16-in. chord and 32-in. span; the trailing edge control surface is 30% of the total wing span centered about 60% span station and is 25% of the chord.[53]

Figure 15.25 shows the finite element structural mesh, whereas Fig. 15.26 depicts the aerodynamic surface mesh of the wing. To demonstrate efficiency of the transpiration method, steady-state Euler solutions were run for Mach 0.77 and zero angle of attack α with 10 deg simulated and also actual flap deflection ($\alpha = 10$ deg). Such results are depicted in Figs. 15.27 and 15.28, amply demonstrating effectiveness of the technique. A flutter prediction comparison is shown in Fig. 15.29. To demonstrate flutter suppression capability of the current procedure a suitable control law (Fig. 15.30) was designed, and Figs. 15.31 and 15.32 depict the results of the related aeroservoelastic analysis.

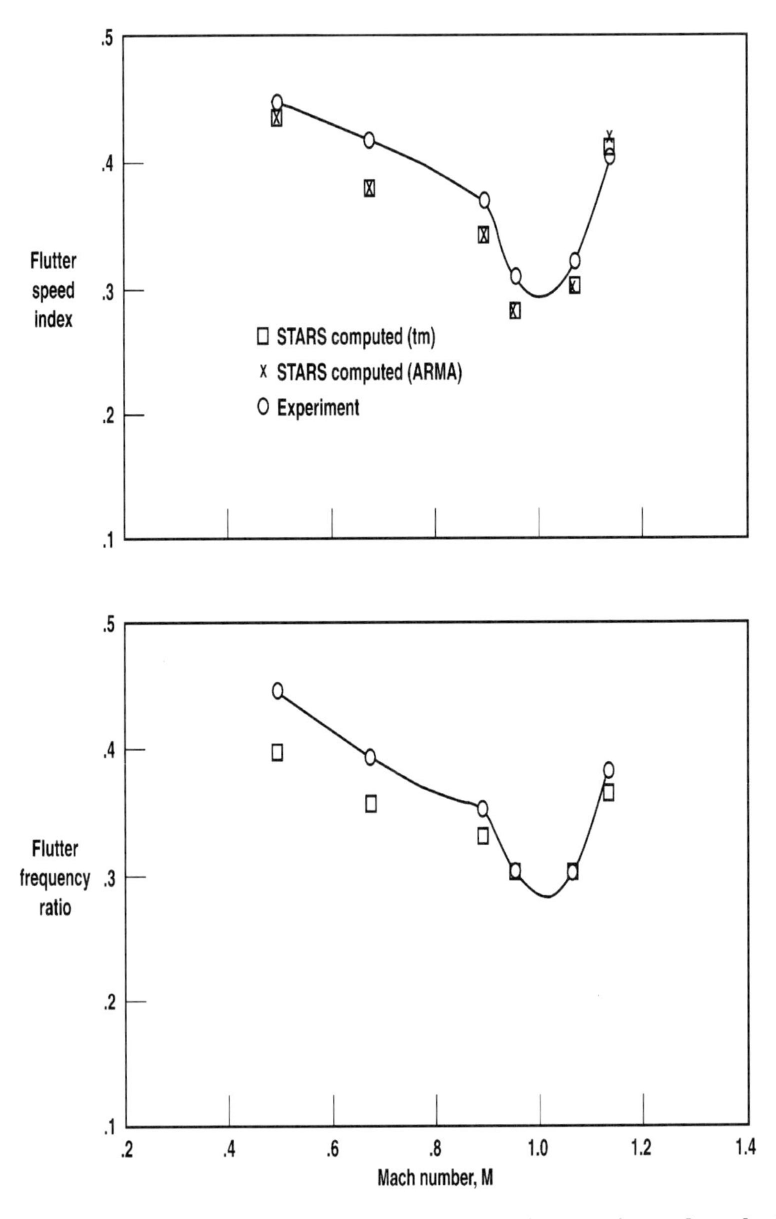

Fig. 15.23 Comparison of STARS flutter predictions with experimental results for the AGARD wing.

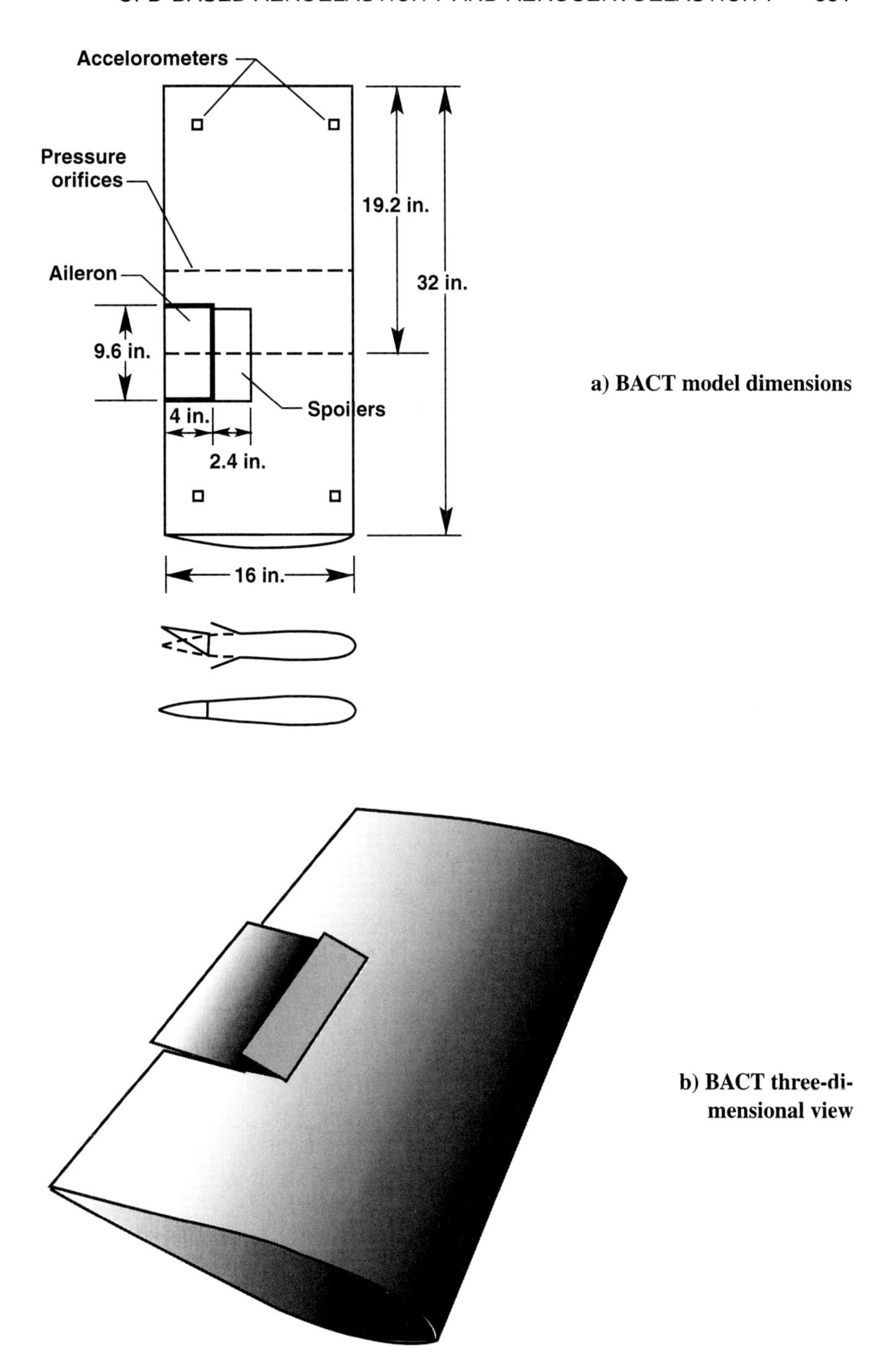

a) BACT model dimensions

b) BACT three-dimensional view

Fig. 15.24 BACT model dimensions.

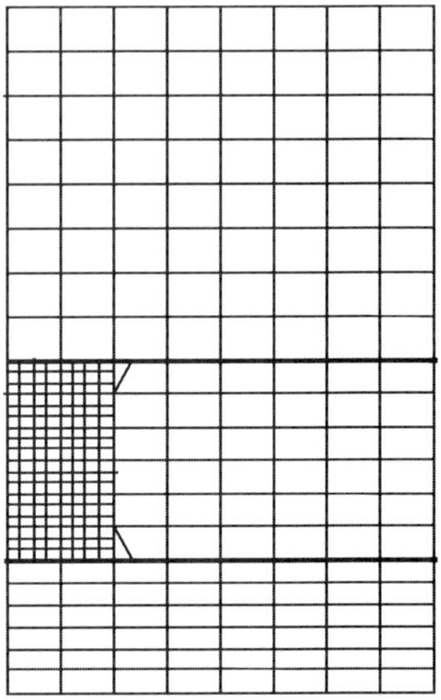

Fig. 15.25 Finite element solids mesh for BACT wing.

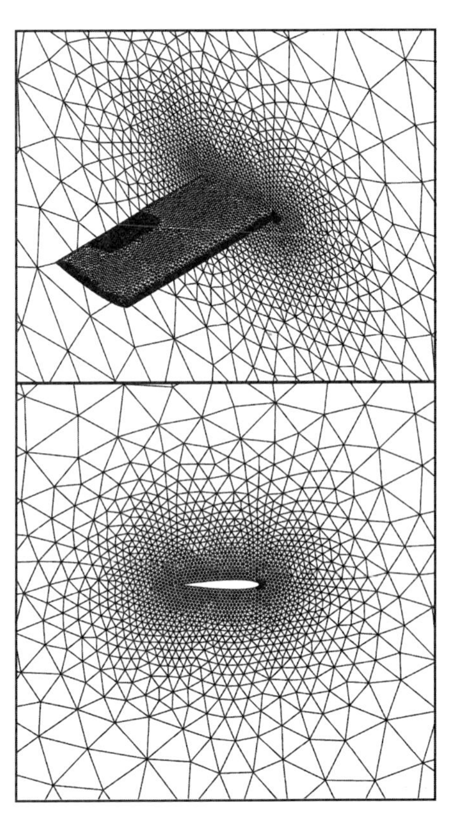

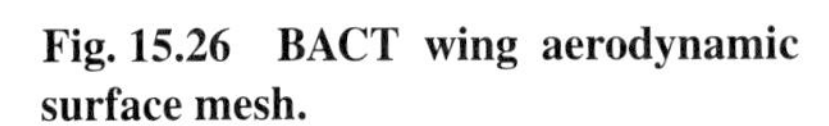

Fig. 15.26 BACT wing aerodynamic surface mesh.

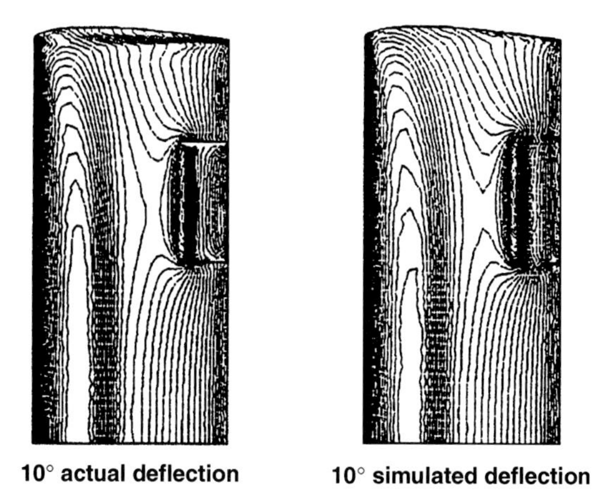

Fig. 15.27 BACT wing surface pressure contours at Mach 0.77, 10-deg control surface deflection.

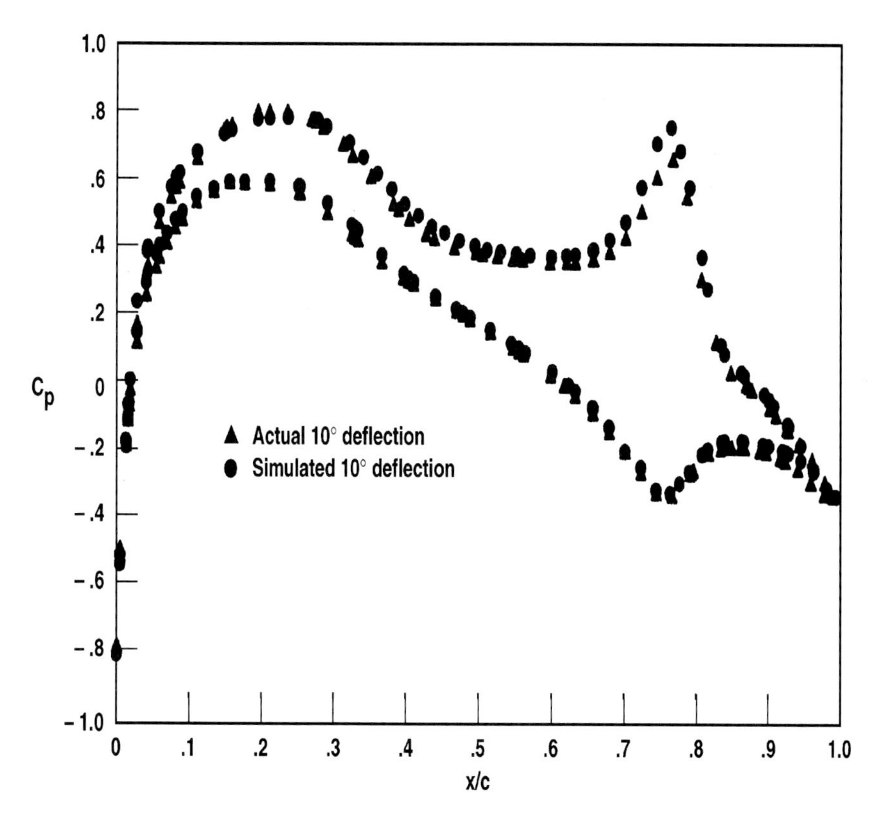

Fig. 15.28 BACT wing comparison of predicted pressure distributions for an actual and simulated 10-deg control surface deflection at Mach 0.77, α = 0-deg.

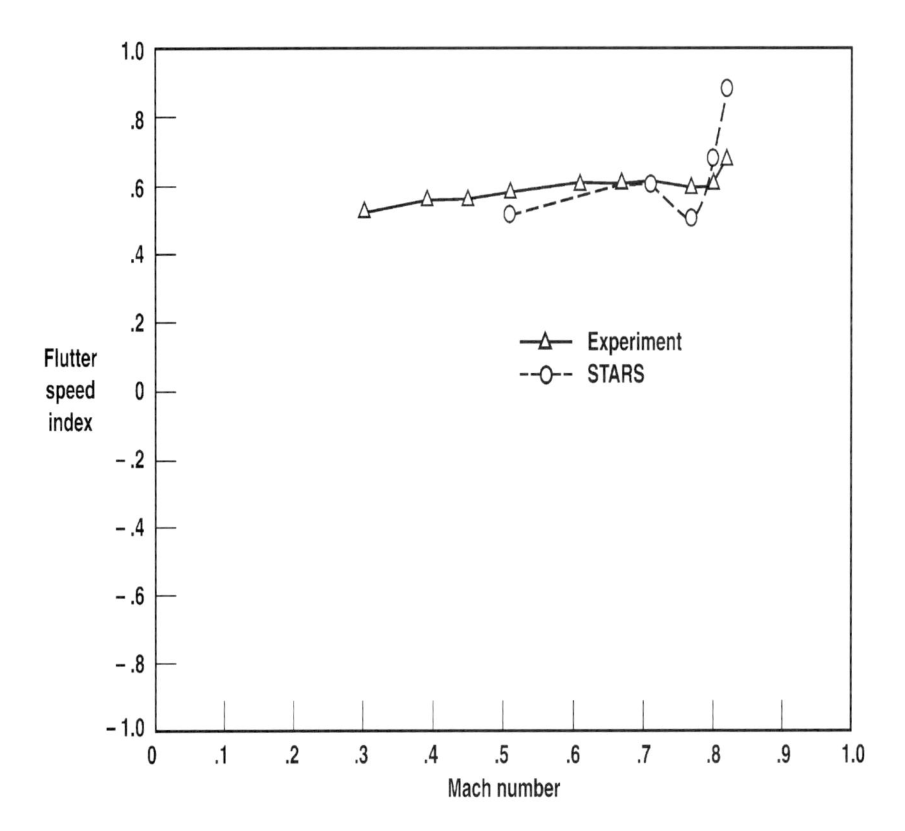

Fig. 15.29 BACT wing flutter prediction comparison with experimental data.

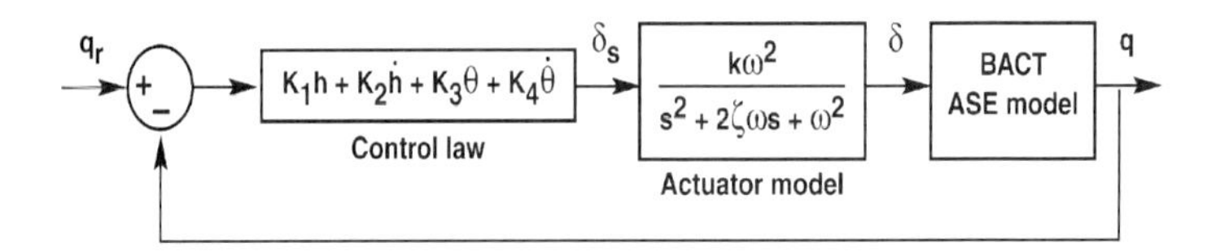

Fig. 15.30 BACT wing block diagram of control implemented into STARS.

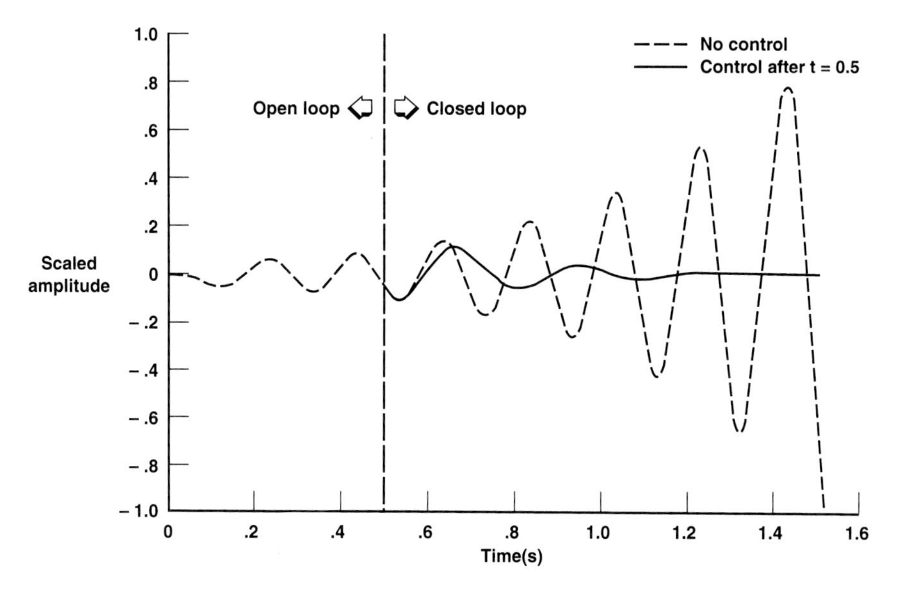

Fig. 15.31 BACT wing time history for plunge (mode 1) response with and without control.

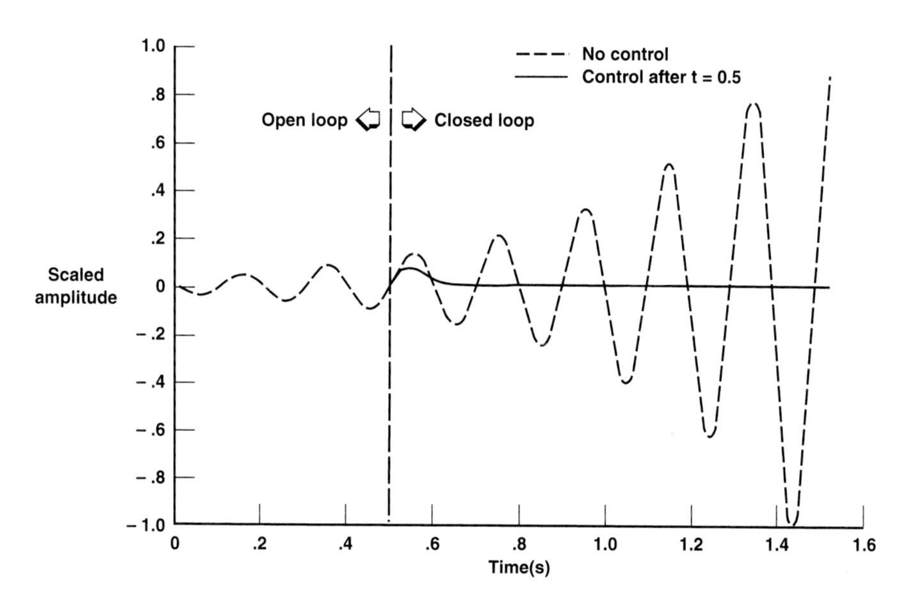

Fig. 15.32 BACT wing time history for pitch (mode 2) response with and without control.

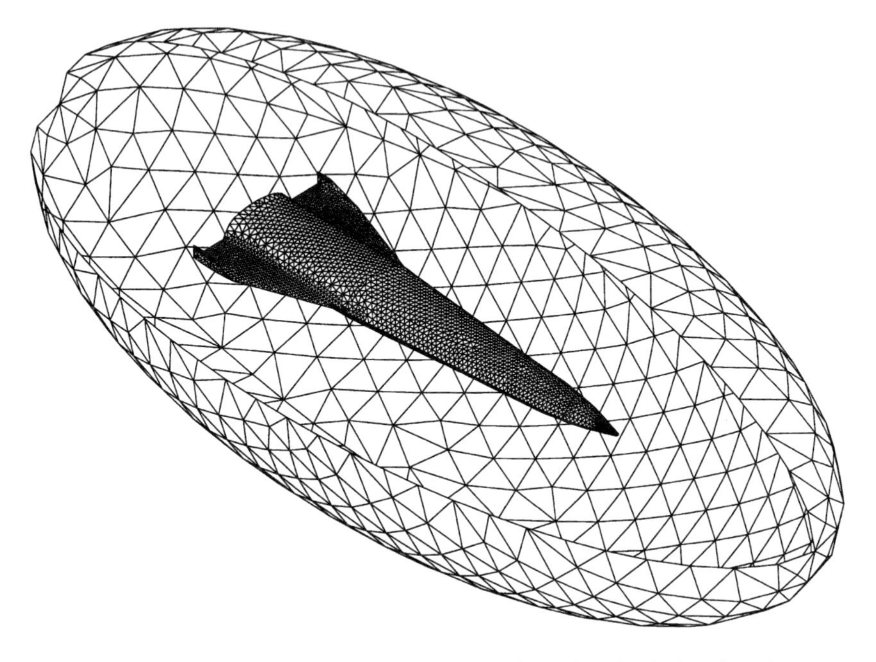

Fig. 15.33 GHV surface mesh for a three-dimensional solution domain.

Example 15.4: Generic Hypersonic Vehicle

In connection with the development of the National Aerospace Plane, a generic hypersonic vehicle (GHV) model in Fig. 9.3 was generated to simulate its flight characteristics. The structural model consist of about 2800 nodes and 5000 elements. First six structural mode shapes are shown in Fig. 9.4. and Tables 9.1 and 9.2 present lists of natural frequencies. Associated three-dimensional CFD unstructured grid consists of about 300,000 elements; Fig. 15.33 depicts the surface mesh of the three-dimensional solution domain. The Euler solution pressure distribution for Mach 2.2 is shown in Fig. 15.34, and results of the aeroelastic analysis[54,55] is shown in Fig. 15.35. Solution time for the entire analysis employing the dtm procedure was about 30 h of CPU time using a 2.8 GHz PC.

Example 15.5 HyperX Flight vehicle

The HyperX stack, Fig. 15.36, consists of a Pegasus booster and the X-43 hypersonic flight test vehicle, which are launched from a B-52 aircraft as shown in Fig. 15.37 (Ref. 56). Once the vehicle is boosted to around 100,000 ft, the X-43 separates from the booster and is designed to perform a hypersonic freeflight at Mach 7 and also at about Mach 10. Stability analysis of the X-43 and the stack at Mach 7 were described in Refs. 57, and 58, respectively. Related analysis of the stack in the transonic flight regime is of critical importance for the success of the project and a representative solution is presented herein. Figure 15.38 presents

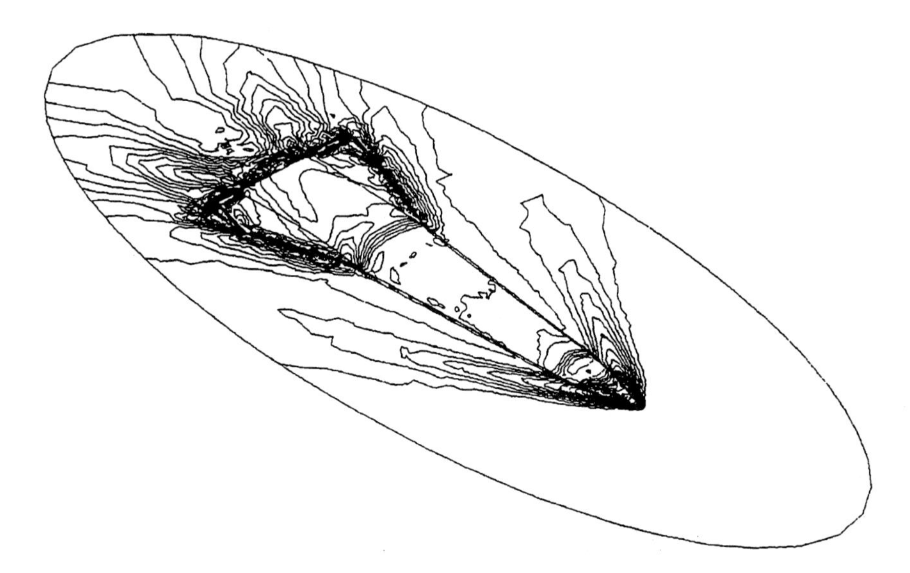

Fig. 15.34 Euler pressure distribution on a GHV for Mach 2.2.

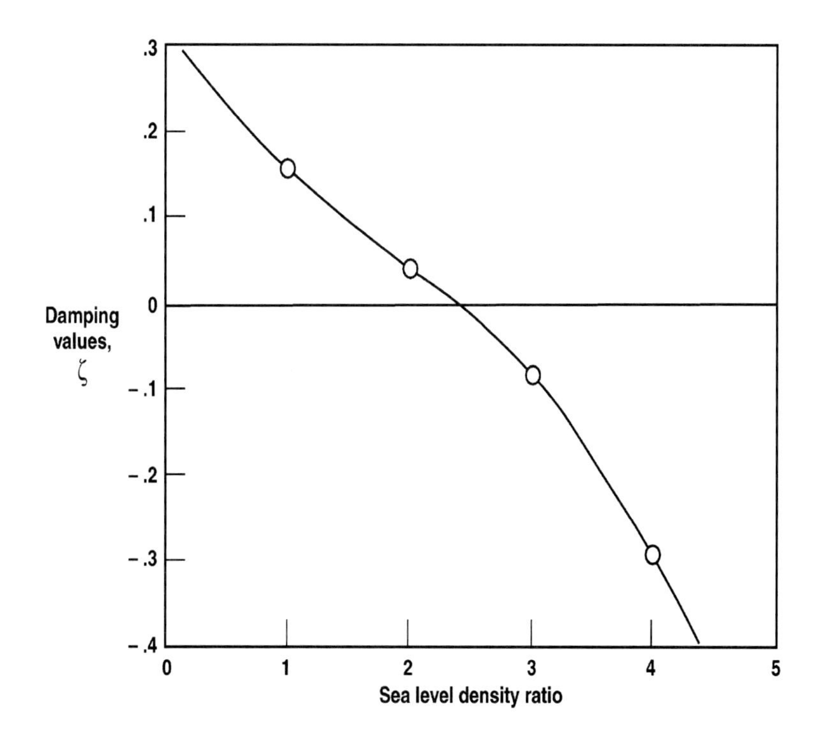

Fig. 15.35 GHV damping plot for Mach 2.2.

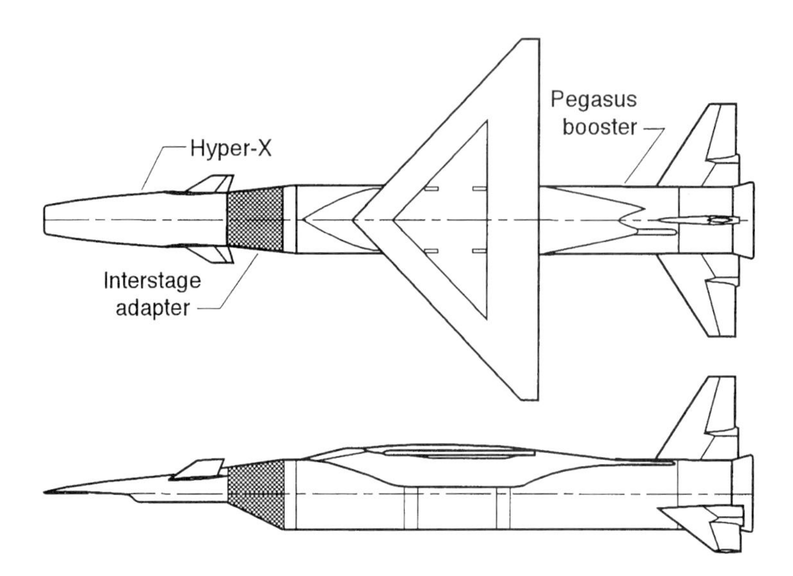

Fig. 15.36 HyperX— flight vehicle.

the first few mode shapes of the free vibration of the HyperX (mode 1 also is depicted in color at the front of this book), and Table 15.3 provides a list of natural frequencies; the angles of attack for the vehicle and the horizontal tails were taken to be 12.43 and −11 deg, respectively. A three dimensional CFD solution domain is shown in Fig. 15.39. Analysis was performed at Mach 0.9, and Mach and pressure distribution are given in Fig. 15.40 (see also the color images at the front of this book). Aeroelastic analysis was performed for Mach 0.9 at altitude 22.5 K, and the distribution of generalized displacements pertaining to elastic modes 1, 2, 4,

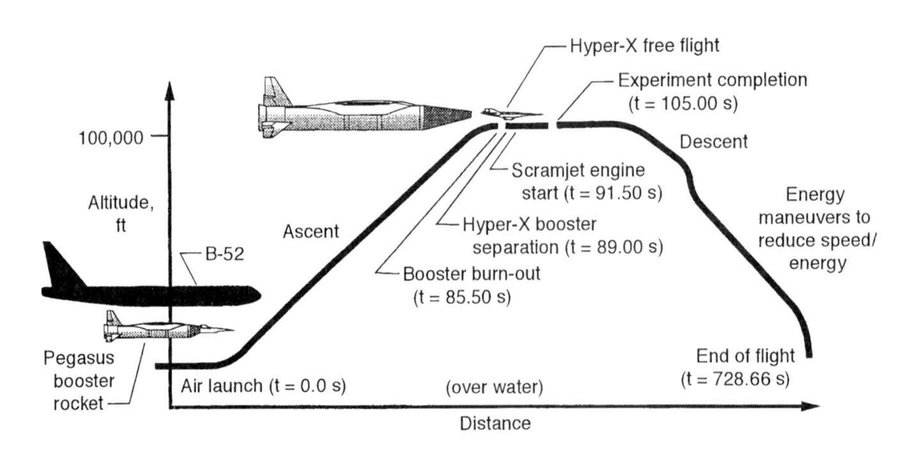

Fig. 15.37 HyperX—flight trajectory.

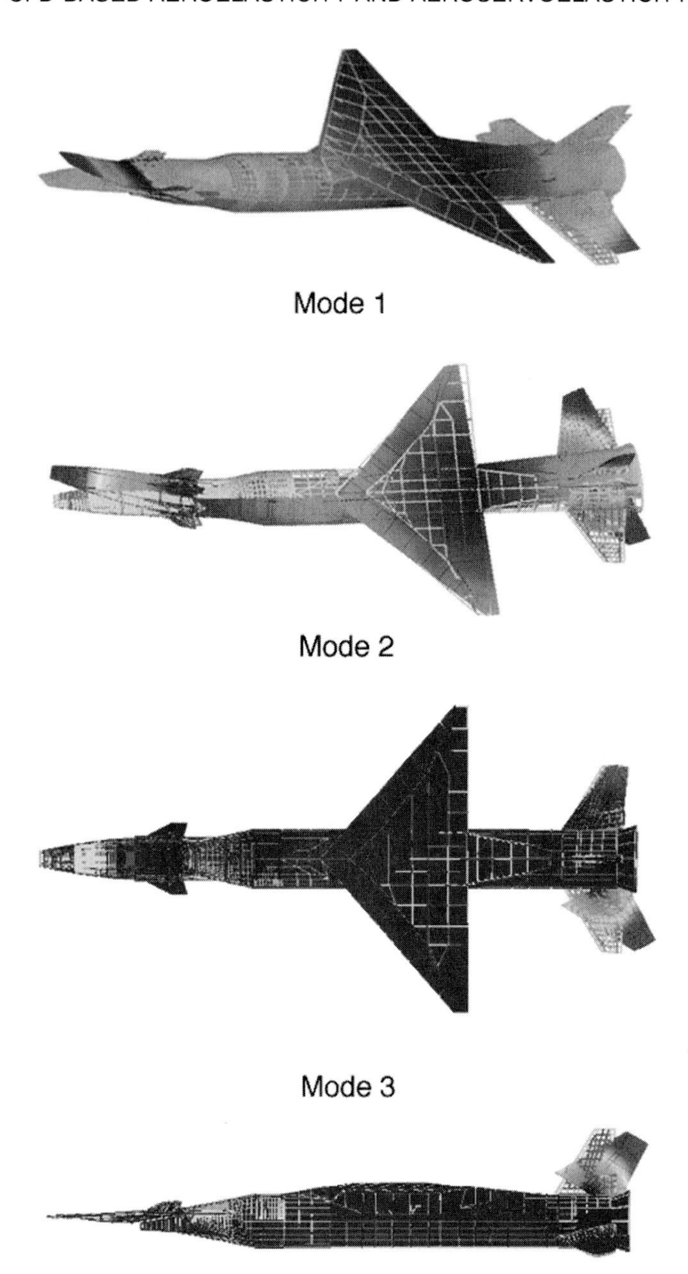

Fig. 15.38 HyperX—Typical mode shapes.

Table 15.3 **HyperX stack free vibration**

Mode	STARS $\alpha_{HT} = -11^O$ frequency, Hz	Mode shape description (major motion)
1–6	0.0	Rigid body modes
7	8.03	F1B Vertical(S) – 1 el
9	9.44	F1B Lateral(A/S) – 2 el
8	11.42	RT FIN F/A – 3 el
10	11.99	LT FIN F/A – 4 el
11	13.21	RT FIN 1B – 5 el
12	14.20	RUDDER 1B – 6 el
13	14.45	LT FIN 1/B – 7 el
14	15.05	RUDDER F/A – 8 el
15	16.38	F2B Vertical(S) – 9 el

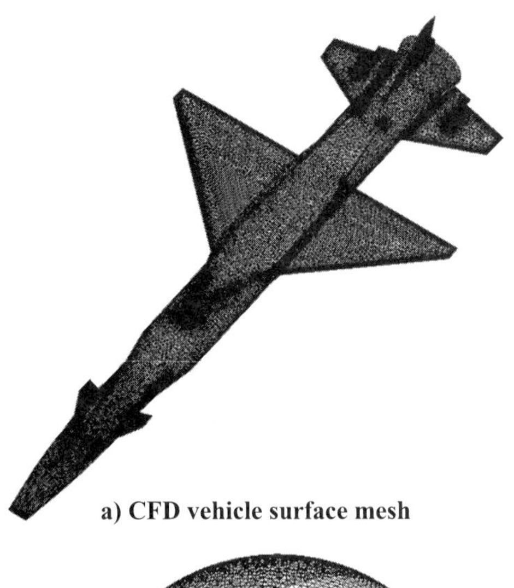

a) CFD vehicle surface mesh

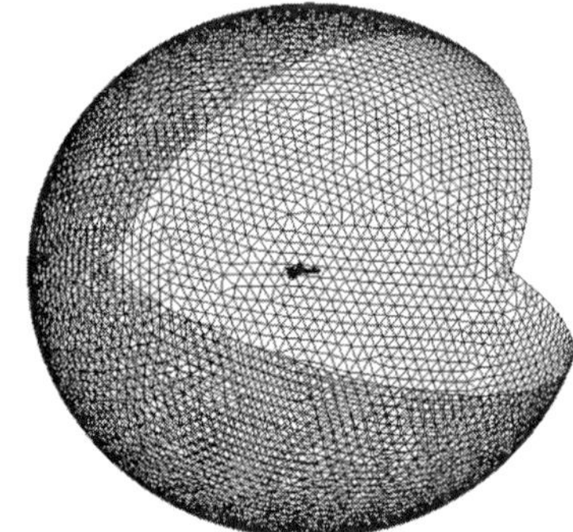

b) CFD mesh for three-dimensional solution domain

Fig. 15.39 HyperX—CFD surface and global mesh.

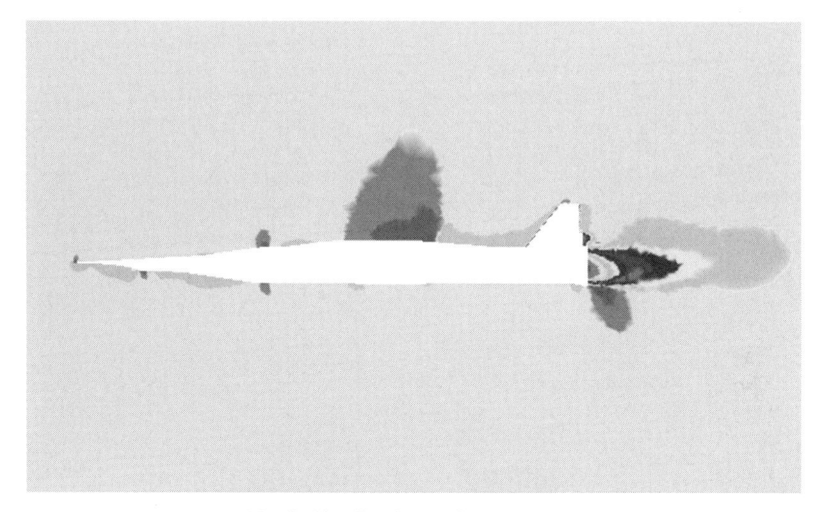

a) Mach distribution—Center cross section

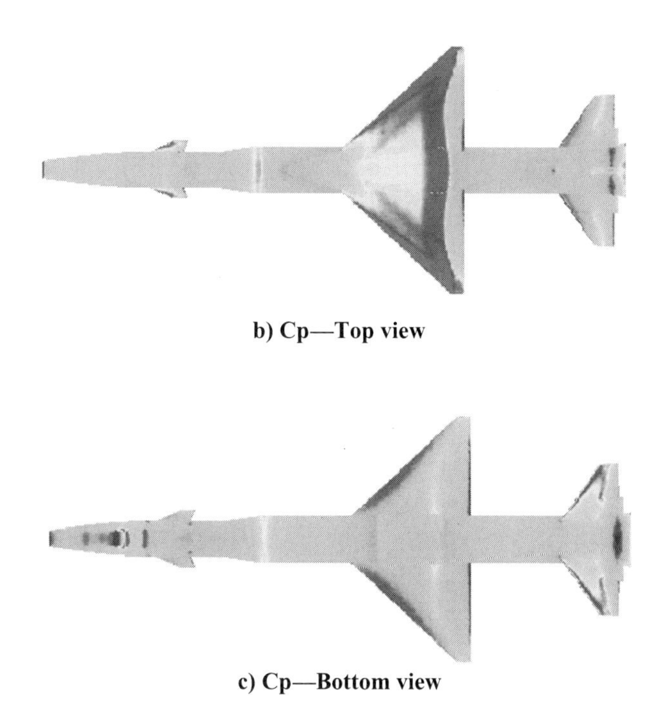

b) Cp—Top view

c) Cp—Bottom view

Fig. 15.40 HyperX—Mach and pressure distribution.

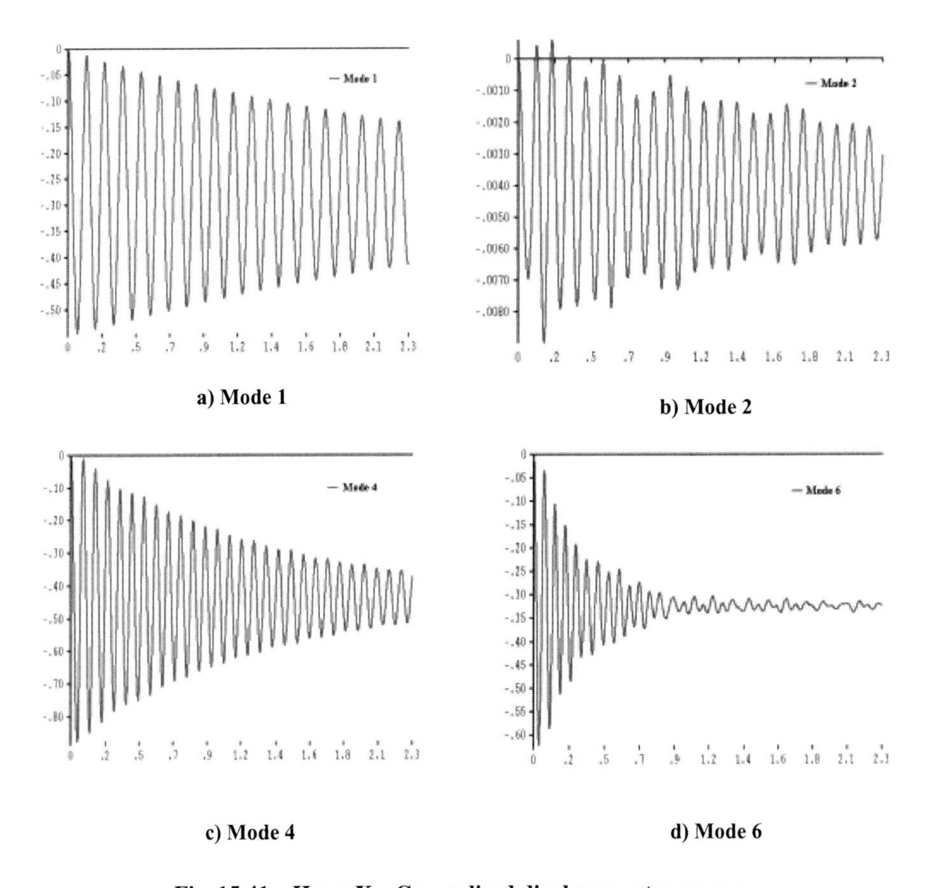

a) Mode 1

b) Mode 2

c) Mode 4

d) Mode 6

Fig. 15.41 HyperX—Generalized displacement response.

and 6 are shown in Fig. 15.41. Typical results of the aeroelastic analysis as damping plots are given in Fig. 15.42.

15.6 Concluding Remarks

Adoption of the common FEM effects accurate transfer of data from fluids to structural domain and vice versa, thereby ensuring accurate representation of their interaction phenomena. Such solution results for the AGARD 45-deg swept-back wing present interesting data on relative efficiencies of the various solution procedures.[59] Thus, Fig. 15.22 depicting STARS analysis[14] and test results,[49] shows good correlation both in subsonic and transonic/supersonic regimes. Table 15.2 provides in tabular form a comparison of flutter test data,[49] with STARS finite element and also finite volume analysis employing structured[6] and unstructured[8] grids, respectively. Whereas all three analysis results correlate rather well in the subsonic area, the current analysis technique also provides excellent correlation with test results in the transonic and supersonic flow regimes. Reference 60

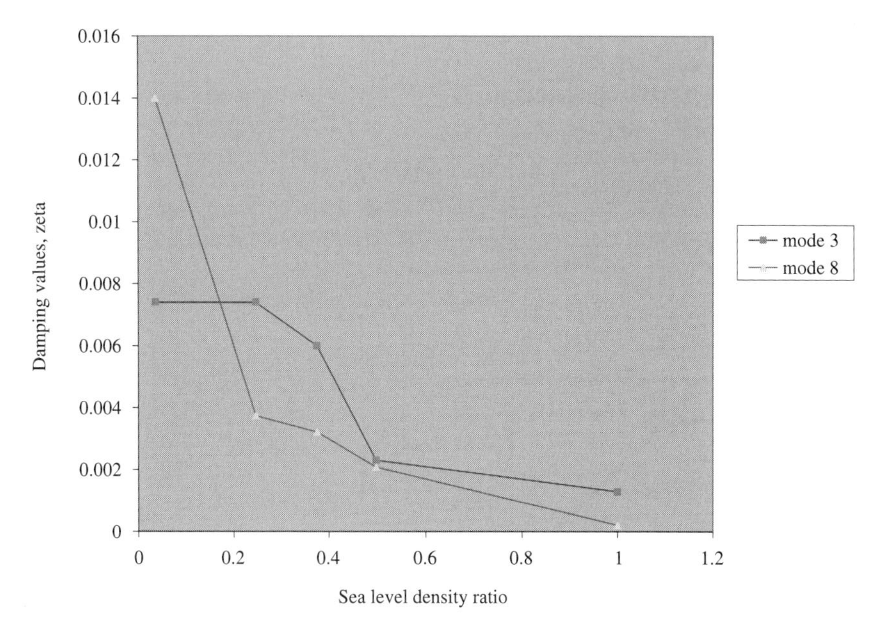

Fig. 15.42 HyperX—typical aeroelastic response.

provides an overview of various analysis methods in aeroelasticity, and a review of CFD-based analyses is given in Ref. 61.

References

[1]Edwards, J. W., and Malone, J. B., "Current Status of Computational Methods for Transonic Unsteady Aerodynamic and Aeroelastic Applications," NASA TM-10491, Dec. 1991.

[2]Ide, H., and Shanker, V. J., "Unsteady Full Potential Aeroelastic Computations for Flexible Configurations," AIAA Paper 87-1238, June 1987.

[3]Guruswamy, G. P., "Time-Accurate Unsteady Aerodynamic and Aeroelastic Calculations of Wings using Euler Equations," AIAA Paper 88-2281, April 1988.

[4]Guruswamy, G. P.," Vortical Flow Computations on a Flexible Blended Wing-Body Configuration," *AIAA Journal*, Vol. 30, No. 10, 1992, pp. 2497–2503.

[5]Robinson, B. A., Batina, J. T., and Yang, H. T. Y., "Aeroelastic Analysis of Wings using the Euler Equations with a Deforming Mesh," *Journal of Aircraft*, Vol. 28, No. 11, 1991, pp. 781–788.

[6]Lee-Rausch, E. M., and Batina, J. T., "Wing Flutter Boundary Prediction using Unsteady Euler Aerodynamic Method," NASA TM-107732, March 1993.

[7]Batina, J. T., "Unsteady Euler Algorithm with Unstructured Dynamic Mesh for Complex-Aircraft Aerodynamic Analysis," *AIAA Journal*, Vol. 29, No. 3, 1991, pp. 327–333.

[8]Rausch, R. D., Batina, J. T., and Yang, H. T. Y., "Three-Dimensional Time-Marching Aeroelastic Analysis using an Unstructured-Grid Euler Method," *AIAA Journal*, Vol. 31, No. 9, 1993, pp. 1626–1633.

[9]Ricketts, R. H., Noll, T. E., Whitlow, W., and Huttsell, L. J., "An Overview of Aeroelasticity Studies for the National Aero-Space Plane," AIAA Paper 93-1313, April 1993.

[10]Bendicksen, O. O., "A New Approach to Computational Aeroelasticity," AIAA Paper 91-0939, April 1991.

[11]Lighthill, M. J., "On Displacement Thickness," *Journal of Fluid Mechanics*, Vol. 4, No. 4, 1958, pp. 383–392.

[12]Sankar, L. N., Malone, J. B., and Schuster, D., "Euler Solutions for Transonic Flow Past a Fighter Wing," *Journal of Aircraft*, Vol. 24, No. 1, 1987, pp. 10–16.

[13]Gupta, K. K., "STARS — An Integrated General-Purpose Finite Element Structural, Aeroelastic and Aeroservoelastic Analysis Computer Program," NASA TM 4795, May 1997.

[14]Gupta, K. K., "Development of a Finite Element Aeroelastic Analysis Capability," *Journal of Aircraft*, Vol. 33, No. 5, 1996, pp. 995–1002.

[15]Moin, P., and Mahesh, K., "Direct Numerical Simulation: a Tool in Turbulence Research," *Annual Review of Fluid Mechanics*, Vol. 30, 1998, pp. 539–578.

[16]Streeter, V. L., *Fluid Dynamics*, McGraw–Hill, New York, 1984.

[17]Donea, J., "A Taylor–Galerkin Method for Convective Transport Problems," *International Journal of Numerical Methods in Engineering*, Vol. 20, No. 1, 1984 pp. 101–119.

[18]Peraire, J., Peiro, J., Formaggia, L., Morgan, K., and Zienkiewicz, O. C., "Finite Element Euler Computations in Three Dimensions," *International Journal of Numerical Methods in Engineering*, Vol. 26, No. 10, 1988, pp. 2135–2159.

[19]Morgan, K., Peraire, J., and Piero, J., "The Computation of Three Dimensional Flows using Unstructured Grids," *Computational Methods in Applied Mechanics and Engineering*, Vol. 87, No. 3, 1991, pp. 335–352.

[20]Gupta, K. K., Petersen, K. L., and Lawson, C. L., "On Some Recent Advances in Multidisciplinary Analysis of Hypersonic Vehicles," AIAA Paper 92-5026, 1992.

[21]Baldwin, B. W., and Barth, T. A. "Thin-Layer Approximation Algebraic Model for Separated Turbulent Flows," AAIA Paper 78-0257, 1978.

[22]Cebeci, P. T., and Smith, A. M. O. "Analysis of Turbulent Boundary Layers," *Series in Applied Mathematics and Mechanics XV*, Academic Press, London, 1974.

[23]Baldwin, B. W., and Barth, T. A. "One-Equation Turbulence Transport Model for High Reynolds Number Wall-Bounded Flows," AIAA Paper 91-0610, 1991.

[24]Spalart, P. R., and Allmaras, S. R. "A One-Equation Turbulence Model for Aerodynamic Flows," AIAA Paper 92-0439, 1992.

[25]Moser, R., Kim, J., and Mansour, N. N "Direct Numerical Simulation of Turbulent Channel Flow up to $Re_t = 590$," *Physics of Fluids*, Vol. 11, 1999, pp. 943–945.

[26]Mansour, N. N., Kim, J., and Moin, P. "Near-Wall $k - \epsilon$ Turbulence Modeling," *AIAA Journal*, Vol. 27, No. 8, 1989, pp. 1068–1073.

[27]Huang, P. G., Coleman, G. N., and Bradshaw, P. "Compressible Turbulence Channel Flows–DNS Results and Modeling," *Journal of Fluid Mechanics* Vol. 305, 1995, pp. 185–218.

[28]Patankar, S. V., and Spalding, D. B. *Heat and Mass Transfer in Boundary Layers, 2nd Ed.,* Intertext, London, 1970.

[29]Rogers, S. E., Kwak, D., and Kiris, C. "Steady and Unsteady Solutions of the Incompressible Navier–Stokes Equations," *AIAA Journal*, Vol. 29, 1991, pp. 603–610.

[30]Spalart, P. R., and Baldwin, B. S., "Direct Numerical Simulation of Turbulent Oscillating Boundary Layer," *Turbulent Shear Flows*, Springer, 1989, pp. 417–440.

[31]Mehta, U. B., and Lavan, Z., "Starting Vortex, Separation Bubbles and Stall: A Numerical Study of Laminar Unsteady Flow Around an Airfoil," *Journal of Fluid Mechanics*, Vol. 67, 1975, pp. 227–256.

[32]Spalart, P. R., Jou, W. J., Strelets, M., and Allmaras, S.R. "Comments on the Feasibility of LES for Wings and on the Hybrid RANS/LES Approach," *Advances in DNS/LES, First AFORSR International Conference on NDS/LES*, Greden Press, 1997.

[33]Leschziner, M. A., and Drikaras, D., "Turbulence Modeling and Turbulent-flow Computation in Aeronautics," *The Aeronautical Journal*, Vol. 106, 2002, pp. 349–384.

[34]Pullian, T. H., and Barton, J. T. "Euler Computations of AGARD Working Group 07 Aerofoil Test Cases," AIAA 23rd Aerospace Sciences Meeting, Jan. 1985.

[35]Castro-Diaz, M. J., Hecht, F., and Mohammadi, B. 'Anisotropic Unstructured Mesh Adaptation for Flow Simulation," *Finite Elements in Fluids Conference*, 1995.

[36]Williams, R., *Adaptive Parallel Meshes with Complex Geometry*, Caltech TR 217-50, California Inst. of Technology, Pasadena, CA, 1991.

[37]Weatherill, N. P., "Mesh Generation in CFD," *Computational Fluid Dynamics*, Von Karman Inst. for Fluid Dynamics, Lecture Series 1989-04, March 1989.

[38]Jameson, A., Baker, T. J., and Weatherill, N. P., "Calculation of Inviscid Transonic Flow over a Complete Aircraft," AIAA Paper 86-0103, 1986.

[39]Baker, T. J., "Automatic Mesh Generation for Complex Three-Dimensional Regions using a Constrained Delaunay Triangulation," *Engineering Computations*, Vol. 5, 1989.

[40]Peraire, J., Vahdati, M., Morgan, K., and Zienkiewicz, O. C., "Adaptive Remeshing for Compressible Flow Computations," *Journal of Computational Physics*, Vol. 72, 1987.

[41]Piradeh, S., "Unstructured Viscous Grid Generation by Advancing-Layers Method," AIAA Paper 93-3453, 1993.

[42]Piradeh, S., "Viscous Unstructured Three-Dimensional Grids by the Advancing-Layers Method," AIAA Paper 94-0417 1994.

[43]Peraire, J., Peiro, J., and Morgan, K., "A 3-D Finite Element Multigrid Solver for the Euler Equations," AIAA Paper 92-0449, 1992.

[44]Lohner, R., "Adaptive H-Refinement on 3-D Unstructured Grids for Transient Problems," AIAA Paper 89-0365, Jan. 1989.

[45]Heeg, J., Zeiler, T., Pototzky, A., Spain, V., and Engelund, W., "Aero Thermoelastic Analysis of a NASP Demonstrator Model," AIAA Paper 93-1366, 1993.

[46]Cowan, T. J. Arena, A. S. and Gupta, K. K., "Accelerating CFD-Based Aeroelastic Predictions using System Identification," AIAA Paper 98-4152, 1998.

[47]Cowan, T.J., Arena, A.S., and Gupta, K.K., "Development of a Discrete Time Aerodynamic Model for CFD-Based Aeroelastic Analysis," AIAA Paper 99-0765, 1999.

[48]Stephens, C. H., Arena, A. S., and Gupta, K. K., "CFD-Based Aeroservoelastic Predictions with Comparisons to Benchmark Experimental Data," AIAA Paper 99-0766, 1999.

[49]Yates, E. C., Jr., Land, N. S., and Fougher, J. T., Jr., "Measured and Calculated Subsonic and Transonic Flutter Characteristics of a 45° Swept-Back Wing Planform in Air and in Freon-12 in the Langley Transonic Dynamics Tunnel," NASA TN D-1616, March 1963.

[50]Yates, E. C., Jr., "AGARD Standard Aeroelastic Configuration for Dynamic Response, Candidate Configuration 1. - Wing 445.6," NASA TM-100492, Aug. 1987; also Proceedings of the 61st Meeting of the Structures and Materials Panel, AGARD-R-765, 1985 pp. 1–73.

[51]Lee-Rausch, E. M., and Batina, J. T., "Calculation of AGARD Wing 445.6 Flutter using Navier–Stokes Aerodynamics," AIAA Paper 93-3479, 1993.

[52]Stephens, C. H., Arena, A. S., and Gupta, K. K., "Application of Transpiration Method for Aeroelastic Prediction using CFD," AIAA Paper 98-2071, April 1998.

[53]Scott, R. C., Hoadley, S. T., Wiesman, C. D., and Durham, M. H., "The Benchmark Active Controls Technology Model Aerodynamic Data," AIAA Paper 97-0829, Jan. 1997.

[54]Gupta, K. K., and Petersen, K. L., *Multidisciplinary Aeroelastic Analysis of a Generic Hypersonic Vehicle*, NASA TM 4544, 1993.

[55]Gupta, K. K., "Development and Application of an Integrated Multidisciplinary Analysis Capability," *International Journal of Numerical Methods in Engineering*, Vol. 40, 1997, pp. 533–550.

[56]Ko, W. L., and Gong, L., "Thermostructural Analysis of Unconventional Wing Structures of a Hyper-X Hypersonic Flight Research Vehicle for the Mach 7 Mission," NASA/TP-2001-210398, 2001.

[57]Gupta, K. K., Voelker, L. S., Bach, C., Doyle, T. and Han, E., "CFD Based Aeroelastic Analysis of the X-43 Hypersonic Flight Vehicle," AIAA Paper 2001-0712, Jan. 2001.

[58]Gupta, K. K., Voelker, L. S., Bach, C., "Finite Element CFD Based Aeroservoelastic Analysis," AIAA Paper 2002-0953, Jan. 2002.

[59]Friedmann, P. P., "Renaissance of Aeroelasticity and its Future," *Journal of Aircraft*, Vol. 36, No. 1, 1999 pp. 105–121.

[60]Dowell, E. H., Crawley, E. F., Jr., H. C., Peters, D. A., Scanlan, R. H., and Sisto, F., *A Modern Course in Aeroelasticity*, 3rd edition, Kluwer Academic, 1995.

[61]Geuzaire, P. Brown, G., and Farhat, C., "Three Field based Nonlinear Aeroelastic Simulation Technology: Status and Application to the Flutter Analysis of an F-16 Configuration," Jan. 2002.

Appendix
Exercises

Chapter 1

1.1 Review existing literature and write a brief history of the beginning of the finite element method as applied to engineering problems.

1.2 Read Courant's paper* and solve the hollow square box torsion problem (Fig. A1) with varying mesh size in appropriate direction to perform a convergence study.

1.3 Review Argyris,† and derive an expression for the stiffness matrix of the rectangular element.

1.4 Examine the plane stress triangular element developed by Turner et. al.‡ and provide a concise summary of its derivation.

Chapter 2

2.1 Extend the line element problem of Example 2.1 (see p. 19) in three dimensions.

2.2 Write an essay summarizing the analysis techniques associated with the finite element procedure for discretizing a physical problem.

Chapter 3

3.1 Derive the material stress–strain matrix for the plane strain isotropic case by setting the appropriate strain components to zero in the three-dimensional strain–stress (compliance) matrix.

Chapter 4

4.1 A 10 × 10 square plate with a circular hole of diameter 1 is subjected to a uniaxial stress field of 100 as shown in Fig. A2. Plot the stress σ_Y along AB and σ_X along CD. Compare results with an available analytical solution.

4.2 A thick cylinder (Fig. A3) with an inner radius of 1 is subjected to internal pressure $p = 10$. Plot the hoop and radial stresses along any radius and compare them with results from the analytical solution.

4.3 A 10 × 2 × 1 cantilever plate (Fig. A4) is subjected to the load condition 1) $m = 100$ and 2) $p = 100$. Calculate the deflection at the center point and at

*Courant, R., "Methods for the Solution of Problems of Equilibrium and Vibrations," *Transactions of the American Mathematical Society*, June 1942, pp. 1–23.

†Argyris, J. H., "Energy Theorems and Structural Analysis Part 1," *Aircraft Engineering*, Vols. 26, 27, Oct. 1954, May 1955.

‡Turner, M. J., Clough, R. W., Martin, H. C., and Topp, L. T., "Stiffness and Deflection Analysis of Complex Structures," *Journal of Aeronautical Sciences*, Vol. 25, No. 9, 1956, pp. 805–823.

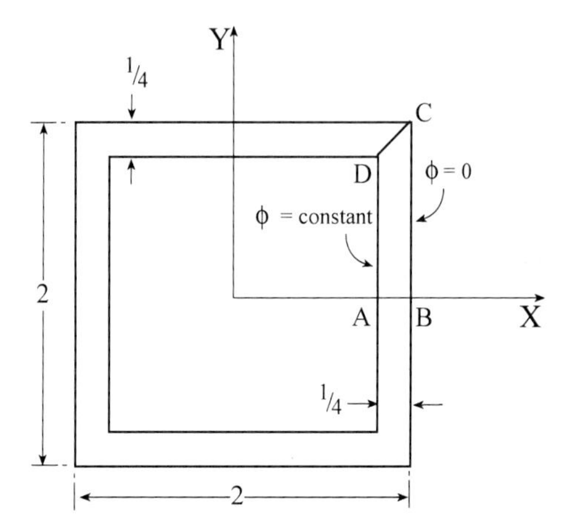

Fig. A1 Square hollow shaft.

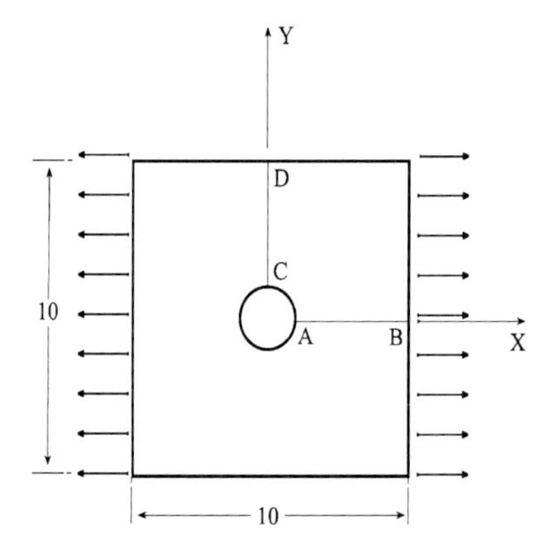

Fig. A2 Square plate with circular hole.

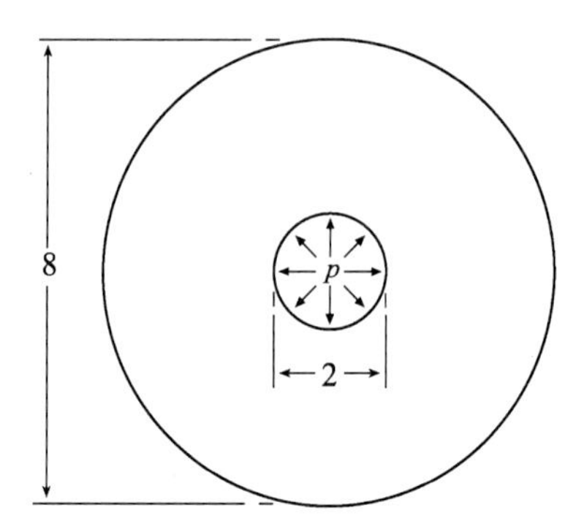

Fig. A3 Thick cylinder subjected to internal pressure.

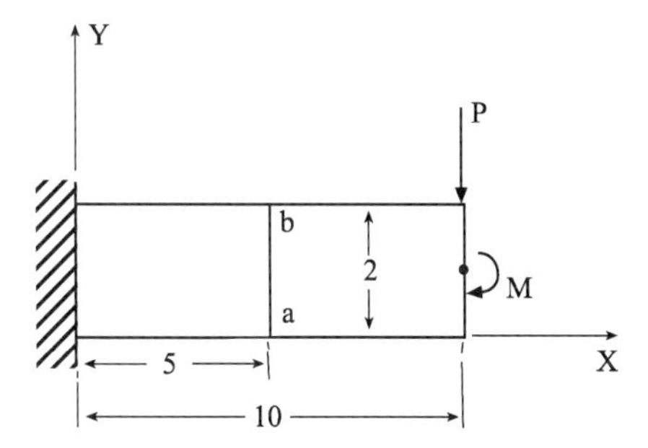

Fig. A4 Cantilever plate with end loading conditions.

the free end. Also plot the distribution of the stresses σ_X and σ_{XY} along the section *a-b*. Compare the calculated deflection and stress values with simple bending theory, allowing for approximation to shear deflection.

4.4 Perform a solution convergence study for a simply supported plate (Fig. 4.23) using constant strain and linear strain elements.

4.5 Perform a similar study as in the preceding exercise for the cantilever cylindrical shell (Fig. 4.24, p. 100) using constant strain, linear strain, and isoparametric triangular shell elements.

Chapter 5

5.1 Derive expressions for Coriolis and centripetal acceleration matrices for a three-dimensional beam element with the following properties: $l = 6$, $A = 1.0$, $E = 30 \times 10^6$, $I_y = 1/12$, $I_z = 1/24$, and $\rho = 0.1666$.

5.2 Find the natural frequencies and mode shapes of a simple cantilever beam (Fig. A5) spinning at the rate of $\Omega_z = 0.1$ Hz and $\Omega_R = 0.1$ Hz ($\Omega_x = \Omega_y = \Omega_z = 0.057735$ Hz), where $l = 60$, $A = 1.0$, $E = 30 \times 10^6$, $I_y = 1/12$, $I_z = 1/24$, and $\rho = 0.1666$.

5.3 For the spinning cantilever square plate example problem (Fig. A6), calculate the natural frequencies and mode shapes for a spin rate of $\Omega_x = 100$ rad/s, where $l_x = l_y = 10$, $t = 0.1$, $E = 10 \times 10^6$, $\nu = 0.3$, and $\rho = 0.259 \times 10^{-3}$.

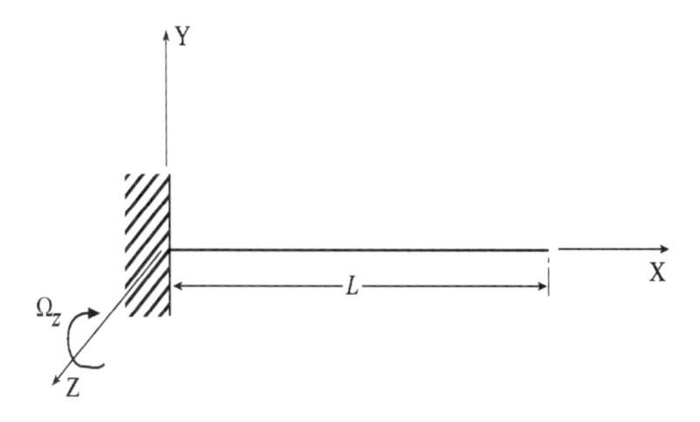

Fig. A5 Spinning cantilever beam.

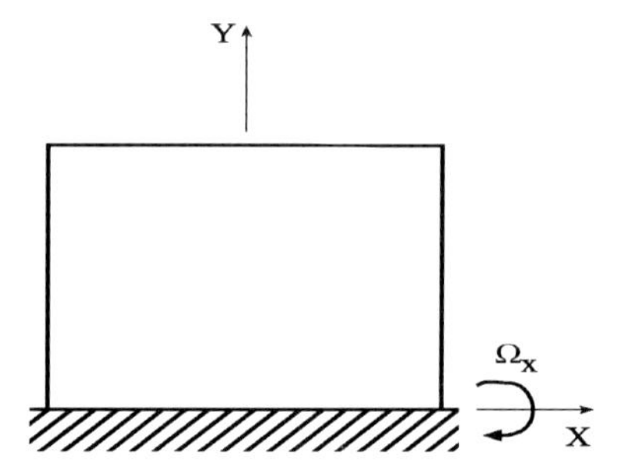

Fig. A6 Spinning cantilever plate.

Chapter 6

6.1 Derive higher-order dynamic correction matrices $\boldsymbol{K}_2$ and $\boldsymbol{M}_4$ for a three-dimensional line element.

6.2 Establish modal convergence characteristics for the cantilever beam of Exercise 5.2 using both finite element and dynamic element methods.

6.3 Calculate the natural frequencies and mode shapes for a simply supported rectangular membrane (Fig. A7) using the finite element and dynamic element method procedures with $\rho = 1$, $t = 1$, and $\sigma = 1$. Vary mesh sizes to study convergence characteristics.

Chapter 7

7.1 Find the minimum bandwidth for the problem of Fig. A8 by renumbering nodes and taking into consideration second-order connectivity.

7.2 For the frame shown in Fig. A9 determine the minimum bandwidth.

7.3 For the space frame with rigid connections, find maximum nodal deflections and element stresses when it is subjected to loads as shown in Fig. A10. The physical parameters for the space frame are Young's modulus $E = 30.24 \times 10^6$, Poisson's ratio $\nu = 0.2273$, cross-sectional area $A = 25.13$, and member length $l = 120$.

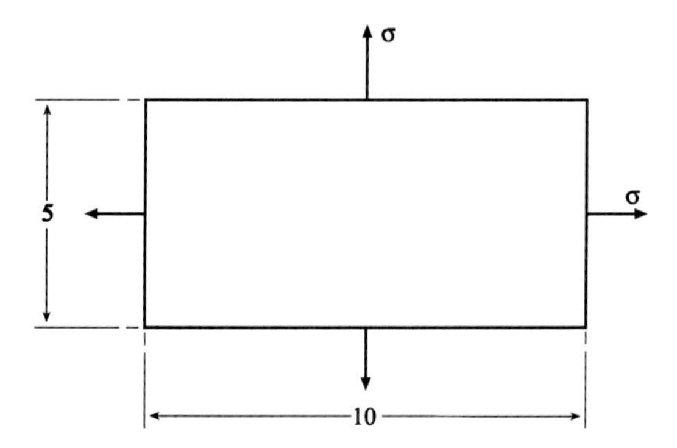

Fig. A7 Rectangular prestressed membrane.

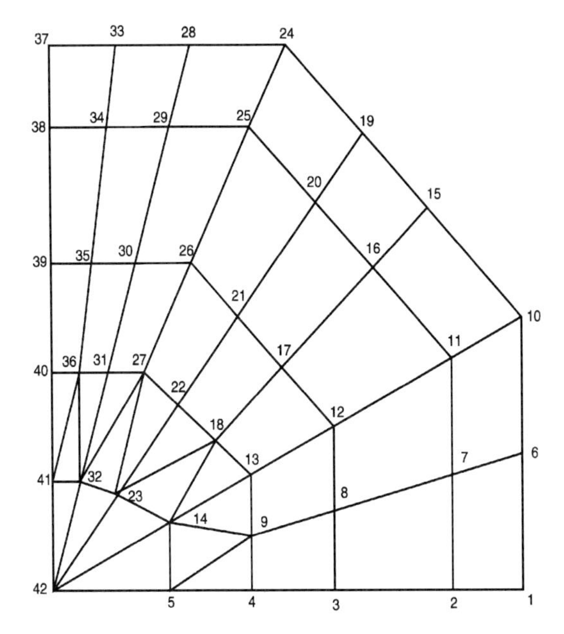

Fig. A8 Simply connected structure.

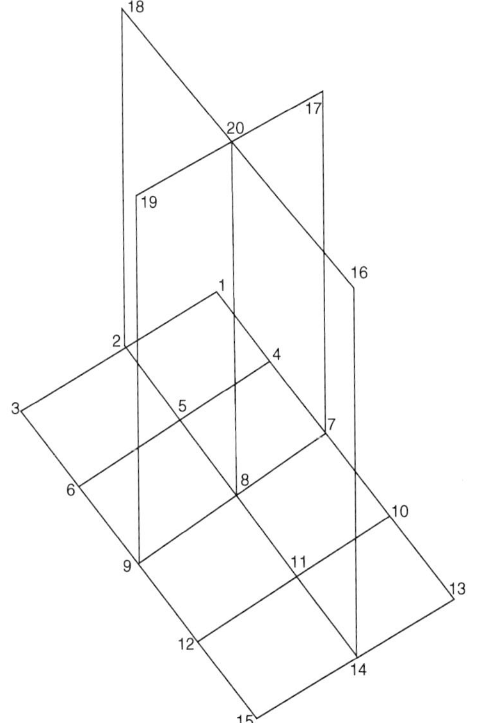

Fig. A9 Space frame structure.

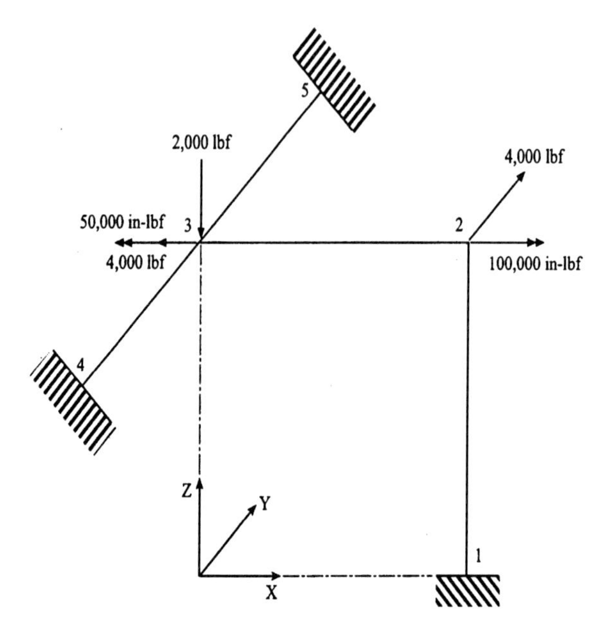

Fig. A10 Space frame structure.

7.4 A uniform cantilever beam with an inclined support at its right end (Fig. A11) is subjected to a concentrated load at its midpoint. Calculate the maximum nodal deformation and element stresses for $\alpha = 30$ deg and $\alpha = 45$ deg. The physical parameters for the beam are as follows:

Young's modulus $E = 30 \times 10^6$
cross-sectional area $A = 1.0$
Moment of inertia:
about the y axis $I_y = 1/12$
about the z axis $I_z = 1/24$
length $L = 10.0$
load $P = 100.0$

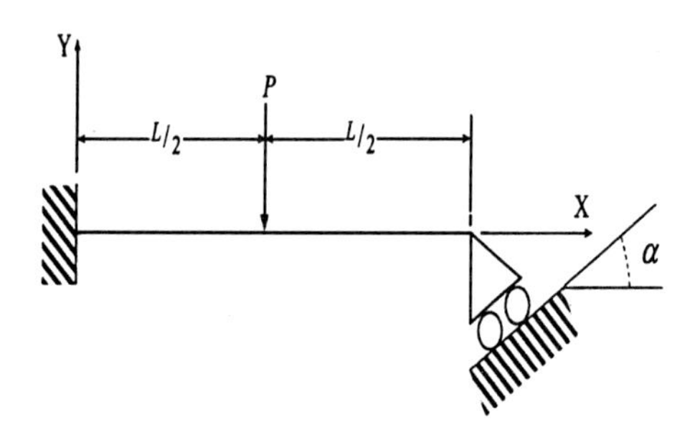

Fig. A11 Cantilever beam with an inclined support.

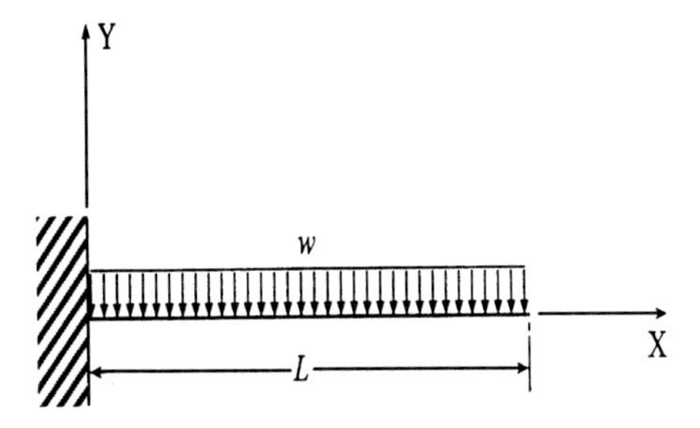

Fig. A12 Cantilever beam subjected to uniformly distributed load.

7.5 A cantilever beam (Fig. A12) is subjected to a uniformly distributed load of 10 lb per unit length. Calculate the tip deflection assuming that the beam has the same geometrical and material properties as in Exercise 7.4.

Chapter 8

8.1 A cantilever beam (Fig. A13) with basic properties as in Exercise 5.2 is subjected to two concentrated loads. Calculate the tip deflection by using Cholesky and also an iterative procedure.

8.2 A cantilever L-beam with basic properties as in Exercise 5.2 is subjected to a concentrated load as shown in Fig. A14. Calculate the tip deflection by employing the Cholesky as well as an iterative method.

8.3 Solve Exercise 7.2 (Fig. A9, p. 371) for minimum fill by using the minimum degree and also the domain decomposition method. Also provide the outlines of the multifrontal method for effective equation solution.

Chapter 9

9.1 Calculate the natural frequencies and mode shapes for Exercise 8.2 assuming a density $\rho = 0.1666$. Use the various eigenproblem solution procedures such as 1) Lanczos and 2) PSI procedures.

9.2 Using the Lanczos as well as the PSI techniques, calculate the frequencies and mode shapes for the beam of Exercise 8.2 when subjected to the following spin rates: $\Omega_y = 0.1$ Hz and $\Omega_z = 0.2$ Hz.

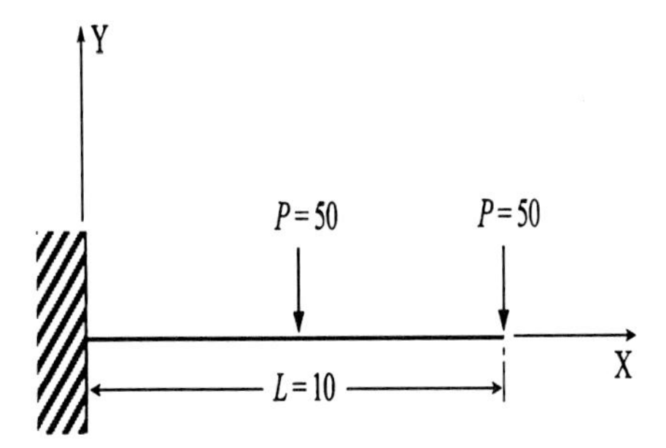

Fig. A13 Cantilever beam subjected to concentrated loads.

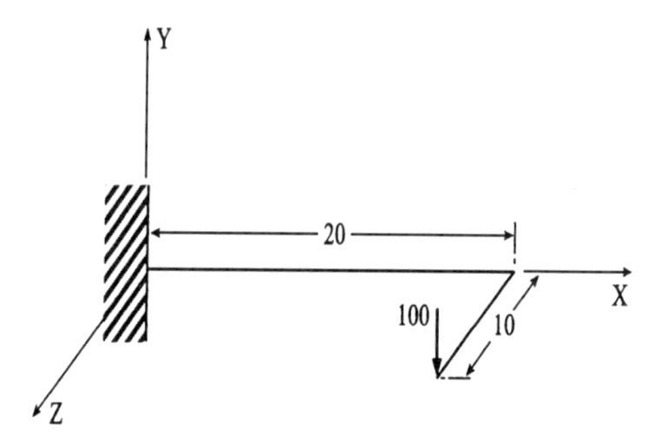

Fig. A14 Cantilever L-beam subjected to tip load.

9.3 Find the natural frequencies and mode shapes of a free-free simple beam having the basic properties of Exercise 5.2.

9.4 A square plate fixed along two opposite edges, and having the same basic properties as in Exercise 5.3, is subjected to uniform temperature loading of 1) 10 deg and 2) 15 deg. Find the vibrational characteristics of the plate using a coefficient of thermal expansion of $\alpha = 6.6 \times 10^{-6}$.

9.5 A square, free-free composite plate has the following basic properties: side length $L = 12$, plate thickness $t = 0.24$, mass density $\rho = 0.1475 \times 10^{-3}$, and composite stacking = [30 deg/−30 deg/−30 deg/30 deg].

The material properties are

$E11 = 30.1169 \times 10^6$	$E12 = 5.5656 \times 10^5$	$E14 = 0.0$
$E22 = 2.6503 \times 10^6$	$E24 = 0.0$	$E44 = 7.84 \times 10^5$
$E55 = 7.84 \times 10^5$	$E56 = 0.0$	$E66 = 7.84 \times 10^6$
$\alpha_X = 0.24 \times 10^{-5}$	$\alpha_Y = 7.84 \times 10^5$	$\alpha_{XY} = 0.0$

Find the first 12 natural frequencies and mode shapes.

9.6 The square composite plate of Exercise 9.5 is subjected to a uniformly varying temperature range along the X axis of 30° at the two edges and the temperature at the center of 200°. Perform a free vibration analysis of the thermally prestressed plate.

9.7 Find the natural frequencies and modes of a cube with the following important data parameters: side length $L = 10$, Young's modulus $E = 1.0 \times 10^7$, Poisson's ratio $\nu = 0.3$, and mass density $\rho = 2.349 \times 10^{-4}$.

9.8 The cantilever beam of Exercise 7.4 (Fig. A11, p. 372) is subjected to a temperature of -1.0°. Compute the buckling load for the beam.

9.9 Perform the buckling analysis of a simple truss (Fig. A15) with the following important data parameters: cross sectional area $A = 0.1$, Young's modulus $E = 1.0 \times 10^3$, and Poisson's ratio $\nu = 0.2$.

9.10 The square plate of Exercise 5.3 is subjected to a uniform unit stress acting along the two edges parallel to the y axis. Calculate the buckling load for the plate.

Chapter 10

10.1 The cantilever beam of Exercise 5.2 is subjected to a time-dependent forcing function (Fig. A16). Perform a dynamic response analysis of the beam.

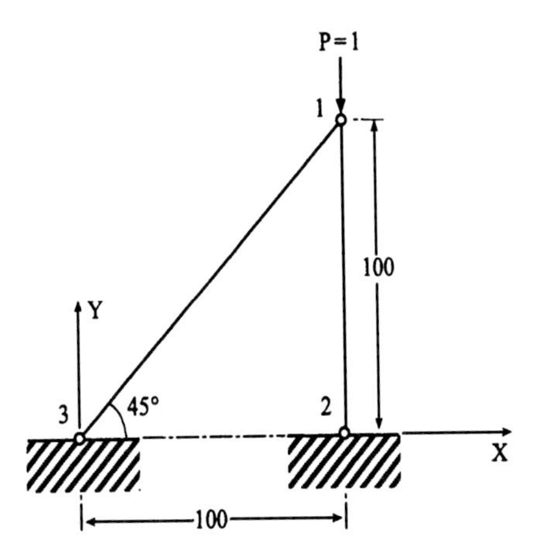

Fig. A15 Simple truss structure.

10.2 A shallow spherical cap (Fig. A17) is subjected to a dynamic load. Perform a linear dynamic analysis using the following important data parameters:

Young's modulus E $= 1.0 \times 10^7$
Poisson's ratio ν $= 0.3$
Thickness t $= 0.01$
Radius R $= 4.0$
Height H $= 0.1$
Length a $= 1.0$
Mass density ρ $= 2.45 \times 10^{-4}$
Dynamic load p_0 $= 10$
Time for load application $\Delta t = 0.5$

Chapter 11

11.1 A beam clamped at both ends (Fig. A18) is subjected to a concentrated load at the center. Perform a nonlinear static and dynamic response analysis

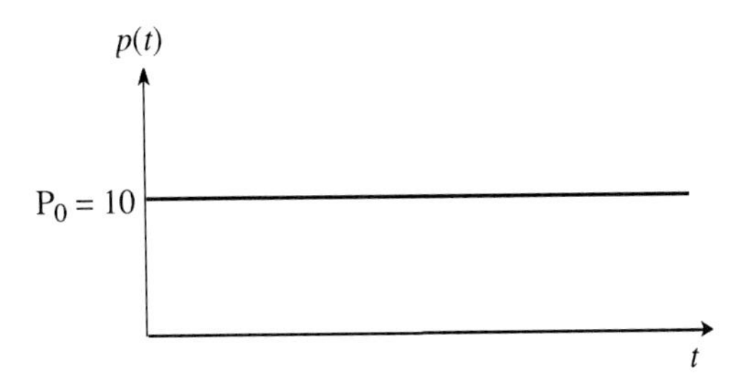

Fig. A16 Beam forcing function.

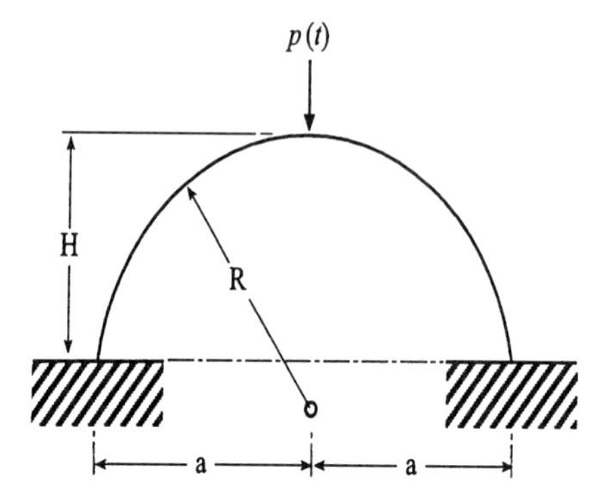

Fig. A17 Clamped spherical cap with apex load.

of the beam involving large displacement and rotation. Important data parameters are

Young's modulus E	$= 30 \times 10^6$
Length a	$= 1.0$
Cross sectional area A	$= 0.125$
Width b	$= 1.0$
Mass density ρ	$= 2.5389 \times 10^{-4}$
Static load p	$= 700$
Δp	$= 10$
Dynamic load p_0	$= 640$
Δt	$= 2.0 \times 10^{-6}$
Load duration t_z	$= 5.0$

11.2 A clamped rectangular plate is subjected to a uniformly distributed load. Perform a nonlinear static analysis and compare with a linear analysis. Important data parameters are Young's modulus $E = 10 \times 10^6$, side lengths a and $b = 20$ and 30, Poisson's ratio $\nu = 0.3$, and uniform pressure $q = 100$, in equal increments of $\Delta q = 4$.

11.3 Perform an elastoplastic analysis of the plate in Exercise 11.2. Assume $\sigma_{yp} = 35 \times 10^3$, $h_p = 0.0$, and $q = 200$.

11.4 For the spherical cap in Exercise 10.2, perform a nonlinear static and dynamic response analysis for the applied load in both cases of magnitude, $p = p_0 = 100$.

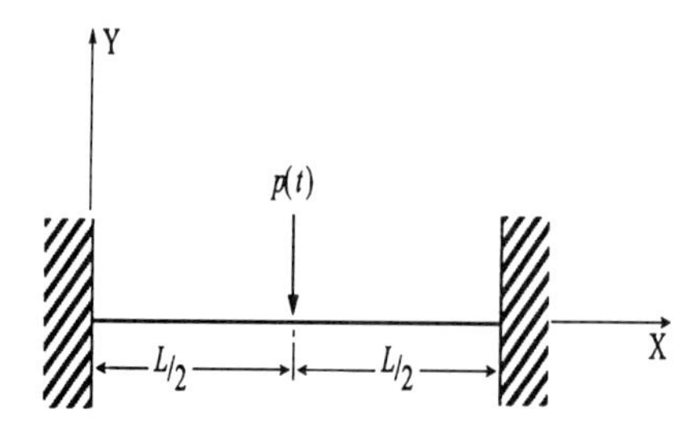

Fig. A18 Clamped beam with concentrated load.

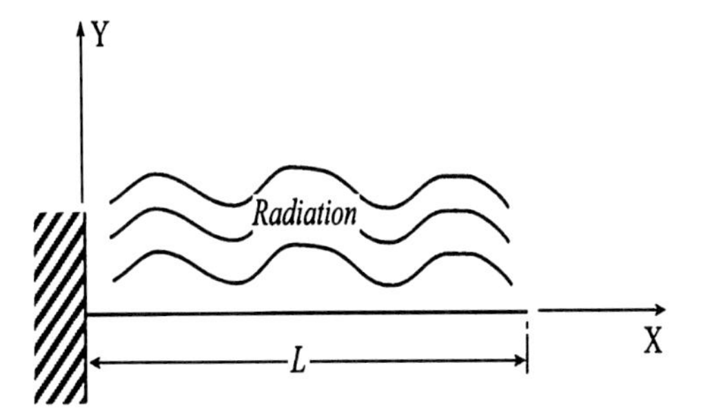

Fig. A19 Cooling fin with radiation.

Chapter 12

12.1 Calculate the maximum stresses in Exercise 11.1.

12.2 For Exercise 11.2, calculate the distribution of maximum and principal stresses.

12.3 Calculate elastoplastic stress distribution for Exercise 11.3.

12.4 Perform a complete stress analysis for Exercise 11.4.

12.5 Solve the problem in Exercise 11.3 for a yield stress reduced to 30,000.

12.6 Using the following constraints, solve the problem in Example 12.2 (p. 271): 1) minimum beam cross sectional area = 0.4 and 2) minimum shell thickness = 0.05.

Chapter 13

13.1 Conduct a heat transfer analysis for a rectangular plate (2×1) subjected to transient heating and having the relevant data parameters as in the numerical example problem of the square plate.

13.2 Perform a nonlinear steady-state radiation analysis of a cooling fin (Fig. A19) using heat transfer line elements. Associated important data parameters are coefficient of conductivity $k = 132.0$, wall temperature $t_w = 1500$, length $L = 1$, area $A = 0.001365$, Stefan–Boltzman constant $\delta = 0.1713 \times 10^{-8}$, and emissivity $\eta = 0.6$.

13.3 Perform a nonlinear transient heat transfer analysis for a composite box with radiation at the top and transient heat flow at the bottom, subjected to a pulse heat flow (Fig. A20). Relevant data parameters are length $L = 1.0$, Stefan–Boltzman constant $\delta = 0.1713 \times 10^{-8}$, emissivity $\eta = 0.6$, and

Material 1	Material 2
Coefficient of conductivity k	Coefficient of conductivity k
$k_{XX} = 10.5$	$k_{XX} = 2.1$
$k_{YY} = 10.5$	$k_{YY} = 2.1$
$k_{ZZ} = 10.5$	$k_{ZZ} = 2.1$
Thickness $t = 0.5$	Thickness $t = 1.0$

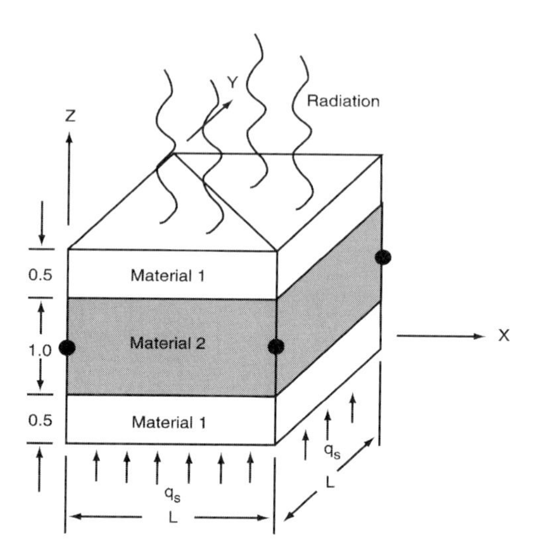

Fig. A20 Composite box subjected to a transient heat flow.

13.4 For the clamped-clamped beam example problem (Fig. 13.7, p. 244) perform a coupled heat transfer and structural analysis with wall temperature $T_{w1} = 10$ and $T_{w9} = 5$, respectively.

13.5 Perform a similar analysis of the plate example problem (Fig. 13.8, p. 245) with nodal temperatures $T_1 = 10.0$ and $T_{81} = 5.0$.

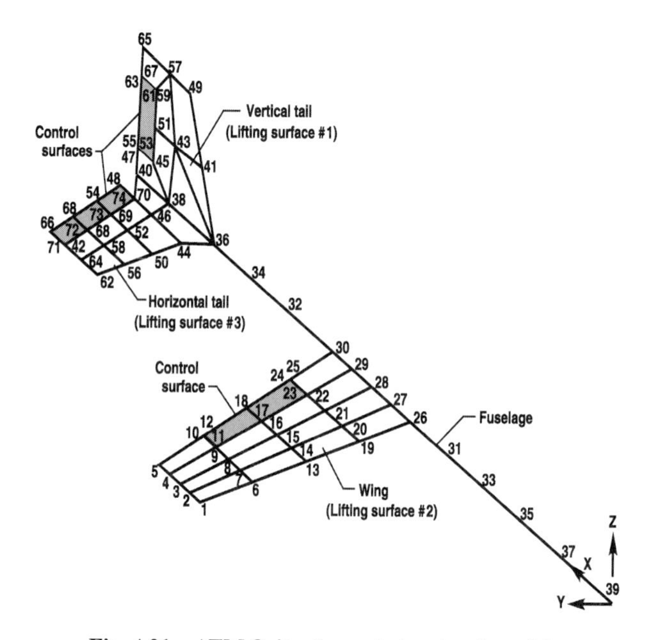

Fig. A21 ATM finite element structural model.

Chapter 14

Perform an aeroelastic and aeroservoelastic analysis of an aircraft test model (ATM) for Mach 0.75. Figure A21 depicts the finite element structural model, the structural data being given in Example A1; associated generalized mass generation data are shown in Example A2. Further, the following figures and examples depict associated aerodynamic and control schemes.

1) Figure A22 shows the direct surface interpolation scheme of modal data from structural to aerodynamic grid points.
2) Figure A23 depicts the aerodynamic boxes.
3) Figure A24 shows the details of aerodynamics panels.
4) Figure A25 depicts the aerodynamic elements.

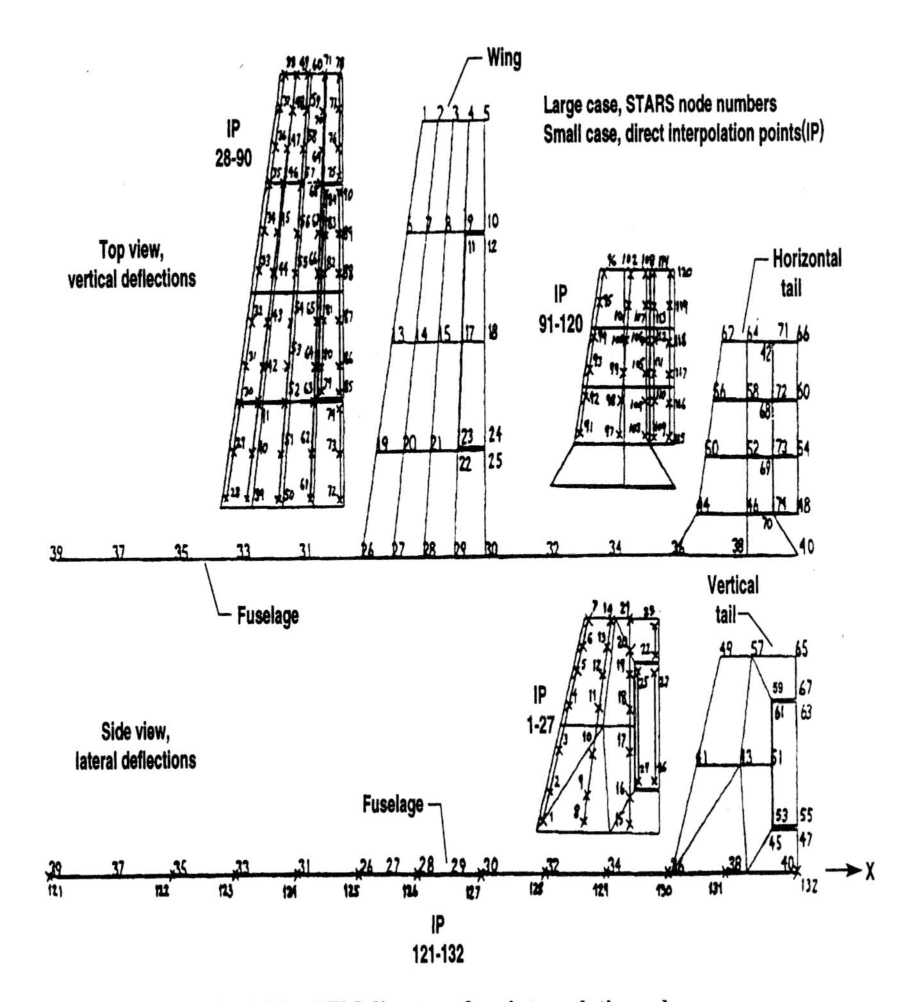

Fig. A22 ATM direct surface interpolation scheme.

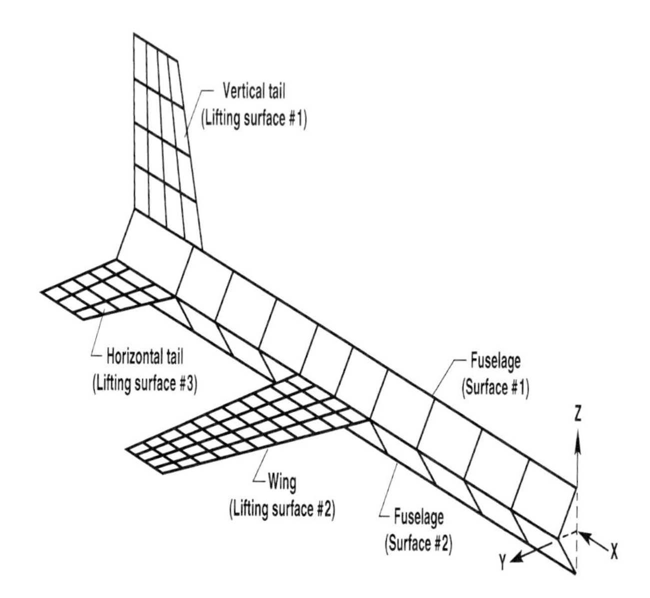

Fig. A23 ATM aerodynamic boxes.

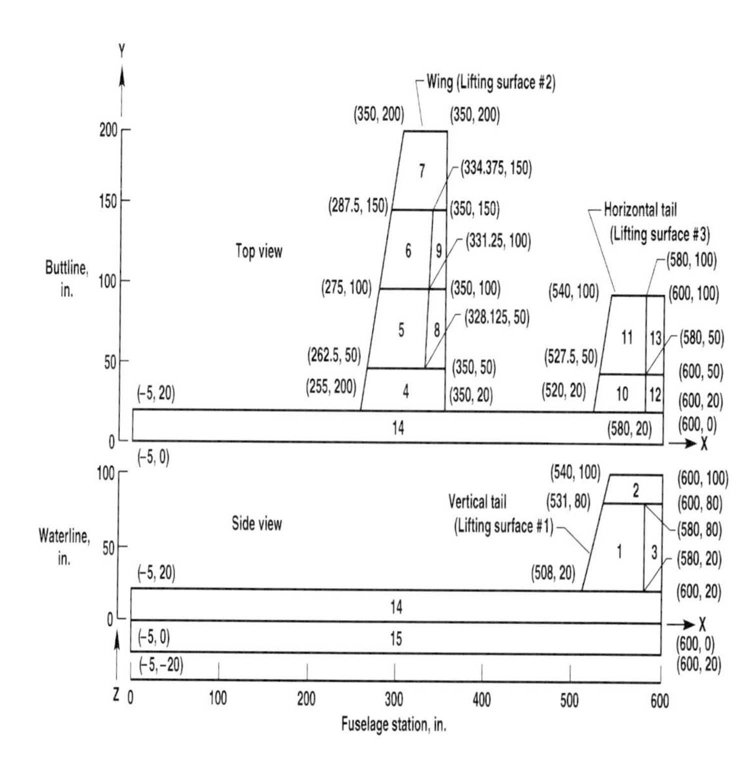

Fig. A24 ATM aerodynamic panels.

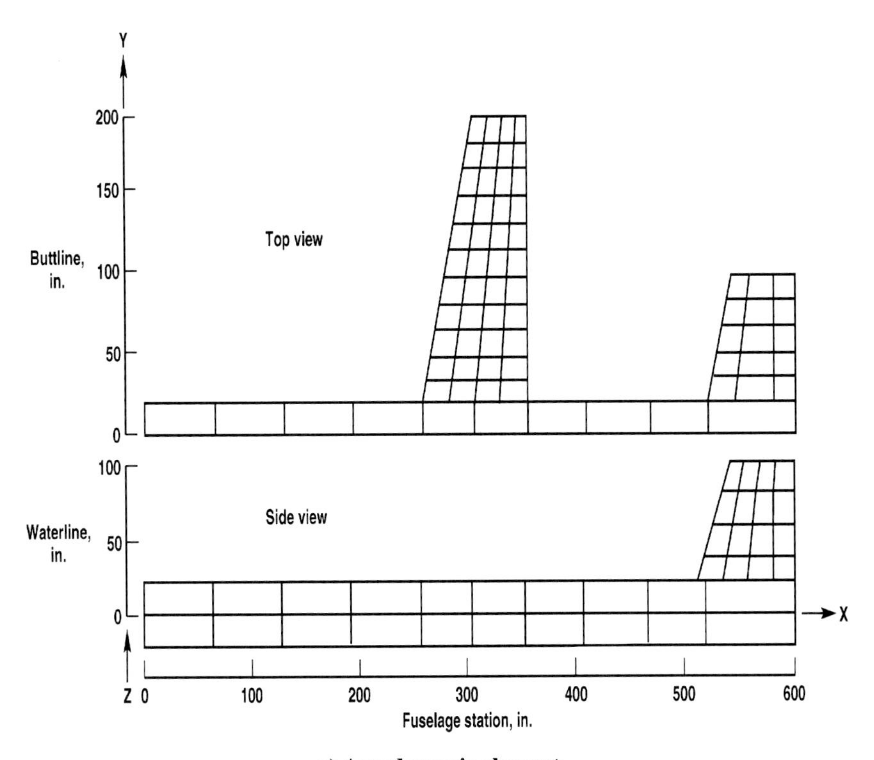

a) Aerodynamic element

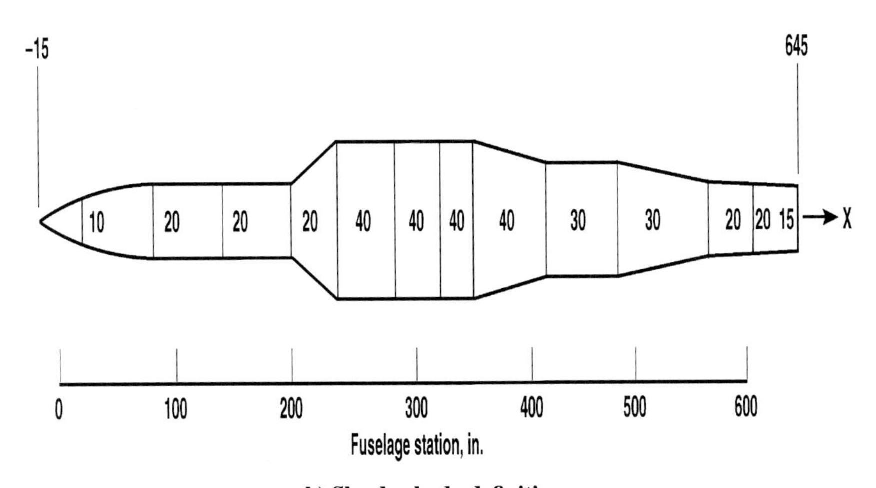

b) Slender-body definitions

Fig. A25 ATM aerodynamic elements.

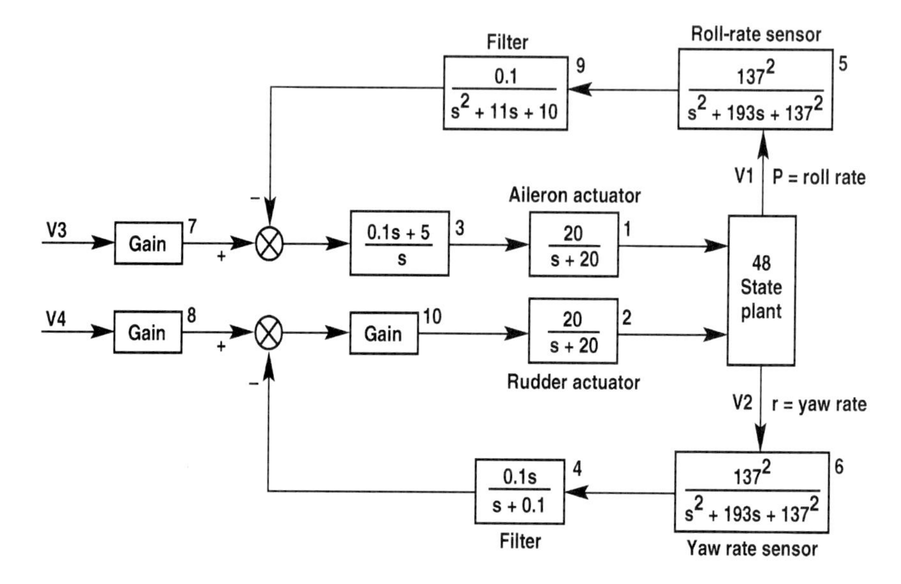

Fig. A26 ATM lateral mode analog control system.

5) Example A3 shows input data for calculation of unsteady aerodynamic forces.
6) Examples A4 and A5 provide details on choice of modes for various analysis.
7) Examples A6 and A7 show Padé polynomials curve fitting input data for flutter and frequency response analysis.
8) Figure A26 shows the analog control system.
9) Examples A8 and A9 show controls input data for open-loop and closed-loop cases, respectively.

A summary of some relevant solution results are given in Tables A1 and A2.

Table A1 Aircraft test model: An aeroelastic antisymmetric analysis using a direct interpolation for AEROS[a] paneling

		ASE	
Mode	Instability number	Velocity, KEAS	Frequency rad/s
Fuselage first bending	Flutter 1	474.1	77.3
Vertical fin first bending	Flutter 2	728.9	136.2
Fin first bending	Divergence 1	651.9	0.0
Fin first torsion	Divergence 2	728.9	0.0

[a]Wind tunnel data.

Table A2 Aircraft test model gain and phase margins

Mode	Phase crossover, rad	Gain margin, dB
Roll	20.20	78.26
Yaw	2.90	−0.80

Example A1: Structural Finite Element Data for an ATM

```
AIRCRAFT TEST MODEL
C PRIMARY STRUCTURAL DATA, MODAL INTERPOLATION AND RIGID BODY/CONTROL MODES
C DATA MAY BE OBTAINED FROM STARS MANUAL
C
C ANTISYMMETRIC HALF MODEL
C IINTP  = 1, DIRECT INTERPOLATION OF MODAL DATA
C
C NCNTRL = 5, FIRST THREE TO GENERATE PERFECT RIGID BODY MODES
C Y TRANSLATION, ROLL AND YAW, PLUS AILERONS AND RUDDER
CONTROL
C MODES.
C/////////////////////////////////////////////////////////////////////////
 No. of Node = 74,  No. of Element = 149,  No. of Material = 1,
 No. of Line Ele. Properties = 22,  No of Shell Thickness  = 5,
 No. of Control Modes = 5 ( 3 Rigid, 2 Control modes),
 No. of Modal Interpolation Points = 132
$ Node    X         Y          Z        ZDBC(Zero Deflection Boundary Condition)
    1  300.0000  200.0000    0.0000        0    0    0    0    0    0
    2  312.5000  200.0000    0.0000        0    0    0    0    0    0
    3  325.0000  200.0000    0.0000        0    0    0    0    0    0
    4  337.5000  200.0000    0.0000        0    0    0    0    0    0
    5  350.0000  200.0000    0.0000        0    0    0    0    0    0
    6  287.5000  150.0000    0.0000        0    0    0    0    0    0
    7  303.1250  150.0000    0.0000        0    0    0    0    0    0
    8  318.7500  150.0000    0.0000        0    0    0    0    0    0
    9  334.3750  150.0000    0.0000        0    0    0    0    0    0
   10  350.0000  150.0000    0.0000        0    0    0    0    0    0
   11  335.3750  149.0000    0.0000        0    0    0    0    0    0
   12  350.0000  149.0000    0.0000        0    0    0    0    0    0
   13  275.0000  100.0000    0.0000        0    0    0    0    0    0
   14  293.7500  100.0000    0.0000        0    0    0    0    0    0
   15  312.5000  100.0000    0.0000        0    0    0    0    0    0
   16  331.2500  100.0000    0.0000        0    0    0    0    0    0
   17  332.2500  100.0000    0.0000        0    0    0    0    0    0
   18  350.0000  100.0000    0.0000        0    0    0    0    0    0
   19  262.5000   50.0000    0.0000        0    0    0    0    0    0
   20  284.3750   50.0000    0.0000        0    0    0    0    0    0
   21  306.2500   50.0000    0.0000        0    0    0    0    0    0
   22  328.1250   50.0000    0.0000        0    0    0    0    0    0
   23  329.1250   51.0000    0.0000        0    0    0    0    0    0
   24  350.0000   51.0000    0.0000        0    0    0    0    0    0
   25  350.0000   50.0000    0.0000        0    0    0    0    0    0
   26  250.0000    0.0000    0.0000        1    0    1    0    1    0
   27  275.0000    0.0000    0.0000        1    0    1    0    1    0
   28  300.0000    0.0000    0.0000        1    0    1    0    1    0
   29  325.0000    0.0000    0.0000        1    0    1    0    1    0
   30  350.0000    0.0000    0.0000        1    0    1    0    1    0
   31  200.0000    0.0000    0.0000        1    0    1    0    1    0
   32  400.0000    0.0000    0.0000        1    0    1    0    1    0
   33  150.0000    0.0000    0.0000        1    0    1    0    1    0
   34  450.0000    0.0000    0.0000        1    0    1    0    1    0
   35  100.0000    0.0000    0.0000        1    0    1    0    1    0
   36  500.0000    0.0000    0.0000        1    0    1    0    1    0
```

```
  37   50.0000     0.0000     0.0000       1   0   1   0   1   0
  38  560.0000     0.0000     0.0000       1   0   1   0   1   0
  39    0.0000     0.0000     0.0000       1   0   1   0   1   0
  40  600.0000     0.0000     0.0000       1   0   1   0   1   0
  41  520.0000     0.0000    50.0000       1   0   1   0   1   0
  42  580.0000   100.0000     0.0000       0   0   0   0   0   0
  43  555.0000     0.0000    50.0000       1   0   1   0   1   0
  44  520.0000    20.0000     0.0000       0   0   0   0   0   0
  45  580.0000     0.0000    20.0000       1   0   1   0   1   0
  46  560.0000    20.0000     0.0000       0   0   0   0   0   0
  47  600.0000     0.0000    20.0000       1   0   1   0   1   0
  48  600.0000    20.0000     0.0000       0   0   0   0   0   0
  49  540.0000     0.0000   100.0000       1   0   1   0   1   0
  50  526.6666    46.6667     0.0000       0   0   0   0   0   0
  51  580.0000     0.0000    50.0000       1   0   1   0   1   0
  52  560.0000    46.6667     0.0000       0   0   0   0   0   0
  53  581.0000     0.0000    21.0000       1   0   1   0   1   0
  54  600.0000    46.6667     0.0000       0   0   0   0   0   0
  55  600.0000     0.0000    21.0000       1   0   1   0   1   0
  56  533.3331    73.3333     0.0000       0   0   0   0   0   0
  57  565.0000     0.0000   100.0000       1   0   1   0   1   0
  58  560.0000    73.3333     0.0000       0   0   0   0   0   0
  59  580.0000     0.0000    80.0000       1   0   1   0   1   0
  60  600.0000    73.3333     0.0000       0   0   0   0   0   0
  61  581.0000     0.0000    79.0000       1   0   1   0   1   0
  62  540.0000   100.0000     0.0000       0   0   0   0   0   0
  63  600.0000     0.0000    79.0000       1   0   1   0   1   0
  64  560.0000   100.0000     0.0000       0   0   0   0   0   0
  65  600.0000     0.0000   100.0000       1   0   1   0   1   0
  66  600.0000   100.0000     0.0000       0   0   0   0   0   0
  67  600.0000     0.0000    80.0000       1   0   1   0   1   0
  68  580.0000    73.3333     0.0000       0   0   0   0   0   0
  69  580.0000    46.6667     0.0000       0   0   0   0   0   0
  70  580.0000    20.0000     0.0000       0   0   0   0   0   0
  71  581.0000   100.0000     0.0000       0   0   0   0   0   0
  72  581.0000    73.3333     0.0000       0   0   0   0   0   0
  73  581.0000    46.6667     0.0000       0   0   0   0   0   0
  74  581.0000    20.0000     0.0000       0   0   0   0   0   0
$ ELEMENT CONNECTIVITY CONDITIONS
 IET  IEN  ND1  ND2  ND3  ND4   IMPP  ITHTH/IEPP
   2    1    6    7    2    1     1     1
   2    2    7    8    3    2     1     1
   2    3    8    9    4    3     1     1
   2    4    9   10    5    4     1     1
   2    5   13   14    7    6     1     1
   2    6   14   15    8    7     1     1
   2    7   15   16    9    8     1     1
   2    8   19   20   14   13     1     1
   2    9   20   21   15   14     1     1
   2   10   21   22   16   15     1     1
   2   11   26   27   20   19     1     1
   2   12   27   28   21   20     1     1
   2   13   28   29   22   21     1     1
   2   14   29   30   25   22     1     1
   2   15   17   18   12   11     1     2
   2   16   23   24   18   17     1     2
   1   17    1    2   39    0     1    11
   1   18    2    3   39    0     1    11
   1   19    3    4   39    0     1     1
   1   20    4    5   39    0     1     1
   1   21    6    7   39    0     1    11
   1   22    7    8   39    0     1    11
   1   23    8    9   39    0     1     1
   1   24    9   10   39    0     1     1
   1   25    6    1   39    0     1    11
   1   26    7    2   39    0     1    11
   1   27    8    3   39    0     1     1
   1   28    9    4   39    0     1     1
   1   29   10    5   39    0     1     1
   1   30   13   14   39    0     1    11
```

```
IET   = Element type
        1  for Line element
        2  for Quad shell
           element
IMPP  = Material property type
ITHTH = Shell element thickness
        type
IEPP  = Line element property
        type
```

```
  129   18.7500    3000.0    1275.0    1275.0
  130   16.5000    2250.0     975.0     975.0
  131   15.0000    1500.0     675.0     675.0
$ SHELL ELEMENT THICKNESSES
$ ITHTH    TH
    1    0.1130
    2    0.0530    ITHTH   =  Element thickness type
    3    0.0900    TH      =  Element thickness
    4    0.0400
    5    0.0100
$ MATERIAL PROPERTIES
$ IMPP  MT
    1    1
      E       POISSON RATIO  AlPHA    RHO
   1.0E+07      0.30          0.    .259E-03
$ NODAL MASS DATA
   IN  IDOF      P      IDOFE
   39    1    0.0195    3
   37    1    0.0389    3
   35    1    0.0584    3
   33    1    0.0972    3  IN  = Node number
   31    1    0.1943    3  IDOF and IDOFE = The start and end degrees of
   26    1    0.2915    3                   freedom assigned with the load P
   28    1    0.2915    3  P    = Nodal load
   30    1    0.2915    3
   32    1    0.2915    3
   34    1    0.2915    3
   36    1    0.2915    3
   38    1    0.2915    3
   40    1    0.1943    3
```

Example A2: Generalized Mass Data Input File

```
$ AERO TEST MODES, ANTISYMMETRIC VERSION
$ ISTMN NLVN GR
    4   39   386.088
note:   ISTMN interger specifying  the starting mode number,  NLVN number of laterally
 vibrating interpolation nodes,  GR  gravitational constant (in/sec**2). Input of a GR
 value is needed to convert generalized mass data into a generalized weight acceptable
 to the AEROs model.
$ LATERALLY MOVING direct interpolation output NODE NUMBERS
    1
    2
    3
    4
    5
    6
    7
    8
    9
   10
   11
   12
   13
   14
   15
   16
   17
   18
   19
   20
   21
   22
   23
   24
   25
   26
   27
```

```
121
122
123
124
125
126
127
128
129
130
131
132
```

Example A3: Input Data for Computation of Unsteady Aerodynamic Forces

```
 AIRCRAFT TEST MODEL - ANTISYMMETRIC CASE - NCNTRL = 5
 SET UP FOR ASE SOLUTION.
 DIRECT SURFACE INTERPOLATION.
 8 ELASTIC MODES FROM 11 SOLIDS (-3 AT GENMASS) + 3 PERF. RIG. + 2 CONTROL
 FILE FOR ASE, FOR FLUTTER USE CONVERT TO EXCLUDE RBC'S, FOR FRESP USE ALL
 MACH NO. = 0.90         ALTITUDE: SEA LEVEL
$ STARS 6.2.2   basic data parameters LC(1) to LC(40)
    1   13    3   10    1    0    0    0    0    0
    1    0    0    0    0    0    0    0    0    0
    1    0    0    0    0    0    0    0    0    0
    0    0    0    0    0    0    0   99    0    0
Note:   LC(1) = 1     state space solution, LC(2)= 13 maximum no of modes, LC(3) = 3
No. of lifting surfaces, LC(4) =10 No. of reduced velocities, LC(5) = 1 No. of air
densities, LC(11) = 1 normalizing mode,  LC(21) = 1 for doublet lattice method,
LC(38) =99  tape unit for data storage.
$ STARS 6.2.3 - Flag to indicate location of vectors and frequencies files
$   INV
    1
$ STARS 6.2.4 - Semichord, Mach.
$     BR       FMACH
     38.89      0.90
$ STARS 6.2.5 - Reduced velocities, LC(4) number.
$ (VBO(I),I=1,LC(4))
   11000.0    1000.0     100.0      50.0       10.0       5.0       1.0
     0.667     0.500      0.25
$ STARS 6.2.11  Maximum and minimum scales for V-g and V-f print plots.
$     GMAX,     GMIN,     VMAX,      FMAX
       .10      -.40     1400.      80.0
$ STARS 6.2.12  Air density ratios, LC(5) number values.
$ (RHOR(I), I=1,LC(5))
       1.0
$ STARS 6.2.18  Reference length and area
$     FL      ACAP
     77.78    15000.
$ STARS 6.2.19 - Doublet lattice/constant pressure methods geometrical
$ paneling and boundary condition parameters
$ NDELT      NP     NB    NCORE   N3  N4   N7
       -1             15       1         1235        0      0      1
Note:  NDELT=-1 index defining aerodynamics antisymmetrical about Y=0, NP=15
 total number of panels on all lifting surfaces, NB=1 number of slender bodies
 used for doublet lattice analysis, NCORE = 1235 problem size aer. elemt.x no. modes,
 N3= 0 print option for pressure influence coefficients, N4 = 0 print option for
 influence coefficients relating downwash on lifting surfaces to  body element
 pressures, N7=1 index specifying the calculation of pressures and generalized
 forces.
$ NP Sets of Surface Paneling Data (Figure A.26)
$ Set number 1 - Surface 1  - Vertical tail (Panel 1), Figure A.26
$ STARS 6.2.21.1    X Y Z  Translates and rotates panels
$       X0        Y0       Z0      GGMAS
                                                 90.0
$ STARS 6.2.21.2  - LCS X and Y coordinates of points defining aerodynamic panel  #1
$    X1         X2         X3         X4          Y1         Y2
     508.0      580.0      532.0      580.0       20.0       80.0
```

```
$ STARS 6.2.21.3  - LCS Z coordinate  and Number of element spanwise and chordwise
$ direction boundaries
$      Z1         Z2     NEBS    NEBC     COEFF
     0.0        0.0       4       4        0.0
Note:  X1 x coordinate of the panel inboard leading edge,  X2 x coordinate of the panel
 inboard trailing edge, X3 x coordinate of the panel outboard leading edge, X4 x
 coordinate of the panel outboard trailing edge,  Y1 y coordinate of the panel inboard
 edge, Y2 y coordinate of the panel outboard edge, Z1 z coordinate of the panel inboard
 edge, Z2 z coordinate of the panel outboard edge, NEBS number of element boundaries in
 the spanwise direction, NEBC number of element boundaries in the chordwise direction.
$ STARS 6.2.21.4  chordwise panel stations in this aero. panel (Figure A.27)
$ (TH(I), I = 1,NEBC)
      0.0     0.3333     0.6666       1.0
$ STARS 6.2.21.5  spanwise panel stations  in this aero. panel (Figure A.27)
$ (TAU(I), I = 1,NEBS)
      0.0     0.3333     0.6666       1.0
$ Set number 2 - Surface 1  - Vertical tail (Panel 2), Figure A.26
$ Repeat STARS 6.2.21.1 to 6.2.21.5
                                      90.0
     532.0      600.0      540.0      600.0       80.0      100.0
      0.0        0.0    2  5           0.0
      0.0     0.2353     0.4705     0.7059         1.0
      0.0        1.0
$ Set number 3 - Surface 1  - Vertical tail (Panel 3), Figure A.26
$ Repeat STARS 6.2.21.1 to 6.2.21.5
                                      90.0
     580.0      600.0      580.0      600.0       20.0       80.0
      0.0        0.0    4  2           0.0
      0.0        1.0
      0.0     0.3333     0.6666       1.0
$ Set number 4 - Surface 2  - Wing (Panel 4), Figure A.26
$ Repeat STARS 6.2.21.1 to 6.2.21.5
                                       0.0
     255.0      350.0      262.5      350.0       20.0       50.0
      0.0        0.0    3  5           0.0
      0.0       0.25       0.50       0.75         1.0
      0.0        0.5        1.0
$ Set number 5 - Surface 2  - Wing (Panel 5), Figure A.26
$ Repeat STARS 6.2.21.1 to 6.2.21.5
                                       0.0
     262.5    328.125      275.0     331.25       50.0      100.0
      0.0        0.0    4  4           0.0
      0.0     0.3333     0.6666       1.0
      0.0       0.34       0.66       1.0
$ Set number 6 - Surface 2  - Wing (Panel 6), Figure A.26
$ Repeat STARS 6.2.21.1 to 6.2.21.5
                                       0.0
     275.0     331.25      287.5    334.375      100.0      150.0
      0.0        0.0    4  4           0.0
      0.0     0.3333     0.6666       1.0
      0.0       0.34       0.66       1.0
$ Set number 7 - Surface 2  - Wing (Panel 7), Figure A.26
$ Repeat STARS 6.2.21.1 to 6.2.21.5
                                       0.0
     287.5      350.0      300.0      350.0      150.0      200.0
      0.0        0.0    4  5           0.0
      0.0       0.25       0.50       0.75         1.0
      0.0       0.34       0.66       1.0
$ Set number 8 - Surface 2  - Wing (Panel 8), Figure A.26
$ Repeat STARS 6.2.21.1 to 6.2.21.5
                                       0.0
   328.125      350.0     331.25      350.0       50.0      100.0
      0.0        0.0    4  2           0.0
      0.0        1.0
      0.0       0.34       0.66       1.0
$ Set number 9 - Surface 2  - Wing (Panel 9), Figure A.26
$ Repeat STARS 6.2.21.1 to 6.2.21.5
                                       0.0
    331.25      350.0    334.375      350.0      100.0      150.0
      0.0        0.0    4  2           0.0
```

```
      0.0       1.0
      0.0       0.34      0.66      1.0

$ Set number 10 - Surface 3  - Horizontal tail (Panel 10), Figure A.26
$ Repeat STARS 6.2.21.1 to 6.2.21.5
                                    0.0
   520.0     580.0     527.5     580.0     20.0     50.0
     0.0       0.0   3  3          0.0
     0.0    0.4167       1.0
     0.0       0.5       1.0
$ Set number 11 - Surface 3  - Horizontal tail (Panel 11), Figure A.26
$ Repeat STARS 6.2.21.1 to 6.2.21.5
                                    0.0
   527.5     580.0     540.0     580.0     50.0    100.0
     0.0       0.0   4  3          0.0
     0.0    0.4167       1.0
     0.0      0.34      0.66       1.0
$ Set number 12 - Surface 3  - Horizontal tail (Panel 12), Figure A.26
$ Repeat STARS 6.2.21.1 to 6.2.21.5
                                    0.0
   580.0     600.0     580.0     600.0     20.0     50.0
     0.0       0.0   3  2          0.0
     0.0       1.0
     0.0       0.5       1.0
$ Set number 13 - Surface 3  - Horizontal tail (Panel 13), Figure A.26
$ Repeat STARS 6.2.21.1 to 6.2.21.5
                                    0.0
   580.0     600.0     580.0     600.0     50.0    100.0
     0.0       0.0   4  2          0.0
     0.0       1.0
     0.0      0.34      0.66       1.0
$ Set number 14 - Fuselage surface 1  (Panel 14), Figure A.26
$ Repeat STARS 6.2.21.1 to 6.2.21.5
                                    0.0
    -5.0     600.0      -5.0     600.0     20.0      0.0
     0.0      20.0   2 11          0.0
     0.0    0.1074    0.2149    0.3223   0.4298   0.5083
  0.5868    0.6777    0.7769    0.8678   1.0000
     0.0       1.0
$ Set number 15 - Fuselage surface 2  (Panel 15), Figure A.26
$ Repeat STARS 6.2.21.1 to 6.2.21.5
                                    0.0
    -5.0     600.0      -5.0     600.0      0.0     20.0
   -20.0       0.0   2 11          0.0
     0.0    0.1074    0.2149    0.3223   0.4298   0.5083
  0.5868    0.6777    0.7769    0.8678   1.0000
     0.0       1.0
$ NB set of slender-body surface Paneling Data
$ Set number 1  slender-body surface
$ STARS 6.2.22.1  Translational value to be added to the LCS  X Y Z coordinate, in.
$     XBO,       YBO,       ZBO
     0.0        0.0        0.0
$ STARS 6.2.22.2    LCS Z and Y coor. Number of slender-body element boundaries. body
$ vibration in Z direction.  Flag of body vibration in Y direction. First and last
$ aerodynamic element on the interference panel associated with this slender body.
$    ZSC        YSC     NF   NZ   NY    COEFF    MRK1   MRK2
    0.0        0.0     14    0    1     0.0      76     95
$ STARS 6.2.22.3   Slender-body element stations
$ F(I) I=1,NF
  -15.000    25.000    85.000   145.000   205.000   245.000
  295.000   335.000   365.000   425.000   485.000   565.000
  605.000   645.000
$ STARS 6.2.22.4   Slender-body radii
$ RAD(i),I=1,NF
     0.0      10.0      20.0      20.0      20.0      40.0
    40.0      40.0      40.0      30.0      30.0      20.0
    20.0      15.0
Note:  ZSC local z coordinate of the body axis, YSC local y coordinate of the body
 axis, NF number of slender-body element boundaries along the axis, NZ flag for body
 vibration in the Z direction, NY flag for body vibration in the Y direction, COEFF
```

entered as 0.,MRK1 index of the last aerodynamic element on the last interference panel associated with this slender body, MRK2 index of the first aerodynamic element of the last interference panel associated with this slender body, F(J) x coordinate of the body station defining a slender-body element in local coordinates, RAD(I) radii of body elements at the stations F(J).

```
$ STARS 6.2.23  General parameters aerodynamics data.
$ NSTRIP    NPR1     JSPECS    NSV    NBV    NYAW
    1        0         0        2      16      1
```
Note: NSTRIP number of chordwise strips of panel elements on all panel (LC(8)=0, set NSTRIP=1), NPR1 print option for pressures, JSPECS index descibing plane aerodynamic symmetry about Z=0, NSV number of strips lying on all vertical panels on the symmetric plane Y=0, NBV number of elements on all vertical panels lying on the plane Y=0, NYAW symmetry flag(1 if NDELT=-1 for antidymmetric about Y=0).
```
$ STARS 6.2.23.1  For NSTRIP =1 a blank card is used.

$ Format STARS 6.2.24.1 to 6.2.29.2 are repeated for LC(3) set of Primary surface Data
$ Lifting surface number 1 - 1st of the LC(3) set - Vertical tail, rotated into X-Y
$ plane
$ STARS 6.2.24.1
$ KSURF    NBOXS    NCS
   T        16       1
```
Note: KSURF flag indicating control surfaces on a primary surface, NBOXS number of elements on this surface, including those elements on control surfaces, NCS number of control surfaces on primary surface.
```
$ STARS 6.2.24.2 - Interpolation data line data on Lifting Surface 1
$ NLINES    NELAXS     NICH     NISP
  4       0      1        1
```
Note: NLINES number of lines along which input modal vector data are prescribed, NELAXS index defining input vector components, NICH index defining the chordwise interpolation/extrapolation from the input vector to aerodynamic elements(0-linear, 1-quadratic, 2-cubic),NISP index defining the spanwise interpolation from the input vector to aerodynamic elements.
```
$ STARS 6.2.25.1 to 6.2.25.2 are repeated for NLINES subsets of data
$ Input modal vector data to be applied to the interpolation of deflections for
$ primary and control surface aerodynamic elements
$ STARS 6.2.25.1
$ NGP   XTERM1       YTERM1   XTERM2    YTERM2  Line 1
   7     502.0         2.0     542.0     100.0
```
Note: NGP number of points on an input vector line, XTERM1 X coordinate specifying the inboard end of an input vector line in the LCS, YTERM1 Y coordinate specifying the inboard end of an input vector line in the LCS, XTERM2 X coordinate specifying the outboard end of an input vector line in the LCS, YTERM2 Y coordinate specifying the outboard end of an input vector line in the LCS.
```
$ STARS 6.2.25.2 - Spanwise coordinate of the interpolation point along
$ the input vector line, inboard to outboard.
$  YGP(I),I=1,NGP   -   Line # 1
   2.0   17.0   37.0   50.0   75.0   85.0   100.0
```
Note: YGP spanwise coordinate of a point along an input vector line, going inboard to outboard in the LCS.
```
$ Line 2 Repeat 6.2.25.1 to 6.2.25.2   -   Line # 2
    7      542.0        2.0      560.0     100.0
   2.0      17.0       37.0       50.0      75.0    85.0   100.0
$ line 3 Repeat 6.2.25.1 to 6.2.25.2   -   Line # 3
    7      578.0        2.0      578.0     100.0
   2.0      17.0       40.0       50.0      73.0    82.0   100.0
$ line 4 Repeat 6.2.25.1 to 6.2.25.2   -   Line # 4
    2      598.0       82.0      598.0     100.0
 82.0     100.0
$ STARS 6.2.27 LCS control surface hinge line coordinate
$    (X1(I), Y1(I), X2(I), Y2(I),    I = 1,NCS)
    580.0      20.0      580.0       80.0
$ STARS 6.2.28 - Number of lines along which input modal vectors are prescribed
$ NLINES NELAXS NICH NISP
   2   0   1   1
$ STARS 6.2.29.1 to 6.2.29.2 Repeated for NLINES subsets of data
$ STARS 6.2.29.1
$ NGP      XTERM1     YTERM1   XTERM2      YTERM2
  2        582.0       22.0    582.0        78.0
$ STARS 6.2.29.2
```

```
$ YGP(I),I=1,NGP
      22.0      78.0
$ Line 2 Repeat 6.2.29.1 to 6.229.2
    2     598.0      22.0     598.0      78.0
      22.0      78.0
$ Lifting Surface number 2    2nd of the LC(3) set = Wing
$ Repeat STARS 6.2.24.1 to 6.2.29.2 - NOTE: data definitions as in lifting surface
$ no. 1
    T   44    1
    6    0    1    1
   11  252.0    2.0    302.0    200.0
  2.0   25.0   50.0     67.0     83.0   102.0   125.0   150.0
       167.0  183.0    200.0
   11  270.0    2.0    313.0    200.0
  2.0   25.0   50.0     67.      83.0   102.0   125.0   150.0
       167.0  183.0    200.0
   11  298.0    2.0    324.0    200.0
  2.0   25.0   50.0     67.0     83.0   102.0   125.0   150.0
       167.0  183.0    200.0
   11  323.0    2.0    336.0    200.0
  2.0   25.0   50.0     67.0     83.0   102.0   125.0   150.0
       167.0  183.0    200.0
    3  348.0    2.0    348.0     48.0
         2.0   25.0     48.0
    4  348.0  152.0    348.0    200.0
       152.0  167.0    183.0    200.0
     328.125   50.0  334.375    150.0
    2    0    1    1
    6  330.0   52.0    336.0    148.0
        52.0   67.0     83.0    102.0   125.0   148.0
    6  348.0   52.0    348.0    148.0
        52.0   67.0     83.0    102.0   125.0   148.0
$ Lifting surface number 3    3rd of the LC(3) set = Horizontal tail
$ Repeat STARS 6.2.24.1 to 6.2.29.2 - NOTE: data definitions as in lifting surface
$ no. 1
    T   15    1
    3    0    1    1
    6     522.0       20.0      542.0      100.0
      20.0      40.0      55.0      72.0      87.0      100.0
    6     550.0       20.0      562.0      100.0
      20.0      40.0      55.0      72.0      87.0      100.0
    6     578.0       20.0      578.0      100.0
      20.0      40.0      55.0      72.0      87.0      100.0
     580.0       20.0      580.0      100.0
    2    0    1    1
    6     582.0       20.0      582.0      100.0
      20.0      40.0      55.0      72.0      87.0      100.0
    6     598.0       20.0      598.0      100.0
      20.0      40.0      55.0      72.0      87.0      100.0
$ 6.2.31 Required when NB > 0.  Repeated NB times.
$ STARS 6.2.31.1  - No. of points on the solid interpolation line for the slender body
$ (Figure A.24, A.26, and A.27)
$ NGP    NSTRIP    IPANEL
  12        2        14
Note: NGP number of points on a slender-body axis at which input vector data are
 prescribed, NSTRIP number of interference panels associated with a slender body,
 IPANEL number of interference panels associated with a slender body, XGP streamwise
 coordinate of each point at which input modal data are prescribed in the LCS.
$ STARS 6.2.31.2 - Streamwise coordinate of each point at which input solid modal data
$ are prescribed in the LCS
$ (XGP(I),I=1,NGP)
      0.0     100.0     150.0     200.0     250.0     300.0
    350.0     400.0     450.0     500.0     560.0     600.0
$ STARS 6.2.32 -  Print option for body aerodynamic elements in the GCS 1-print  0-no
$ print
$ KLUGLB
    0
```

Example A4: ASE Convert Data Input File for Flutter Analysis

```
$ CONVERT FILE FOR ASE FLUTTER AND DIVERGENCE SOLUTION
$ NM total number of desired modes to form reduced generalized matrices
   8
$ MODAL SELECTION AND ORDERING
$ IOLD    INEW
   1,1
   2,2
   3,3
   4,4
   5,5
   6,6
   7,7
   8,8
```

Example A5: ASE Convert Data Input File for Frequency Response Analysis

```
$ CONVERT FILE FOR ASE SOLUTION
$ NM total number of desired modes to form reduced generalized matrices
   13
$ MODAL SELECTION AND ORDERING
$ IOLD    INEW
   9,      1
  10,      2
  11,      3
   1,      4
   2,      5
   3,      6
   4,      7
   5,      8
   6,      9
   7,     10
   8,     11
  12,     12
  13,     13
```

Example A6: Padé Input Data for Flutter Analysis

```
$ ATM ASE FLUTTER ANALYSIS, 0.9 MACH AT SEA LEVEL - AERO analysis
$ NRM,NEM,NCM,NG,NS,NK,NA, RHOR,    VEL,     CREF, IWNDT, NQD
    0,  8,  0,  0, 0, 10, 2, 1.0,  1004.79,    3.2,   0,     69
Note: NRM  number of rigid-body modes,  NEM  number of elastic modes,  NCM number of
 control modes, NG  number of gusts,  NS  number of sensors, NK number of sets of input
 data at discret reduced frequencies,  NA  order of Pade equation,  RHOR  relative
 aerodynamic density with respect to sea level,  VEL  true airspeed(ft/sec),  CREF
 reference chord(ft),  IWNDT wind tunnel correction index, NQD  number of velocities
 for flutter and divergence analysis, to be set to 0 for ASE analysis.
$ TENSION COEFFICIENTS
$ (BETA(I),I=1,NA)  Pade approximation dat
      0.4     0.2
$ GENERALIZED MASS
$ (GMASS(I,J),J=I,NM),I=1,NM) the upper symmetric half of the generalized mass matrix
  .2543E+00  0.0  0.0  0.0  0.0  0.0  0.0  0.0
  .7310E+01  0.0  0.0  0.0  0.0  0.0  0.0
  .1446E+01  0.0  0.0  0.0  0.0  0.0
  .1882E+01  0.0  0.0  0.0  0.0
  .6351E+01  0.0  0.0  0.0
  .1488E+01  0.0  0.0
  .1003E+00  0.0
  .7438E+01
```

```
$ GENERALIZED DAMPING
$ (DAMP(I),I=1,NM) generalized damping data.
  .00000000E+00  .00000000E+00  .00000000E+00  .00000000E+00
  .00000000E+00  .00000000E+00  .00000000E+00  .00000000E+00
$ Natural Frequencies (radians)
$ (OMEGA(I),I=1,NM) Modal frequency data.
  .63715276E+02  .78224005E+02  .92261642E+02  .18064940E+03
  .18729576E+03  .20387943E+03  .22453529E+03  .32130318E+03
$ VELOCITIES FOR FLUTTER AND DIVERGENCE ANALYSIS (ft/sec)
$ (VEL(I),I=1,NQD) airspeed values to be used to calculate the frequency and damping
    1.0
  100.0
  200.0
  300.0
  400.0
  500.0
  600.0
  700.0
  800.0
  900.0
 1000.0
 1100.0
 1200.0
 1210.0
 1220.0
 1230.0
 1240.0
 1250.0
 1260.0
 1270.0
 1280.0
 1290.0
 1300.0
 1400.0
 1500.0
 1600.0
 1700.0
 1800.0
 1900.0
 2000.0
 2050.0
 2100.0
 2150.0
 2200.0
 2250.0
 2300.0
 2350.0
 2400.0
 2450.0
 2500.0
 2550.0
 2600.0
 2650.0
 2700.0
 2710.0
 2730.0
 2740.0
 2750.0
 2760.0
 2780.0
 2790.0
 2800.0
 2850.0
 2875.0
 2900.0
 2950.0
 3000.0
 3050.0
 3100.0
 3150.0
```

```
3200.0
3250.0
3300.0
3350.0
3400.0
3450.0
3500.0
3550.0
3600.0
```

Example A7: Padé Input Data for Frequency Response Analysis

```
$ ATM ASE ANALYSIS, 0.9 MACH AT 40K FEET - VERSION II DATA
$ NRM,NEM,NCM,NG,NS, NK,NA, RHOR,    VEL,     CREF, IWNDT, NQD
  3, 8,   2, 0, 2, 10, 2, 0.24708, 871.27,  3.2,    0,    0
Note: NRM  number of rigid-body modes,  NEM  number of elastic modes,  NCM number of
 control modes, NG  number of gusts,  NS  number of sensors, NK number of sets of input
 data at discrete reduced frequencies,  NA  order of Pade equation,  RHOR  relative
 aerodynamic density with respect to sea level,  VEL  true airspeed(ft/sec),  CREF
 reference chord(ft),  IWNDT wind tunnel correction index, NQD  number of velocities
 for flutter and divergence analysis, to be set to 0 for ASE analysis.
$ TENSION COEFFICIENTS
$ (BETA(I),I=1,NA)  Pade approximation dat
     0.4     0.2
$ GENERALIZED MASS
$ (GMASS(I,J),J=I,NM),I=1,NM) the upper symmetric half of the generalized mass matrix
0.7875E+02  0.0  0.0  0.0  0.0  0.0  0.0  0.0  0.0  0.0  0.0  0.0  0.0
0.4701E+04  0.0  0.0  0.0  0.0  0.0  0.0  0.0  0.0  0.0  0.0  0.0
0.1831E+05  0.0  0.0  0.0  0.0  0.0  0.0  0.0  0.0  0.0  0.0
0.2543E+00  0.0  0.0  0.0  0.0  0.0  0.0  0.0  0.0  0.0
0.7310E+01  0.0  0.0  0.0  0.0  0.0  0.0  0.0  0.0
0.1446E+01  0.0  0.0  0.0  0.0  0.0  0.0  0.0
0.1882E+01  0.0  0.0  0.0  0.0  0.0  0.0
0.6351E+01  0.0  0.0  0.0  0.0  0.0
0.1488E+01  0.0  0.0  0.0  0.0
0.1003E+00  0.0  0.0  0.0
0.7438E+01  0.0  0.0
0.3996E+01  0.0
0.4419E+00
$ GENERALIZED DAMPING
$ (DAMP(I),I=1,NM) generalized damping data.
0.00000000E+00 0.00000000E+00 0.00000000E+00 0.00000000E+00
0.00000000E+00 0.00000000E+00 0.00000000E+00 0.00000000E+00
0.00000000E+00 0.00000000E+00 0.00000000E+00 0.00000000E+00
0.00000000E+00
$ Natural Frequencies (radians)
$ (OMEGA(I),I=1,NM) Modal frequency data.
          0.0              0.0              0.0  .63715276E+02
 .78224005E+02  .92261642E+02  .18064940E+03  .18729576E+03
 .20387943E+03  .22453529E+03  .32130318E+03            0.0
          0.0
$ PHI,THETA, PSI,     US,  VS,  WS,  PS,  QS,  RS, PHID, THAD, PSID, NDOF
  0.0, 0.0,  0.0, 871.27, 0.0, 0.0, 0.0, 0.0, 0.0, 0.0,  0.0,  0.0,  -3
Note:  PHI  roll angle(deg); THETA pitch angle(deg);  PSI  yaw angle(deg); US, VS, WS
 are body axis velocities;  PS, QS, Rs  are angular rates; PHID, THAD, PSID  are Euler
 angle rates;  NDOF number of aircraft degrees of fredom( a negative sign indicates
 antisymmetric case).
$ SENSOR DATA
$ IFLSI =0 for a non-GVS case
 0
$ XS, YS, ZS  -  coordinate of the sensor.
   300.00         0.0        50.0
$ LX, MY, NZ, THX, THY, THZ
  0.0  0.0  0.0  1.0  0.0  0.0
Notes:  LX, MY, NZ are direction cosine for an accelerometer normal in X, Y, and Z;
 THX, THY, THZ are the dirction cosine for the pitch axis about X, Y, Z axis.
   300.00         0.0        50.0
      0.0         0.0         0.0         0.0         0.0         1.0
```

Example A8: Control Data Input for Open-Loop Case

```
$ ATM ANTI-SYMMETRIC 3 RIGID, 8 ELASTIC, AND 2 CONTROL MODES- OPEN LOOP CASE LOOP OPEN
$ BETWEEN BLOCKS 3 AND 9 AS WELL AS BETWEEN 4 AND 10
$ NX   NY  NU  NV  NXC   DELTAT  TDELAY  MAXBC  MAXPO
  48,  12,  4, 4,  58,    0.0,     0.0,    3,     3
$ NB  NYBTUV  IADDRA  IADDCB IADDRC NLST NDRESP IRP ITRP
  10    4       10      6      10    4     2     1    1
$ MAXA   MAXB   MAXC   NBSS
   2      1      1      2
Notes:  NX number of states in the plant A matrix,  NY number of the row of the plant
 C matrix, NU  number of column of the plant B matrix,  NV  number of external inputs
 to the system,  NXC total number of continuous states( plant  plus all elements),
 DELTAT sample time for digital elements,  TDELAY  system time delay,  MAXBC  maximum
 number of block connectivity,  MAXPO maximum polynomial order plus one, NB  number of
 blocks of anlog and digital elements in the system, NYBTUV is NYTOV+NBTOU,  IADDRA
 additional rows of A caused by the augmentation of control elements, IADDCB additional
 columns of B caused by the augmentaion of control elements,  IADDRC additional rows of
 C caused by augmentation of control elements,  NLST total number of frequency range
 specifications for frequency-response computations, NDRESP  number of times the loops
 are broken for open-loop response evaluation, IRP frequency-response problem number to
 be evaluated, ITRP total number of frequency-response cases, MAXA maximum row
 dimension of the direct input state-space A matrix for the control element, MAXB
 maximum column dimension of the direct input B matrix, MAXC maximum row dimension of
 the direct input C matrix,  NBSS  number of control blocks that has state-space
 matrices form input instead of transfer function.
$ BLOCK CONNECTIVITY
$ IBN ICN1  ICN2  ICN3   IEXI   IELPCL  ISLPCL
   1   3     0     0      0       0       0
   2  10     0     0      0       0       0
   3   7     0     0      0       0       0
   4   6     0     0      0       0       0
   5   0     0     0      1       0       0
   6   0     0     0      2       0       0
   7   0     0     0      3       0       0
   8   0     0     0      4       0       0
   9   5     0     0      0       0       0
  10   8     0     0      0       0       0
Notes:  IBN  integer defining the block number; ICN1, ICN2, ICN3 connecting block
 numbers.  If IBN has been augmented then the ICN1 block is next being augmented to
 the plant in connection to block IBN. If IBN has not been augmented to the plant
 then it will be augmented to the plant in connection with ICN1 block.
Note:  IEXI integer defining the block connected directly to the plant( related with
 IYTOV2 down below in the data),  IELPCL  integer defining the closing block (feedback
 signal of the closed loop system, ISLPCL integer defining the starting block of the
 closed-loop system.
$ TRANSFER FUNCTION DESCRIPTIONS- as order of polynomials for each block
$ IBN    ICNP    ICDP    ICROW/
$          NA     NCB     NCR
    1            1         2          0
    2            1         2          0
    3            2         2          0
    4            2         2          0
    5            2         1          1
    6            1         3          0
    7            1         1          0
    8            1         1          0
    9            2         1          1
   10            1         1          0
Note:  ICNP  integer defining the number of coefficients in the numerator polynomial /
 NA size of A matrix; ICDP integer defining the number of coefficients in the
 denominator polynomial/ NCB number of columns of the B matrix;  ICROW integer
 defining the number of coefficients in the denominator polynomial/  NCR number of row
 for the C matrix( for the direct state-space matrix input option ICROW is not 0)
$ LISTING OF POLYNOMIAL COEFFICIENTS
$ IBN     (POLCON(I),I=1,MAXPO)
    1 .2000E+02 .0000E+00 .0000E+00
$ IBN     (POLCOD(I),I=1,MAXPO)
    0 .2000E+02 .1000E+01 .0000E+00
    2 .2000E+02 .0000E+00 .0000E+00
    0 .2000E+02 .1000E+01 .0000E+00
```

```
    3 .5000E+01 .1000E+00 .0000E+00
    0 .0000E+00 .1000E+01 .0000E+00
    4 .0000E+00 .1000E+01 .0000E+00
    0 .1000E+00 .1000E+01 .0000E+00
    6 .1877E+05 .0000E+00 .0000E+00
    0 .1877E+05 .1930E+03 .1000E+01
    7 .1000E+01 .0000E+00 .0000E+00
    0 .1000E+01 .0000E+00 .0000E+00
    8 .1000E+01 .0000E+00 .0000E+00
    0 .1000E+01 .0000E+00 .0000E+00
   10 .1000E+01 .0000E+00 .0000E+00
    0 .1000E+01 .0000E+00 .0000E+00
Note:  POLCON  numerator coefficients  of the transfer function for block IBN, POLCOD
 denominator coefficients of the transfer function for control block IBN are placed in
 the next row next to the numerator coefficients, one block at a time.  NB-NBSS sets of
 data are required.
$ Listing of state space matrices for NBSS control blocks
$ IBN   NA  NCB  NCR
   5    2    1    1
$ ((A(I,J),J=1,NA,B(I,J),J=1,NCB),I=1,NA)   for control block IBN
        0.            1.           0.
   -18770.         -193.        18770.
$ ((C(I,J),J=1,NA,D(I,J),J=1,NCB),I=1,NCR)   for control block IBN
        1.000        0.            0.
        9            2             1         1
        0.           1.            0.0
      -10.         -11.            0.1
        1.           0.             .0
Notes:  NA is the size of matrix A of control block IBN,  NCB is the column of matrix B
 of the control block IBN,  NCR is the row number of matrix C of the control block IBN;
 A, B, C, D  are the state space matrices of the control block IBN.   NBSS sets of data
 are required for input.
$ GAIN INPUTS FOR EACH BLOCK
$ IBN    GAIN -  NB sets of data are required for input.
  1 .1000E+01  2 .1000E+01  3 .1000E+01  4 .1000E+00  5 .1000E+01
  6 .1000E+01  7 .1000E+01  8 .1000E+01  9 .1000E+01 10 .1000E+01
$ SPECIFICATION FOR SYSTEM OUTPUTS
$ ISO1  ISO2   .....      ISONYB
  7  8   0   0   0   0   0   0   0   0   0   0   0   0   0
  0  0   0   0   0   0
Notes:  ISOI desired output from any sensor( the corresponding row of the C matrix for
 the plant) and any control element (augmented thereafter).  NYB = NY+NB number of data.
$ SPECIFICATION FOR SYSTEM INPUTS
$ ISI1   ISI2   ....  ISINUV
      7    8    0    0    0    0    0    0
Note:  ISII plant input( the correspinding column of the B matrix for the plant) and
 external input.   NUV = NU+NY number of data.
$ CONNECTION DETAILS FROM PLANT TO BLOCKS
$ NYTOV   NBTOU   NBTOK
    2       2       0
$ IYTOV1    IYTOV2
      7         1
      8         2
$ IBTOU1    IBTOU2
      1         1
      2         2
Note:  NYTOV number of connections from plant outputs to external inputs, NBTOU number
 of block outputs connected to plant inputs,  NBTOK number of digital element outputs
 connected to analog element inputs,  IYTOV1  row number of the plant C matrix to the
 feedback control system IYTOV2 appears in the IEXI column in the block connectivities
 section above,  IBTOU1 block number to be connected to the plant input column IBTOU2
 of B matrix.
$ FREQUENCY RANGE SPECIFICATIONS - NLST sets of frequency data required
$ FREQI   FREQF   NFREQ
   0.1      .9      200
   1.0      9.      200
   10.     90.      200
  100.    500.      100
Note:  FREQI initial frequency, FREQF final frequency,  NFREQ  number of frequencies
 within this range, logarithmically spaced.
```

```
$ LOOP DEFINITIONS  - out block  inp block  outblock row  input block row
$ ILOOP  IPRINT
    1      0
$ NBRAK1  NBRAK2  NOUTBL   NINBL
     4       8       1        1
     9       7       1        1
Note:  ILOOP  integer defining the loop type( 0- closed-loop response case,
 1-open-loop response case, 2- eigensolution of the system), IPRINT eigensolution print
 option for the closed-loop case ( 0- print eigenvalues only, 1- prints eigenvalues
 and vectors),  NBRAK1 block having the output signal( 999 identified the plant),
 NBRAK2 block having the input signal, NOUTBL output of block NBRAK1,  NINBL input of
 block NBRAK2. Required NDRESP sets of data.
```

Example A9: Control Data Input for Closed-Loop Case

```
$ ATM ANTI-SYMMETRIC 3 RIGID, 8 ELASTIC, AND 2 CONTROL MODES- OPEN
$ LOOP CASE. LOOP CLOSED
$ BETWEEN BLOCKS 3 AND 9 AS WELL AS BETWEEN 4 AND 10
 NX   NY  NU  NV  NXC    DELTAT   TDELAY MAXBC  MAXPO
 48,  12,  4,  4,  58,     0.0,     0.0,     3,     3
$ NB  NYBTUV  IADDRA  IADDCB IADDRC  NLST NDRESP IRP ITRP
  10    4       10       6     10      4     2    1    1
$ MAXA   MAXB   MAXC   NBSS
    2      1      1     2
$ BLOCK CONNECTIVITY
$ IBN ICN1  ICN2  ICN3   IEXI  IELPCL  ISLPCL
    1   3     0     0      0       0       0
    9   0     0     0      0       5       3
    3   7     0     0      0       0       0
    5   0     0     0      1       0       0
    6   0     0     0      2       0       0
    7   0     0     0      3       0       0
    8   0     0     0      4       0       0
    2  10     0     0      0       0       0
    4   0     0     0      0       6      10
   10   8     0     0      0       0       0
$ TRANSFER FUNCTION DESCRIPTIONS- as order of polynomials for each block
$ IBN   ICNP   ICDP   ICROW/
$  NA    NCB    NCR
      1         1          2         0
      2         1          2         0
      3         2          2         0
      4         2          2         0
      5         2          1         1
      6         1          3         0
      7         1          1         0
      8         1          1         0
      9         2          1         1
     10         1          1         0
$ LISTING OF POLYNOMIAL COEFFICIENTS
$ IBN     (POLCON(I),I=1,MAXPO)
    1 .2000E+02 .0000E+00 .0000E+00
$ IBN     (POLCOD(I),I=1,MAXPO)
    0 .2000E+02 .1000E+01 .0000E+00
    2 .2000E+02 .0000E+00 .0000E+00
    0 .2000E+02 .1000E+01 .0000E+00
    3 .5000E+01 .1000E+00 .0000E+00
    0 .0000E+00 .1000E+01 .0000E+00
    4 .0000E+00 .1000E+01 .0000E+00
    0 .1000E+00 .1000E+01 .0000E+00
    6 .1877E+05 .0000E+00 .0000E+00
    0 .1877E+05 .1930E+03 .1000E+01
    7 .1000E+01 .0000E+00 .0000E+00
    0 .1000E+01 .0000E+00 .0000E+00
    8 .1000E+01 .0000E+00 .0000E+00
    0 .1000E+01 .0000E+00 .0000E+00
   10 .1000E+01 .0000E+00 .0000E+00
    0 .1000E+01 .0000E+00 .0000E+00
```

```
$ Listing of state space matrices for NBSS control blocks
$ IBN   NA  NCB  NCR
     5       2       1       1
$ ((A(I,J),J=1,NA,B(I,J),J=1,NCB),I=1,NA)    for control block IBN
            0.             1.          0.
   -18770.       -193.      18770.
            1.             0.          0.
     9       2       1       1
            0.             1.          0.
         -10.           -11.           0.1
            1.             0.          0.
$ GAIN INPUTS FOR EACH BLOCK
$ IBN    GAIN -  NB sets of data are required for input.
  1 .1000E+01  2 .1000E+01  3 .1000E+01  4 .1000E+00   5 .1000E+01
  6 .1000E+01  7 .1000E+01  8 .1000E+01  9 .1000E+01  10 .1000E+01
$ SPECIFICATION FOR SYSTEM OUTPUTS
$ IS01  IS02   .....     ISONYB
   7   8   0   0   0   0   0   0   0   0   0   0   0   0   0   0
   0   0   0   0   0   0
$ SPECIFICATION FOR SYSTEM INPUTS
$ ISI1   ISI2   ....  ISINUV
   7    8    0    0    0    0    0    0
$ CONNECTION DETAILS FROM PLANT TO BLOCKS
$ NYTOV   NBTOU   NBTOK
       2               2               0
$ IYTOV1     IYTOV2
      7               1
      8               2
$ IBTOU1    IBTOU2
      1               1
      2               2
$ FREQUENCY RANGE SPECIFICATIONS - NLST sets of frequency data required
$ FREQI   FREQF   NFREQ
      0.1          .9          200
      1.0         9.           200
     10.         90.           200
    100.        500.           100
$ LOOP DEFINITIONS  - out block  inp block  outblock row  input block
$ row
$ ILOOP  IPRINT
   0    0
$NBRAK1  NBRAK2  NOUTBL    NINBL
   999       8       8       1
   999       7       7       1
```

Chapter 15

Figure A.27 shows a cantilever plate with a NACA 0012 airfoil and solution domain. Important data parameters are as follows: wing span = 2.0178, wing chord length = 1.0089, Mach number = 0.6, angle of attack = 0 deg and also 10 deg, and speed of sound at infinity = 340.29. The Structural data are Young's modulus $E = 6.8947 \times 10^{10}$, Poisson's ratio $\nu = 0.3$, and density $\rho = 2764.925$. Perform a CFD-based aeroelastic analysis and determine flutter parameter corresponding to the zero damping case.

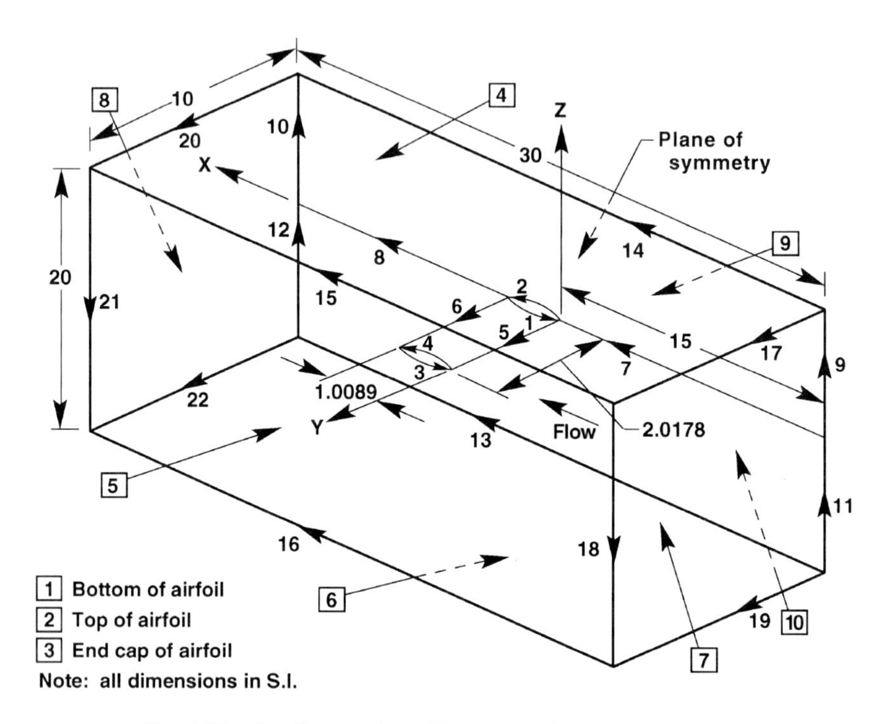

Fig. A27 Cantilever wing with aeroelastic solution domain.

Index

TEXTS PUBLISHED IN THE AIAA EDUCATION SERIES

Finite Element Multidisciplinary Analysis, Second Edition
K. K. Gupta and John L. Meek 2003
ISBN 1-56347-580-4

Flight Testing of Fixed-Wing Aircraft
Ralph D. Kimberlin 2003
ISBN 1-56347-564-2

The Fundamentals of Aircraft Combat Survivability Analysis and Design, Second Edition
Robert E. Ball 2003
ISBN 1-56347-582-0

Analytical Mechanics of Space Systems
Hanspeter Schaub and John L. Junkins 2003
ISBN 1-56347-563-4

Introduction to Aircraft Flight Mechanics
Thomas R. Yechout with Steven L. Morris, David E. Bossert, and Wayne F. Hallgren 2003
ISBN 1-56347-577-4

Aircraft Design Projects for Engineering Students
Lloyd Jenkinson and James Marchman 2003
ISBN 1-56347-619-3

Elements of Spacecraft Design
Charles D. Brown 2002
ISBN 1-56347-524-3

Civil Avionics Systems
Ian Moir and Allan Seabridge 2002
ISBN 1-56347-589-8

Helicopter Test and Evaluation
Alastair K. Cooke and Eric W. H. Fitzpatrick 2002
ISBN 1-56347-578-2

Aircraft Engine Design, Second Edition
Jack D. Mattingly, William H. Heiser, and David T. Pratt 2002
ISBN 1-56347-538-3

Dynamics, Control, and Flying Qualities of V/STOL Aircraft
James A. Franklin 2002
ISBN 1-56347-575-8

Orbital Mechanics, Third Edition
Vladimir A. Chobotov, Editor 2002
ISBN 1-56347-537-5

Basic Helicopter Aerodynamics, Second Edition
John Seddon and Simon Newman 2001
ISBN 1-56347-510-3

Aircraft Systems: Mechanical, Electrical, and Avionics Subsystems Integration
Ian Moir and Allan Seabridge 2001
ISBN 1-56347-506-5

Design Methodologies for Space Transportation Systems
Walter E. Hammond 2001
ISBN 1-56347-472-7

Tactical Missile Design
Eugene L. Fleeman 2001
ISBN 1-56347-494-8

Flight Vehicle Performance and Aerodynamic Control
Frederick O. Smetana 2001
ISBN 1-56347-463-8

Modeling and Simulation of Aerospace Vehicle Dynamics
Peter H. Zipfel 2000
ISBN 1-56347-456-5

Applied Mathematics in Integrated Navigation Systems
Robert M. Rogers 2000
ISBN 1-56347-445-X

Mathematical Methods in Defense Analyses, Third Edition
J. S. Przemieniecki *2000*
ISBN 1-56347-396-6

Finite Element Multidisciplinary Analysis
Kajal K. Gupta and John L. Meek *2000*
ISBN 1-56347-393-3

Aircraft Performance: Theory and Practice
M. E. Eshelby *1999*
ISBN 1-56347-398-4

Space Transportation: A Systems Approach to Analysis and Design
Walter E. Hammond *1999*
ISBN 1-56347-032-2

Civil Jet Aircraft Design
Lloyd R. Jenkinson, Paul Simpkin, and Darren Rhodes *1999*
ISBN 1-56347-350-X

Structural Dynamics in Aeronautical Engineering
Maher N. Bismarck–Nasr *1999*
ISBN 1-56347-323-2

Intake Aerodynamics, Second Edition
E. L. Goldsmith and J. Seddon *1999*
ISBN 1-56347-361-5

Integrated Navigation and Guidance Systems
Daniel J. Biezad *1999*
ISBN 1-56347-291-0

Aircraft Handling Qualities
John Hodgkinson *1999*
ISBN 1-56347-331-3

Performance, Stability, Dynamics, and Control of Airplanes
Bandu N. Pamadi *1998*
ISBN 1-56347-222-8

Spacecraft Mission Design, Second Edition
Charles D. Brown *1998*
ISBN 1-56347-262-7

Computational Flight Dynamics
Malcolm J. Abzug *1998*
ISBN 1-56347-259-7

Space Vehicle Dynamics and Control
Bong Wie *1998*
ISBN 1-56347-261-9

Introduction to Aircraft Flight Dynamics
Louis V. Schmidt *1998*
ISBN 1-56347-226-0

Aerothermodynamics of Gas Turbine and Rocket Propulsion, Third Edition
Gordon C. Oates *1997*
ISBN 1-56347-241-4

Advanced Dynamics
Shuh-Jing Ying *1997*
ISBN 1-56347-224-4

Introduction to Aeronautics: A Design Perspective
Steven A. Brandt, Randall J. Stiles, John J. Bertin, and Ray Whitford *1997*
ISBN 1-56347-250-3

Introductory Aerodynamics and Hydrodynamics of Wings and Bodies: A Software-Based Approach
Frederick O. Smetana *1997*
ISBN 1-56347-242-2

An Introduction to Aircraft Performance
Mario Asselin *1997*
ISBN 1-56347-221-X

Orbital Mechanics, Second Edition
Vladimir A. Chobotov, Editor *1996*
ISBN 1-56347-179-5

Thermal Structures for Aerospace Applications
Earl A. Thornton *1996*
ISBN 1-56347-190-6

Structural Loads Analysis for Commercial Transport Aircraft: Theory and Practice
Ted L. Lomax *1996*
ISBN 1-56347-114-0

Spacecraft Propulsion
Charles D. Brown 1996
ISBN 1-56347-128-0

Helicopter Flight Dynamics: The Theory and Application of Flying Qualities and Simulation Modeling
Gareth D. Padfield 1996
ISBN 1-56347-205-8

Flying Qualities and Flight Testing of the Airplane
Darrol Stinton 1996
ISBN 1-56347-117-5

Flight Performance of Aircraft
S. K. Ojha 1995
ISBN 1-56347-113-2

Operations Research Analysis in Test and Evaluation
Donald L. Giadrosich 1995
ISBN 1-56347-112-4

Radar and Laser Cross Section Engineering
David C. Jenn 1995
ISBN 1-56347-105-1

Introduction to the Control of Dynamic Systems
Frederick O. Smetana 1994
ISBN 1-56347-083-7

Tailless Aircraft in Theory and Practice
Karl Nickel and Michael Wohlfahrt 1994
ISBN 1-56347-094-2

Mathematical Methods in Defense Analyses, Second Edition
J. S. Przemieniecki 1994
ISBN 1-56347-092-6

Hypersonic Aerothermodynamics
John J. Bertin 1994
ISBN 1-56347-036-5

Hypersonic Airbreathing Propulsion
William H. Heiser and David T. Pratt 1994
ISBN 1-56347-035-7

Practical Intake Aerodynamic Design
E. L. Goldsmith and J. Seddon 1993
ISBN 1-56347-064-0

Acquisition of Defense Systems
J. S. Przemieniecki, Editor 1993
ISBN 1-56347-069-1

Dynamics of Atmospheric Re-Entry
Frank J. Regan and Satya M. Anandakrishnan 1993
ISBN 1-56347-048-9

Introduction to Dynamics and Control of Flexible Structures
John L. Junkins and Youdan Kim 1993
ISBN 1-56347-054-3

Spacecraft Mission Design
Charles D. Brown 1992
ISBN 1-56347-041-1

Rotary Wing Structural Dynamics and Aeroelasticity
Richard L. Bielawa 1992
ISBN 1-56347-031-4

Aircraft Design: A Conceptual Approach, Second Edition
Daniel P. Raymer 1992
ISBN 0-930403-51-7

Nonlinear Analysis of Shell Structures
Anthony N. Palazotto and Scott T. Dennis 1992
ISBN 1-56347-033-0

Orbital Mechanics
Vladimir A. Chobotov, Editor 1991
ISBN 1-56347-007-1

Critical Technologies for National Defense
Air Force Institute of Technology 1991
ISBN 1-56347-009-8

Space Vehicle Design
Michael D. Griffin and James R. French 1991
ISBN 0-930403-90-8

Defense Analyses Software
J. S. Przemieniecki *1990*
ISBN 0-930403-91-6

Inlets for Supersonic Missiles
John J. Mahoney *1990*
ISBN 0-930403-79-7

Introduction to Mathematical Methods in Defense Analyses
J. S. Przemieniecki *1990*
ISBN 0-930403-71-1

Basic Helicopter Aerodynamics
J. Seddon *1990*
ISBN 0-930403-67-3

Aircraft Propulsion Systems Technology and Design
Gordon C. Oates, Editor *1989*
ISBN 0-930403-24-X

Boundary Layers
A. D. Young *1989*
ISBN 0-930403-57-6

Aircraft Design: A Conceptual Approach
Daniel P. Raymer *1989*
ISBN 0-930403-51-7

Gust Loads on Aircraft: Concepts and Applications
Frederic M. Hoblit *1988*
ISBN 0-930403-45-2

Aircraft Landing Gear Design: Principles and Practices
Norman S. Currey *1988*
ISBN 0-930403-41-X

Mechanical Reliability: Theory, Models and Applications
B. S. Dhillon *1988*
ISBN 0-930403-38-X

Re-Entry Aerodynamics
Wilbur L. Hankey *1988*
ISBN 0-930403-33-9

Aerothermodynamics of Gas Turbine and Rocket Propulsion, Revised and Enlarged
Gordon C. Oates *1988*
ISBN 0-930403-34-7

Advanced Classical Thermodynamics
George Emanuel *1987*
ISBN 0-930403-28-2

Radar Electronic Warfare
August Golden Jr. *1987*
ISBN 0-930403-22-3

An Introduction to the Mathematics and Methods of Astrodynamics
Richard H. Battin *1987*
ISBN 0-930403-25-8

Aircraft Engine Design
Jack D. Mattingly, William H. Heiser, and Daniel H. Daley *1987*
ISBN 0-930403-23-1

Gasdynamics: Theory and Applications
George Emanuel *1986*
ISBN 0-930403-12-6

Composite Materials for Aircraft Structures
Brian C. Hoskin and Alan A. Baker, Editors *1986*
ISBN 0-930403-11-8

Intake Aerodynamics
J. Seddon and E. L. Goldsmith *1985*
ISBN 0-930403-03-7

The Fundamentals of Aircraft Combat Survivability Analysis and Design
Robert E. Ball *1985*
ISBN 0-930403-02-9

Aerothermodynamics of Aircraft Engine Components
Gordon C. Oates, Editor *1985*
ISBN 0-915928-97-3

Aerothermodynamics of Gas Turbine and Rocket Propulsion
Gordon C. Oates *1984*
ISBN 0-915928-87-6

Re-Entry Vehicle Dynamics
Frank J. Regan *1984*
ISBN 0-915928-78-7